Springer Series in Optical Sciences Volume 38

Edited by Theodor Tamir

Springer Series in Optical Sciences

1 **Solid-State Laser Engineering**
By. W. Koechner

2 **Table of Laser Lines in Gases and Vapors**
3rd Edition
By R. Beck, W. Englisch, and K. Gürs

3 **Tunable Lasers and Applications**
Editors: A. Mooradian, T. Jaeger, and P. Stokseth

4 **Nonlinear Laser Spectroscopy**
By V. S. Letokhov and V. P. Chebotayev

5 **Optics and Lasers**
An Engineering Physics Approach
By M. Young

6 **Photoelectron Statistics**
With Applications to Spectroscopy and Optical Communication
By B. Saleh

7 **Laser Spectroscopy III**
Editors: J. L. Hall and J. L. Carlsten

8 **Frontiers in Visual Science**
Editors: S. J. Cool and E. J. Smith III

9 **High-Power Lasers and Applications**
2nd Printing
Editors: K.-L. Kompa and H. Walther

10 **Detection of Optical and Infrared Radiation**
2nd Printing
By R. H. Kingston

11 **Matrix Theory of Photoelasticity**
By P. S. Theocaris and E. E. Gdoutos

12 **The Monte Carlo Method in Atmospheric Optics**
By G. I. Marchuk, G. A. Mikhailov, M. A. Nazaraliev, R. A. Darbinian, B. A. Kargin, and B. S. Elepov

13 **Physiological Optics**
By Y. Le Grand and S. G. El Hage

14 **Laser Crystals** Physics and Properties
By A. A. Kaminskii

15 **X-Ray Spectroscopy**
By B. K. Agarwal

16 **Holographic Interferometry**
From the Scope of Deformation Analysis of Opaque Bodies
By W. Schumann and M. Dubas

17 **Nonlinear Optics of Free Atoms and Molecules**
By D. C. Hanna, M. A. Yuratich, D. Cotter

18 **Holography in Medicine and Biology**
Editor: G. von Bally

19 **Color Theory and Its Application in Art and Design**
By G. A. Agoston

20 **Interferometry by Holography**
By Yu. I. Ostrovsky, M. M. Butusov, G. V. Ostrovskaya

21 **Laser Spectroscopy IV**
Editors: H. Walther, K. W. Rothe

22 **Lasers in Photomedicine and Photobiology**
Editors: R. Pratesi and C. A. Sacchi

23 **Vertebrate Photoreceptor Optics**
Editors: J. M. Enoch and F. L. Tobey, Jr.

24 **Optical Fiber Systems and Their Components**
An Introduction. By A. B. Sharma, S. J. Halme, and M. M. Butusov

25 **High-Peak-Power Nd: Glass Laser Systems**
By D. C. Brown

26 **Lasers and Applications**
Editors: W. O. N. Guimaraes, C. T. Lin, and A. Mooradian

27 **Color Measurement** Theme and Variations
By D. L. MacAdam

28 **Modular Optical Design**
By O. N. Stavroudis

29 **Inverse Problems of Lidar Sensing of the Atmosphere** By V. E. Zuev and I. E. Naats

30 **Laser Spectroscopy V**
Editors: A. R. W. McKellar, T. Oka, and B. P. Stoicheff

31 **Optics in Biomedical Sciences**
Editors: G. von Bally and P. Greguss

32 **Fiber-Optic Rotation Sensors** and Related Technologies
Editors: S. Ezekiel and H. J. Arditty

33 **Integrated Optics: Theory and Technology**
By R.G. Hunsperger

34 **The High-Power Iodine Laser**
By G. Brederlow, E. Fill, and K. J. Witte

35 **Engineering Optics** By K. Iizuka

36 **Transmission Electron Microscopy**
By L. Reimer

37 **Opto-Acoustic Molecular Spectroscopy**
By V.S. Letokhov and V.P. Zharov

38 **Photon Correlation Techniques** in Fluid Mechanics
Editor: E. O. Schulz-DuBois

39 **Optical and Laser Remote Sensing**
By D. K. Killinger and A. Mooradian

Photon Correlation Techniques

in Fluid Mechanics

Proceedings of the 5th International Conference
at Kiel-Damp, Fed. Rep. of Germany, May 23–26, 1982

Editor: E. O. Schulz-DuBois

With 218 Figures

Springer-Verlag Berlin Heidelberg GmbH 1983

Sponsored by
Deutsche Forschungsgemeinschaft · State Government of Schleswig-Holstein

ISBN 978-3-662-13529-7 ISBN 978-3-540-39493-8 (eBook)
DOI 10.1007/978-3-540-39493-8

Library of Congress Cataloging in Publication Data. Main entry under title: Photon correlation techniques in fluid mechanics. (Springer series in optical sciences ; v. 38) Proceedings of the 5th International Conference on Photon Correlation Techniques. Includes bibliographical references and index. 1. Fluid mechanics - Congresses. 2. Photon correlation - Congresses. I. Schulz-Dubois, E. O. II. International Conference on Photon Correlation Techniques (5th : 1982 : Kiel, Germany) III. Series. QC138.P48 1983 532'.05 83-459

Originally published by Springer-Verlag Berlin Heidelberg New York in 1983.
Softcover reprint of the hardcover 1st edition 1983

2153/3130-543210

Preface

Photon correlation is a kind of spectroscopy designed to identify optical frequency shifts and line-broadening effects in the range of many MHz down to a few Hz. The optical intensity is measured in terms of single photon-detection events which result in current pulses at the output of photomultiplier tubes. This signal is processed in real time in a special-purpose parallel processor known as a correlator. The resulting photon correlation function, a function in the time domain, contains the desired spectral information, which may be extracted by a suitable algorithm.

Due to the non-intrusive nature and the sound theoretical basis of photon correlation, the phenomena under study are not disturbed, and the parameters in question can be precisely evaluated. For these reasons photon correlation has become a valuable and in many instances indispensable technique in two distinct fields. One of these is velocimetry in fluid flow. This includes hydro- and aerodynamic processes in liquids, gases, or flames where the velocity field may be stationary, time periodic, or turbulent, and may range from micrometers per second for motion inside biological cells to one kilometer per second for supersonic flow. The other major field is stochastic particle propagation due to Brownian motion. Typically one studies an aqueous suspension of solid spheres of an inorganic or organic nature, ordered macromolecular structures such as vesicles or micelles, constituents of biological cells such as chromatin or DNA macromolecules, emulgated droplets, or suspended bubbles. The measurements give the size distribution of the particles and changes in it with time which result from processes like polymerization or aggregation.

In little more than a decade, photon correlation has rapidly developed from humble beginnings. Its progress was made possible by advances in photomultiplier technology, the availability of stable laser light sources and, perhaps most significantly, by advances in integrated circuits for digital electronics. These advances in hardware were accompanied by a growing theoretical understanding of the statistical nature of the measurement process, which led to improved algorithms for the preprocessing and postprocessing of data. Major contributions to what is now photon correlation were made Dr. E.R. PIKE and his associates at the Royal Signals and Radar Establishment at Malvern, UK. This and other British laboratories still play an important role in advancing this technique, although in the last few years significant contributions have also come from many other industrialized countries.

This book contains the Proceedings of the 5th International Conference on Photon Correlation Techniques, which took place May 23 – 26, 1982, at Damp 2000, a modern resort on the Baltic Sea not far from Kiel, F.R. Germany. Two summer schools, in 1973 and 1976 on the island of Capri, Italy, and four conferences, two at Cambridge, UK, and one each in Stockholm, Sweden, and Stanford, California, were previously devoted to this field. The lecture notes of both summer schools were compiled in book form (see Refs. 2 and 3 of paper 1) and still represent the most valuable surveys of the field. The proceedings of the first and fourth conferences are available in limited editions only, while those of the second and third ones were published in Physica Scripta Vol. *19* (1979) and Optica Acta Vol. *27* (1980) respectively. For the present, fifth conference, an attempt was made to achieve reasonably complete and representative coverage of the field. To this end, the members of the Advisory Board and the Technical Programme Committee encouraged contributions on virtually all important aspects and all typical applications of photon correlation techniques. Invited papers were solicited for an introduction to the major topics in this field and for surveys on the state of the art. Thus this book gives an authoritative account of the field, of what has been achieved thus far, and the directions in which current research is heading. The book should be useful to students, scientists and engineers who want to enter the field, but also to specialists already active in it who are looking for an up-to-date synopsis.

It is a pleasure to thank all those who contributed to the conference and, by the same token, to this volume. I wish first and foremost to thank the authors of the papers collected in this book. This includes several colleagues who because of adverse financial or political conditions in their respective countries were not able to deliver their papers in person at the conference. The contributions to this book were written after the conference, and this enabled the authors to utilize ideas and results that had been exchanged there.

Thanks are due to the members of the Advisory Board and the Technical Programme Committee for helping to select timely subjects and knowledgeable authors. As a Topical Conference of the European Physical Society, the meeting was held under the auspices of the "Präsident der Christian-Albrechts-Universität zu Kiel" and the "Deutsche Physikalische Gesellschaft, Arbeitsgemeinschaft Quantenoptik." I am grateful to Dr. H. Lotsch for offering to publish this volume in the Springer Series in Optical Sciences. Of the members of the Institute of Applied Physics at the University of Kiel who helped to run the conference, I am particularly grateful to Mrs. Hildegard Kock, who carried a remarkable work load efficiently and cheerfully.

Kiel, December, 1982 *E.O. Schulz-DuBois*

Contents

Part 3 Photon Correlation Spectroscopy of Brownian Motion: Polydispersity Analysis and Studies of Particle Dynamics

Overview

Erich O. Schulz-DuBois

Institut für Angewandte Physik der Universität Kiel, Olshausenstraße 40-60
D-2300 Kiel, Fed. Rep. of Germany

The table of contents - like the conference programme - shows three major parts. The first one deals with methods of photon correlation velocimetry and their applications in various, sometimes uninviting environments. The second one is devoted to correlation techniques as such and their application to signals other than photoelectron pulses. The third one discusses photon correlation studies of Brownian motion and the associated data reduction schemes for evaluating the particle size distribution, together with the application of these techniques to the study of reacting and interacting particle systems.

In the following paragraphs, the individual papers in this book are briefly commented on. This should help the reader to find those papers which best serve his need and curiosity.

Paper 1 by SCHULZ-DUBOIS is intended for readers without previous exposure to the subject matter of this book; it views photon correlation techniques as a branch of spectroscopy and introduces theoretical background, technical concepts and typical applications.

Papers 2 to 21 are devoted to photon correlation velocimetry; more specifically, 2 to 13 discuss laser Doppler or cross-beam velocimetry; of these, 8 to 13 describe applications of this technique under difficult or "hostile" environmental conditions; papers 14 to 19 are concerned with laser transit or laser two-focus velocimetry; and finally papers 20 and 21 propose methods for the direct measurement of components of the velocity gradient tensor.

Paper 2 by ABBISS is both an introduction to and a survey of laser Doppler velocimetry by photon correlation; it reviews photoelectron statistics and techniques such as clipping in the experimental realization of photon correlation functions; it finally gives a rather complete survey on how to extract velocity distribution data from measured correlograms.

Paper 3 by BROWN reviews both laser Doppler and transit velocimetry; its primary concern is the practitioner's desire to obtain most reliable and accurate data on velocity distribution from experimental correlograms; the argument is supported by many measured and simulated results.

Paper 4 by INMAN and BRADBURY addresses the question of how one may obtain accurate turbulence data even if the laser beam cross section is not precisely Gaussian as assumed in many data extraction algorithms.

Paper 5 by CATALANO et al. gives a fluid dynamic characterization of the turbulent wake behind a cylindrical obstacle as measured by photon correlation velocimetry in a "two-dimensional" air tunnel.

Paper 6 by BRDLIK et al. reports experimental data on the turbulent boundary layer inside a vertical tunnel of rectangular cross section under conditions of convection which is either free - due to thermal effects only - or forced - driven by a fan - or both in combination.

Paper 7 by CATALANO et al. presents velocity and turbulence data for a geometry of considerable importance in aerodynamics: the ejector is a duct connecting the lower and upper side of an air foil, and in operation it leads to increased thrust and lift.

Paper 8 by DANIELSSON and PIKE lists design principles for long-range (about 100 m) laser wind velocimeters and gives optimization strategies; such instruments are in demand in connection with wind power development and airport surveillance; several existing systems are described.

Paper 9 by SOMMER and PFEIFER shows advantages for pulsed as opposed to c. w. laser operation in long-range velocimetry under realistic operating conditions.

Paper 10 by DURST and RICHTER describes an operating long-range wind velocimetry system and discusses the underlying design considerations; a sample of data measured at a distance of 106 m is given.

Paper 11 by NORRIS shows the necessity to know the flow field just forward of the propeller of full-scale operating ships; it describes the technical difficulties of measurements in this environment, their solution and sample results.

Paper 12 by KUX and STÖHRMANN, originally offered in the discussion period, presents similar data obtained with different equipment on other ships.

Paper 13 by MAX and FERNSTRÖM shows velocity and turbulence data of the gas flow inside the cylinder of a running diesel motor; the latter was run without fuel injection and without combustion.

Paper 14 by ERDMANN is a thorough probabilistic treatment of turbulent particle flow in the laser transit velocimeter and the resultant consequences for the correlation function; corrective algorithms based on these results are expected to yield highly accurate velocity, turbulence and other data.

Paper 15 by BROWN and RICHARDS discusses an interesting aspect of flow measurements inside a transonic cascade; evidence for vortex shedding at the Strouhal frequency is found in the wake close to the blade trailing edge by what amounts to an observation of rate correlation.

Paper 16 by GILL et al. presents measurements inside a rapidly turning turbocharger and points to a difficulty having to do with relating measurements obtained at different angles.

Paper 17 by KELLEY et al. gives evidence for a scaling law in the performance of transit velocimeters; based on the empirical Junge size distribution of aerosol particles, it allows predictions for velocimeter data rates if the range and optical system parameters are altered.

Papers 18 and 19 by SMART and MAYO, available only as abstracts, indicate some results that were obtained with an advanced 5 ns correlator and sophisticated data reduction schemes.

Paper 20 by HANSON proposes to measure velocity gradients by a combination of Fourier optical components with traditional detection and cross-correlation equipment.

Paper 21 by KEVELOH and STAUDE describes a velocity gradient measuring system; it is based on the speckle system due to embedded scattering particles; the speckle pattern moves due to the gradient flow field, and this motion is identified by cross-correlation.

Papers 22 to 27 report on advances of correlation techniques per se; this refers to alternative schemes of signal preprocessing and processing in the correlator; it also includes the use of signals other than photon detection pulses in correlation techniques.

Paper 22 by SCHÄTZEL reviews the correlation of several signal types including that of photoelectron pulses under the unifying concept of rate correlation; it treats traditional and novel schemes of preprocessing for format reduction of correlator input signals; and it discusses the structurator currently being developed which seems to become a competitor to existing correlators.

Paper 23 by ADRIAN shows that fixed level clipping does not lead to distortion of the correlation function if the input signal is dominated by a constant background light intensity leading to Poisson fluctuations of photoelectron pulses.

Paper 24 by PFISTER et al. presents results obtained by the cross-beam rate correlation technique on the onset and frequency spectra, of singly or multiply time-periodic laminar and weakly turbulent flows in the so-called Taylor geometry.

Paper 25 by KLEWE et al. reports on the eddy size, i. e. the turbulent microscale, in a flame; this information is deduced from non-Gaussian light intensity fluctuations resulting from the fluctuating index of refraction.

Paper 26 by BRYANSTON-CROSS et al. is concerned with the dynamics of index gradients as measured by schlieren correlation techniques; a study of transonic flow in a linear turbine section and around an air foil model reveal the vortex shedding frequency.

Paper 27 by MERZ et al. reports on optical phase and phase difference statistics that were obtained by newly developed interferometers on a plume of hot, rising air; the data can be fitted by the so-called gamma and K distributions, respectively.

Papers 28 to 43 deal with photon correlation studies of particles undergoing Brownian motion. More specifically, papers 28 to 33 address the problem of polydispersity: how to get best the particle size distribution out of a measured correlation function; there seem to be different, sometimes controversial, answers to this question; papers 34 to 43 treat a variety of problems of current research interest; this includes contributions to photon correlation spectroscopy techniques as such and applications of these techniques to the analysis of specific types of particle dynamics.

Paper 28 by OSTROWSKY and SORNETTE gives a critical review of existing methods for polydispersity analysis and practical hints for their application to obtain an approximate particle size distribution.

Paper 29 by BERTERO and PIKE recognizes the evaluation of particle size distribution functions from measured correlograms to be an "ill-posed inversion problem" in the sense of Hadamard and argues in favor of a linear eigenfunction analysis.

Paper 30 by FORD and CHU gives practical advice on iterative approaches for evaluating particle size distributions; the number of parameters which characterize the distribution is increased until the experimental signal-to-noise ratio of the correlogram does not permit the extraction of further information.

Paper 31 by DANOVICH and SERDYUK proposes regularization as the preferred method for the computation of particle size distributions from measured correlograms and reports experiences with computer simulations.

Paper 32 by PROVENCHER describes a versatile and well-tested computer method for the evaluation of particle size distribution functions from measured correlograms; it makes use of regularization and reasonable additional constraints such as positivity and least curvature of the result function.

Paper 33 by STELZER et al. reports experimental data that were evaluated with the help of PROVENCHER's programme.

Paper 34 by PUSEY is a tutorial introduction to the dynamics of interacting Brownian particles; it outlines the current understanding of these phenomena and relates it to experimental observations by means of photon correlation spectroscopy; distinctions of identical and non-identical charged particles and of large or small scattering wave vectors are essential for the interpretation of results.

Papers 35 and 36 by GRÜNER et al., available only as abstracts, report measurements on the Brownian motion of a macromolecule in the dilute, the semidilute, and the concentrated regimes, respectively, and on the dynamics of latex spheres in solution whose arrangement is liquid-like at low and crystalline-like at high concentration.

Paper 37 by KOPS-WERKHOVEN et al. deals with a system of Brownian particles where, by varying the solvent refractive index, a smaller or larger part is made visible. This allows the experimental identification of two modes of diffusion, i. e. collective and self-diffusion, which should be distinguished at high particle concentration.

Paper 38 by RANDLE and RARITY deals with aggregating latex particles; as the particles grow by aggregation, their number decreases; by incoherent detection and multichannel analysis, the scattered light intensity distribution function is obtained and this, in turn, gives information on the number of scatterers.

Paper 39 by WILTZIUS et al. (abstract only) reports on fibrin polymerization, i. e. the process of blood coagulation; static and dynamic light scattering data reveal detailed information on the process by which bleeding is stopped in nature.

Paper 40 by CORTI and DEGIORGIO deals with mixed micelles, i. e. micelles made out of two rather different molecular constituents. Static and dynamic light scattering data support a simple formula for the composition of the mixed micelles.

Paper 41 by HILLER and SIMONSOHN reports on dynamic light scattering of thermal fluctuations in a liquid; the data give an accurate value of the heat conductivity without the need for a macroscopic thermal gradient; a small inconsistency in the data allows several tentative explanations and stimulates further work.

Paper 42 by DHONT analyses the effects of multiple scattering on measured light intensity and correlation functions; an iterative scheme is given which allows the calculation of single scattering data from measured data provided only double scattering effects need be corrected for.

Paper 43 by KASK et al. deals with fluorescence correlation; for the selected fluorescent molecules, their dynamics may be studied in complex environments; this basically involves number fluctuations since the fluorescent emission is incoherent; some results and problems of this technique are discussed.

Introductory Survey

1. High-Resolution Intensity Interferometry by Photon Correlation

Erich O. Schulz-DuBois

Institut für Angewandte Physik der Universität Kiel, Olshausenstraße 40-60
D-2300 Kiel, Fed. Rep. of Germany

In this tutorial introduction, photon correlation is introduced as a method to achieve very high resolution in optical spectroscopy through detection and processing of light signals in the time domain. The underlying basic theorems are reviewed: the Wiener-Khintchine theorem which gives the relation between spectrum and amplitude autocorrelation function; the Siegert relation which gives the connection between amplitude and intensity autocorrelation functions; and the Mandel formula which describes the Poisson (or shot noise) character of the photodetection process and gives rise to the factorial moment equations.

The experimental determination of correlation functions requires digital processing, mostly in special hardware parallel processors known as correlators. The working principle is briefly described for three correlator types. These include the original one-bit correlator where the required multiplication is replaced by the fast logical AND operation; the fast correlator for sparse signals in which the time differences between one-bit events are stored in a multichannel analyzer; the four-bit by four-bit correlator (and structurator) in which multiplication (or differencing and squaring) is done by fast table look-up in a read-only-memory.

As an application of photon correlation techniques, the intensity interferometry of HANBURY-BROWN and TWISS is cited in which the diameter of a star is obtained from the loss of coherence in the correlogram with increasing distance between photodetectors. Another important application, laser velocimetry, involves a Doppler frequency shift. It can be detected only by beating with a second frequency, for example in a reference beam setup or in the crossbeam geometry. The other important application is the measurement of the diffusion constant in photon correlation studies of Brownian motion. In spectroscopy terms, this is a line-broadening mechanism which corresponds to a Lorentzian line with a typical width in the kHz range.

Finally it is pointed out that the availability of fast digital correlators makes it possible to determine also correlation functions of signals other than photoelectron pulses. This should open the way to further applications of correlation techniques in other fields of science and technology.

1. Introduction

Spectroscopists often wish to enhance the resolution of their instruments. They do so not so much to get better numbers for known effects. Rather they often may observe entirely new effects. The field of laser spectroscopy, to mention an important example, has given previously inaccessible information on the interaction of atomic systems with monochromatic radiation through greatly enhanced resolution.

Sometimes the spectroscopic information of interest is a line shift. Consider, for example, the Doppler shift due to a scatterer moving with velocity ν which is of order $\Delta\omega \quad \omega\nu/c$ where ω is the light frequency and c the velocity of light. Practically interesting velocities range between 1 μm/s for fluid motion inside biological cells and 1 km/s for supersonic aerodynamics. For visible light the corresponding Doppler shifts range between 1 Hz and 1 GHz. Apparently only the top end of this frequency range may be resolved by conventional optical means, for frequency shifts up into the MHz range alternative spectroscopic methods are required.

In other cases one is interested in a line-broadening mechanism. Consider the Brownian motion of spherical particles of 1 μm radius in aqueous suspension at room temperature. The light scattered off these particles gives rise to Lorentzian line broadening by an amount up to 1 kHz which depends on the direction of observation. Here again conventional optical means would not allow sufficient resolution, in fact the stability of most laser sources would not be adequate.

Let us discuss the limiting resolution of conventional optical dispersing instruments. For a prism spectrometer one has to compare the diffraction limited spot size of monochromatic radiation with the transverse displacement of this spot when the wavelength is changed. One obtains for the resolution

$$R = \lambda/\Delta\lambda = \nu/\Delta\nu = B\, dn/d\lambda \;,$$

where λ and ν are optical wavelength and frequency, and n and B index of refraction and baselength of the prism, respectively. For λ = 500 mm, B = 10 cm and $dn/d\lambda = 10^{-4}\ \text{nm}^{-1}$ one finds $R = 10^4$ and $\Delta\nu$ = 60 GHz. A similar consideration for a grating spectrometer leads to

$$R = \nu/\Delta\nu = mD/g = mN \;,$$

where N is the total number of lines, D the total width, g the grating constant of the grating and m the order of the spectrum (m is a small integer). For λ = 500 nm, D = 10 cm, g = 1 μm, $N = 10^5$ and $m = 1$ one finds $R = 10^5$ and $\Delta\nu$ = 6 GHz. While these numbers are no firm upper limits, it is at least expensive to do better. For an interferometer the resolution is given by the ratio of frequency ν to frequency width $\Delta\nu$ of the transmission maximum

$$R = \nu/\Delta\nu = \nu\, Q/\Delta f = \nu Q 2L/c \;;$$

here $\Delta f = c/2L$ is the free spectral range, Q the quality factor or finesse, and L the distance between mirrors. For λ = 500 nm, L = 10 cm, and Q = 250 one finds $R = 10^8$ and $\Delta\nu$ = 6 MHz.

One may look at the dispersion in these optical instruments as an effect which takes place in the frequency domain. The frequency resolution is between about 1 MHz and 100 GHz. Slower effects or smaller frequency displacements cannot be resolved whereas faster processes and larger shifts can of course be measured.

Now let us discuss signal recording and processing in the time domain. Suppose the signal amplitude is sampled in increments of the sampling period Δt. In view of SHANNON's sampling theorem, the spectral content of the signal may be obtained from these samples up to the limiting frequency $1/2\ \Delta t$. If the signal contains higher frequencies than that, the true spectrum cannot be obtained from the sampled signal. A spectrum evaluated under these conditions is subject to systematic errors known as aliasing. If the signal is sampled and recorded for a total time span τ, the frequency resolution in the spectrum is $1/2\ \tau$, a very small limit in all cases of interest.

Thus, in contrast to signal processing in the frequency domain, processing in the time domain is characterized by a highest resolvable frequency. It should be applied only if the sampling period Δt can be made small enough to be compatible with the fastest variations in the signal. Since the frequency resolution improves with total sampling time span τ, it depends primarily on economic considerations such as the capacity of the storage medium or the size of the computer used for Fourier transformation.

The discussion of the preceding two paragraphs referred to a signal amplitude that should be sampled. Now, in the optical spectral range, it is not possible to measure the instantaneous value of the electrical field amplitude, at least not if rapid periodic repetitions of the measurement are envisaged.

What can be measured rapidly and repeatedly, on the other hand, if only with statistical scatter, is the optical power or intensity. The question arises, therefore, whether and what spectral information may be obtained from sampled intensity values. The answer to this question forms the basis of the intensity interferometry of HANBURY-BROWN and TWISS [4]. It also forms the basis of photon correlation techniques [1, 2, 3] as discussed in this volume. In this technique, the statistical nature of the photon detection process used for intensity measurements is explicitly taken into consideration. The photon correlation technique was pioneered by E. R. PIKE and several members of his laboratory at the Royal Signals and Radar Establishment at Malvern, UK. The formal relation between the spectrum and sampled intensity estimates in terms of photon counts will be treated in the third section. Prior to that, in the second section a simple model is discussed which should reveal the pertinent facts in a rather intuitive way.

2. Simplified model discussion of intensity interferometry

Consider the signal amplitude described by equation (1) and illustrated in fig. 1

$$A(t) = A_o \sum_k E(t-t_k) \exp\left(-\gamma(t-t_k)\right) \exp\left(i\omega_o(t-t_k)\right) . \tag{1}$$

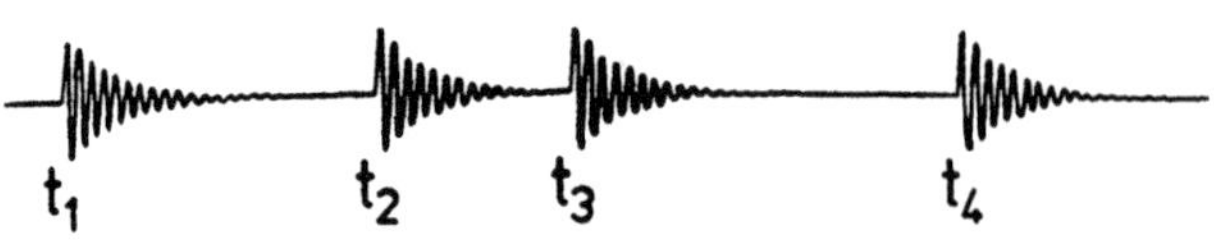

Fig. 1. Signal amplitude given by equation (1)

Here $E(t)$ is the unit step, $E(t) = 0$ for $t < 0$ and $E(t) = 1$ for $t > 0$, A_o, γ, and ω_o are amplitude, decay constant and center frequency of exponentially damped oscillations, respectively, whose starting times t_k are statistically distributed.

This type of signal amplitude may be interpreted as a simplified semi-classical model of spontaneous emission of radiation from an ensemble of excited atoms.

The spectrum associated with such radiation could be obtained experimentally by conventional spectroscopy or interferometry. To be specific, this would be amplitude interferometry since interference is obtained by the superposition of amplitudes, and it yields the power spectrum, i. e., the dependence of power on frequency. From equation (1) one obtains the spectrum given in equation (2),

$$I^{(1)}(\omega) = A_o^2 \; \gamma^2/\left((\omega-\omega_o)^2 + \gamma^2\right) , \tag{2}$$

with the spectral maximum normalized to A_o^2, and it is graphically shown in fig. 2. The spectrum is a Lorentzian line of the center frequency ω_o and half width at half power of γ.

Fig. 2. Spectrum of the signal shown in fig. 1

The intensity corresponding to the above amplitude is given by its absolute square in equation (3) *)

$$I(t) = A^*(t)\,A(t) = A_o^2 \sum_k E(t-t_k)\,\exp\left(-2\gamma(t-t_k)\right) , \tag{3}$$

where crossterms are neglected since their contribution vanishes on average. A plot of the intensity is shown in fig. 3. Since the intensity is necessarily positive, it contains an average d. c. level.

One may ask for the spectrum of the intensity. By Fourier transformation one obtains formally equation (4),

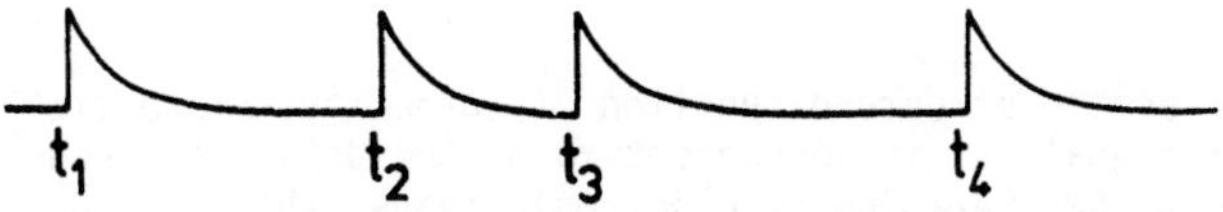

Fig. 3. Intensity signal corresponding to the signal amplitude shown in fig. 1

*) The star * denotes the conjugate complex

$$I^{(2)}(\omega) = A_o^4 \left(\delta(\omega) + 4\gamma^2/(\omega^2+4\gamma^2)\right) . \tag{4}$$

This function is plotted in fig. 4. The first term, the δ-function spike at zero frequency, is the spectral representation of the d. c. level. The second term shows, again, a Lorentzian line, however with center frequency zero and half width at half maximum of 2γ.

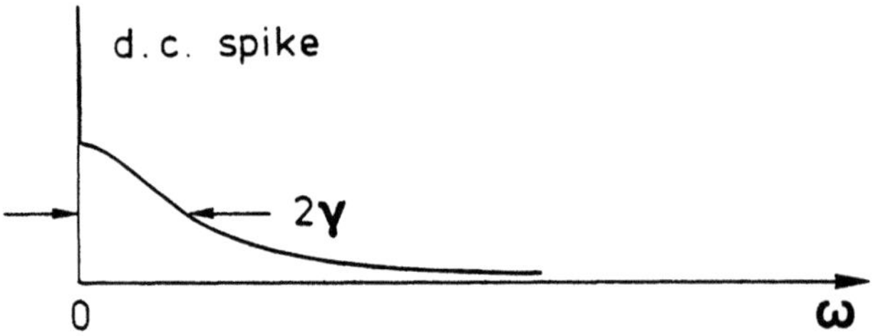

Fig. 4. Intensity spectrum corresponding to the intensity signal shown in fig. 3

Note that $I^2(\omega)$ contains the field amplitude A_o in the fourth power. Thus it is not a quantity that may be measured directly by conventional optical means. It could be measured by two-photon fluorescence or similar nonlinear effects, but these methods require rather high intensity and will not be discussed here.

But how can intensity interferometry be done? All interferometric methods have in common that an instantaneous and a delayed signal are superposed and the superposition is then squared and averaged. In conventional amplitude interferometry, for example, one essentially forms

$$S^{(1)}(\tau) = \lim_{T\to\infty} \frac{1}{T} \int_o^T \left| A(t) - A(t+\tau) \right|^2 dt , \tag{5}$$

where the absolute squaring is inherent in the photodetection process and integral and limit operations imply averaging. As far as intensity interferometry is concerned, one may take advantage of the fact that photodetection produces electrical signals which are - at least statistically - proportional to instantaneous intensity $I(t)$. It is therefore possible to do the necessary operations - superposition of actual and delayed signal followed by squaring and averaging - on the electrical replicas of optical intensity. In other words, one has to form

$$S^{(2)}(\tau) = \lim_{T\to\infty} \frac{1}{T} \int_o^T \left(I(t) - I(t+\tau)\right)^2 dt . \tag{6}$$

This expression is known as the structure function of intensity. Since $I(t)$ represents an electrical signal, it is convenient to do the delay, the subtraction, the squaring and the averaging by electronic means. This will be possible with the aid of an instrument currently under development, the digital structurator.

The square in equation (6) may be expanded giving two squares and a mixed product before averaging. As it turns out, for the most important type of signals, i. e., stationary signals, the squares asymptotically become constants. This means that the interesting spectral information of the structure function may alternatively be obtained by forming the intensity autocorrelation function

$$G^{(2)}(\tau) = \lim_{T\to\infty} \frac{1}{T} \int_{0}^{T} I(t)\, I(t+\tau)\, dt \,. \tag{7}$$

A number of digital electronic instruments known as correlators are on the market which permit the real-time evaluation of the correlation function.

Summarizing this section, for the proposed simple model signal, the conventional power spectrum $I^{(1)}(\omega)$ is centered at the optical frequency ω_o and has halfwidth γ, whereas the intensity spectrum $I^{(2)}(\omega)$ has a d. c. term and a half line centered at $\omega = o$ with halfwidth 2γ. The next section will give a more general and rigorous formulation of these facts. It was also shown that interferometry in general and intensity interferometry in particular call for the superposition of actual and delayed signals followed by squaring and averaging. This naturally leads to the definition of the structure and correlation functions which may be formed by digital electronic structurators or correlators.

3. Basic relations of intensity interferometry

In this section we cite three important theorems which are basic to intensity interferometry. A full, albeit somewhat simplified derivation is given in each case since these theorems are not easy to find in standard textbooks.

The line of reasoning in presenting these theorems is as follows. Suppose, firstly, we were able to measure instantaneous values of the optical field amplitude as a function of time. Then, in order to get the power spectrum out of measurements of amplitude fluctuations we need the WIENER-KHINTCHINE theorem. But we appreciate, secondly, that only estimates of intensity may be measured as a function of time. We then need the SIEGERT relation to translate the measured information on intensity fluctuations into results on amplitude fluctuations. This is possible under certain conditions. We finally realize that intensity measurements in the optical range are carried out by the detection of individual photons which is a process governed by the statistical laws of quantum theory. We therefore need the MANDEL equation and some of its consequences to assess the impact of these statistics on experimental estimates of quantities such as the intensity autocorrelation function.

3.1 The Wiener-Khintchine theorem

This theorem was derived independently by WIENER [6] at M.I.T. in the USA and by KHINTCHINE [7], member of the famous Moscow school of theorists, in the 1930s. It belongs to the mathematics of stochastic variables and correlation functions which then was developed in both places.

Consider a stationary stochastic variable $A(t)$ (which may be complex) and its autocorrelation function defined by

$$G^{(1)}(\tau) = \lim_{T\to\infty} \frac{1}{T} \int_{0}^{T} A^{*}(t)\, A(t+\tau)\, dt \,. \tag{8}$$

We form its Fourier transform

$$\lim_{T'\to\infty} \frac{1}{T'} \int_0^{T'} G^{(1)}(\tau) \exp(i\omega\tau)\, d\tau$$

$$= \lim_{T'\to\infty} \lim_{T\to\infty} \frac{1}{T'} \frac{1}{T} \iint A^*(t) \exp(-i\omega t)\, A(t+\tau) \exp\big(i\omega(t+\tau)\big)\, dt\, d\tau$$

$$= \Big(\lim_{T\to\infty} \frac{1}{T} \int A^*(t) \exp(-i\omega t)\, dt\Big) \Big(\lim_{T'\to\infty} \frac{1}{T'} \int A(s) \exp(i\omega s)\, ds\Big)$$

$$= A^*(\omega)\, A(\omega) = I^{(1)}(\omega) \; . \qquad (9)$$

This is the WIENER-KHINTCHINE theorem which states that the Fourier transform of the amplitude autocorrelation function is the power spectrum. Its significance is obvious: The spectrum may be computed from amplitude measurements in the time domain.

In the above derivation, the gratuitous factor $\exp(-i\omega t)\exp(i\omega t)$ is placed into the definition of the Fourier transform, and $s = t+\tau$ is introduced as a new variable; this causes some complications with the limits of one of the integrals, which, however, are inconsequential; the expression may finally, by changing the order of integrations and limits, be written as the product of conjugate Fourier-transformed amplitudes.

3.2 The SIEGERT relation

We next ask for the connection between amplitude and intensity autocorrelation functions. The desired relation is not universally valid but applies only to a particular class of signals, the so-called Gaussian signals. We consider the Gaussian signal amplitude

$$A(t) = \sum_k a_k \exp\big(i\varphi_k(t)\big) \qquad (10)$$

where the summation extends, in the limit, over infinitely many terms and the phases φ_k are distributed with equal probability between $-\pi$ and $+\pi$; the amplitudes a_k may or may not depend on time. If they do, the fluctuations of a_k and φ_k should be uncorrelated. The signal $A(t)$ may be thought to represent the coherent superposition of light scattered by many particles numbered by the index k and having scattering amplitudes and optical phase propagation delay, a_k and φ_k, respectively.

We first form the amplitude autocorrelation function defined by

$$G^{(1)}(\tau) = \langle A^*(t)\, A(t+\tau) \rangle$$

$$= \sum_{k,l} \langle a_k\, a_l \exp\big(i\varphi_l(t+\tau) - i\varphi_k(t)\big) \rangle \; . \qquad (11)$$

Here the averaging sign < > stands for the combination of integration and limit operation used in equations (5) to (8). Examining equation (11), for $k \neq l$ the phase difference $\varphi_l(t+\tau) - \varphi_k(t)$ has any value between $-\pi$ and $+\pi$ (modulo 2π) with equal likelihood. The exponential function represents a point on the circumference of the unit circle in the complex plane for any choice of l, k, t and τ. But since all points on the circumference are equally probable, the expectation of the exponential function collapses to the point of origin, that is it vanishes for $k \neq l$. Therefore only terms with $k = l$ contribute to the correlation function and we obtain for a Gaussian signal

$$G^{(1)}(\tau) = \sum_k a_k^2 < \exp\left(i\varphi_k(t+\tau) - i\varphi_k(t)\right) > . \tag{12}$$

The intensity is defined as

$$I(t) = A^*(t)\, A(t)$$
$$= \sum_{k,l} a_k\, a_l \exp\left(i\varphi_l(t) - i\varphi_k(t)\right) . \tag{13}$$

Hence the intensity correlation function is

$$G^{(2)}(\tau) = < I(t)\, I(t+\tau) >$$
$$= < \sum_{klmn} a_k\, a_l\, a_m\, a_n \exp i\left(\varphi_l(t) - \varphi_k(t) + \varphi_n(t+\tau) - \varphi_m(t+\tau)\right) > . \tag{14}$$

A similar argument as above shows that the expectation of the exponential function vanishes unless the indices k, l, m, n are pairwise equal as follows:

a) Let $k = n$ and $l = m$ but $k \neq l$; this gives the most significant contribution

$$(N^2-N) < a_k^2 > < \exp\left(i\varphi_k(t+\tau) - i\varphi_k(t)\right) > < a_l^2 > < \exp\left(i\varphi_l(t) - i\varphi_l(t+\tau)\right) >$$

$$= \left(1-\frac{1}{N}\right) \left| \sum_k a_k^2 < \exp\left(i\varphi_k(t+\tau) - i\varphi_k(t)\right) > \right|^2$$

$$= \left(1-\frac{1}{N}\right) \left| G^{(1)}(\tau) \right|^2 .$$

b) Let $k = l$ and $m = n$ but $k \neq m$; this gives a contribution

$$(N^2-N) < a_k^2 > < a_m^2 > = \left(1-\frac{1}{N}\right) \left(G^{(1)}(0)\right)^2 .$$

c) Let $k = l = m = n$; this gives a contribution

$$N < a_k^4 > .$$

Summing these contributions in the Gaussian limit $N \to \infty$ gives the result

$$G^{(2)}(\tau) = \left(G^{(1)}(0)\right)^2 + \left| G^{(1)}(\tau) \right|^2 . \tag{15}$$

This formula was derived by SIEGERT [8] during World War II in connection with work on the statistical interpretation of radar return signals.

Two comments are indicated. The Siegert relation is not generally valid but only for Gaussian signals. In many research situations one may not be certain whether or not the signal is Gaussian. Then uncritical use of intensity interferometry and, implicitly, of the Siegert relation may lead to erroneous results.

In cases where the amplitude correlation function is a genuine complex quantity, its phase gets lost in the intensity correlation function. Thus there are situations where intensity interferometry is not able to yield as much information as amplitude interferometry.

3.3 The MANDEL formula and factorial moments

In the optical region intensity cannot be measured directly. The photomultiplier, for example, responds to single photons by a current pulse at its output. A response is not obtained for all photons impinging on the photocathode, but only to a fraction given by the quantum efficiency α. Thus the detection of photons is a statistical process. It is governed by Poisson statistics, a fact which has its deep roots in quantum electrodynamics and may be demonstrated by means of GLAUBER's coherent states and P representation [10].

If I is a constant intensity, α the quantum efficiency and T the sampling time, then the number n of photodetection pulses should have the expectation

$$< n > = \alpha IT . \tag{16}$$

Since n is Poisson distributed, the probability of finding n detected photons in the time T is

$$p(n;\ T) = \frac{(\alpha IT)^n}{n\ !} \exp\ (-\alpha IT) . \tag{17}$$

MANDEL's formula [9] is a generalization of equation (17) for the case that the intensity itself is not known but only its probability distribution $P(I)$. While it seems plausible, its full derivation makes use of quantum theory. It reads

$$p(n;\ T) = \int \frac{(\alpha IT)^n}{n\ !} \exp\ (-\alpha IT)\ P(I)\ dI . \tag{18}$$

We now use this formula to derive the expectation values $< n >$ and $< n(n-1) >$ which are also known as the first two factorial moments

$$\begin{aligned} < n > &= \sum n\ p(n;\ T) \\ &= \int \sum n\ \frac{(\alpha IT)}{n\ !} \exp\ (-\alpha IT)\ P(I)\ dI \\ &= \int \alpha IT \sum \frac{(\alpha IT)^{n-1}}{(n-1)\ !} \exp\ (-\alpha IT)\ P(I)\ dI \\ &= \int \alpha IT \sum \frac{(\alpha IT)^m}{m\ !} \exp\ (-\alpha IT)\ P(I)\ dI \\ &= \int \alpha IT \exp\ (\alpha IT) \exp\ (-\alpha IT)\ P(I)\ dI \\ &= \int \alpha IT\ P(I)\ dI \\ &= \alpha T < I > \end{aligned} \tag{19}$$

$$< n(n-1) > = \sum n(n-1)\, p(n;\, T)$$

$$= \int \sum n(n-1) \frac{(\alpha IT)^n}{n\,!} \exp(-\alpha IT)\, P(I)\, dI$$

$$= \int (\alpha IT)^2 \sum \frac{(\alpha IT)^{n-2}}{(n-2)\,!} \exp(-\alpha IT)\, P(I)\, dI$$

$$= \int (\alpha IT)^2 \sum \frac{(\alpha IT)^m}{m\,!} \exp(-\alpha IT)\, P(I)\, dI$$

$$= \int (\alpha IT)^2 \exp(\alpha IT) \exp(-\alpha IT)\, P(I)\, dI$$

$$= \int (\alpha IT)^2\, P(I)\, dI$$

$$= \alpha^2 T^2 < I^2 > . \qquad (20)$$

In both derivations, we started with the definition of these expectations; we then eliminated a common factor n or $n(n-1)$ in the numerator and the factorial; the summation index was renamed and the series representation of the exponential function was recognized. Generalizing this approach, the kth factorial moment becomes

$$< n(n-1) \ldots (n-k+1) > = \alpha^k T^k < I^k > . \qquad (21)$$

Several conclusions of relevance to photon correlation techniques may be drawn from these equations. Equation (19) shows that the intensity may be estimated without error from photon counts. The relative variance of the photon count becomes

$$\frac{< n^2 > - < n >^2}{< n >^2} = \frac{< I^2 > - < I >^2}{< I >^2} + \frac{1}{\alpha T < I >} . \qquad (22)$$

Thus, even if the intensity is constant with zero variance, the photon count variance is finite, but decreases with longer sampling time. For two correlated intensities a derivation similar to equation (19) has to make use of the concept of conditional probability and leads to

$$< n_1\, n_2 > = \alpha^2 T^2 < I_1\, I_2 > . \qquad (23)$$

Equation (23) shows that intensity auto- or crosscorrelation functions may be obtained in terms of photon count correlations. An exception is the "zeroeth channel" in autocorrelation. From equations (20) and (19) one finds

$$< n_1^2 > = \alpha^2 T^2 < I_1^2 > + \alpha T < I_1 > . \qquad (24)$$

The second term on the right-hand side is often called the shot noise term since it also occurs in the theory of shot noise. This term would lead to a discontinuous step in experimental correlograms between channels zero and one. In fact, because of this fact, but also because it would be difficult to implement channel zero, it is not offered in commercial correlators.

4. Experimental determination of photon correlation functions by digital correlators

The mathematical definition of the intensity correlation function was (see equations (7) and (14))

$$G^{(2)}(\tau) = \lim_{T\to\infty} \frac{1}{T} \int_0^T I(t)\, I(t+\tau)\, dt \,. \qquad (25)$$

Several modifications suggest themselves in order to arrive at a definition which is suitable for experimental realization by means of digital electronics.

a) Evidently the limit operation $T \to \infty$ is unrealistic in experimental procedures. The use of a finite limit T means that one determines an estimator instead of the asymptotic correlation function.

b) The normalizing factor $1/T$ is omitted. Then the absolute magnitude of the correlation function increases with increasing measurement time T. This makes it possible to determine the correlation function by a continuous accumulation process in the correlator.

c) In equation (25), time is a continuous variable. In order to make digital processing possible, a discrete time basis is introduced in integral multiples of a sampling time Δt. In view of SHANNON's sampling theorem, this does not lead to erroneous spectral information if $I(t)$ has no faster frequency contents than with $1/2\ \Delta t$. Let the discrete time basis be defined by

$$t = i\ \Delta t, \quad \tau = j\ \Delta t, \quad T = N\ \Delta t, \quad j_{max} = J \,.$$

d) The intensity $I(t)$ is a continuous variable whereas the number of photoelectrons n' counted in the time sample i, that is between times $i\Delta t$ and $(i+1)\ \Delta t$, is an integer $n'(i)$.

e) Depending on the choice of Δt and the capabilities of the correlator, the true number $n'(i)$ of detected photoelectrons may be too large for processing in the correlator. A reduced number $n(i)$ may be obtained then by division

$$n(i) = n'(i)/M_1 \,,$$

where M_1 is an integer. The division is implemented by continuous counting up to M_1-1. Every time the counter overflows, a unit is added to $n(i)$. Since there is no additional resetting of the counter, the division does not leave a rest but rather transfers it to the next sampling time. If the number $n'(i)$ has a significant lower limit M_2, $n'(i)$ may also be reduced by subtraction

$$n(i) = n'(i) - M_2 \,.$$

Subtraction is implemented by a counter which is reset to $-M_2$ at the beginning of every sampling time so that at its end the content is $n(i)$. Division and subtraction may also be used in combination.

f) An even more drastic reduction of format to one-bit accuracy is required for the delayed photon count $n(i+j)$ in the original Malvern correlator.

(Some other correlator models require a reduction to a four-bit digital number). This reduction is done as follows:

$$m(i+j) = \begin{cases} 1 & \text{if } n(i+j) > n_o \\ 0 & \text{if } n(i+j) \leq n_o . \end{cases}$$

If, in this decision process, n_o is a fixed number, the process is called clipping and n_o is the clip level. If n_o varies statistically with equal likelihood between zero and the maximal value of $n(i+j)$, the process is called random clipping. A kind of random clipping is called scaling. It is implemented by continuous counting of the photon counts with an upper limit equal to the maximum of $n(i+j)$. A one-bit signal is generated whenever the counter overflows.

Of these methods, clipping is simple to implement but may lead to distortion of the correlation function unless the signal has certain statistical properties. Random clipping is ideal in that it does not cause any distortion of the correlation function; but it tends to be expensive because fast random number generators are needed. Scaling is a more economic way of doing random clipping, but it works without distortion only with sufficiently random signals, for example those of photodetection.

With these modifications, the digital photon correlation function may be written

$$G^{(2)}(j) = \sum_{i=1}^{N} n(i)\, m(i+j) \; . \qquad 1 \leq j \leq J \; . \tag{26}$$

In typical correlators, the number of channels J is of order 10^2, for example 24 to 512, and in typical experiments the number of time samples, N, may range up to 10^8 or 10^{10}.

The operating principle of the original one-bit correlator [2, 3] is illustrated in fig. 5. Photoelectron pulses, perhaps reduced in number by division, are fed to a string of AND gates and simultaneously to a clipping or scaling unit which issues a one or zero signal for every sampling time. These one-bit signals enter a shift register which is controlled, together with the clipping or scaling unit, by the sample time clock. The content of every shift register cell controls the adjacent AND gate: photo-

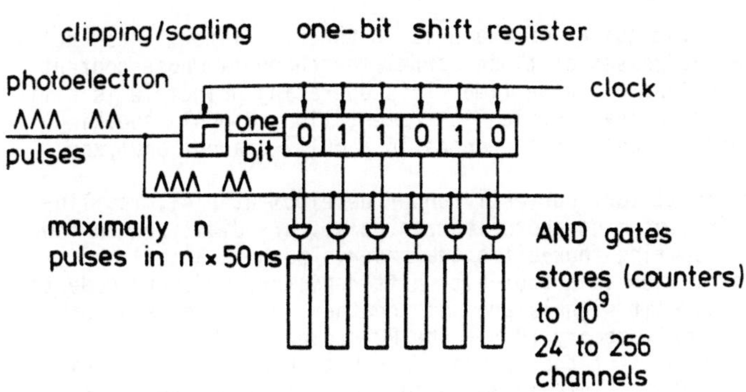

Fig. 5. Scheme of the original one-bit Malvern correlator

electron pulses enter the counters acting as stores if a one is in the shift register cell, but do not if it is a zero. This correlator is fast because the multiplication inherent in the definition of the correlation function (see equation (26)) is replaced by the logical AND operation. The latter is both fast and economic to implement compared to a genuine multidigit multiplication.

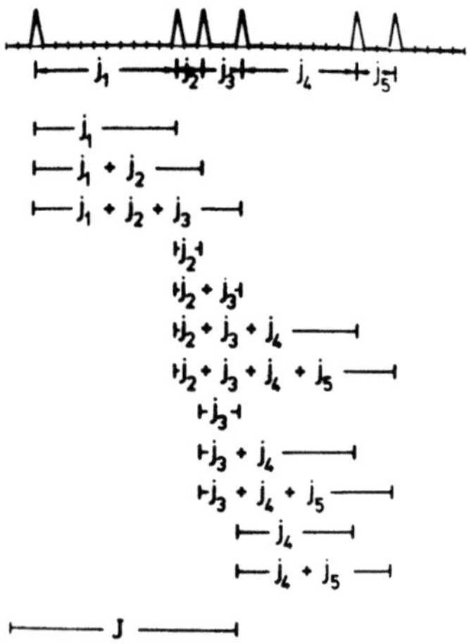

Fig. 6. Illustrating the function of the sparse signal correlator. Typical sparse signal is shown in top line, below all possible delay differences computed from it up to the upper limit J

Another type of correlator [11, 12, 13] is an order of magnitude faster, but works only on sparse signals. As shown in fig. 6, a sparse signal is one where in most sampling times the signal is zero, and only occasionally there is one bit or pulse. Then every non-vanishing term in the sum definition, equation (26), of the correlation function is equal to one. For evaluation of the correlation function, the contents of the store with number j should be incremented by one bit if $j\,\Delta t$ is the elapsed time between single photon counts. In the correlator, the elapsed times between consecutive photon counts are measured in units of Δt and stored as a string of integers j_k, j_{k+1}, j_{k+2} ... in a fast buffer. A computer program then forms all possible delay differences

$$j_k\ ,\quad j_k + j_{k+1}\ ,\quad j_k + j_{k+1} + j_{k+2}\ ,\ \ldots.$$

$$j_{k+1}\ ,\ j_{k+1} + j_{k+2}\ ,\ \ldots.$$

up to a maximum of J, the number of channels in the correlator. These delay differences serve as addresses of those correlator channels whose content will be incremented by one bit at a time. This operating principle is similar to methods used in nuclear physics instrumentation. Roughly speaking it involves the recording of delay differences in a multi-channel analyzer.

A third type of correlator, currently under development [14], is illustrated in fig. 7. The incoming photoelectron pulses are digitized to four-bit format and the resulting four-bit signals serve both as actual and delayed signals, the latter in a four-bit shift register. At every node the actual and delayed four-bit signals are put together and serve as an eight-bit address to a read-only memory (ROM). The ROMs contain the multiplication table and yield the product of the current four-bit inputs, i. e., an eight-bit number, at their output. This is added to the contents of the store. An advantage of this approach is the fact that as an alternative to the corre-

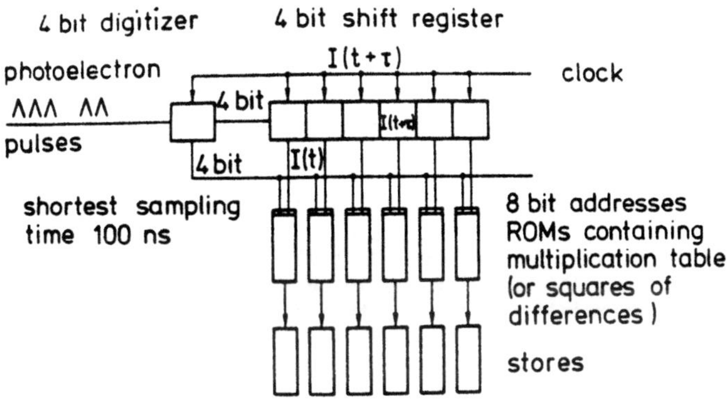

Fig. 7. Scheme of four-bit by four-bit correlator and structurator

lation function

$$G^{(2)}(j) = \sum_{i=1}^{N} n(i)\, n(i+j) \ ,$$

the structure function

$$S^{(2)}(j) = \sum_{i=1}^{N} \left(n(i) - n(i+j)\right)^2 \qquad (27)$$

may be obtained at no extra cost. This is so because commercial ROMs are laid out for a nine-bit input. The ninth bit can then be used to indicate the choice of correlation or structure function, and in the memory of the ROM there is space for both the multiplication table and the table of squares of differences. The essential feature of this system is the replacement of cumbersome and time consuming operations like multiplication or squaring of differences by a fast table look-up.

This brief discussion of several typical correlators is not meant to be exhaustive. In any event they may be characterized as special purpose hardware parallel processors. It should be added, however, that with sufficiently slow or sparse signals, the correlation or structure function may be formed by means of a general purpose computer or minicomputer with special software and serial processing. If table look-up is employed in the computation, the structure function can be conveniently formed with a much smaller table since for M distinguishable signal levels the number of possible products in the multiplication table is M^2 whereas the number of squares is only M. These so-called software correlators or structurators are several orders of magnitude, typically two to four, slower than the hardware processors.

5. Scientific and technical applications of photon correlation techniques

After having discussed the theoretical basis of photon correlation techniques in section 3 and the electronic processing of correlation functions in section 4, this section is devoted to three rather different applications of photon correlation techniques which belong, respectively, to the fields of astronomy, fluid dynamics, and the borderline area of chemistry and biology.

5.1 Determination of stellar diameters by intensity interferometry

When the papers by HANBURY-BROWN and TWISS [4] on intensity interferometry appeared in the 1950's, they originally stirred up doubts. How should intensities interfere or more precisely, why should the correlation function show a nontrivial structure? The work of Hanbury-Brown and Twiss was defended by Nobel Laureate PURCELL [5] who pointed out that the shape of intensity correlation functions follows from both classical and quantum considerations. The peak at zero delay in fact is equivalent to photon bunching, a consequence of the boson nature of photons. In the present context it is convenient to discuss the measurement of stellar diameters by intensity interferometry in the terminology of photon correlation which, historically, developed one or two decades later.

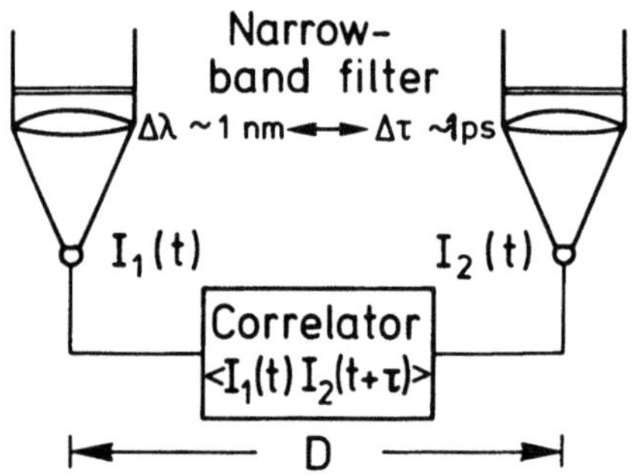

Fig. 8. Experimental setup for stellar interferometery according to HANBURY-BROWN and TWISS

The experimental setup is sketched in fig. 8. Two telescopes, separated by a distance D, each equipped with an optical narrow-band filter and a photomultiplier at the focus, are directed at the same stellar object. The electrical detector signals $I_1(t)$ and $I_2(t)$ are fed to an electronic correlator and the intensity cross-correlation function $\langle I_1(t)\, I_2(t+\tau) \rangle$ is formed.

Geometrical coherence considerations show that two terrestrial observers, separated by a lateral distance D, receive coherent light if

$$D < \lambda/\beta = \lambda L/2R \ ,$$

where λ is the optical wavelength, β and $2R$ are the stellar diameter in angle and absolute units, respectively, and L is the distance between star and earth. Before discussing the intensity cross correlation, consider the amplitude cross-correlation function $\langle A_1^*(t)\, A_2(t+\tau) \rangle$. It could be measured in a conventional interferometer, where light from both telescopes is superposed. The measured amplitude correlogram would have the shape illustrated in fig. 9. The correlogram as a whole is largest for $D = 0$, is reduced to about

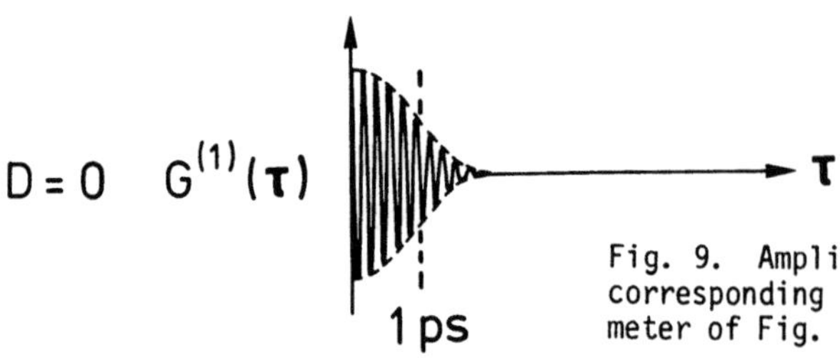

Fig. 9. Amplitude correlation function corresponding to the stellar interferometer of Fig. 8

half that size for $D = \lambda/\beta$ and vanishes for much larger D. As a function of delay time, the contrast tapers off, to half of its value at $\tau = 0$, for a time given solely by the bandwidth of the radiation, that is, of the optical filter. For a filter of $\Delta\lambda$ = 1 nm width at λ = 500 nm, this characteristic time is about 1 ps.

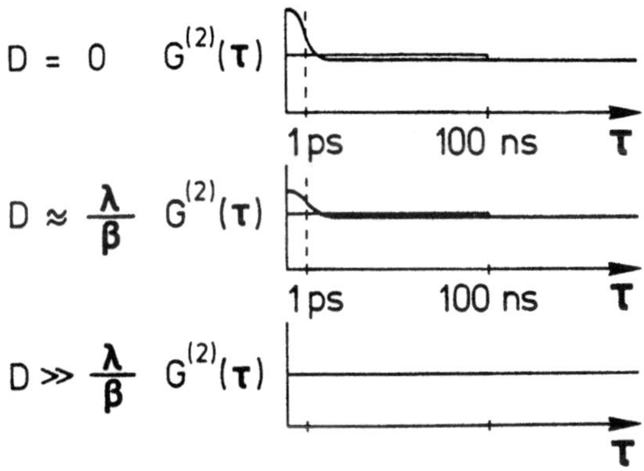

Fig. 10. Intensity correlation functions of the stellar interferometer of Fig. 8

From the amplitude correlation function of fig. 9, the intensity correlation function is obtained via the Siegert relation, equation (15). The result is shown in terms of heavy lines in fig. 10a, b, c for the cases $D = 0$, $D \approx \lambda/\beta$, and $D \gg \lambda/\beta$. For $D = 0$, $G^{(2)}(0)$ is predicted to have twice the value of $G^{(2)}(\tau)$ for larger values of τ. Indeed this is the formal expression of photon bunching which holds for ordinary, so-called Gaussian or chaotic light. Note, however, that such correlation functions could only be obtained with electronic correlators that have a time resolution in the fractional picosecond range. In reality, there is a severe mismatch in time resolution. The correlator may have a sampling or resolving time of 100 ns, that is five orders of magnitude more than the optical filter. As a result, the sharp enhancement of the correlogram at zero delay is reduced by averaging over a 10^5 times larger sampling time. Thus, at $D = 0$ the contents of the first two channels in the correlator are not in a ratio of 2:1, but as symbolically suggested by the horizontal line, in a ratio of $(1+10^{-5})$:1. The stellar diameter is then related to the distance D between telescopes which reduces this ratio to $(1+0.5\cdot10^{-5})$:1.

It speaks for the experimental skill of HANBURY-BROWN and TWISS that they were able to make reproducible diameter measurements on the basis of ratios which are so close to one. From our discussion it is apparent, however, that larger ratios and hence better measurements should be possible if the mismatch in time resolution of the optical and electronic devices is reduced. This calls for sharper optical filters and a faster correlator.

5.2 Laser Doppler velocimetry by photon correlation

This subject [15] is treated extensively in other papers of this volume so that the present discussion may be brief. We restrict ourselves primarily to a negative result: we shall show how laser velocimetry cannot be done.

Consider the setup sketched in fig. 11. Small particles are carried along by a fluid moving with velocity $\bar{u}$ and passing a laser beam. Some of the scattered light is received coherently by a photodetector. Through BRAGG's formula we introduce the scattering wave vector $\bar{q}$ of magnitude

$$q = (2\pi/\lambda)\ 2 \sin \vartheta/2 \tag{28}$$

whose direction is also shown in fig. 11 together with the scattering angle ϑ. Let x be a coordinate measured, and u be the component of $\bar{u}$, in the direction of $\bar{q}$.

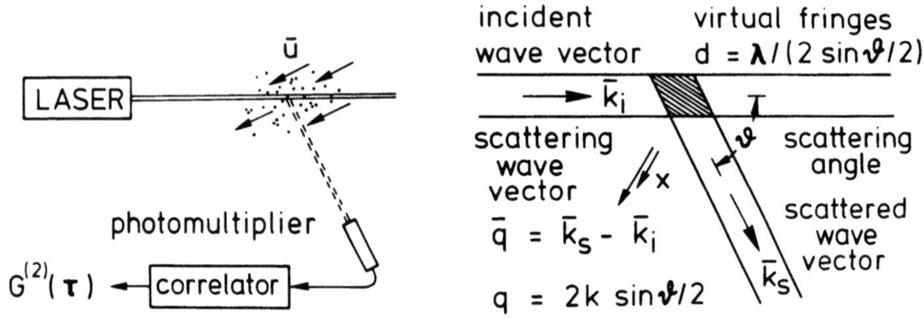

Fig. 11. Laser light scattering experiment which cannot be used for velocity measurements. At left, sketch of the setup; at right, details of the scattering geometry.

A consideration of optical phase delay shows that the amplitude of scattered light reaching the detector surface can be written

$$A(t) = \sum_k a_k \exp\left(iqx_k(t) + i\varphi_{ko}\right) . \tag{29}$$

The summation extends over particles which are both illuminated by the laser and in view of the detector. The amplitude correlation function becomes

$$G^{(1)}(\tau) = \left\langle \sum_{k,l} a_k a_l \exp\left(iqx_l(t+\tau) - iqx_k(t)\right) \right\rangle . \tag{30}$$

The argument previously used in going from equation (11) to (12) applies here, too. Cross terms with $k \neq l$ do not contribute to the expectation since the particles are distributed randomly in space, so that

$$G^{(1)}(\tau) = \sum a_k^2 \left\langle \exp\left(iqx_k(t+\tau) - iqx_k(t)\right) \right\rangle . \tag{31}$$

But in uniform flow with velocity component u, the displacement during time τ is

$$x_k(t+\tau) - x_k(t) = u\tau .$$

This leads to the amplitude correlation function

$$G^{(1)}(\tau) = \sum a_k^2 \exp(iqu\tau) \tag{32}$$

which is necessarily complex. For photon correlation we need the intensity correlation function which from the SIEGERT relation (15) is

$$G^{(2)}(\tau) = 2\ (\sum a_k^2)^2 = \text{const.} \qquad (33)$$

Evidently from a constant no information can be extracted. One has to conclude that velocity measurements are not possible with a single laser beam as in the setup of fig. 11. We could also arrive at the same conclusion more directly. At the photodetector the light frequency is shifted to ω_o+qu from the original laser frequency ω_o. But since there is only this one frequency present at the photodetector surface, no frequency information remains after photodetection.

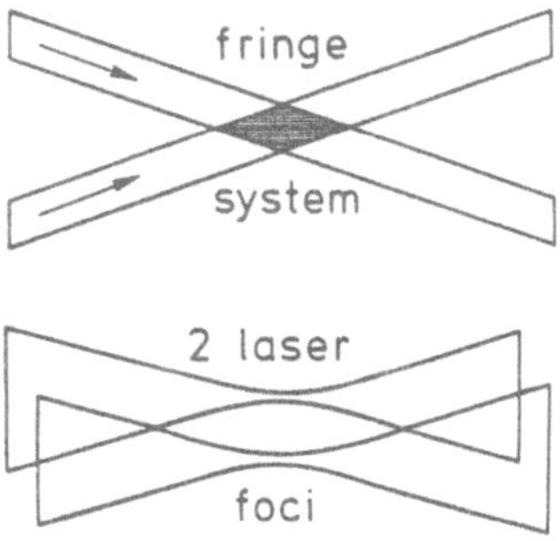

Fig. 12. Laser beam geometries for velocimetry

In fact, as is well known, two laser beams are required for laser velocimetry. The most frequently used arrangements are, as shown in fig. 12, the cross-beam geometry and the laser transit or two-focus system. Much of their early development goes to the credit, respectively, of PFEIFER at ISL and SCHODL at DFVLR. In either case, fluid velocities in the direction and location indicated by the arrows are measured by analyzing light scattered off particles which move with the fluid. The function of these systems may be explained in several ways.

a) Both focal regions in the laser transit system may be viewed as light barriers. The velocity measurement is then a time-of-flight measurement for passing particles.

b) As indicated in the figure, there are interference fringes in the overlap region of the cross-beam geometry. These may be considered as a multiple light barrier which gives rise to a multiply peaked scattering signal.

c) Light of the first beam is scattered off a moving particle and hence shifted in frequency. The same happens to the second beam. At the detector, the difference between these shifted frequencies is obtained.

d) In view of the two laser beams, the amplitude autocorrelation function contains two terms similar to equation (32) with different values for qu. After forming the absolute square called for in the Siegert relation, the intensity correlation function contains the difference frequency term.

e) The superposition of two light signals, whether they are scattered or direct laser beams, implies optical heterodyning.

Of course, these explanations are more or less equivalent so that a choice is a matter of personal preference.

5.3 Measurement of diffusion constant by photon correlation [16]

The setup is similar to that shown in fig. 11 except that the particles under study are contained in a cell where they move only by Brownian motion. The amplitude autocorrelation function equation (31) as derived in the preceding paragraph is obtained here, too. Its exponent contains the displacement. According to EINSTEIN's theory of Brownian motion [17], the displacement $x_k(t+\tau) - x_k(t)$ is normally distributed with zero mean and mean square

$$< \left(x(t+\tau) - x(t)\right)^2 > = 2D\tau \ ; \qquad (34)$$

here D is the diffusion constant. Note that the mean-square displacement is just the structure function of the instantaneous x coordinate. Next we apply a result due to BLOCH [18], according to which for a normally distributed quantity φ with variance σ^2 one has

$$< \exp(i\varphi) > = \exp(-\sigma^2/2) \ . \qquad (35)$$

Hence we find

$$G^{(1)}(\tau) = \sum a_k^2 \exp(-q^2 D\tau) \qquad (36)$$

and from the Siegert relation we have

$$G^{(2)}(\tau) = (\sum a_k^2)^2 \left(1 + \exp(-2q^2 D\tau)\right) \ . \qquad (37)$$

Thus a measurement of $G^{(2)}$ yields the diffusion constant D which for spherical particles gives the particle radius r through the STOKES-EINSTEIN [17] relation

$$D = kT/6\pi\eta r \ . \qquad (38)$$

Here η is the viscosity of the medium in which the particles are suspended, k the Boltzmann constant and T the absolute temperature.

Photon correlation has proved a fast and accurate method for the measurement of the diffusion constant which, in addition, is non-intrusive. Thus, for example, the change of particle radius due to chemical reactions may be conveniently studied in vitro. This and the above derivation apply to monodisperse systems where all particles are identical. For polydisperse systems, equation (36) is replaced by a superposition of exponentials with the appropriate diffusion constants. In that case it is difficult to obtain the size distribution of the particles out of the measured correlation function with reasonable accuracy. The reason is, loosely speaking, that decaying exponentials are rather similar so that it is difficult to tell them apart in the analysis.

6. Operating conditions for photon correlation experiments and correlation measurements with other types of signals

Like any other experimental method, photon correlation techniques have a preferred operating window. Limitations exist with respect to the time resolution and to the optical power level. The time resolution, given by the shortest sampling time of most commercial correlators, is 50 or 100 nanosesonds and may be as short as a few nanoseconds for the sparse signal correlator mentioned in connection with fig. 6. Of course, all correlators can run with longer sampling time, but there is no point in using a correlator with sampling time larger than say 1 millisecond because a suitably programmed general purpose minicomputer offers competitive performance. The highest intensity that can be handled is of the order 10^8 photons per second. For higher intensity the individual photon count pulses cannot be resolved by fast pulse electronics. Photon pile-up and dead-time, i. e., the inability to process two photo detector pulses one immediately after the other, lead to systematic error in photon correlation measurements at very high photon rates. But there is also a lower limit on the number of photon counts. It is of the order 10^4 photons per second if one wants to get reasonably accurate measurements in a time not exceeding several hours.

The existence and availability of digital correlators made experimentalists consider other uses for these powerful instruments. Some alternative correlator applications will be mentioned here while others will likely be proposed in the future.

a) ADRIAN [19] has considered the case that the photon count rate is too high for time-resolved processing of individual photoelectron pulses. Then it is more reasonable to accumulate the photoelectronic charge in every sampling time and use these charge values as estimates of instantaneous intensity. In fact, these data tend to be more reliable since dead-time effects are largely eliminated. ADRIAN has shown that charge may be used in correlators in place of photon numbers. The problem is essentially one of interfacing. The charge data represent sampled analog values, and it is necessary to convert them into the digital format appropriate for the respective correlator.

b) By means of a light barrier and threshold detection of scattered light, ERDMANN [20, 21] gained a sequence of pulses which represents the arrival rate of particles carried along in an air flow, and feeding this so-called rate signal into a correlator obtained the rate correlation function. If the particles are statistically distributed in volume, the average rate is proportional to velocity. Thus the rate correlation is proportional to the velocity correlation function. From such measurements one obtains directly the period (or periods) of oscillating flows or the time scales of turbulence.

c) A variant of rate correlation was devised by SCHÄTZEL [22] at the author's laboratory. The laser crossbeam fringe system is used. A tracker [23] with free-wheeling feature produces a synthetic Doppler signal of constant amplitude whose instantaneous frequency is synchronized with the Doppler signal received at the photodetector whenever it has adequate amplitude due to the presence of scattering particles. The rate signal, in this case, is the sequence of Doppler cycles. As in Erdmann's original rate correlation, here, too, the result is proportional to the velocity correlation function. However, due to the narrow-band filtering and noise reduction functions of the tracker, the correlogram is obtained with significantly greater accuracy, and, because of the cross-beam geometry, the correlation function refers to Cartesian velocity components.

Other signals which seem suited for correlator applications, either in auto- or cross-correlation mode, include force or displacement signals from strain gauges, pressure in rotating machinary or transonic aerodynamics, and optical phase in gas and combustion dynamics.

The author acknowledges stimulating discussions with a great number of individuals, in fact too many to mention their names. Among them are colleagues in many countries, most notably Dr. E. R. PIKE and his coworkers of the Royal Signal and Radar Establishment at Malvern, UK, who pioneered much of what today is known as photon correlation. He also likes to acknowledge his coworkers at Kiel and, last not least, his students who forced him to look for simple explanations.

References

1 There is no introductory text book on photon correlation techniques; the summer school proceedings [2] and [3] present an excellent synoptic survey of the field.

2 H. Z. Cummins and E. R. Pike, eds., Photon correlation and light spectroscopy, Plenum Press 1974. In particular consult the papers by Cummins, Jakeman, Oliver, Pike, and Pusey.

3 H. Z. Cummins and E. R. Pike, eds., Photon correlation spectroscopy and velocimetry, Plenum Press 1977.

4 R. Hanbury-Brown and R. Q. Twiss, Nature 177, 27 (1956).

5 E. M. Purcell, Nature 178, 1449 (1956).

6 N. Wiener, Acta Math. 55, 117 (1930).

7 Khintchine, Math. Ann. 109, 604 (1934).

8 A. J. F. Siegert, MIT Rad. Lab. Report No. 465 (1943).

9 L. Mandel, Proc. Phys. Soc. 74, 233 (1959).

10 R. J. Glauber, Phys. Rev. 131, 2766 (1963).

11 A. E. Smart and W. T. Mayo, Jr., eds., Proceedings of the fourth international conference on photon correlation techniques in fluid mechanics; see papers by Mayo and Smart for description of a commercial correlator of this type.

12 S. Chopra and L. Mandel, Rev. Sci. Instrum. 43, 1489 (1972); 44, 466 (1973). Describes laboratory models of correlators for second- and third-order intensity correlation measurements.

13 J. D. Cresser, J. Häger, G. Lenchs, M. Rateika, H. Walther in: Dissipative systems in quantum optics, R. Bonifacio, ed., Springer 1982. Describes use of very fast cross correlation measurements for the demonstration of anticorrelated photons.

14 E. O. Schulz-DuBois and I. Rehberg, Appl. Phys. 24, 323 (1981).

15 For an introduction to laser velocimetry, see, for example, T. S. Durrani and C. A. Greated, Laser systems in flow measurements, Plenum 1977; F. Durst, A. Melling and J. H. Whitelaw, Principles and practice of laser-Doppler-anemometry, Academic Press 1976.

16 For an introduction to this technique, see, for example, B. Chu, Laser light scattering, Academic Press 1974; B. J. Berne and R. Pecora, Dynamic light scattering, John Wiley 1976.

17 A. Einstein, Ann. Phys. 17, 549 (1905); 19, 371 (1906).

18 The theorem may be proved by Fourier transforming the normal distribution and setting the new variable equal to one.

19 R. J. Adrian, in: Photon correlation techniques in fluid mechanics, proceedings from the 4th international conference, W. T. Mayo, Jr. and A. E. Smart, eds., Stanford University 1980.

20 J. C. Erdmann and R. I. Gellert, J. Opt. Soc. Am. 68, 787 (1978).

21 J. C. Erdmann and R. I. Gellert, Physica Scripta 19, 396 (1979).

22 K. Schätzel, Optica Acta 27, 45 (1980).

23 B. S. Fedders and A. Köneke, J. Phys. E (Sci. Instrum.) 12, 766 (1979).

Part I

Photon Correlation Velocimetry: Assessment of Methods and Selected Examples of Technical Applications

2. Theoretical Aspects of Photon Correlation Methods in Doppler-Difference Anemometry

J.B. Abbiss

Royal Aircraft Establishment, Farnborough, Hampshire, UK

Summary

In this review, the developments in the theory of photostatistics and photon correlation spectroscopy which are of particular relevance to the design and operation of digital clipping correlators are first summarised. A discussion then follows of the mathematical forms which the scattered-light signal and its autocorrelation function take, in laminar and turbulent flow, for the Doppler-difference type of laser anemometer, which represents a highly practical and efficient form of the apparatus. Various methods which have been proposed for the interpretation of the correlator output are briefly described. An analysis is also made of the information contained in the autocorrelation function resulting from a transit of the scattering region by a single particle in the presence of optical noise. Finally the properties of laser radiation which are especially important in the design of laser anemometry systems are discussed.

List of Symbols

E_{phot}	energy of a photon (4×10^{-19} J approximately for green light)
$g^{(1)}(\tau)$	normalised amplitude autocorrelation function
$g^{(2)}(\tau)$	normalised intensity autocorrelation function
$g_k^{(2)}(\tau)$	single-clipped intensity autocorrelation function
$g_{00}^{(2)}(\tau)$	double-clipped intensity autocorrelation function
$G^{(2)}(\tau), G(\tau)$	unnormalised intensity autocorrelation function
$\hat{G}(\tau)$	estimate of the unnormalised intensity autocorrelation function
$H(\tau)$	intensity autocorrelation function for turbulent flow
$I(t)$	classical intensity of the incident light beam
$\underset{\sim}{k}_s$	wave vector for the scattered radiation
$\underset{\sim}{k}_{01}, \underset{\sim}{k}_{02}$	wave vectors for the incident radiation
m	fringe visibility
n	number of photocounts in a sample time
n_j	number of photocounts in the jth sample time
$\bar{n}$	mean number of photocounts per sample time
N	total number of sample times in an experiment

N_c	number of correlator channels
N_{tot}	total number of pulses received during an experiment
$p_u(u), p_v(v)$	probability density distributions for u and v components of velocity
$p_{uv}(u,v)$	joint probability density distribution for u and v components of velocity
r_d	distance of particle from optical axis at closest approach
r_0	beam radius in measurement volume
R	radius of curvature of wave front
s	fringe spacing
t	time
T_d	dead-time associated with the detector system
T_e	total time occupied by an experiment
T_r	beam-radius transit time
T_s	sample time
u,v,w	components of velocity in an orthogonal coordinate system
$U(t,T_s)$	integrated intensity (over a sample time)

Greek:

α	quantum efficiency of detector system
Δf	Doppler-difference frequency (Hz)
$\Delta\nu_1, \Delta\nu_2$	Doppler shifts in frequency of scattered radiation
ζ	ratio of beam amplitudes
η_u, η_v	turbulence intensities associated with u and v components of velocity (standard deviation divided by mean)
θ	beam intersection angle
λ	wavelength
σ	radius of a laser beam
σ_0	radius of a laser beam at waist
σ_u, σ_v	standard deviations of fluctuations in u and v components of velocity
τ	correlation delay time
ϕ	beam divergence

1. Introduction

The invention of the laser has made possible one of the most significant advances in instrumentation for research in fluid mechanics for many decades. The coherence and directional intensity of the laser beam, which cannot be matched by any conventional light source, can be exploited to yield information on fluid-dynamical variables, including velocity, temperature and density, without introducing any material devices into the flow under study. Here we shall be concerned with the use of laser light for the determination of gas velocity, a technique which is now generally known as laser anemometry.

The fundamental source of information in this case is the Doppler-shifted radiation scattered out of the incident beams by small particles borne along with the fluid. Laser anemometry has now been the subject of intensive development for over a decade, but severe difficulties can still be encountered in some applications, both in equipment design and in the interpretation of the experimental data. This is particularly true, for example, where long optical paths in high-speed enclosed flows are involved. However, the essentially non-intrusive character of these laser-based measurement techniques makes them especially attractive candidates for the quantitative exploration of flowfields in which the instrumentation previously available may suffer from serious limitations; for example, in wake flows and vortices, or the interactions of shock waves and boundary layers. Laser anemometers in which photon correlation is used for the initial acquisition and processing of the data possess, in addition, the advantages of the maximum possible sensitivity, since they respond to individual photoelectric events, coupled with the accuracy inherent in any entirely digital arrangement.

Before considering the nature of the scattered-light signal and the form which the output of a correlator may take under various flow conditions, we shall discuss the developments in the theory of photostatistics and photon correlations which have made possible the construction of an efficient and practical instrument.

2. Photon Counting and Photon Correlation Spectroscopy

The rapid development of the laser in the early nineteen-sixties stimulated the application in the field of optical spectroscopy of techniques already familiar in the radio and microwave regions of the electromagnetic spectrum. In particular, intensity-fluctuation and heterodyne-detection methods initiated an era of ultra-high-resolution studies of scattered laser radiation. The signal processing following detection of this radiation can in principle be carried out equally well by analysis in the frequency or time domain, since the autocorrelation function of the signal and its power spectrum are, by the Wiener-Khintchine theorem, a Fourier transform pair. Signal autocorrelation is in general a more adaptable and convenient procedure, however, and for very weak light fields, when individual photoelectric events can be detected, digital correlation techniques become especially attractive.

We introduce first some of the basic physical ideas, together with the relevant mathematical formulae, which underlie the methods of what is now known as photon correlation spectroscopy. General references for the following discussion include the Proceedings of the two Advanced Study Institutes held in Capri, Italy [1,2], especially the lectures by JAKEMAN and OLIVER in the earlier of these, and the books by CHU [3] and CROSIGNANI *et al.* [4]. For accounts of early photocounting studies with thermal and laser sources, see [5 and 6].

2.1 Light Detection and Photostatistics

The phenomenon on which all photocounting experiments are based is the photoelectric effect, which results from the interaction of optical radiation with a light-sensitive material. The hypothesis that the radiation is absorbed in discrete quanta (photons) was originally proposed by EINSTEIN [7] in order to explain certain experimentally observed features of the interactive process which appeared to conflict with the predictions of classical physical theory. To each absorption of a photon there corresponds the liberation, within the detector material, of a single electron. A complete description of the interaction requires quantisation of the incident radiation field as well as of the electronic levels in the atom. For the present purposes, however, the photoelectric effect can be treated entirely by means of the semiclassical approximation, in which the light incident on the detector is regarded as a classical electromagnetic field and the interaction is with electrons in their quantised states. (See [8] for an extended discussion of this topic.) The electronic transitions are then from the bound states in the solid material of the detector cathode to a quasi-free state outside the cathode. It is through the agency of these electrons that 'photon counting' becomes an experimental possibility.

It has been shown [9] that, using the semiclassical approximation, the probability $p_e(t)$ of emission of an electron in a time Δt is proportional to the classical intensity $I(t)$ of the incident light:

$$p_e(t)\Delta t = \alpha I(t)\Delta t \quad , \tag{2-1}$$

where α is the quantum efficiency of the detection process, which depends on the surface area and other characteristics of the detector, and the intensity is defined as the squared modulus of the positive frequency component of the optical field at the detector [10]. (Since this is equal to the square of the envelope function of the field, such a definition of intensity is entirely consistent with the fact that the response time of the detector is much longer than the inverse of the optical frequency.) It is assumed that the light arrives as a plane quasi-monochromatic wave incident normally on the surface, and that Δt is much longer than the inverse of the mean optical frequency, but much less than the coherence time of the light (the inverse of the frequency spread). In photocounting experiments we are interested in calculating the probability $P(n,T_s)$ of counting n photoelectrons in some interval of time from t to $t + T_s$, T_s being the sample time. It follows from equation (2-1) that, provided the detections are statistically independent (which is a physically reasonable assumption for a large assembly of atoms), the distribution is Poisson [9,11]:

$$P(n,T_s) = \frac{1}{n!}\left\{\alpha U(t,T_s)\right\}^n \exp\left\{-\alpha U(t,T_s)\right\} \tag{2-2}$$

where $$U(t,T_s) = \int_t^{t+T_s} I(x)dx \tag{2-3}$$

is the integrated intensity.

If the intensity is constant, as for example in the case of a laser operating well above threshold, the dependence on t disappears and equation (2-2) reduces to

$$P(n,T_s) = \frac{1}{n!}(\alpha I T_s)^n \exp(-\alpha I T_s) \quad . \tag{2-4}$$

It follows that the mean value of n, which we denote by $\bar{n}$, is now equal to $\alpha I T_s$. Hence, for the constant intensity case, we have the alternative form

$$P(n,T_s) = \frac{\bar{n}^n}{n!}e^{-\bar{n}} \quad . \tag{2-5}$$

If, on the other hand, $I(t)$ is a statistically stationary random variable, the distribution resulting from an experiment occupying a period of time which is long compared with the slowest fluctuations of the light beam will be

$$P(n,T_s) = \frac{1}{n!}\left\langle(\alpha U)^n \exp(-\alpha U)\right\rangle , \qquad (2\text{-}6)$$

where the angle brackets denote an ensemble average with respect to the sample times. In terms of the distribution function for the integrated intensity, $p_U(U)$ say, this expression becomes

$$P(n,T_s) = \int_0^\infty \frac{(\alpha U)^n}{n!} \exp(-\alpha U) p_U(U) dU . \qquad (2\text{-}7)$$

This result is now usually referred to as MANDEL'S formula [11]. An inverse formula, by means of which $p_U(U)$ can be recovered from a known distribution $P(n,T_s)$, has been given by WOLF and MEHTA [12].

When I, and hence U, are deterministic functions, $p_U(U)$ can be redefined as the distribution governing the values of U for sample times taken at random from the set constituting the experiment; the photocounting distribution is then once again given by (2-7). Experimental photocounting distributions agreeing closely with the predictions of MANDEL'S formula were first obtained by FRAY *et al.* [13], using a modulated laser as light source. General expressions for the distributions associated with arbitrarily modulated radiation sources have been obtained by DIAMENT and TEICH [14].

Generally, the mean value $\bar{n}$ of the number of photocounts is, using (2-7),

$$\bar{n} = \sum_{n=0}^{\infty} nP(n,T_s)$$

$$= \sum_{n=0}^{\infty} n \int_0^\infty \frac{(\alpha U)^n}{n!} \exp(-\alpha U) p_U(U) dU = \alpha \int_0^\infty U p_U(U) dU .$$

Hence, if $\bar{U}$ denotes the mean value of the integrated intensity (expressed in units of the energy of a photon),

$$\bar{n} = \alpha \bar{U} . \qquad (2\text{-}8)$$

Higher-order statistics for the integrated intensity can be conveniently obtained from the measured $P(n,T_s)$ by using the factorial moments of the distribution, the mth factorial moment being defined as

$$\left\langle\frac{n!}{(n-m)!}\right\rangle = \sum_{n=m}^{\infty} n(n-1) \ldots (n-m+1) P(n,T_s) . \qquad (2\text{-}9)$$

Using again MANDEL'S formula, we find for example that

$$\langle n(n-1)\rangle = \alpha^2 \overline{U^2}$$

and therefore, from (2-8),

$$\langle n^2\rangle = \alpha\bar{U} + \alpha^2 \overline{U^2} \ . \qquad (2\text{-}10)$$

Excellent agreement between measurement and theory has been reported by OLIVER [1] up to the sixth factorial moment, for both coherent and incoherent sources; examples of these would be, respectively, a laser well above thresh-threshold and a thermal source, such as a tungsten lamp. For a polarised thermal source the electromagnetic field can be represented as a stationary Gaussian process, and it can be shown (see for example BERTOLOTTI in [1]) that, for sample times very much less than the coherence time of the light (roughly, the inverse of the spectral width of the source), the photocounting distribution is of BOSE-EINSTEIN (geometric) form:

$$P(n,T_s) = \frac{\bar{n}^n}{(1+\bar{n})^{n+1}} \ . \qquad (2\text{-}11)$$

For sample times much greater than the coherence time the photocount numbers are governed by the Poisson distribution, as for the coherent source.

In a real detector system, the output of a pulse corresponding to a photodetection is always followed by a brief interval, the dead time, during which the system is incapable of responding to any further detections, and the measured photocount distribution will differ from its ideal form. An exact expression for the modification due to dead-time effects has been obtained by BÉDARD [15]. If the dead time T_d is very much less than the sample time T_s , the observed distribution is approximately

$$P(n,T_s,T_d) = \frac{(\alpha U)^n}{n!} \exp(-\alpha U)\left\{1 + n(\alpha U - n + 1)\frac{T_d}{T_s}\right\} \ . \qquad (2\text{-}12)$$

Studies of the effect of dead time for a Poisson distribution of photocounts from a thermal light source have been made by JOHNSON *et al.* [16]. Their results agree closely with the theoretical predictions. (It should be noted that dead times for currently available equipment are of the order of 10 nanoseconds, and their effects are only significant at the shortest sample times.)

An exact inversion formula by means of which the undistorted distribution can be recovered from experimental data has been given by MANDEL [17].

2.2 Photon Correlations and the Spectrum of the Incident Light

Useful information on the spectrum of simple optical fields can be obtained by varying the sample time and observing the changes induced in the photocount distribution; for a thermal source (such as the tungsten lamp), for example, the form of $P(n,T_s)$ changes from geometric to Poisson as T_s is increased from much less to much greater than the source coherence time. It was soon realised, however, that the autocorrelation function of the photocount numbers encoded spectral information in a particularly convenient and compact form, and we now summarise briefly the theoretical basis for this relationship.

Because optical frequencies are so high (some 10^{14} Hz), the lowest-order correlation function of the light field, supposed stationary and ergodic, that is accessible to measurement is the second-order (intensity) correlation function, defined in normalised form as

$$g^{(2)}(\tau) = \langle I(0)I(\tau)\rangle \Big/ \langle I\rangle^2 \quad , \tag{2-13}$$

where the angle brackets denote time or ensemble averages:

$$\langle I\rangle = \lim_{T\to\infty} \frac{1}{T} \int_{-T/2}^{T/2} I(t)dt \tag{2-14}$$

and

$$\langle I(0)I(\tau)\rangle = \lim_{T\to\infty} \frac{1}{T} \int_{-T/2}^{T/2} I(t)I(t+\tau)dt \quad . \tag{2-15}$$

For a Gaussian optical field (more precisely, if the Fourier field amplitudes are statistically independent), the first-order (amplitude) autocorrelation function $g^{(1)}(\tau)$ can be expressed directly in terms of $g^{(2)}(\tau)$ by means of the Siegert relationship [18]:

$$g^{(2)}(\tau) = 1 + |g^{(1)}(\tau)|^2 \quad . \tag{2-16}$$

Thus measurement of the intensity correlation function can yield information on the optical spectrum; for example, for Gaussian light having a Lorentzian spectrum, the line-width of the source can be determined from the decay rate of the (exponential) second-order function. When the field is non-Gaussian, the formula connecting $g^{(1)}(\tau)$ and $g^{(2)}(\tau)$ will be more complicated, and heterodyne (Doppler) methods are commonly used (see, for example, JAKEMAN in [1]). We shall see later that all the relevant spectral information in laser Doppler-difference anemometry, where two incident beams intersect in the scattering region, is carried by the intensity autocorrelation function. We shall be concerned there with estimates of the unnormalised form of this function, defined by

$$G^{(2)}(\tau) = \langle I(0)I(\tau)\rangle \quad . \tag{2-17}$$

We shall also drop the upper suffix 2 in those discussions.

The determination of the second-order autocorrelation function involves in practice samples taken over a finite time T_s. We therefore introduce a modified function:

$$g^{(2)}(\tau,T_s) = \langle U(0,T_s)U(\tau,T_s)\rangle \Big/ \langle U\rangle^2 , \qquad (2\text{-}18)$$

where the integrated intensity U is defined by (2-3). By generalising MANDEL'S formula to cover the case of joint distributions of photocounts contained in sample times separated by a delay τ, it can be shown [19] that

$$g^{(2)}(\tau,T_s) = \frac{\langle n(0,T_s)n(\tau,T_s)\rangle}{\bar{n}^2} , \qquad \tau \geqslant T_s \qquad (2\text{-}19)$$

where $\bar{n} = \alpha\langle U\rangle$ is the mean number of counts per sample time over the whole experiment or ensemble of experiments. At zero delay the equation requires a statistical correction [19], given by (2-10):

$$g^{(2)}(0,T_s) = \frac{\left\langle\left\{n(0,T_s)\right\}^2\right\rangle}{\bar{n}^2} - \frac{1}{\bar{n}} , \qquad \tau = 0 . \qquad (2\text{-}20)$$

In unnormalised form we have

$$G^{(2)}(\tau,T_s) = \begin{cases} \langle n(0,T_s)n(\tau,T_s)\rangle , & \tau \geqslant T_s \\ \left\langle\left\{n(0,T_s)\right\}^2\right\rangle - \bar{n} , & \tau = 0 . \end{cases} \qquad (2\text{-}21)$$

Thus the autocorrelation function of the intensity fluctuations can be obtained directly from measurements of the photocount distribution made over finite sample times.

2.3 Double- and Single-Clipped Autocorrelation Functions

Eq. (2-21) is in principle a very convenient result, since the photocount numbers $n(t,T_s)$ are already digitised. However, the computation of the full autocorrelation function with the desired time resolution (of the order of 10 nanoseconds) is beyond the capabilities of available electronic circuitry. A solution to this problem is to be found in a theorem discovered by VAN VLECK [20,21] in 1943, which states that if a Gaussian signal is represented by one or zero according to whether its instantaneous value is above or below its mean, the autocorrelation function of this 'clipped' version is equal to $2/\pi$ times the arc sine of the full autocorrelation function. Although the theorem cannot be taken over directly into the optical regime, analogous forms have been derived [22] which are applicable to the photocount sequences obtained from stationary Gaussian light fields where the sample time is much less than the source coherence time. For this purpose we define the clipped count number by the relations

$$n_k = \begin{cases} 1 , & n > k \\ 0 , & n \leqslant k . \end{cases}$$

For clipping at zero (k = 0), the autocorrelation function of the sequence of numbers n_k is given by [22]

$$g_{00}^{(2)}(\tau) = \frac{(1 + \bar{n})2\bar{n} + (1 - \bar{n}^2)g^{(2)}(\tau)}{1 + 2\bar{n} + 2\bar{n}^2 - \bar{n}^2 g^{(2)}(\tau)} . \tag{2-22}$$

The cross-correlation function of the unclipped signal with the delayed signal clipped at some arbitrary level k (the single-clipped autocorrelation function) is

$$g_k^{(2)}(\tau) = \frac{\bar{n} - k + (1 + k)g^{(2)}(\tau)}{1 + \bar{n}} . \tag{2-23}$$

Note that the form of $g^{(2)}(\tau)$ is unaltered by single clipping, and that if $k = \bar{n}$, the clipped and unclipped functions are identical.

These equations provide the theoretical basis for the design of a parallel-processing correlator capable of operating at high speed in the single- or double-clipping mode; it was fortuitous, perhaps, that the electronic logical elements available in the late sixties, when this development occurred, provided an upper limit to the spectral resolution range (about 10 MHz) which so neatly overlapped the lower limit of classical interferometric methods [23].

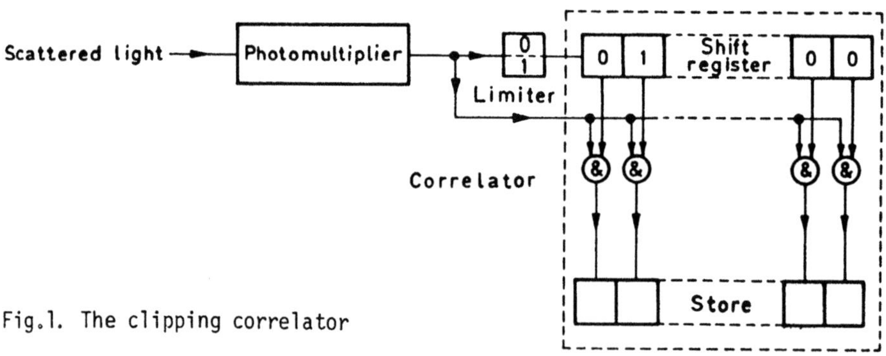

Fig.1. The clipping correlator

The essential features of the instrument developed at the Royal Signals and Radar Establishment at Malvern are shown in Fig.1. The delay register consists of a one-bit shift register carrying a series of 0's and 1's representing the clipped form of the delayed signal. Each sample-time clock-pulse shifts the contents of this register one place to the right and enters the new clipped count from the clipping gate. To form the single-clipped correlation function these delayed clipped data must be multiplied by the unclipped input; this multiplication is performed by AND gates, as shown in the diagram. The simplicity of the multiplication process makes possible the simultaneous computation of the numbers which are to be added to the individual output storage channels. Each channel consists of an independent scaler. The double-clipped autocorrelation function can be generated by feeding the AND gates from the output, rather than the input, of the clipping gate. Further details of the theory and design of this correlator can be found in [1].

The results quoted in (2-22) and (2-23) are only strictly true for the Gaussian optical field, and clipping of a non-Gaussian signal will in general result in some distortion of the spectrum. Formulae connecting the clipped and unclipped autocorrelation functions for certain non-Gaussian fields have been obtained by BENDJABALLAH [24,25]. It can be shown, however, that by varying the clipping level in a uniformly random manner during the course of the experiment the true autocorrelation function can still be recovered, whatever the statistics of the field (see JAKEMAN [1]); the penalty, of course, is usually a considerable loss of efficiency. The varying clipping level can be derived either from a random number generator or from a ramp generator [26], provided that the ramp frequency does not coincide with a significant frequency component in the signal. Alternatively, an effect equivalent to random clipping can be achieved by the technique of scaling [27], in which the signal is correlated with a 'scaled' version of itself. The latter takes the value one on the arrival of every sth count, and zero otherwise, the scaling level s being set sufficiently high that the probability of registering more than one scaled count in a sample time is negligibly small. The question of the statistical accuracy of measurements using digital correlation of photocounts, including the effects of clipping and scaling, is discussed at length in the lectures by JAKEMAN and OLIVER in [1], together with the implications for correlator operation.

The Doppler-difference signal from a single scatterer, which is the basis for the laser anemometry methods considered later, is of course fundamentally deterministic in structure and very far from the Gaussian case. At high light levels distortion of the correlation function is therefore to be expected - at low light levels, however, the true autocorrelation function can still be recovered under appropriate operating conditions, whatever the statistics of the signals. We consider such an experiment in the next section.

2.4 The Single-Clipped Autocorrelation Function for Weak Non-Gaussian Light Fields

We suppose that the correlator is operating in the single-clipping mode at a clipping level of zero and that the value of $\bar{n}$, the mean number of counts per sample time, is much less than one. Suppose also that the experiment lasts for a total of N sample times, each of duration T_s. Denote the number of counts recorded in the jth sample time by n_j and the corresponding clipped count number, which is either 0 or 1, by n_{0j}. (Dead-time effects in the detector system, which are significant only in the first few channels at the shortest sample times, will be neglected.) If N is much larger than the total number of correlator channels N_c, so that end effects can be neglected, the content of the kth channel at the end of the experiment will be

$$G_k = \sum_{j=1}^{N} n_{0j} n_{j+k} \qquad (k = 1,2,\ldots,N_c). \qquad (2\text{-}24)$$

Since the clipping level is zero, the probability that a specific n_{0j} has the value 1 is equal to the probability of absorption of one or more photons by the detector. At sufficiently low light levels the photodetections become independent and Poisson distributed, whatever the statistics of the incident field, the information concerning the intensity fluctuations now being carried by the time-dependent mean value of the Poisson process. Furthermore, from (2-2), the expectation value for the number of photodetections is αU_j, where α is the quantum efficiency of the detector and U_j is the integrated intensity for the jth sample time:

$$U_j = \int_{t=(j-1)T_s}^{t=jT_s} I(t)dt .$$

For a Poisson process with mean αU_j, the probability that no photodetection occurs in this period is $\exp(-\alpha U_j)$ and hence

$$\mathrm{Prob}\{n_{0j} = 1\} = 1 - \exp(-\alpha U_j) .$$

The expectation value αU_j is typically less than 0.1 for $\bar{n} \ll 1$, and we can take, as a reasonable approximation,

$$\mathrm{Prob}\{n_{0j} = 1\} = \alpha U_j . \qquad (2\text{-}25)$$

For sufficiently small T_s (much smaller than the period of the most rapid fluctuation),

$$\mathrm{Prob}\{n_{0j} = 1\} = \alpha T_s I(jT_s) . \qquad (2\text{-}26)$$

These expressions also represent the probability that the unclipped count n_j is non-zero. If the light field is sufficiently weak ($\bar{n} \ll 1$), n_j is very unlikely to be greater than one. (For a Poisson distribution with $\bar{n} = 0.1$, $n_j > 1$ for only one sample in 200.) Hence with little error we can assume that both clipped and unclipped counts can take only the values 0 and 1, and that

$$\mathrm{Prob}\{n_j = 1\} = \mathrm{Prob}\{n_{0j} = 1\} = \alpha T_s I(jT_s) \qquad (2\text{-}27)$$

for sufficiently small T_s.

For weak light fields we can also take the n_j to be independent random variables, each governed only by the Poisson statistics of the detection process. Hence the probability that the product $n_{0j}n_{j+k}$ has the value 1 will be simply the product of the probabilities that the clipped and unclipped counts are separately 1:

$$\mathrm{Prob}\{n_{0j}n_{j+k} = 1\} = (\alpha T_s)^2 I(jT_s) I\{(j + k)T_s\} . \qquad (2\text{-}28)$$

If the experiment lasts sufficiently long for all the significant characteristics of the fluctuating field to have appeared many times, or, equivalently, if a large number of short experiments are carried out at random intervals, the expected value for the content of the kth channel will be closely approximated by the formula

$$G_k = (\alpha T_s)^2 \sum_{j=1}^{N} I(jT_s) I\{(j+k)T_s\} \quad . \tag{2-29}$$

(If the sample time were insufficiently short the summation would be over the products of integrated intensities U_j .)

But the estimate of the unnormalised autocorrelation function of the continuous function $I(t)$ is, after N sample times,

$$\hat{G}(\tau) = \frac{1}{NT_s} \int_{-NT_s/2}^{NT_s/2} I(t) I(t+\tau) dt \quad . \tag{2-30}$$

Eq.(2-29) thus establishes the result that for sufficiently weak light fields and sufficiently short sample times, a single-clipping correlator will generate, at a clipping level of zero, an unnormalised and sampled but essentially undistorted estimate of the classical function $G(\tau)$, whatever the actual statistics of the incident field.

3. The Doppler-Difference Signal and its Autocorrelation Function

The first paper to be published on the measurement of fluid flow by means of the Doppler effect at optical frequencies [28] described an optical arrangement in which part of the output beam of the laser bypasses the flow and acts as a local oscillator at the detector, where it is combined with radiation scattered out of the flow; thus the beat signal carries the Doppler information. Reference-beam or heterodyne arrangements of this sort, with coherent detection, are generally used for laser Doppler studies wherever the scatterers are spatially highly concentrated. However, for experiments in which the scattering particles are sparsely distributed throughout the medium and appear in the measurement region only rarely, as in most fluid-mechanical investigations, the maximum possible optical efficiency is usually required. Largely for this reason, the Doppler-difference arrangement [29] is now much more commonly used in these applications.

3.1 The Doppler-Difference Optical Arrangement

In a Doppler-difference system (Fig.2), two beams of mutually coherent radiation of wavelength λ are incident from different directions characterised by the wave vectors $\underset{\sim}{k}_{01}$ and $\underset{\sim}{k}_{02}$ on the scattering centre, which is moving with some velocity $\underset{\sim}{V}$. The Doppler shifts in the frequency of the radiation scattered in some direction corresponding to the wave vector $\underset{\sim}{k}_s$ are, in units of radians per second,

$$\Delta\nu_1 = (\underset{\sim}{k}_s - \underset{\sim}{k}_{01}) \cdot \underset{\sim}{V} \quad \text{and}$$

$$\Delta\nu_2 = (\underset{\sim}{k}_s - \underset{\sim}{k}_{02}) \cdot \underset{\sim}{V} \quad .$$

(Note that $\underset{\sim}{k}_{01}$, $\underset{\sim}{k}_{02}$, $\underset{\sim}{k}_s$ and $\underset{\sim}{V}$ are not necessarily coplanar.) The Doppler-difference frequency is:

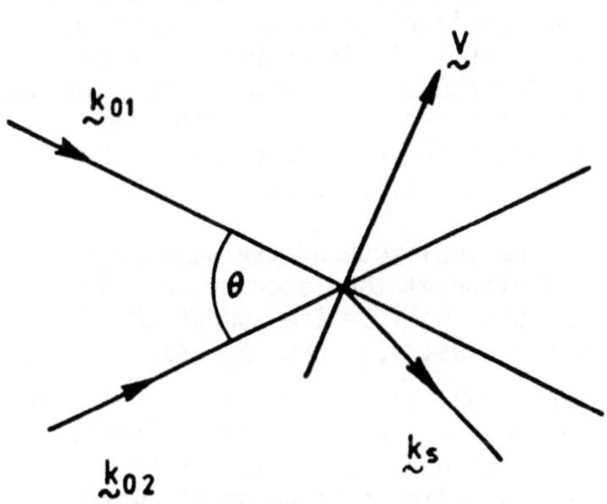

Fig. 2. Vector diagram

$$\Delta\nu_1 - \Delta\nu_2 = (\underset{\sim}{k}_{02} - \underset{\sim}{k}_{01}) \cdot \underset{\sim}{V}$$

which is independent of the direction of the vector $\underset{\sim}{k}_s$. This in turn implies that the receiving lens aperture is not restricted by coherence requirements - the light-collecting efficiency can thus be very high for such an arrangement. If u is the component of the velocity in the direction $(\underset{\sim}{k}_{02} - \underset{\sim}{k}_{01})$ - *ie* normal to the bisector of the angle θ between the beams - then the equation reduces to

$$\Delta\nu_1 - \Delta\nu_2 = \frac{2\pi}{\lambda} 2u \sin\frac{\theta}{2} . \qquad (3\text{-}1)$$

Another interpretation of this result can be made by considering the following experiment. Suppose a scattering screen is set up in the plane normal to the included bisector of the incident beams and passing through the point of intersection of the beam axes. Fringes will be observed on the screen which arise from the interference of the two beams. For plane wave fronts it can easily be shown that the fringe spacing s will be given by the formula

$$s = \lambda \Big/ \left(2 \sin\frac{\theta}{2}\right) . \qquad (3\text{-}2)$$

If the screen is now translated in its own plane with velocity component u normal to the fringes, the frequency (in Hertz) with which the fringes pass a fixed point on the screen will be

$$\Delta f = \frac{u}{s} = \left(2u \sin\frac{\theta}{2}\right) \Big/ \lambda \qquad (3\text{-}3)$$

in agreement with (3-1). However, the widely used 'fringe' model of the Doppler-difference system should not be taken too literally; for example, it cannot account for the reduced modulation of the received signal which can occur when a non-isotropic scatterer is viewed from particular directions. But we can speak, metaphorically, of 'fringes' as if they actually existed in the region where the beams cross and it is in the sense of the experiment with the scattering screen that the word should be understood.

3.2 The Scattered-Light Signal

In practice, the signal at the fundamental frequency Δf will be amplitude-modulated as the scattering particle moves across the beams. In order to calculate the form of this modulation we shall use the more general geometry of Fig.3. Here the included bisector of the beam axes is taken to be coincident with the Z axis; the plane containing the laser beams is taken as the X-Z plane. It is assumed that the divergence of the beams in the overlap region is so small that the beam radii, both supposed equal to r_0, are essentially constant.

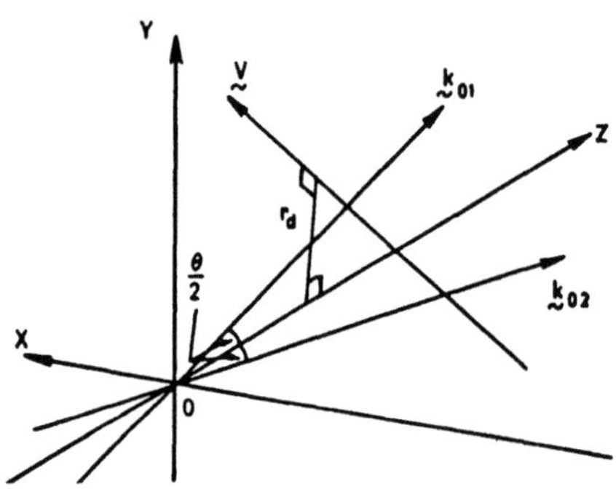

Fig 3 Scattering geometry

The amplitude of the scattered radiation at the detector, at any instant, will be the sum of the amplitudes, $\underset{\sim}{E}_1$ and $\underset{\sim}{E}_2$ say, of the individual scattered fields. A square-law device such as a photomultiplier tube responds to the intensity I of the radiation falling on it, where

$$I = |\underset{\sim}{E}_1 + \underset{\sim}{E}_2|^2 ,$$

and, by expanding the expression on the right-hand side, the manner in which the detected intensity varies with time for a single small particle traversing the scattering region, as shown diagrammatically in Fig.3, can be determined. Suppose time is measured from the instant at which the particle has reached an arbitrary point on its trajectory, having coordinates (x_0, y_0, z_0), and that (u,v,w) are the components of the velocity vector $\underset{\sim}{V}$. Then if ζ is the ratio of the beam amplitudes, it can be shown [30] that

$$I(t) = I_0 \exp\left\{-\frac{2}{r_0^2}(y_0+vt)^2\right\}\left\{\exp\left[-\frac{2}{r_0^2}\left\{(x_0+ut)\cos\frac{\theta}{2} - (z_0+wt)\sin\frac{\theta}{2}\right\}^2\right]\right.$$

$$+ \zeta^2 \exp\left[-\frac{2}{r_0^2}\left\{(x_0+ut)\cos\frac{\theta}{2} + (z_0+wt)\sin\frac{\theta}{2}\right\}^2\right]$$

$$+ 2\zeta \exp\left[-\frac{2}{r_0^2}\left\{(x_0+ut)^2\cos^2\frac{\theta}{2} + (z_0+wt)^2\sin^2\frac{\theta}{2}\right\}\right] \times$$

$$\left.\times \cos\left\{\frac{4\pi}{\lambda}(x_0+ut)\sin\frac{\theta}{2}\right\}\right\} \qquad (3\text{-}4)$$

where I_0 is some constant for the particular transit which depends on the beam amplitudes and on scattering and collecting efficiencies; it is assumed that the scattering efficiency of the particle is constant over the range of scattering angles concerned. For the purposes of this analysis any background radiation has been ignored.

The frequency of the cosine term is seen to be

$$2u\left(\sin\frac{\theta}{2}\right)\Big/\lambda$$

in agreement with (3-3). Figs.4 to 6 illustrate the behaviour of the Doppler-difference signal of (3-4) for transits parallel to the X axis at different distances from the centre of the scattering volume. Here it has been assumed that the beams have the same amplitude, so that $\zeta = 1$ and (3-4) can be written

$$I(t) = 2I_0 \exp\left[-\frac{2}{r_0^2}\left\{(z_0+wt)^2\sin^2\frac{\theta}{2} + (x_0+ut)^2\cos^2\frac{\theta}{2} + (y_0+vt)^2\right\}\right]$$

$$\times \left[\cosh\left\{\frac{4}{r_0^2}(z_0+wt)(x_0+ut)\sin\frac{\theta}{2}\cos\frac{\theta}{2}\right\} + \cos\left\{\frac{4\pi}{\lambda}(x_0+ut)\sin\frac{\theta}{2}\right\}\right]. \qquad \ldots (3\text{-}5)$$

The unit of length ℓ_0 in these figures is the distance along the Z axis at which the e^{-2} beam contours intersect. This is the conventional measure of the longitudinal extent of the scattering volume. v and y_0 have been taken to be zero in all cases. The figures are drawn to the same scale.

The double-peak characteristic of transits well away from the centre arises from the hyperbolic cosine form. If this can be effectively removed by high-pass filtering the amplitude of the remaining cosine term will fall off ellipsoidally with distance from the centre; the contours of constant amplitude of the Doppler-difference signal are ellipsoids.

Eq.(3-4) can be simplified considerably if certain experimental conditions are met. We assume that transits for which w is large are rare and that the collecting optics impose an upper limit on z_0, and therefore on wt, of a few beam radii. If, in addition, the angle θ is sufficiently small, all terms involving $(z_0 + wt) \sin \frac{\theta}{2}$ can be neglected, $\cos \frac{\theta}{2}$ can be replaced by unity, and the equation becomes

$$I(t) = I_0 \exp\left[-\frac{2}{r_0^2}\left\{(x_0 + ut)^2 + (y_0 + vt)^2\right\}\right]$$

$$\times \left[1 + \zeta^2 + 2\zeta \cos\left\{\frac{4\pi}{\lambda}(x_0 + ut) \sin \frac{\theta}{2}\right\}\right] . \tag{3-6}$$

Fig 4

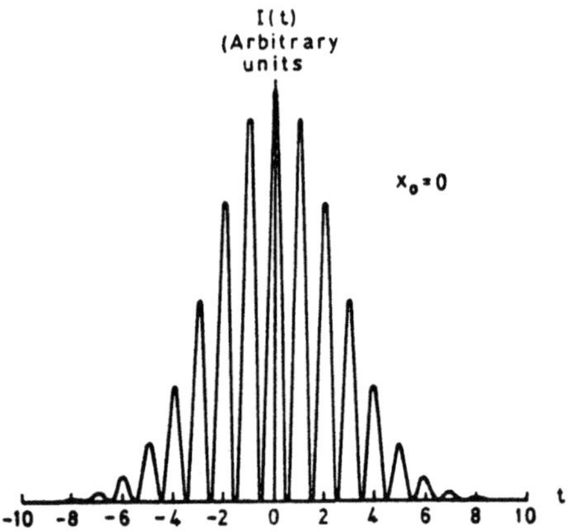

Fig 5

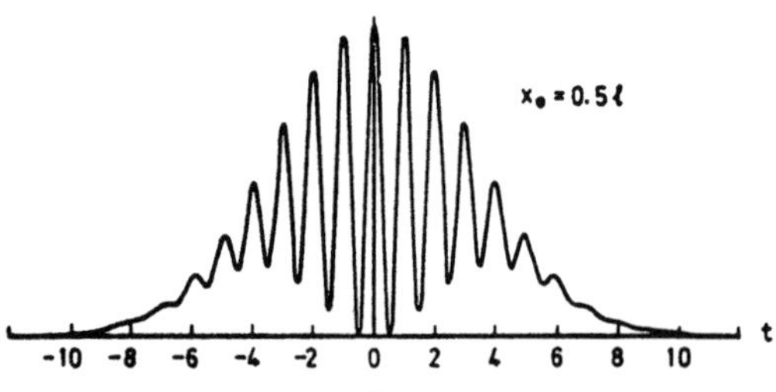

Fig 6

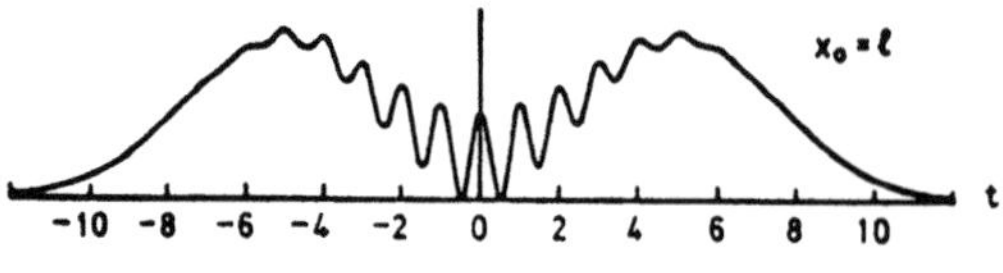

Figs. 4-6. Doppler-difference signals for transits at various distances from the centre of the scattering volume

By taking (x_0, y_0, z_0) as the point of closest approach to the Z axis, I(t) can be expressed in terms of r_d, the distance from the Z axis to this point. For r_d must be perpendicular to the particle trajectory, so that

$$\underset{\sim}{r}_d \cdot \underset{\sim}{V} = 0 \ ,$$

giving

$$x_0 u + y_0 v = 0 \ .$$

Hence, writing m for the quantity $2\zeta/(1+\zeta^2)$ and s for $\lambda/2 \sin\frac{\theta}{2}$ we find

$$I(t) = I_0(1+\zeta^2)\exp\left(-\frac{2r_d^2}{r_0^2}\right)\exp\left\{-\frac{2}{r_0^2}(u^2+v^2)t^2\right\}\left[1+m\cos\left\{\frac{2\pi}{s}(x_0+ut)\right\}\right] \ . \qquad \ldots (3\text{-}7)$$

This equation describes the dependence on time of the response of an ideal classical square-law detector when exposed to the scattered light fields under the stated geometrical restrictions. However, in the semi-classical approximation (see section 2.1), the detection process is assumed to consist of the absorption of individual quanta from the incident field. In this process statistical features play a vital part, and (3-7) now governs the way in which the probability of photodetection is distributed in time; the probability of absorption of a photon is greatest around the peaks of the cosinusoidal term. When the scattered light is extremely weak, so that the output of the detector becomes a train of clearly separated pulses (corresponding to the individual photodetection events at the cathode), the frequency components in the signal can still be extracted if the contributions from a number of successive particles are added together in some appropriate way. Autocorrelation of the pulse train is an efficient method of accomplishing this task, since successive signals are added in the correct relative phase, and signal frequencies are unchanged. We have also seen that, under the correct operating conditions, the clipped autocorrelation function is a sampled but essentially undistorted version of the autocorrelation function calculated as if the detected signal were a continuous representation of the incident light intensity. Hence by basing the calculation on the intensity distribution as given by (3-7) we can determine explicitly the functional form of the output of the correlator when operating in laminar flow; these results can then be used to predict the form of the output in turbulent conditions.

3.3 The Autocorrelation Function in Laminar and Turbulent Flow

After an experimental time T, the unnormalised output $\hat{G}(\tau)$ of an ideal autocorrelator operating on the input signal I(t) would be (see section 2.4)

$$\hat{G}(\tau) = \frac{1}{T}\int_{-T/2}^{T/2} I(t)I(t+\tau)dt$$

where the instant at which the particle reaches the point (x_0, y_0, z_0) on its trajectory is now taken as the mid-point of the experiment.

Since the envelope of $I(t)$ is Gaussian, the total energy in the signal from a single scatterer is finite. Hence we drop the factor $1/T$, allow T to increase without limit and obtain, as the limiting form to which the autocorrelation function will tend,

$$G(\tau) = \int_{-\infty}^{\infty} I(t)I(t+\tau)dt .$$

Inserting the expression in (3-7) for $I(t)$, we find

$$G(\tau) = \frac{a_0}{\sqrt{u^2+v^2}} \exp\left\{-\frac{(u^2+v^2)\tau^2}{r_0^2}\right\}\left[1 + 2m \cos\frac{\pi u\tau}{s} \exp\left(-\frac{\pi^2 r_0^2}{4s^2}\right) + \tfrac{1}{2}m^2 \exp\left(-\frac{\pi^2 r_0^2}{s^2}\right) + \tfrac{1}{2}m^2 \cos\frac{2\pi u\tau}{s}\right]$$

where $$a_0 = \frac{\sqrt{\pi}}{2} r_0 I_0^2 (1+\zeta^2)^2 \exp\left(-\frac{4r_d^2}{r_0^2}\right) .$$

(If more than one scatterer contributes to $G(\tau)$, the quantity r_d has to be given some appropriate mean value.) Provided that $r_0 > 2s$, say, the middle two terms in the square bracket are negligible compared with the other two, and we find, to a very good approximation,

$$G(\tau) = \frac{a_0}{\sqrt{u^2+v^2}} \left[\exp\left\{-\frac{(u^2+v^2)\tau^2}{r_0^2}\right\}\right]\left(1 + \tfrac{1}{2}m^2 \cos\frac{2\pi u\tau}{s}\right) . \qquad (3\text{-}8)$$

This result holds for laminar flow under the simplifying assumptions mentioned above. The form of the autocorrelation function corresponding to a general optical geometry, when the variation of the intensity in time and space is described by (3-4), can be found in [30].

Now suppose we are dealing with a three-dimensional turbulent flow, so that successive contributions to the output of the correlator are characterised in general by different velocity vectors. In the following argument, all particles are assumed to have the same scattering properties and to be completely randomly distributed in the fluid, which will be treated as incompressible. It will also be assumed that not more than one scatterer is present at any instant and that the dimensions of the scattering region are small compared with the length scale of the turbulence.

With the specified limitation on θ, the scattering volume (the spatial region defined now by the field of view of the collecting optics) is approximately cylindrical, the length 2ℓ being always considerably greater than the diameter D of the beams in the experimental arrangements considered here. The cross-sectional area A presented to the flow by the scattering volume is therefore approximately $2D\ell \sin\xi$, where ξ is the angle between the axis of the cylinder (the direction of the w component) and the velocity vector (u,v,w). Hence

$$A \;=\; 2D\ell\sqrt{(u^2 + v^2)\Big/(u^2 + v^2 + w^2)} \quad .$$

If n is the number of scatterers per unit volume of the fluid, the number of transits per unit area per unit time is

$$n\sqrt{u^2 + v^2 + w^2}$$

and the number of transits per unit time across A is

$$2nD\ell\sqrt{u^2 + v^2}$$

which is independent of the w component. Now in a three-dimensional turbulent flow the velocity distribution is characterised by some joint probability density function $p(u,v,w)$. If the duration of the experiment, say T_e , is sufficiently long to include all the components of the turbulence frequency spectrum, the fractional time for which the flow velocity lies in the range (u,v) to $(u+\delta u,\ v+\delta v)$ is

$$\left(\int_{-\infty}^{\infty} p(u,v,w)dw\right)\delta u\delta v$$

and the total number of transits of A during the course of the experiment by particles with velocity vectors in the same range will be

$$2nD\ell T_e\sqrt{u^2 + v^2}\left(\int_{-\infty}^{\infty} p(u,v,w)dw\right)\delta u\delta v \quad .$$

Hence the composite autocorrelation function will be proportional to

$$\int_{-\infty}^{\infty}\int_{-\infty}^{\infty}\int_{-\infty}^{\infty}\sqrt{u^2 + v^2}\; p(u,v,w)G(\tau)dudvdw$$

and the expected output of the correlator in turbulent conditions becomes

$$H(\tau) \;=\; a_1\int_{-\infty}^{\infty}\int_{-\infty}^{\infty}\int_{-\infty}^{\infty} p(u,v,w)\,\exp\left\{-\,\frac{(u^2 + v^2)\tau^2}{r_0^2}\right\}\left(1 + \tfrac{1}{2}m^2\cos\frac{2\pi u\tau}{s}\right)dudvdw \qquad \ldots (3\text{-}9)$$

where a_1 depends on a_0 , the number of scatterers and the duration of the experiment.

Note that the factor $(u^2+v^2)^{-\frac{1}{2}}$ appearing in (3-8) is exactly cancelled by the reciprocal factor arising from the instantaneous rate at which transits occur. This convenient effect, which eliminates a potentially significant source of bias in the data, is due essentially to the fact that for small θ the scattering volume is approximately cylindrical in shape.

Integration over the w component in (3-9) can be carried out immediately, since the kernel of the integral is independent of w . Denoting the joint probability density function for the u and v components by $p_{uv}(u,v)$, we have

$$p_{uv}(u,v) = \int_{-\infty}^{\infty} p(u,v,w)dw$$

and hence

$$H(\tau) = a_1 \int_{-\infty}^{\infty}\int_{-\infty}^{\infty} p_{uv}(u,v)\left[\exp\left\{-\frac{(u^2+v^2)\tau^2}{r_0^2}\right\}\right]\left(1+\tfrac{1}{2}m^2\cos\frac{2\pi u\tau}{s}\right)dudv \quad .$$

... (3-10)

It should be remarked here that if frequency-shifting devices are used in the optical system, a constant offset will be added to the Doppler-difference frequency u/s , and the frequency of the cosinusoid in (3-10) will be changed by the same amount. However, none of the arguments leading up to the result will be affected. Note in particular that the exponential term in (3-8) and (3-10), which arises from the (common) Gaussian field distribution across the beams, is unmodified; it is of course intuitively clear that the distribution will be unaffected by frequency modulation of either beam.

Thus (3-10) represents the fundamental mathematical model for the operation of the correlator. Its quantitative interpretation for various types of flow forms the topic of the next section.

4. Interpretation of the Correlator Output

The following notes on data-reduction techniques are based on a review [31] of the various procedures which are in current use or which have been proposed for processing the autocorrelation function. The output of the correlator under real experimental conditions will of course include a noise component, which usually arises mainly from stray radiation present in the environment of the experiment. Provided that this noise is uncorrelated with the signal or with itself, and that the signal is of adequate strength, its effect will merely be to add a fairly constant contribution to the contents of each channel, whose mean level is the value to which $G(\tau)$ will tend as τ tends to infinity; this quantity is normally included in the data-reduction procedures as another unknown whose magnitude has to be determined. For the purposes of the present discussion it will be assumed that this operation has been carried out. It will also be assumed that the necessary geometrical conditions by means of which (3-6) was derived are satisfied, so that the simplified formulae of (3-8) and (3-10), for the laminar and turbulent cases respectively, are applicable.

4.1 Laminar Flows

Here (3-8) applies and data-reduction can proceed by straightforward curve-fitting. Alternatively, the problem can be regarded as a special case of the more general one of estimating mean velocity in turbulent flow and a single computer programme can be designed to cover both conditions. Purely laminar flows are in any case rarely, if ever, encountered in practice and an estimate of the turbulence level present is usually wanted.

4.2 Turbulent Flows

Here (3-10) applies, and we are immediately presented with an impasse, for we cannot recover the two-dimensional distribution p_{uv} from an experimental function of a single variable. The problem centres essentially on the exponential term, since this factor couples both u and v components into the measured correlation function, and it is obvious that some extra information is required if progress is to be made at all. If an explicit and complete mathematical model for p_{uv} is available, data reduction can again be carried out by curve-fitting to $H(\tau)$. This will not be the case, however, in most problems of real intrinsic interest (*ie* the majority of experiments). We shall therefore describe a number of methods, involving generally weaker assumptions, by means of which useful information about the velocity field can still be recovered from the experimental function $H(\tau)$.

4.2.1 p_{uv} Separable

We write

$$p_{uv}(u,v) = p_u(u)p_v(v) .$$

Note that this implies that the stress term $\overline{u'v'}$ must be zero; primes denote the zero-mean fluctuating parts of the velocity components and capital letters their mean values:

$$u = U + u' , \qquad v = V + v' .$$

For, by definition,

$$\overline{uv} = \int_{-\infty}^{\infty}\int_{-\infty}^{\infty} uvp_{uv}(u,v)dudv$$

$$= \int_{-\infty}^{\infty} up_u(u)du \int_{-\infty}^{\infty} up_v(v)dv$$

$$= UV .$$

But

$$\overline{uv} = \overline{(U + u')(V + v')}$$

$$= UV + \overline{u'v'} .$$

Hence

$$\overline{u'v'} = 0 .$$

Eq.(3-10) can be written

$$H(\tau) = a_1 \int_{-\infty}^{\infty} p_v(v) \exp\left(-\frac{v^2\tau^2}{r_0^2}\right) dv \int_{-\infty}^{\infty} p_u(u)\left[\exp\left(-\frac{u^2\tau^2}{r_0^2}\right)\right]\left(1 + \tfrac{1}{2}m^2 \cos\frac{2\pi u\tau}{s}\right)du. \quad \text{... (4-1)}$$

Both integrals are unknown functions of τ, so that progress cannot be made in determining the characteristics of either p_u or p_v without further assumptions.

(a) u and v components both normally distributed .

We put

$$p_u(u) = \frac{1}{\sigma_u\sqrt{2\pi}} \exp\left\{-\frac{(u-U)^2}{2\sigma_u^2}\right\} \tag{4-2}$$

where σ_u is the standard deviation of the fluctuations in u , with a similar expression for $p_v(v)$.

In this case the integrals can be evaluated explicitly, and we find, to a very good approximation,

$$H(\tau) = a_1\left[\exp\left\{-\frac{(U^2+V^2)\tau^2}{r_0^2}\right\}\right]\left[1+\tfrac{1}{2}m^2\exp\left(-\frac{2\pi^2\sigma_u^2\tau^2}{s^2}\right)\cos\frac{2\pi U\tau}{s}\right] \tag{4-3}$$

for values of τ such that

$$\tau^2 \ll \min\left\{\frac{1}{2}\frac{r_0^2}{\sigma_u^2}, \frac{1}{2}\frac{r_0^2}{\sigma_v^2}\right\} . \tag{4-4}$$

(With these restrictions, the result can also be shown to hold if p_{uv} is a bivariate normal distribution, so that a shear stress term can be accommodated. In this case, p_{uv} is not separable.) If the values of the experimental parameters r_0 and s are known, σ_u, U and V can now be derived from (4-3) by curve-fitting.

It is often more convenient to work in terms of the turbulence intensities η_u and η_v , the ratios of the standard deviations to the means, rather than in terms of the standard deviations themselves. The inequality then becomes

$$\tau^2 \ll \min\left\{\frac{1}{2}\frac{r_0^2}{\eta_u^2U^2}, \frac{1}{2}\frac{r_0^2}{\eta_v^2V^2}\right\} .$$

Suppose the autocorrelation function contains q cycles, each corresponding to the transit of one interference fringe, and that there are n fringes contained in a beam diameter. Thus the fringe spacing is $2r_0/n$ and the maximum value of τ for the u component would be

$$\frac{2r_0q}{nU} .$$

In order that the inequality is satisfied we require that

$$\eta_u^2 \ll \frac{n^2}{8q^2} \tag{4-5}$$

with a similar result for η_v .

Fairly typical values are $n = 20$, $q = 10$, for which the condition becomes

$$\eta_u^2 \ll 0.5 \quad .$$

Hence η_u (and η_v) should not be more than, say, 0.25 (25%). The restriction is unnecessarily stringent, however, if only a few cycles are visible in the record. If, for example, the effective value of q is 4, then, for the same value of n , 50% turbulence would be tolerable. On the other hand, at this level the exponential factor damping the cosine term in (4-3) has fallen from its initial value of unity to 0.01 in only one cycle, so that the velocity information contained in the signals is no longer being coded in an efficient manner.

(b) Hermite polynomial expansion for $p_u(u)$: v component normally distributed

This approximation is intended for applications where the distribution of one velocity component may be highly non-Gaussian and rests, as stated in [32], on the assumption that $p_v(v)$ is Gaussian and that $p_u(u)$ can be written in the form of a Gram-Charlier expansion:

$$p_{uv}(u,v) = \frac{1}{2\pi\sigma_u\sigma_v} \exp\left\{-\frac{(u-U)^2}{2\sigma_u^2} - \frac{(v-V)^2}{2\sigma_v^2}\right\} \sum_{k=0}^{2L-1} A_k He_k\left(\frac{u-U}{\sigma_u}\right)$$

where $A_k = \frac{1}{k!}\left\langle He_k\left(\frac{u-U}{\sigma_u}\right)\right\rangle$

and He_k is the Hermite polynomial defined by the relation

$$He_k(z) = (-)^k \exp(\tfrac{1}{2}z^2) \frac{\partial^k}{\partial z^k}\left[\exp(-\tfrac{1}{2}z^2)\right] \quad .$$

The choice of L in the upper limit $(2L-1)$ of the sum will depend on the amount of information required from the calculation. The first few polynomials have the following forms:

$$He_0(z) = 1 , \qquad He_1(z) = 2z , \qquad He_2(z) = 4z^2 - 2 ,$$

$$He_3(z) = 8z^3 - 12z , \qquad He_4(z) = 16z^4 - 48z^2 + 12 .$$

The double integral can again be evaluated explicitly and, using again the approximations based on (4-4), takes the form

$$H(\tau) = a_1\left[\exp\left\{-\frac{(U^2+V^2)\tau^2}{r_0^2}\right\}\right]\left[1 + \tfrac{1}{2}m^2 \exp\left(-\frac{2\pi^2\sigma_u^2\tau^2}{s^2}\right)\right]\sum_{k=0}^{L}(-)^k\left(\frac{2\pi\sigma_u\tau}{s}\right)^{2k} \times$$
$$\times\left[A_{2k}\cos\left(\frac{2\pi U\tau}{s}\right) + A_{2k+1}\left(\frac{2\pi\sigma_u\tau}{s}\right)\sin\left(\frac{2\pi U\tau}{s}\right)\right] \quad . \qquad (4\text{-}6)$$

The A_k are now determined by curve-fitting. Fig.3 of [32] illustrates the way in which a better approximation can be obtained to a skewed velocity distribution using the Gram-Charlier method than with the purely Gaussian model of the last section.

(c) Eigenfunction expansion for $p_u(u)$: v component assumed negligibly small

For a one-dimensional flow with velocity vector normal to the fringe system, (3-10) reduces to

$$H(\tau) = a_1 \int_{-\infty}^{\infty} p_u(u) \left[\exp\left(-\frac{u^2\tau^2}{r_0^2}\right)\right]\left(1 + \tfrac{1}{2}m^2 \cos\frac{2\pi u\tau}{s}\right) du \quad .$$

It is convenient at this point to introduce an important practical consideration. Because the kernel of the integral is an even function of u, any negative values of u which may occur will contribute to $H(\tau)$ in exactly the same way as equal positive ones. Hence if an undistorted form of $p_u(u)$ is to be recovered, the u component must always be of the same sign; *ie* the flow must be non-reversing. (In addition, the analysis summarised in this section rests on Hilbert-Schmidt theory, which requires that the kernel be symmetric in u and τ.) Let us suppose then that u is always positive, so that the lower limit of the integral can be set to zero:

$$H(\tau) = a_1 \int_{0}^{\infty} p_u(u) \left[\exp\left(-\frac{u^2\tau^2}{r_0^2}\right)\right]\left(1 + \tfrac{1}{2}m^2 \cos\frac{2\pi u\tau}{s}\right) du \quad . \qquad (4\text{-}7)$$

The procedure now is to try to find a set of eigenfunctions ϕ_i and eigenvalues λ_i which satisfy the associated Fredholm equation

$$\int_0^\infty \phi_i(u) K(u\tau) du = \lambda_i \phi_i(\tau)$$

where $$K(u\tau) = \left[\exp\left(-\frac{u^2\tau^2}{r_0^2}\right)\right]\left(1 + \tfrac{1}{2}m^2 \cos\frac{2\pi u\tau}{s}\right) \quad .$$

Suppose also that the ϕ_i form a complete orthogonal set. Then a function such as $p_u(u)$ can be expanded in a series of the form

$$p_u(u) = \sum_{j=0}^{\infty} c_j \phi_j(u)$$

where the c_j are unknown coefficients which have to be determined. Substituting this expression into the equation for $H(\tau)$ and proceeding formally, we have

$$H(\tau) = a_1 \int_0^\infty \left[\sum_{j=0}^{\infty} c_j \phi_j(u)\right] K(u\tau) du = a_1 \sum_{j=0}^{\infty} c_j \lambda_j \phi_j(\tau) \quad .$$

Using the orthogonality of the ϕ_i, we have

$$\int_0^\infty H(\tau)\phi_k(\tau) d\tau = a_1 \int_0^\infty \left[\sum_{j=0}^{\infty} c_j \lambda_j \phi_j(\tau)\right] \phi_k(\tau) d\tau = a_1 c_k \lambda_k \quad ,$$

so that

$$c_k = \frac{1}{a_1 \lambda_k} \int_0^\infty H(\tau)\phi_k(\tau)d\tau \quad ,$$

and finally

$$p_u(u) = \frac{1}{a_1} \sum_{j=0}^{\infty} \frac{1}{\lambda_j} \phi_j(u) \left[\int_0^\infty H(\tau)\phi_j(\tau)d\tau \right] \quad . \tag{4-8}$$

A complete set of eigenfunctions and their associated eigenvalues for the Doppler-difference kernel has been found [33]. Further developments of the method are discussed in [34].

4.2.2 p_{uv} not separable

Consider again (3-10):

$$H(\tau) = a_1 \int_{-\infty}^{\infty} \int_{-\infty}^{\infty} p_{uv}(u,v) \left[\exp\left\{ - \frac{(u^2 + v^2)\tau^2}{r_0^2} \right\} \right] \left(1 + \tfrac{1}{2}m^2 \cos \frac{2\pi u\tau}{s} \right) du dv \quad .$$

We discuss two important cases, distinguished by the behaviour of the exponential term.

(a) Exponential term approximately unity

If there are sufficient fringes across the beam ($r_0 \gg s$) the exponential term will be almost unity over the observed length of the autocorrelation function and we obtain immediately the approximation

$$H(\tau) \simeq a_2 \int_{-\infty}^{\infty} p_u(u) \cos \frac{2\pi u\tau}{s} du + a_1 \tag{4-9}$$

since

$$p_u(u) \triangleq \int_{-\infty}^{\infty} p_{uv}(u,v)dv \quad .$$

Note that it is unnecessary to make assumptions now about the shape of the function p_{uv} and that high turbulence levels in the v component will not affect the result. Again we remark that since the kernel in (4-9) is even, the correct form for the distribution $p_u(u)$ can only be obtained if the u component is of constant sign. Here we assume that if necessary, a constant known bias u_s is imposed in order to make $u + u_s$ positive for each particle transit by introducing a frequency-shifting device into one or both beams[1]. We can then without loss of generality again replace the lower limit of the integral by zero:

$$H(\tau) = a_2 \int_0^\infty p_u(u) \cos \frac{2\pi u\tau}{s} \, du + a_1 \quad . \qquad (4\text{-}10)$$

Hence $p_u(u)$ can be recovered by performing a Fourier cosine transform on the data, after removal of the additive background constant a_1. The latter can be carried out automatically by the data-reduction programme.

The requirement that $r_0 \gg s$ implies a lower limit to the spatial resolution, characterised by the beam diameter $2r_0$, which can be achieved in high-speed flows, if the effect of the exponential term is to be negligible. (In nearly laminar conditions frequency shifting could conceivably be used to bring the effective velocity down.) However, for low-speed flows it is very often possible to arrange the experiment appropriately. Note also that if the speed is sufficiently low, frequency-shifting methods can now be used to accommodate very high levels of turbulence, for which flow reversals will occur from time to time, or to achieve fine spatial resolution.

(b) Exponential term varying significantly: narrow distributions of the velocity components

These conditions may be met with, for example, in supersonic boundary layers. In this case, since u and v are assumed to lie within a narrow range, the exponential term in (3-10) varies much more slowly within the integral (for a given range of τ) than the cosine term and can be replaced to a good approximation by the value it takes when u and v are given their mean values, U and V. It can then be taken outside the integral and the integration over the v component again carried out, yielding the approximate expression.

$$H(\tau) = a_1 \left[\exp\left\{ - \frac{(U^2 + V^2)\tau^2}{r_0^2} \right\} \right] \int_{-\infty}^{\infty} p_u(u) \left(1 + \tfrac{1}{2} m^2 \cos \frac{2\pi u\tau}{s} \right) du \quad .$$

An estimate of the accuracy of this approximation can be made by considering again the case when the u and v distributions are both Gaussian. The

[1] In [35] a method is described for recovering $p_u(u)$ in the case of reversing flows, without the use of frequency shifting. Polarisation devices are included in the transmitting optics and arranged to form two virtual fringe systems, displaced by one quarter of a fringe spacing. Two orthogonally polarised detector systems are used. The output from one is autocorrelated in the usual way and also cross-correlated with the output of the second. By adding and subtracting the Fourier transforms of the two correlation functions, the complete distribution $p_u(u)$ can be recovered. It should be remarked, however, that the result will in general be distorted if there are insufficient fringes across the beam diameter.

equation above then reduces to the result obtained in section 4.2.1(a), which under typical conditions was shown to be a good approximation for turbulence intensities at least up to 25%. For these relatively low turbulence levels the lower limit can be replaced by zero, and finally we have

$$H(\tau) = a_1 \left[\exp\left\{ - \frac{(U^2 + V^2)\tau^2}{r_0^2} \right\} \right] \left[1 + \tfrac{1}{2}m^2 \int_0^\infty p_u(u) \cos \frac{2\pi u \tau}{s} \, du \right] . \qquad (4\text{-}11)$$

The computer programme can be designed to determine $r_0/\sqrt{U^2 + V^2}$ as a free parameter, together with any residual background contribution to the correlation function. The magnitude of the exponential term can now be calculated for each value of τ and a modified form of $H(\tau)$ derived from which the damping effect of the exponential has been eliminated. A second Fourier transform will now give an improved estimate for $p_u(u)$ from which η_u, together with a refined value for U, can be determined. A similar calculation can be carried out to obtain $p_v(v)$, and hence η_v, from an orthogonal fringe system.

4.3 Data-Reduction Using Fourier Transforms

Clearly the most satisfactory experimental technique is to arrange wherever possible that there are very many fringes contained within the beam diameter, since analysis can then proceed with the minimum of *a priori* assumptions about the flow. In addition, we have seen in (4-10) that the problem of data-reduction collapses to that of carrying out a Fourier transform on the experimental data, which is a simple and rapid computational procedure. We remark here that the fast Fourier transform (FFT) is not usually the most suitable technique for general applications to turbulent flow since it lacks flexibility and, at low turbulence levels, the necessary resolving power. However, it has been successfully used in difficult engineering applications [36], and if sufficient signal is available from each scatterer, so that the correlator can be operated in a burst mode, the FFT is the obligatory method for processing each autocorrelation function [37].

4.3.1 The Fourier Transform for Laminar and Low-Turbulence Flows

The limited number of channels provided in available clipping correlators significantly restricts the number of cycles in the record obtained from laminar or low-turbulence flows. The rectangular data window produces an effect in the transform plane which is exactly equivalent to instrumental broadening in classical spectroscopy. The delta function which would ideally be present in laminar flow is replaced by a function having a $\sin x/x$ behaviour, although still centred at the same point on the u axis. Thus an accurate estimation of the velocity can be made simply by finding the position of the peak of the Fourier transform. (The perturbation arising from the associated $\sin x/x$ function centred at the symmetric point on the negative u axis is usually negligible.) The $\sin x/x$ structure can also be used to decide whether, within the limits of experimental accuracy, the flow is laminar [31].

At sufficiently high turbulence levels, the autocorrelation function will have decayed away effectively to zero by the end of the record. For a 48-

channel correlator operating with typical settings on a flow having a fairly symmetric unimodal velocity distribution, the level below which truncation effects are significant is about 4 or 5%; broadening of the distribution due to the finite record length then results in an overestimate of the turbulence intensity. Mean velocity estimates are very little affected at such low levels, but the accurate determination of the turbulence intensity begins to pose problems. If a good model of the flow is available for curve-fitting purposes, it should be possible to obtain accurate estimates down to the laminar condition; for example, a Gaussian model may often be acceptable.

The problem of instrumental broadening is of course very commonly encountered - for example when a recording or a spectroscopic measurement is being made - and a number of procedures for achieving resolution within the instrumental line-width have in recent years been proposed. Methods based on analytic continuation of band-limited signals (such as those considered here) have been intensively studied, particularly in the field of optical imaging [38]. Applications of these and other extrapolation techniques to photon correlation anemometry have been discussed at previous conferences in this series [39-41].

One particularly powerful and highly practical method of analytic continuation, usually referred to as GERCHBERG'S method [42], is based on iterative Fourier transform procedures and can be adapted for application to photon correlation data. The essential requirement in the latter case is that the bounds within which the velocity distribution must lie are known explicitly. The actual procedure then consists, at each iteration, of first truncating the latest estimate of the distribution to these limits and calculating its Fourier transform over the desired extrapolation length. The portion of this function lying between the origin and the maximum delay time of the measured autocorrelation function is then replaced by the original data and the resulting composite function again Fourier-transformed to form the new estimate of the velocity distribution. In the absence of noise this procedure has been shown to converge to the true distribution [43]. A stabilised non-iterative version of the method has been developed for use when the signals are contaminated by noise and has been successfully applied to photon correlation anemometry data [44].

4.4 The Autocorrelation Function for a Single Transit in the Presence of Noise

It has been noted before that an experimental autocorrelation function usually contains a reasonably constant additive term, attributable mainly to the stray radiation present in the test environment. If it can also be assumed that only one scatterer has contributed to the data, additional information about the circumstances of the experiment can be extracted in the following way. (An analysis given in [31] has been extended here to include fringe visibility as a free parameter.)

Given certain simplifying assumptions, (3-4) describing the scattered-light intensity for a single particle was found to reduce to the form of (3-7):

$$I(t) = I_0(1+\zeta^2)\left\{\exp\left(-\frac{2r_d^2}{r_0^2}\right)\right\}\left[\exp\left\{-\frac{2(u^2+v^2)t^2}{r_0^2}\right\}\right]\left(1+m\cos\frac{2\pi ut}{s}\right) . \qquad \ldots (4\text{-}12)$$

(The phase factor $2\pi x_0/s$ appearing in (3-7), which is not significant in the following argument, has been ignored.)

Suppose then as a first approximation that the background illumination contributes a constant term, c_2 say, to the intensity of the radiation incident on the detector. Writing

$$c_1 = I_0(1 + \zeta^2) \exp\left(-\frac{2r_d^2}{r_0^2}\right) ,$$

which is a constant for a given transit, and

$$T_r = r_0 \Big/ \sqrt{u^2 + v^2}$$

(the beam-radius transit time), we have

$$I(t) = c_1\left\{\exp\left(-\frac{2t^2}{T_r^2}\right)\right\}\left(1 + m \cos\frac{2\pi ut}{s}\right) + c_2 \quad . \tag{4-13}$$

Hence, if the experiment lasts for a time $T_e \gg T_r$, a good approximation to the correlation function will generally be

$$G(\tau) = \int_{-\infty}^{\infty} I(t)I(t+\tau)dt$$

$$= c_1^2 \frac{\sqrt{\pi}}{2} T_r\left\{\exp\left(-\frac{\tau^2}{T_r^2}\right)\right\}\left(1 + \tfrac{1}{2}m^2 \cos\frac{2\pi u\tau}{s}\right)$$

$$+ c_1c_2\sqrt{2\pi}\, T_r\left\{1 + \tfrac{1}{2}m \exp\left(-\frac{\pi^2u^2T_r^2}{2s^2}\right)\right\} + c_2^2T_e \quad .$$

The second exponential will be quite negligible in any real experiment (for which T_r is always considerably greater than s/u) and we have the result

$$G(\tau) = c_1^2 \frac{\sqrt{\pi}}{2} T_r\left\{\exp\left(-\frac{\tau^2}{T_r^2}\right)\right\}\left(1 + \tfrac{1}{2}m^2 \cos\frac{2\pi u\tau}{s}\right) + c_1c_2\sqrt{2\pi}\, T_r + c_2^2T_e \quad . \tag{4-14}$$

Note that for a detector of unit area, the units of $I(t)$ are energy per unit time and of $G(\tau)$ energy squared per unit time. If the statistical fluctuations of the detection process are ignored for the present purpose, the total count rate arising from signal and background (*ie* the input to the correlator in pulses per second) can be taken to be directly proportional to the incident intensity. The constant of proportionality will be the ratio of the detector quantum efficiency, α, to the energy of a photon, E_{phot}. Thus (4-13) for $I(t)$ can also be used to represent the time dependency of the count-rate during a particle transit, with c_1 and c_2 replaced by d_1 and d_2, where

$$d_i = \left(\frac{\alpha}{E_{phot}}\right)c_i \quad , \qquad i = 1,2.$$

The expression for $G(\tau)$ in (4-14) will be similarly modified and, by integration over the sample time T_s, can be used to calculate the expected contents of the correlator channels. Provided that T_s is very much less than T_r and the transit time s/u for one fringe spacing this integration reduces, for the kth channel, to replacing τ by kT_s and multiplying each term by T_s:

$$G(kT_s) = d_1^2 \frac{\sqrt{\pi}}{2} T_r T_s \left[\exp\left\{-\left(\frac{kT_s}{T_r}\right)^2\right\}\right]\left(1 + \tfrac{1}{2}m^2 \cos\frac{2\pi ukT_s}{s}\right)$$
$$+ d_1 d_2 \sqrt{2\pi}\, T_r T_s + d_2^2 N T_s^2 \ . \qquad (4\text{-}15)$$

(Note that T_e has been replaced by NT_s, where N is the number of sample times in the experiment.)

Conversely, given the values of certain experimental parameters, approximate values for d_1 and d_2 can be recovered from the correlation function if it can also be assumed that only one particle transit is involved. Suppose the maximum and minimum channel contents near the origin ($k = 0$) are G_{max} and G_{min}; these correspond to values of the cosine term of approximately +1 and -1 respectively. Then, with the usual assumption that the beam radius is significantly greater than the fringe spacing, (4-15) gives

$$G_{max} - G_{min} = \frac{\sqrt{\pi}}{2} T_r T_s m^2 d_1^2$$

$$G_{max} + G_{min} = \sqrt{\pi}\, T_r T_s d_1^2 + 2\sqrt{2\pi}\, T_r T_s d_1 d_2 + 2NT_s^2 d_2^2 \ .$$

The total number of pulses received during the experiment, N_{tot} say, is obtained by integrating the expression for the intensity:

$$N_{tot} = \int_{-\frac{1}{2}NT_s}^{\frac{1}{2}NT_s} I(t)dt = \sqrt{\frac{\pi}{2}}\, T_r d_1 + NT_s d_2 \ .$$

Hence the numbers of signal and noise counts are $\sqrt{\pi/2}\, T_r d_1$ and $NT_s d_2$ respectively.

From these formulae, expressions for d_1, d_2 and m can be derived in terms of known quantities. Putting

$$G_{max} - G_{min} = h_1$$

and

$$G_{max} + G_{min} = h_2 ,$$

we find

$$d_1 = \left[\frac{Nh_2 - 2N_{tot}^2}{\sqrt{\pi}\, T_r(NT_s - \sqrt{\pi}\, T_r)}\right]^{\frac{1}{2}}$$

$$d_2 = \frac{N_{tot}}{NT_s} - \frac{1}{NT_s}\left[\frac{\sqrt{\pi}\, T_r\left(Nh_2 - 2N_{tot}^2\right)}{2(NT_s - \sqrt{\pi}\, T_r)}\right]^{\frac{1}{2}}$$

$$m = \left[\frac{2h_1(NT_s - \sqrt{\pi}\, T_r)}{h_2NT_s - 2N_{tot}^2T_s}\right]^{\frac{1}{2}} .$$

In Fig. 7 an autocorrelation function for a single transit is displayed. The Fourier transform of this record yields a value of 18.8 m s^{-1} for the particle velocity. The fringe size for this experiment was 12.5 microns and the beam diameter 420 microns. Hence the beam radius transit time is 11.17 μs. The sample time setting was 0.1 μs and the total number of samples 10^3. The monitor channel value for N_{tot} was 152. From the figure we take G_{max} and G_{min} as approximately 50 and 30 respectively. Then the formulae give

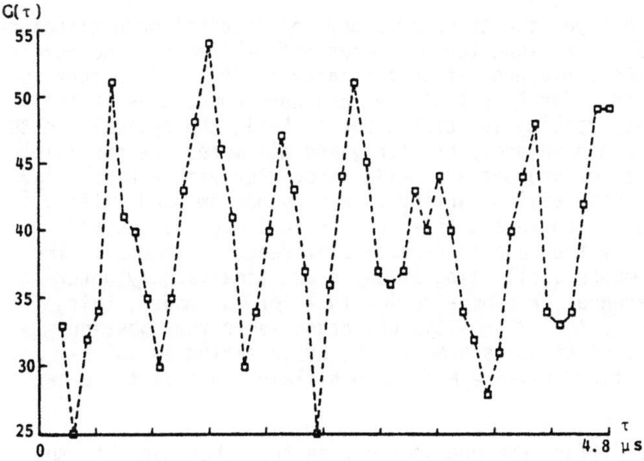

Fig. 7. Autocorrelation function for a single particle transit

$d_1 = 4.613 \times 10^6\ s^{-1}$

$d_2 = 0.874 \times 10^6\ s^{-1}$

$m = 0.974$.

The signal and noise counts are thus approximately 65 and 87 respectively. The peak photodetection rates would have been about $9.1 \times 10^6\ s^{-1}$ for the signal and $10^7\ s^{-1}$ for signal plus background. These rates are in fact approaching the level at which the detector system would become saturated and the output pulse train cease.

5. Lasers and Laser Beams

We consider finally the properties of laser radiation which are of importance in the design of laser anemometry systems and introduce formulae by means of which the characteristics of the light beams incident on the scattering particle can be calculated.

The basic resonant cavity of a laser consists simply of a pair of mirrors, one of which transmits a small amount of the radiation incident upon it, aligned so that the light wave passes repeatedly through the lasing medium. In the case of the commonly used helium-neon or argon gas lasers, the source of energy is an electric discharge in the gas itself. The argon-ion laser is capable of operating simultaneously at a number of wavelengths in the visible spectrum (and also in the ultraviolet) and is for this reason particularly widely used, since multicomponent measurements can be made with a single source. If only one wavelength is required, the output can be concentrated into a single line by using an intra-cavity dispersing prism in combination with the rear mirror.

The spatial and temporal coherence properties of the laser beam are often of great importance in laser anemometry. The spatial coherence is associated with the phase across the output beam; if this is highly uniform, the beam can be focussed effectively down to the diffraction limit. Temporal coherence manifests itself in the length of time for which the output wave train maintains its phase structure. This property is clearly of direct relevance to experiments in which the Doppler effect is exploited.

We are thus led to consider the transverse and longitudinal mode structure of the laser output [45]. The longitudinal modes differ from one another in the number of wavelengths contained within the laser cavity; this number is necessarily an integer in order to satisfy the boundary conditions at the mirrors. Their frequency spacing is $c/2L$, where L is the optical length of the cavity and c is the velocity of light, and all modes are possible which lie within the Doppler-broadened profile associated with a particular transition. An intra-cavity etalon, acting as a very narrow band filter, is generally used if operation in a single longitudinal mode is desired, though this is not usually the case in Doppler-difference anemometry. The number of longitudinal modes oscillating at any given time largely determines the temporal coherence, or coherence length - for n modes, this length is approximately $2L/n$. For a typical argon laser the coherence length under multimode conditions is some 10 cm; by inserting an intra-cavity etalon and permitting only one mode to oscillate, this distance can become tens of metres.

The transverse modes differ from one another, on the other hand, in the way in which the electromagnetic field varies in amplitude and phase across

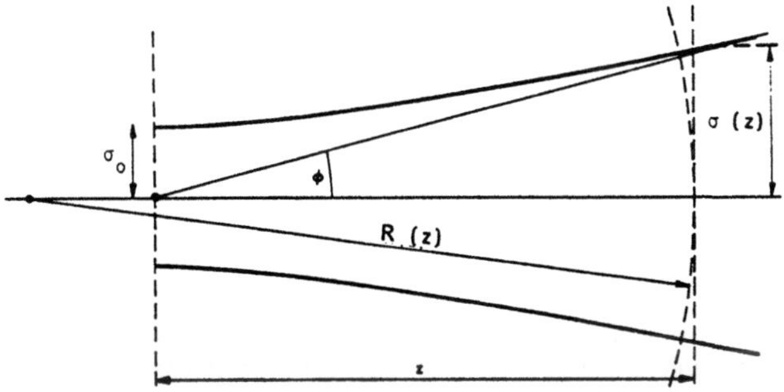

Fig. 8. The Gaussian beam

the wavefront, although it should be noted that provided the transverse mode structure is unchanging with time, the laser output possesses full spatial coherence in all cases. However, it is generally preferable to operate the laser in the fundamental TEM_{00} (uniphase) mode, since the energy in the beam is concentrated closer to the axis of propagation for this mode than for any other.

The TEM_{00} mode has a Gaussian intensity profile and a hyperbolic beam contour; see Fig 8. Only at the waist, where the beam diameter is a minimum, is the wavefront planar; the beam radius σ is defined by KOGELNIK and LI [45] as the radial distance at which the field amplitude is e^{-1} times its value on the axis. The axial field amplitude decays in a Lorentzian fashion with distance z from the waist and beam radius σ increases according to the formula

$$\sigma^2(z) = \sigma_0^2\left[1+\left(\frac{\lambda z}{\pi\sigma_0^2}\right)^2\right] \tag{5-1}$$

where σ_0 is the value at the waist. The radius of curvature of the wave front at z is

$$R(z) = z\left[1+\left(\frac{\pi\sigma_0^2}{\lambda z}\right)^2\right] . \tag{5-2}$$

The radius of curvature has a minimum value when $z = \pi\sigma_0^2/\lambda$. The beam divergence is defined as the angle ϕ between the axis and the hyperbola asymptotes and therefore

$$\phi = \frac{\lambda}{\pi\sigma_0} . \tag{5-3}$$

This is the far-field diffraction angle of the Gaussian beam; measurement of the rate of growth of the beam diameter (spot size) in the far field is thus a convenient way of estimating approximately the diameter at the waist.

If a lens of focal length f is placed in such a Gaussian beam, the radius of the wavefront is transformed according to the usual lens formula:

$$\frac{1}{R_{in}} - \frac{1}{R_{out}} = \frac{1}{f} \tag{5-4}$$

and if the lens is of sufficient power the emergent beam, also Gaussian, will converge to a new waist. It can be proved that if the waist of the incident beam lies in the back focal plane of the lens, a new waist will be formed in the front focal plane, although the radii will in general be different. In general, however, the new waist will not be at the focal distance from the lens; its size and position can be calculated by inserting the appropriate values for the focal length and position of the lens into (5-1), (5-2) and (5-4). It is also an interesting and useful property of the emergent beam that its radius at a distance f beyond the lens, σ_f say, is related to the divergence of the incident beam by the simple formula

$$\sigma_f = f\phi \ . \tag{5-5}$$

It follows from these considerations that particular care should always be taken in selecting and positioning the optical components used for providing beams of the required diameter. Special attention should be paid to ensuring that any variation in beam diameters in the measurement region is negligibly small, since it is only then that the equiphase fronts can be considered plane and the fringe spacing constant throughout this region, its value being given by (3-2). The plane-wavefront requirement is in fact only met at the waists or at positions infinitely far from them, and for experiments in which the spatial resolution, as determined by the beam diameters, is not an important factor, the safest procedure is to use only long-focus lenses, if a limited degree of control over beam size is still necessary, or to omit them entirely.

6. Concluding Remarks

The preceding discussion has been intended to demonstrate that provided certain operational and geometrical criteria are met, photon correlation techniques in Doppler-difference laser anemometry can be placed on a firm theoretical basis. However, the important question of the overall accuracy of a specific measurement at a point in the fluid has not been considered. The problem is in practice a complicated one, since, in addition to the statistical character of the signals themselves, many experimental factors are involved, including, for example, the light-scattering and fluid following properties of the seeding particles, the level of optical noise with which the signals are contaminated, the length of time over which the data have been acquired and the accuracy with which the various optical and geometrical constants are known. Further complications arise if the experimental arrangements do not meet the simplifying criteria discussed in section 3.2. Broadly speaking, however, acceptable accuracy can generally be achieved by the use of small beam-intersection angles and by integrating the signals over a sufficiently long period. The fact that the ratio of signal to background can usually be improved continuously by integration is of course a major advantage of photon correlation techniques, provided that time-averaged quantities, such as the first and second moments of the velocity distribution, are all that is required.

If, however, an estimate of velocity is required from each scattering particle (*ie*, if the anemometer is to be operated in the 'burst' mode), a fundamental limitation on the accuracy attainable will be imposed by the statistical nature of the photodetection process itself. A comparison of various methods of analysing photon-resolved signals generated by single-particle transits has been carried out by OLIVER [46], who showed that in Doppler-difference anemometry the highest accuracy is achieved by fitting the formula for a single transit, (3-8), to the measured autocorrelation function; equivalently, the Fourier transform of (3-8) can be fitted to the power spectrum of the signal. The latter procedure reduces in the 'many-fringe' case to the location of a peak in the spectrum which, together with the Fourier transform stage, could be realised relatively simply in solid-state form. With such a signal-processing system, valuable information on the important components of the turbulent fluctuations, for many types of flow, would become directly accessible, although concomitant improvements in the frequency response of the photon-counting photomultiplier tubes and their associated circuitry, perhaps by an order of magnitude, would also be highly desirable. As the range of applications of photon correlation laser anemometry continues to widen, these and other improvements in apparatus and technique can be confidently anticipated.

References

1 H.Z. Cummins, E.R. Pike (eds): *Photon correlation and light beating spectroscopy*. NATO Advanced Study Institute Series, Vol.B3, Plenum Press, New York & London (1974)

2 H.Z. Cummins, E.R. Pike (eds): *Photon correlation spectroscopy and velocimetry*. NATO Advanced Study Institute Series, Vol.B23, Plenum Press, New York & London (1977)

3 Benjamin Chu: *Laser light scattering*. Academic Press, New York & London (1974)

4 B. Crosignani, P. Di Porto, M. Bertolotti: *Statistical properties of scattered light*. Academic Press, New York & London (1975)

5 F.A. Johnson, T.P. McLean, E.R. Pike: Proceedings, International Conference on the Physics of Quantum Electronics, Puerto Rico, 28-30 June 1965. (P.L. Kelley, B. Lax and P.E. Tannenwald, eds) McGraw Hill, New York (1966)

6 F.T. Arecchi: Phys. Rev. Letters, 15, 24, 912-6 (1965)

7 A. Einstein: Ann. Physik, 17, 132 (1905)

8 L. Mandel: 'The case for and against semiclassical theory.' In *Progress in Optics*, Vol.XIII (ed. E. Wolf), North-Holland Publishing Company, Amsterdam & Oxford (1976)

9 L. Mandel, E.C.G. Sudarshan, E. Wolf: Proc. Phys. Soc., 84, 435-444 (1964)

10 R.J. Glauber: Phys. Rev. Letters, 10, 84-6 (1963)

11 L. Mandel: Proc. Phys. Soc., 74, 3, 233-243 (1959)

12 E. Wolf, C.L. Mehta: Phys. Rev. Letters, 13, 705 (1964)

13 S. Fray, F.A. Johnson, R. Jones, T.P. McLean, E.R. Pike: Phys. Rev. 153, 2, 357-9 (1967)

14 P. Diament, M.C. Teich: J. Opt. Soc. Am., 60, 5, 682-9 (1970)

15 G. Bédard: Proc. Phys. Soc., 90, 131-141 (1967)

16 F.A. Johnson, R. Jones, T.P. McLean, E.R. Pike: Phys. Rev. Letters, 16, 13, 589-92 (1966)

17 L. Mandel: J. Opt. Soc. Am., 70, 7, 873-4 (1980)

18 R.J. Glauber: Phys. Rev., 131, 2766-88 (1963)

19 E. Jakeman: J. Phys. A., 3, 201-215 (1970)

20 J.H. Van Vleck: Harvard University Radio Research Lab Report, No.51 (1943)

21 J.H. Van Vleck, D. Middleton: Proc IEEE, 54, 2-18 (1966)

22 E. Jakeman, E.R. Pike: J. Phys. A., Series 2, 2, 411-2 (1969)

23 E.R. Pike: 'Photon correlation spectroscopy.' In *Very High Resolution Spectroscopy: a Symposium.* (ed. R.A. Smith), Academic Press, London & New York (1976)

24 C. Bendjaballah: J. Phys. A., 6, 837 (1973)

25 C. Bendjaballah: Optics Communications, 34, 2, 164-6 (1980)

26 P. Tartaglia, T.A. Postal, S.H. Chen: J. Phys. A., 6, L35 (1973)

27 P.N. Pusey, W.I. Goldberg: Phys. Rev. A3, 766-76 (1971)

28 Y. Yeh, H.Z. Cummins: Applied Physics Letters, 4, 10, 176-8 (1964)

29 H.J. Pfeifer, H.D. vom Stein: Technical Report No.T12/67, Institut Franco-Allemand, St Louis, France (1967)

30 J.B. Abbiss: 'Photon correlation velocimetry in aerodynamics.' (in Ref 2)

31 J.B. Abbiss: Physica Scripta, 19, 388-395 (1979)

32 A.D. Birch, D.R. Brown, J.R. Thomas: J. Phys. D: Appl. Phys., 8, 438-447 (1975)

33 J.G. McWhirter, E.R. Pike: Physica Scripta, 19, 417-425 (1979)

34 J.G. McWhirter, E.R. Pike: Optica Acta, 27, 1, 83-105 (1980)

35 R.G.W. Brown, E.R. Pike: 'Laser Doppler anemometry in reversing flows without frequency shifting.' Proceedings, 5th National Quantum Electronics Conference, Hull, UK, September 1981

36 C.J. Moore, A.E. Smart: J. Phys. E., 9, 977-981 (1976)

37 R.G.W. Brown, R.D. Callan, W. Jenkins, R. Jones, I. Miller, C.J. Oliver, E.R. Pike, D.J. Watson, M. Wedd: Physica Scripta, 19, 365-368 (1979)

38 B.R. Frieden: 'Evaluation, design and extrapolation methods for optical signals, based on the use of the prolate functions.' In *Progress in Optics*, Vol.IX (ed. E. Wolf), North-Holland Publishing Company, Amsterdam & London (1971)

39 J.B. Abbiss: 'Inversion and extrapolation of experimental autocorrelation data.' Proceedings, Conference on Photon Correlation Techniques in Fluid Mechanics, Churchill College, Cambridge, UK, 6-7 April 1977

40 P.S. Stone: Physica Scripta, 19, 402-410 (1979)

41 J.B. Abbiss, C. De Mol, M.A. Fiddy, G. Ross: 'The superresolution problem in Doppler-difference laser anemometry.' Proceedings, Fourth Conference on Photon Correlation Techniques in Fluid Mechanics, Stanford University, Stanford, California, USA, 25-27 August 1980

42 R.W. Gerchberg: Optica Acta, 21, 9, 709-20 (1974)

43 P. De Santis, F. Gori: Optica Acta, 22, 8, 691-5 (1975)

44 J.B. Abbiss, C. De Mol, H.S. Dhadwal: 'Regularised iterative and non-iterative procedures for object restoration from experimental data.' (To be published in Optica Acta)

45 H. Kogelnik, T. Li: Applied Optics, 5, 10, 1550-67 (1966)

46 C.J. Oliver: J. Phys. D: Appl. Phys., 13, 7, 1145-59 (1980)

3. Velocimetry Signals and Data Reduction in Simulation and Practice

Robert G.W. Brown

Royal Signals and Radar Establishment, St. Andrews Road
Malvern, Worcestershire, UK

1. Introduction

It is 10 years since the first applications of photon correlation techniques to laser Doppler velocimetry. In that time many hundreds of papers have been published on the subject and it is therefore out of the question to review all the developments in detail. Instead, we ask a series of questions and attempt to answer them, hoping to provide a critical assessment of the state of the art. Problems will therefore dominate the discussion and the many valuable achievements will be suppressed. Available space requires the assumption that the reader is familiar with the basic principles of laser Doppler-difference velocimetry [1,2,3] (LDV), laser transit velocimetry [4] (LTV) and photon correlation [5].

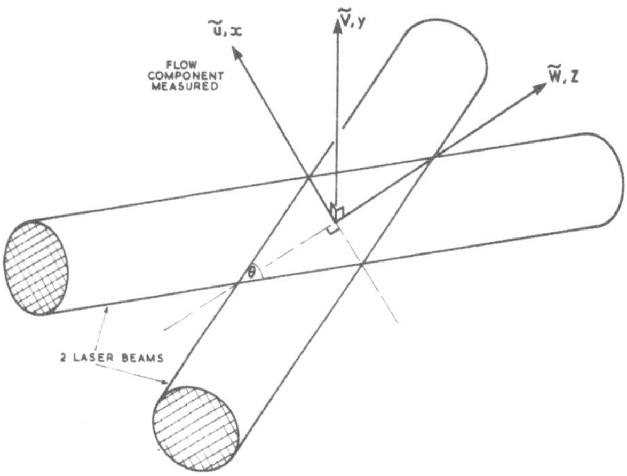

Fig. 1. Laser Doppler-difference velocimetry geometry

In this review, the velocity component that we measure will be termed u. In LDV, referring to Fig.1, u is the velocity component in the plane of the two beams that intersect to form the apparent fringes in the sampling or measurement volume, and perpendicular to the bisector of the angle between the beam propagation directions. The u, v and w components are mutually orthogonal, the w component being in the direction of the angle bisector. The first part of the review comprises four questions concerning LDV.

2. Laser Doppler Velocimetry, LDV

2.1 What are the forms of LDV signals?

The dynamic range of the signals can be very large, depending on the laser power, scattering geometry, optical geometry and scattering particle or seed characteristics. Photon correlation signal processing requires a random pulse train corresponding to individual photodetections, usually only a few per Doppler cycle. Alternative techniques [3] require orders of magnitude more photodetections which are filtered and smoothed to give quasi-analogue Doppler signals. Using natural atmospheric seeding, the signal strengths from occasional large particles cause photodetection pile-up because of finite pulse width, but the bulk of the signals should be discrete, resolvable pulses. If this is not the case, then either experimental improvements or longer experiment times are necessary to gain better statistics, or may be (rarely) photon correlation is inappropriate.

To calculate signal strengths for a given particle we use an equation due to PIKE [1] which gives the mean number of photodetections due to a particle transit through the sampling volume

$$\langle N_s \rangle = \frac{I_o \cdot N_{ph} \cdot \Omega \cdot \eta_e \cdot \eta_f \cdot i_1 \cdot T \cdot (\pi a^2)}{\pi \cdot r_o^2} , \qquad (2.1.1)$$

where I_o is the laser power, r_o the sampling volume radius, N_{ph} the number of photons per watt, Ω the solid angle of the detector's field of view, η_f and η_e optical filter and system efficiencies, T the residence time of the particle in the sampling volume, a is the particle radius and i_1 is its Mie scattering intensity function. Calculation of i_1 is straightforward for spherical dielectric scatterers but not for irregular shapes and complex refractive indices. Graphs useful for laser velocimetry are shown in Figs. 2, 3 and 4; note the similarity between side and full backscatter strengths.

Simulation of the structure of Doppler difference signals is based on the equation for the intensity of scattered radiation at a point in space due to an arbitrary transit of a particle through the sampling volume (ABBISS, in [1])

$$I\,(x,y,z) = 2I_o \exp\left\{\frac{-2}{r_o^2}\left(x^2 \, Cos^2(\tfrac{\theta}{2}) + y^2 + z^2 \, Sin^2(\tfrac{\theta}{2})\right)\right\} \times$$
$$\left[Cosh\left(\frac{4}{r_o^2}\, x \cdot z \cdot Sin(\tfrac{\theta}{2}) Cos(\tfrac{\theta}{2})\right) + Cos\left(\frac{4\pi}{\lambda}\, x \, Sin(\tfrac{\theta}{2})\right)\right] . \qquad (2.1.2)$$

This equation is the photodetection rate modulating function. It is straightforward to convolve a given particle size with this equation and the previous one to yield, step by step, a series of photodetection numbers and positions from Poisson number generators. This procedure is valid for one particle only in the sampling volume at any instant in time. By placing the phototube rise and fall times on each pulse we can assemble a picture of the signal due to a particle. Examples of varying signal magnitude are shown in Figs. 5a to 5f, in which the particle size is progressively increased. It has been shown that correlation and reduction of a signal such as Fig.5a can give 1% accuracy in velocity estimation.

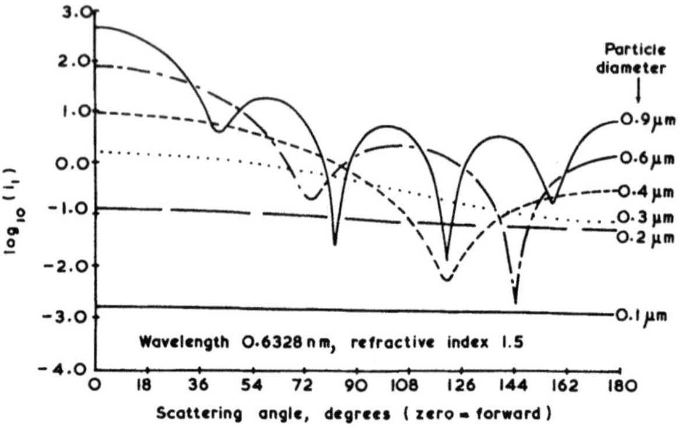

Fig.2. Mie scattering curves for laser velocimetry

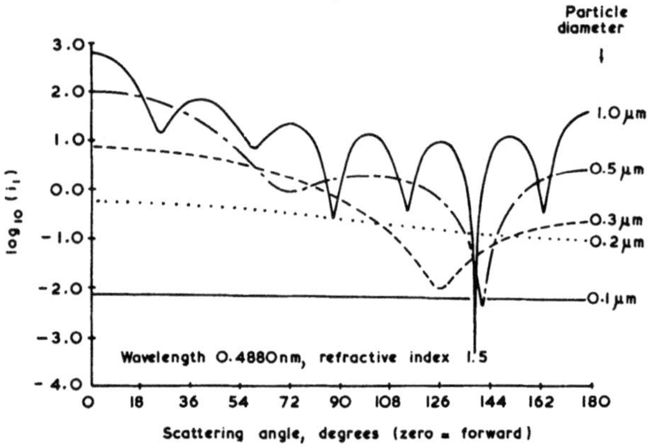

Fig.3. Angular Mie scattering curves for laser velocimetry

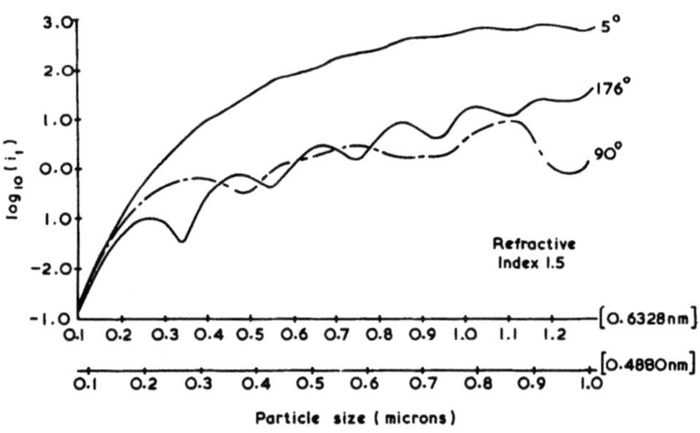

Fig.4. Mie scattering variations with particle size

Complete simulation of a velocimetry experiment has been described by MAYO [6,7]. It is a triply stochastic Poisson process, the processes being the turbulent flow, the particle arrivals and the photodetections. To proceed to full simulation of correlation functions, we select velocities and particle sizes at random from suitable distributions, computing signals for each particle and constructing a signal sequence. Discrimination, scaling or clipping [5] and autocorrelation of the signal is then performed. Noise signals may be mixed into the velocimeter signals but accurate prediction of noise levels is very difficult, though reasonable attempts may be made with full

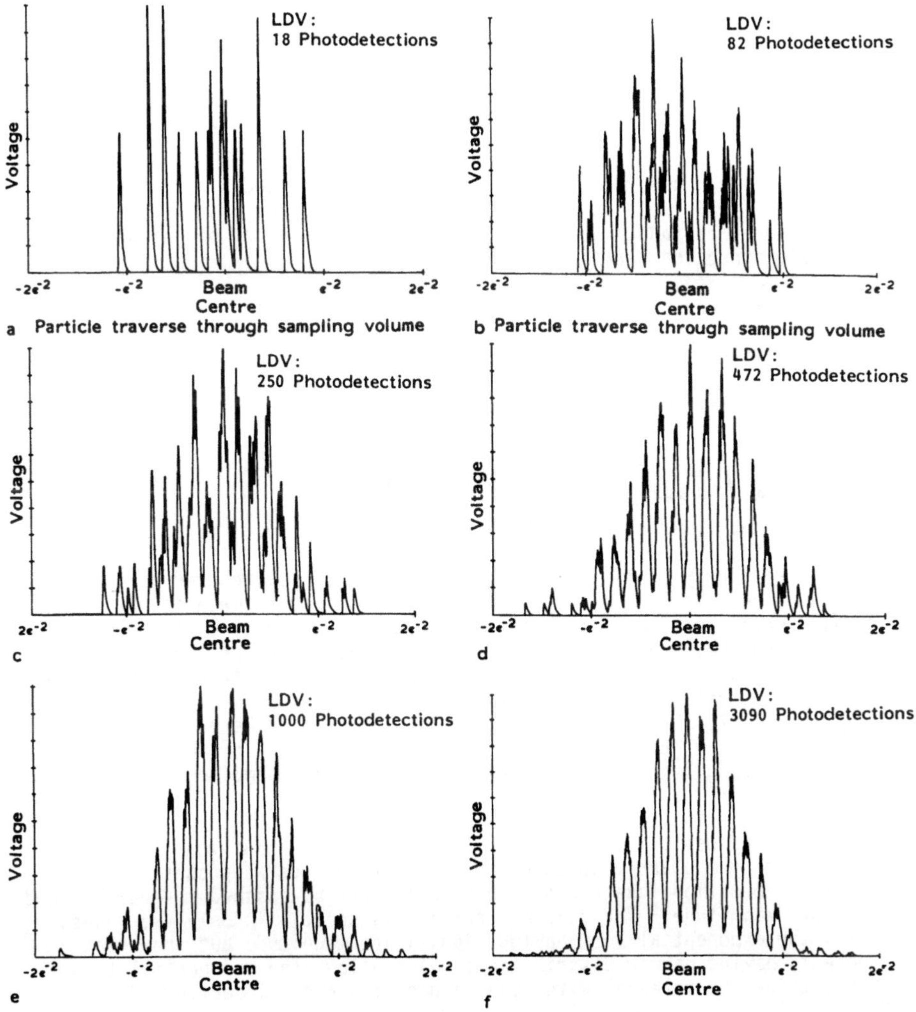

Fig. 5a-e

backscatter geometries and images of defocussed beams. The simulation of just one correlogram demands very long calculations, even with a large, fast computer.

2.2 How should we process our LDV data?

The main problem is the form of the LDV correlogram. The passage of a single particle through a laser Doppler-difference sampling volume gives rise to an autocorrelogram [1]

$$G_s(\tau) = \frac{a}{(u^2+v^2)^{\frac{1}{2}}} \exp\left\{\frac{-(u^2+v^2)\tau^2}{r^2}\right\}\left[1 + \frac{m^2}{2}\,\mathrm{Cos}\left(\frac{2\pi u\tau}{s}\right)\right] \qquad (2.2.1)$$

which is velocity biased by the square root term. In the turbulent, many-particle case the equation becomes

$$G(\tau) = \iiint_{-\infty}^{+\infty} p(u,v,w) \exp\left\{\frac{-(u^2+v^2)\tau^2}{r^2}\right\}\left[1 + \frac{m^2}{2}\,\mathrm{Cos}\left(\frac{2\pi u\tau}{s}\right)\right] dudvdw \qquad (2.2.2)$$

which is unbiased, assuming random particle positions and a cylindrical sampling volume. This equation is insoluble unless further assumptions and approximations are made, but note that the w component can be integrated immediately since the kernel is w independent.

Most assumptions [8] concern the separability of the u and v components, $v \ll u$ and that flow reversals are non-existent or frequency shifted to remove their ambiguity.

The correlogram form collapses to

$$G(\tau) = a \int_0^{\infty} p_u(u) \exp\left(\frac{-u^2\tau^2}{r^2}\right)\left[1 + \frac{m^2}{2}\,\mathrm{Cos}\left(\frac{2\pi u\tau}{s}\right)\right] du \qquad (2.2.3)$$

and an increasingly sophisticated series of methods may be used to process the data. These include a simple, rapid 3-turning-point method [1,9] with virtually noise free data, a 6-parameter Gaussian velocity distribution model fitting technique, a Hermite polynominal expansion [10] for the u distribution with v assumed Gaussian distributed (often unstable in practice), Fourier cosine transformation and the information-theoretic approaches of PIKE and McWHIRTER [11,12,13]. None of the techniques discussed here includes corrections for velocity gradients in the sampling volume [14] or possible corrections to low turbulence estimates due to finite axial sampling volume extent [15].

These techniques all require _a priori_ assumptions about the forms and magnitudes of the u and v components. However, unless we possess such _a priori_ information, the u and v components must be assumed inseparable and of unknown magnitude. The general equation is soluble in only two cases, firstly the exponential term having significant magnitude but the u and v components having narrow distributions which allows the exponential to be taken outside the integral with mean values of u and v inserted [12]

$$G(\tau) = a\left[\exp\left\{\frac{-(\bar{u}^2+\bar{v}^2)\tau^2}{r^2}\right\}\right] \times \int_0^{\infty} p_u(u)\left\{1 + \frac{m^2}{2}\,\mathrm{Cos}\left(\frac{2\pi u\tau}{s}\right)\right\} du \quad . \qquad (2.2.4)$$

Fourier cosine transformation of the correlogram yields a velocity distribution slightly distorted by the exponential term, but iterations reduce this problem.

To avoid completely assumptions concerning u and v it is necessary to reduce the general equation to a pure Fourier cosine transformation by placing a very large number of fringes in the sampling volume to make the exponential effects negligible. In practice this is generally not possible, due either to processor speeds or optical geometry and spatial resolution requirements. Thus, _all_ data processing techniques require approximations and assumptions concerning the flow in question. A further problem with the Fourier cosine transform ideal is that to have many fringes we require a large angle between the laser beams unless spatial resolution is to be sacrificed through large sampling volumes. The cylindrical sampling volume assumption that causes the unbiasing of the correlogram form is now weakened to some extent.

All correlograms have finite length, transformations of them incur spurious, unphysical negative oscillations. The truncation represents a velocity resolution limit. To overcome simultaneously truncation oscillations and to increase velocity resolution, some people pursue techniques such as extrapolation of the correlogram, use of maximum entropy ideas and other methods of forcing the velocity distribution to be positive. Extrapolation is a dangerous philosophy involving invention of data to extend the correlogram length. Any quantity of noise renders such techniques ill conditioned, it is then difficult to justify the choice of one from an infinite number of possible solutions. Noise affects the "fine structure" of velocity distributions and much spurious resolution can be introduced into the results without the ability to detect or control it. Maximum entropy approaches are equivalent to a form of extrapolation and, furthermore, use models that are entirely inappropriate for laser anemometry. Simply forcing the velocity distribution to be positive usually entails making the negative regions of oscillation positive and of slightly reduced magnitude, which is of no great value, just aesthetically satisfying.

With all these problems in mind, it is appropriate to discuss the information theoretic approaches of McWHIRTER and PIKE [11,12,13]. They are centred on the first kind Fredholm integral equation

$$\dot{g}(\tau) = \int_0^\infty K(u\tau)\, p(u)\, du \quad . \tag{2.2.5}$$

For LDV the kernel is

$$K(u,\tau) = \exp\left(\frac{-u^2\tau^2}{r^2}\right)\left[1 + \frac{m^2}{2}\,\mathrm{Cos}\left(\frac{2\pi u\tau}{s}\right)\right] ; \tag{2.2.6}$$

the equation is ill-conditioned and has limited validity due to approximations discussed earlier. To solve this equation one tries to find a set of functions ϕ_i and factors λ_i, eigenfunctions and eigenvalues which satisfy the associated Fredholm equation

$$\int_0^\infty \phi_i(u)K(u,\tau)du = \lambda_i\phi_i(\tau) \quad . \tag{2.2.7}$$

The eigenfunctions are basic elements of information which retain their identity under the action of the integral operator, but are scaled in magnitude by the eigenvalue λ_n. The larger λ_n, the more efficient the transmission of the corresponding information element through the integral. Elements for which λ_n is sufficiently small are transmitted so weakly that they are lost in noise. The faster the eigenvalue spectrum decays to the noise level in an experiment, the less the information (resolution) that can be extracted. Information or resolution to an amount determined by the noise level is the maximum available. Having found all the ϕ_i and λ_i possible, the velocity distribution is constructed and, usually the first two moments calculated reliably, higher moments depending on noise level.

In practice, in the absence of a quantitative measure of the noise, we compute a number of solutions for the data, each requesting more eigenvalues until such time as the reconstructed velocity distribution is affected by the noise magnitude, though this can be avoided if the noise level is known. The penultimate iteration represents the maximum resolution result obtainable. Early implementations of this approach were computationally lengthy, but recently, McWHIRTER has developed a competitively rapid cubic B spline model approach [17] taking about 45 seconds of computation on a DG Nova computer compared to (say) the Abbiss-Sharpe Fourier Cosine transform software which takes about 30 seconds per correlogram [8,16]. The inclusion of a smooth model for the velocity distribution makes the problem well conditioned and stable. It overcomes problems of ill-conditioning, sampling and truncation in a justifiable way and allows a resolution to a level determined by all these criteria. It is worth noting that in laser anemometry, resolution even up to the Nyquist limit is generally not possible, because the presence, always, of the exponential term of the kernel causes a more rapid decay of the eigenvalue spectrum than one obtains for a pure Fourier cosine transform relation. In the limits of a vanishing exponential term, the spline model approach and Fourier cosine method are virtually identical. The new spline model approach allows positivity of the velocity distribution to be included as a constraint, but only at considerable computational expense.

To illustrate data reduction performance in practice we choose sets of experimental data acquired with 10 nano-second sample time photon correlators [17]. In a collaborative study with Mary Gill of Cranfield Institute [18] we used a variety of data processing software packages on a wide range of quality and type of correlogram. Here, we select just two sets of data and use 6 parameter Gaussian model fitting, the Abbiss-Sharpe Fourier cosine transform software and McWHIRTER's new spline model software.

Figure 6 shows a correlogram with more than 40 fringes in the sampling volume diameter and the "best fit" 6-parameter Gaussian model fit, showing clearly that the velocity distribution is non-Gaussian and exhibits skewness because the oscillations persist for longer than a Gaussian distribution and the oscillation frequency changes across the correlogram [19]. The inversion of the spline model result is shown in Figure 7, and clearly fits the correlogram almost perfectly. The velocity distribution computed by the spline approach is shown in Figure 8. Note the details near the peaks and position of the baseline. Analysis of this data by Fourier cosine transformation is shown in Figures 9 to 11. Figure 9 shows the same data following division by an estimate of the exponential term, in preparation for transformation, and correction of the G(0) and G(1) values to define a velocity distribution baseline and correct for phototube deadtime effects. Figure 10 shows the velocity distribution following transformation. The mean velocities computed by spline and cosine transform methods agree to better than 1/20%

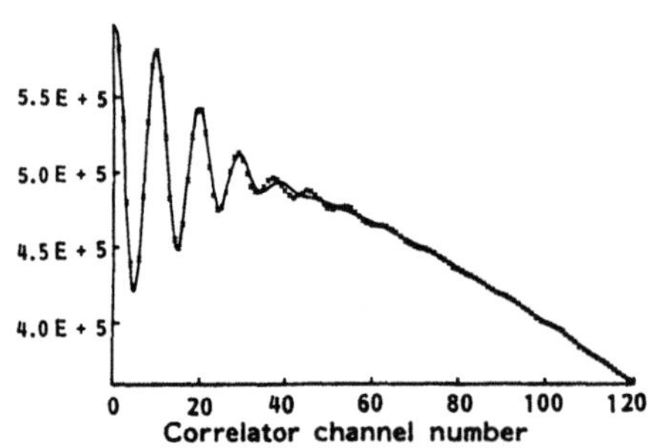

Fig.6. 6-Parameter Gaussian model fit

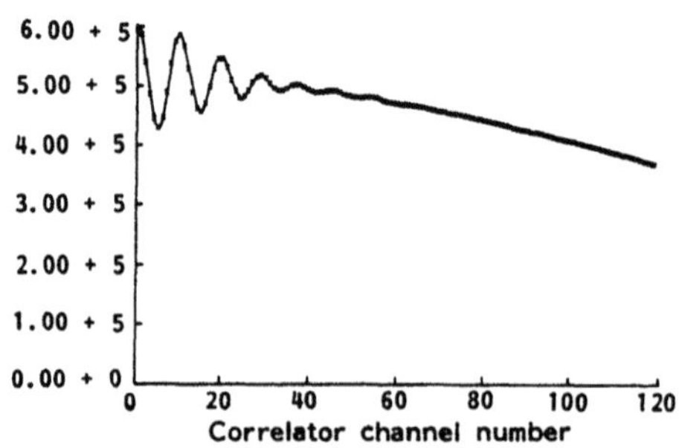

Fig.7 Inversion of spline result

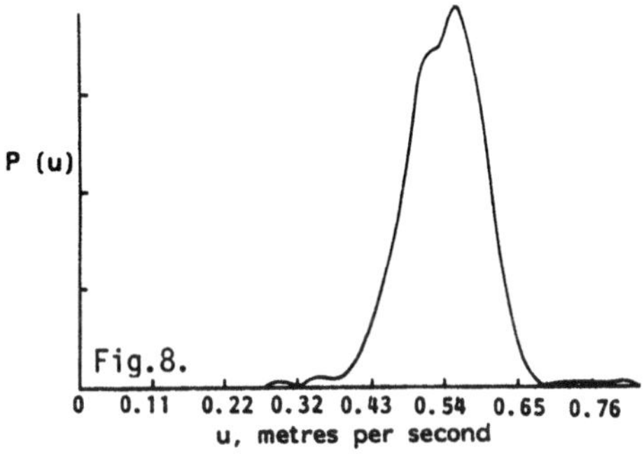

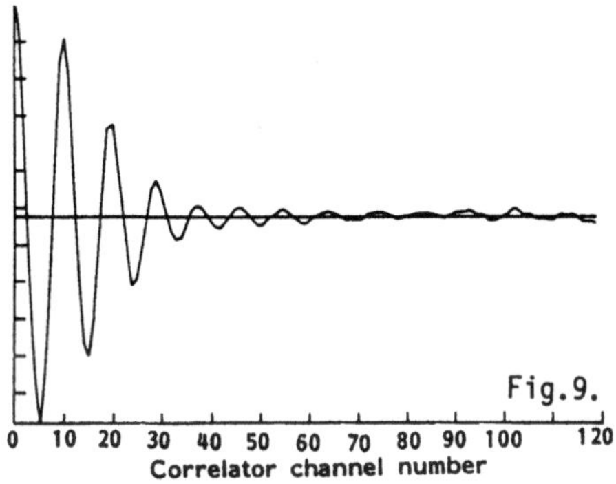

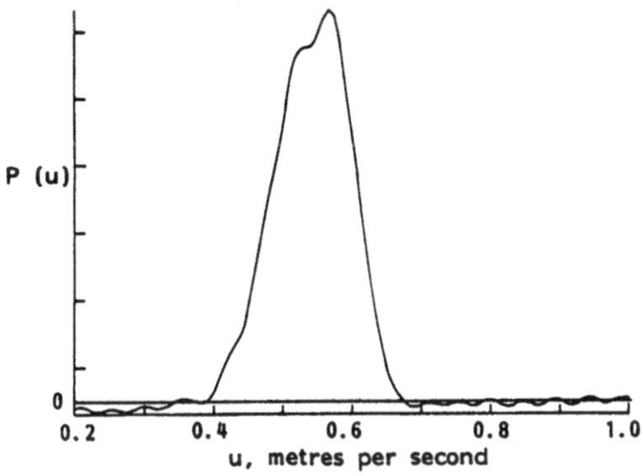

Fig.8 P(u) Reconstructed from cubic B spline model

Fig.9 FCT data after beam profile removal

◄Fig.10. P(u) from FCT of data

and turbulence intensities were 10.94% and 9.79% respectively. Figure 11 shows the inverse cosine transform of that range of the velocity distribution used to compute the moments, showing good agreement but slight radius error (cf. Figure 7, the spline inverse).

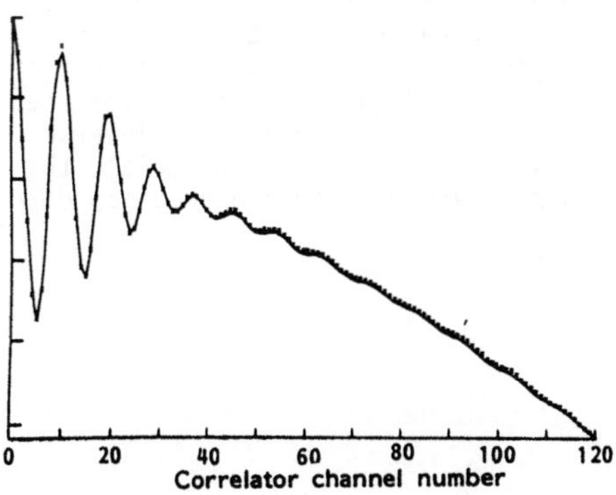

Fig.11. Inverse FCT of P(u)

The second set of data is drawn from experiments in an axisymmetric jet, where $v \ll u$. Figure 12 shows 12 fringes in the sampling volume and the 6-parameter Gaussian model "best fit". Again, skewness is evident. Figure 13 shows the inverse of the spline result showing an almost perfect fit to the

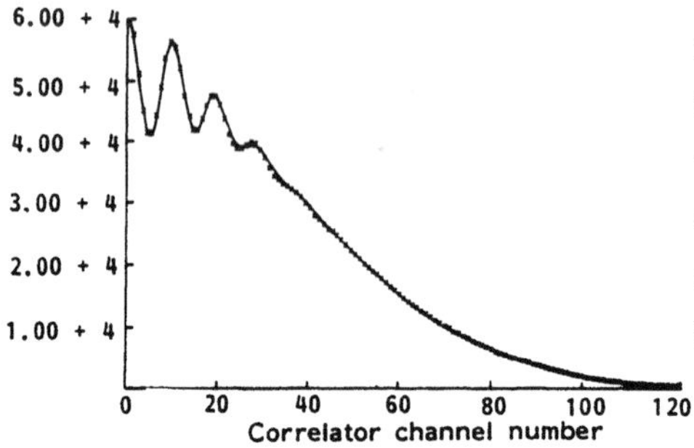

Fig.12. 6-Parameter Gaussian model fit

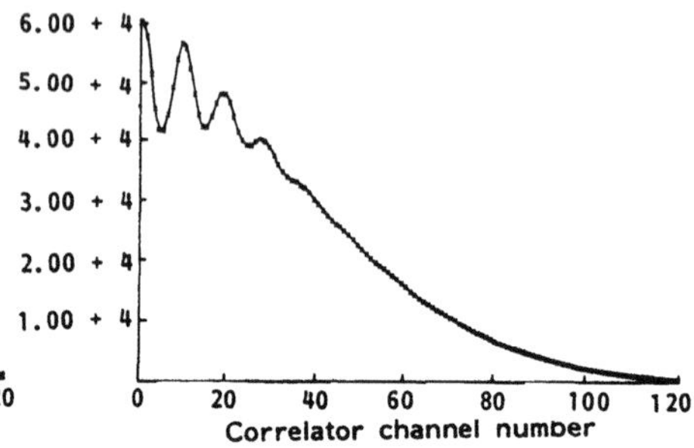

Fig.13. Inversion of spline result

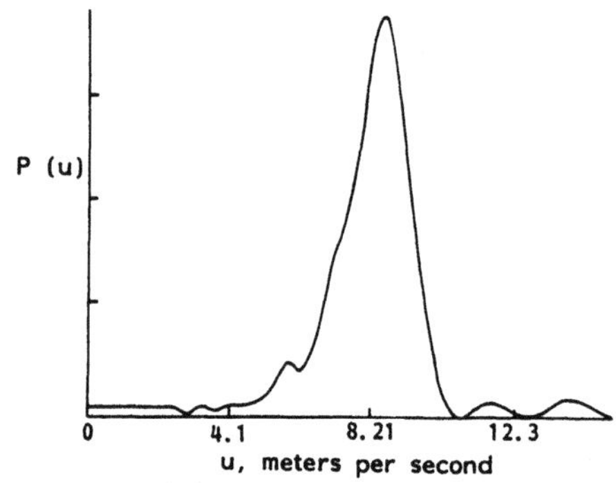

Fig.14. P(u) reconstructed from spline model

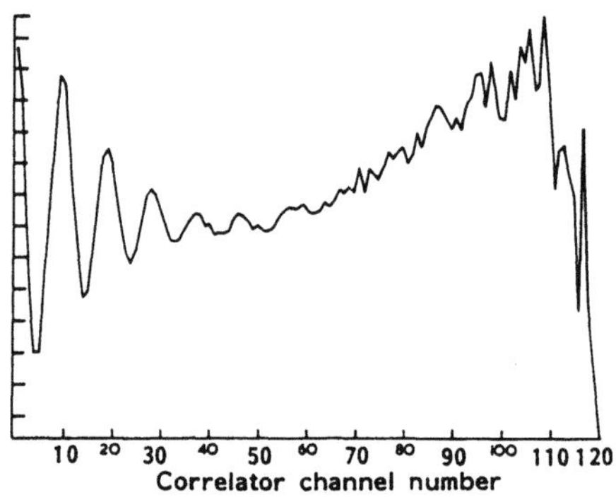

Fig.15. FCT data after beam profile removal

data, the spline model velocity distribution result being shown in Figure 14; note the skewness, as expected. Analysis of this data by cosine transformation is shown in Figures 15 to 18. Figure 15 shows the data following division by the estimate of the exponential term in preparation for transformation, the same exponential term that is part of the fit in the spline approach. A slight error in the beam radius estimate has caused high distortion at the high τ end of the correlogram. Truncation of the data to 68 channels was used to remove this problem, yielding the velocity distribution shown in Figure 16, with consequent truncation oscillations and loss of the negative skewness information shown previously to be present. The inverse transform of that region of the velocity distribution used to calculate the moments is shown in Figure 17, showing poor agreement with the original data.

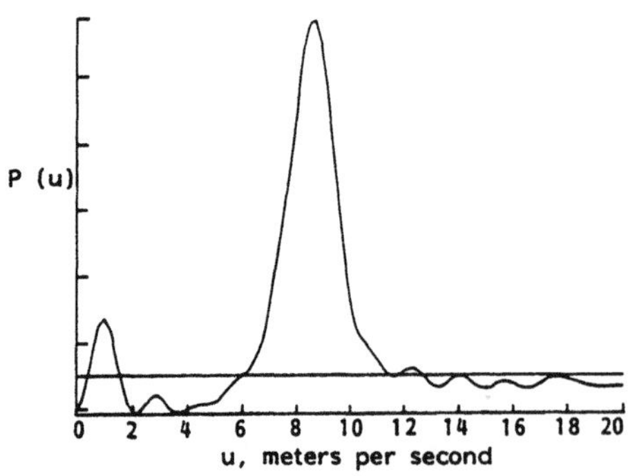

Fig.16. FCT P(u)

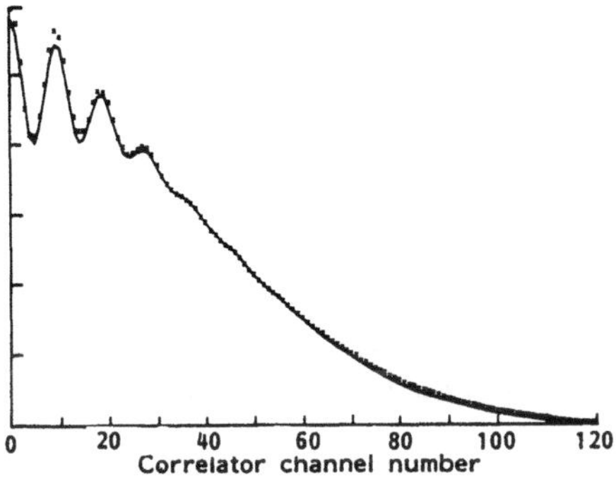

Fig.17. Inverse FCT of P(u)

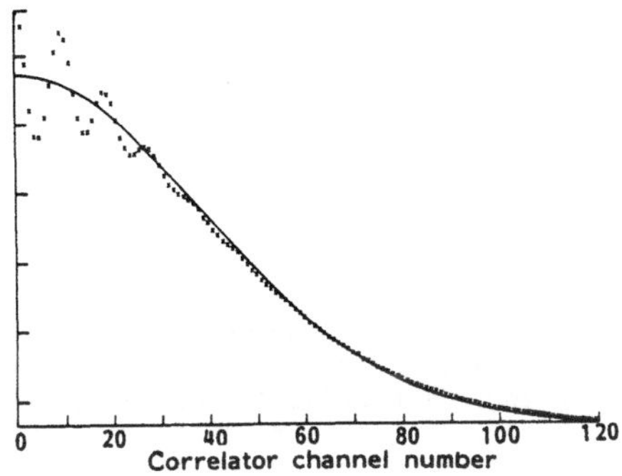

Fig.18. FCT beam radius estimate compared to data

The discrepancy between velocities calculated by spline and cosine transformation was 3½% in this instance, the turbulence intensities were 12.8% and 9.74% respectively. The slight error in beam radius that caused the cosine transform problems is shown against the original data in Figure 18.

In practice, we find that the cosine transform approach requires preferably more than 20 fringes in the sampling volume radius but has advantages of computation speed and simplicity, and can process very noisy data and data with less than 3 or 4 coefficients per Doppler cycle, under which circumstances fitting procedures have great difficulty in convergence.

Under the limitations cited, the information-theoretic approach may be the most sophisticated data processing possible; in terms of generally applicable processing there may yet be room for further development, particularly of the case where the exponential is taken outside the integral with narrow u and v distributions assumed. Taking extra terms of the expansion of the exponential term may lead to wider applicability, i.e. less restrictive u and v criteria, through an interative correction term technique.

2.3 Can we predict and estimate the accuracy of our LDV results?

Virtually no realistic analytical prediction of experimental accuracy has been made because the signal and noise distributions in laser velocimetry are usually highly non-Gaussian. We can, however, process simulated correlograms and compare the results to the models used for simulation, or we can multidimensionally integrate [20], but both approaches require extensive computation and neglect other problems such as particle slip with respect to the fluid.

Clipping of non-Gaussian signals introduces distortions into the correlogram, but for certain types of non-Gaussian signal and low mean count rates per sample time such distortions are negligible [21].

The estimate of fringe spacing limits the absolute accuracy obtainable. It is time consuming and difficult to measure fringe spacings to better than ½ to 1% accuracy, although it is possible [22].

Spatially non-random particle concentrations also affect accuracy, large errors being possible where flows of differing particle concentrations are mixing [23]. Thus we always try to avoid the use of artificial seeding generators and seeding dispersion problems.

Accuracies of better than 1% on mean velocity estimates are claimed in many circumstances, turbulence estimates can be considerably less accurate, depending on data processing method and accuracy of the beam radius estimate. Finally, we note that correlogram validity depends on the experiment statistics being stationary.

2.4 What are the basic limits of data retrieval by burst correlation?

The concept of burst correlation is to process in realtime correlograms formed of the signals from single particles, in a similar manner to burst counting. The principal advantage is sensitivity, i.e. lower laser powers are required. Burst correlation is required to obtain either turbulence spectra or a continuous record of flow velocity. There are many factors involved in attempting either measurement.

There are a variety of basic limits to information rate and accuracy in laser velocimetry. The fundamental photon statistics limits have been calculated by OLIVER [24], irrespective of limitations due to particle slip and fringe spacing accuracy. Some 30 photodetections are required for a 1% accurate velocity estimate in the presence of no noise [1], and experimental verification of good accuracy from 34 signal pulses in the presence of 708 noise detections has been achieved [8]. However, we have seen that this number of pulses from an LDV particle transit can cause some pulse "pile-up" due to high instantaneous photodetection rates, and consequent loss of information and accuracy. Reducing laser power to cure this problem causes a reduced rate of observed particles and rapidly lengthening experiment times.

If we require particle concentrations sufficient to allow continuous "tracking" of real time velocity fluctuations, it must be considered that for spatially random particles, the distribution of times between particle arrivals at the sampling volume is exponential, the probability of absence of particles for longer than time Δt is

$$p(t > ts + \Delta t) = \exp - \bar{N}(1 + \frac{\Delta t}{ts}) \quad , \qquad (2.4.1)$$

where ts is the transit time of the scatterer through the sampling volume, and $\bar{N}$ the average number of particles in the volume. If $ts \simeq \Delta t$, to obtain negligible signal loss or "dropout", $\bar{N} \gtrsim 3$ or 4 at least. For negligible probability of two or more particles simultaneously present in the sampling volume, $\bar{N} \ll 1$.

But it is well known and easily verified that the quasi-continuous Doppler signal from many particles in the sampling volume simultaneously is randomly phase modulated. The random errors in phase (phase noise, Doppler ambiguity or transit time broadening) occur on a time scale of the beam transit time and take values between $\pm \pi$. Thus, instantaneous velocity estimates acquired during a period of one transit time can be in error by as much as approximately the reciprocal of the number of fringes in the sampling volume. The particle concentration requirements for negligible signal dropout and accuracy are conflicting. The phase autocorrelation function can significantly distort the Doppler correlogram unless precautions are taken. The phase noise problem need not be of concern in the measurement of spectra since that information may be found by random sampling a flow seeded so lightly that the usual single-particle criteria is met [25], or if the cross-correlation of two Doppler signals from spatially adjacent measurement volumes within the flow microscale is performed [26].

For the situation of a single measurement volume filled with many scatterers, phase noise is present whether either incoherent or coherent detection of the Doppler signals is used, because the signal de-correlates continuously during a transit time as the particle set changes completely. With incoherent detection of scattering from many particles, many speckles fill the receiver's aperture and the Doppler oscillations are superimposed. As the particle number density increases, so the background "d.c." level increases but the Doppler oscillation amplitude decreases, thus the Doppler signal disappears and in any case DC saturation will occur of any sensitive detection system such as a photon counting tube.

Coherent detection of the Doppler-difference signals implies the addition of optical field amplitudes instead of intensities. Now, the signal strength increases with photon and particle intensities. However, coherent detection implies an aperture limitation that may render signal strengths unusable unless forward-scatter and moderate laser powers (some tens of milliwatts) are employed.

To track velocity continuously by burst correlation it is therefore necessary to:

(i) Estimate the spatial and temporal flow microscales.

(ii) Set the correlogram integration time less than the temporal microscale to measure an instantaneous velocity.

(iii) Set the sampling or measurement volume smaller than the spatial microscale dimension to avoid averaging.

(iv) Place many fringes in the sampling volume to reduce phase noise to an acceptable level of error.

(v) Arrange for many particles to be in the sampling volume simultaneously unless the microscales are so large that one can guarantee a single particle velocity estimation every microscale. The regularly sampled, contiguous velocity record will therefore be uncorrelated with the flow fluctuations and particle arrivals.

(vi) Coherently detect the Doppler signals with possibly consequent increased laser powers.

A burst correlator has now been built [17] and successfully demonstrated, however, there is some further theoretical analysis to be done, particularly in the proper treatment of signal and noise levels in terms of an inverse root relative variance [27]. This work is currently in progress at RSRE and will form the basis of a future publication.

3. Laser Transit Velocimetry, LTV

This velocimetry mode comprises two parallel laser beams and two photodetectors assigned to detect scattered light from different laser beams, Figure 19. Again, the u component is the one that is sought. The LTV technique has been developed [28,29,30] principally for applications inside machinery, because of its capability of very high spatial resolution.

We ask four questions about LTV.

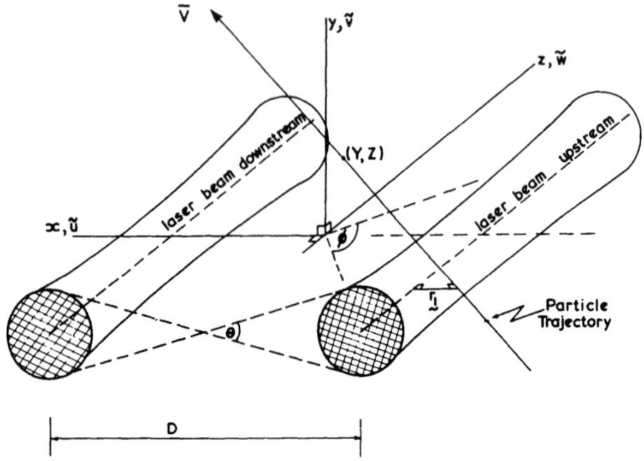

Fig.19. Laser transit velocimetry geometry

3.1 How do LTV signals compare with those of LDV?

Again, the dynamic signal range can be large, but now, because of the small beam waist dimensions usually employed and consequent increased power densities, the existence of photon "pile-up" pulses is frequent. It is for this reason that pulse centre estimation techniques [31,32] are often employed to provide one pulse for each particle and yield particle-event correlograms in place of photopulse correlograms.

We simulate such signals in the same manner as LDV simulation but using a three-dimensional Gaussian laser beam intensity expression as the rate modulating function. Simulated signal pictures are shown in Figures 20a and b. Pulse centre estimations using too few photodetections will be in error about the true beam centre, a discrimination level is necessary to reduce this problem.

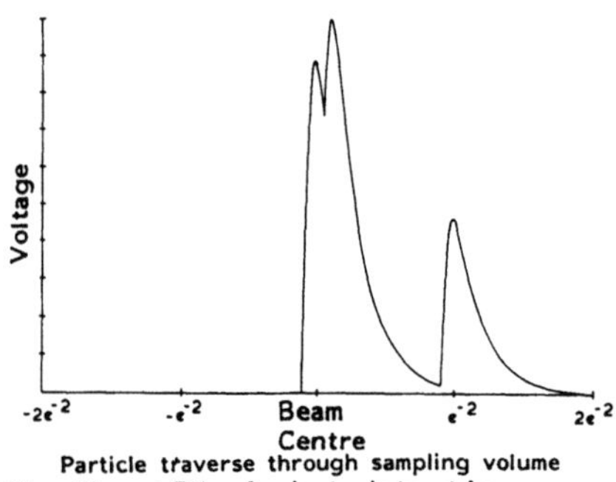

Fig.20a. LTV: 4 photodetections

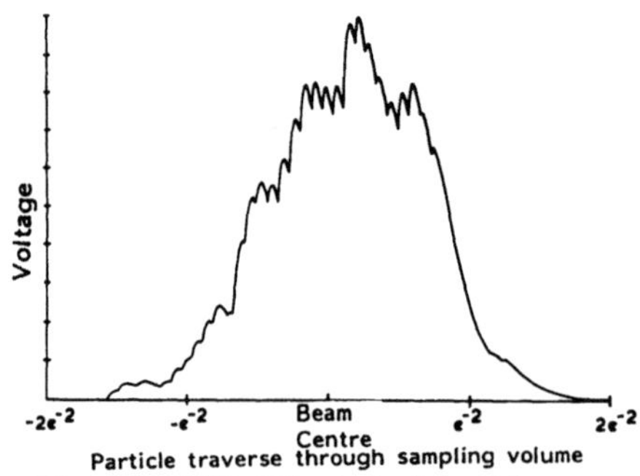

Fig.20b. LTV 250 photodetections

In terms of usage of natural aerosol particles, LTV can use signals from the naturally more abundant smaller particles in the atmosphere because of higher power densities in the sampling volume. Figure 21 illustrates this with an example taken from a design study for 200 ms^{-1} flow velocity, 100 mW laser power and 15^{0} optical transmitter/receiver ray cone in full backscatter. The limits drawn around particle regions relate to usable particle concentrations, fluid tracking capability and desirable data rates. We must not be deceived by thoughts of potential one or two orders of magnitude data rate

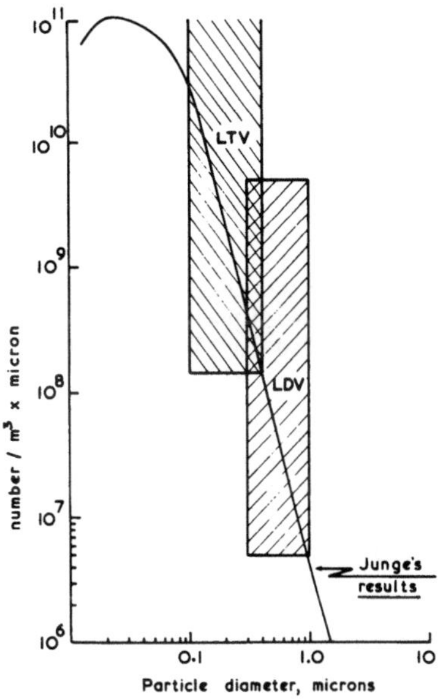

Fig.21. Comparison of uses of aerosol by LDV and LTV for arrangements described

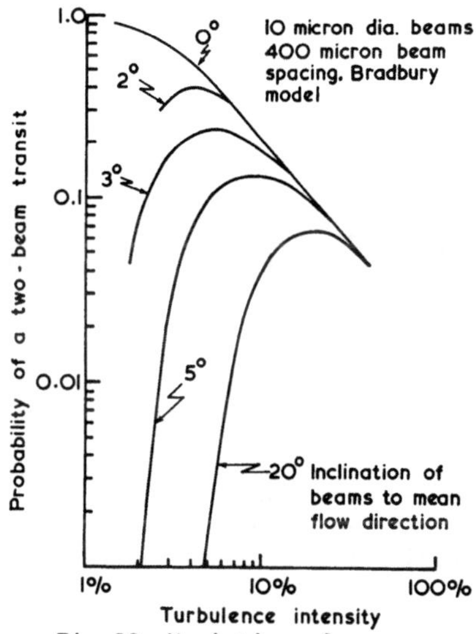

Fig.22. Variation of successful two-beam transit probability with turbulence intensity and inclination of beam plane of the flow direction

advantage of LTV over LDV, which may be construed from this graph. Such advantages may not exist in practice for four reasons: firstly, lack of high efficiency, diffraction limited optics; secondly, the smaller cross-section of the sampling volume causes interception of fewer particles; thirdly, the laser beam intensity cross-section is Gaussian and not a cylindrical step function and fourthly, and most significantly, the turbulence intensity and angle of the beam plane to the mean flow direction can cause dramatic reductions in the probability that a particle will intercept both laser beams and contribute to the correlogram positively.

To illustrate the fourth reason we may use a cylindrical beam and Gaussian turbulence model of the LTV situation proposed by BRADBURY (unpublished). The probability of a successful two-beam transit is

$$p(\theta,\phi) = \frac{1}{(2\pi\sigma^2)^{3/2}} \int_0^{\infty} e^{-\frac{(u-\bar{u})^2}{2\sigma^2}} \int_{-u\tan\theta/2}^{u\tan\theta/2} e^{-\frac{(v-\bar{v})^2}{2\sigma^2}} \int_{-u\tan\phi/2}^{u\tan\phi/2} e^{-\frac{w^2}{2\sigma^2}} \, du\,dv\,dw \quad (3.1.1)$$

Results are shown in Figure 22. With low turbulence and small angles of the beam plane to mean flow direction the two-beam transit probability is greatly reduced, at high turbulence intensities the probability becomes independent of angle. This is in accord with experience. Data rates depend strongly on alignment and turbulence intensity.

3.2 Should we use a correlator or an MCA to form correlograms?

The multichannel analyser is a "single stop" correlator [5], in contrast to standard correlators which are "all stop". It has been shown theoretically [33] that with MCA processing it is possible for signals due to seeding concentration and/or noise level to produce spurious, false transit time "stop" pulses and thus to distort the correlogram base line and degrade the accuracy of estimates. Such effects have been quantified and it has been concluded that the likelihood of erroneous measurements occurring using natural aerosol particles is extremely small, but if using entrained, high density artificial seeding the chance is significant. Unfortunately, noise signals are indistinguishable from particle signals and thus in high noise levels, eg LTV operation close to surfaces, MCA distortion effects are likely to be seen unless high thresholds are used to provide signal discrimination. These effects are illustrated in Figure 23, where it can be seen that for the example mentioned earlier, if the mean signal photodetections per particle are 25 times the mean noise rate, a threshold may be set so that true transit signals dominate false, noise-produced signals. With a signal to noise ratio of 4:1 the same is true, but much of the true data is lost due to the high threshold necessary, and for a signal to noise ratio of only 2:1 the domination of true signals by threshold increase is not possible at all. Supportive experimental evidence exists for these conclusions [34].

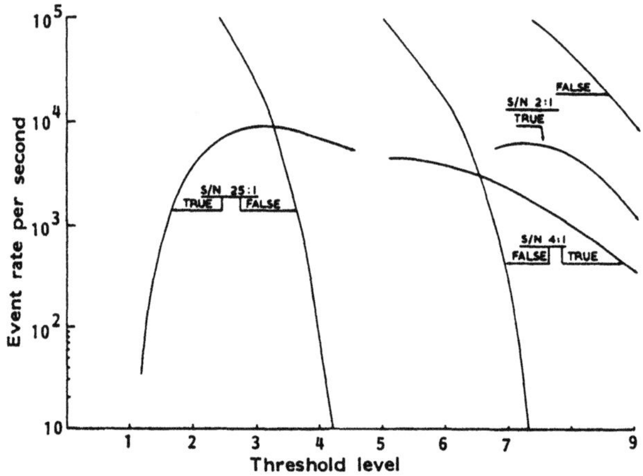

Fig.23.
MCA true and false signal rates

Disadvantages with currently available photon correlators are minimum sample time and limited correlogram length, particularly for low turbulence estimates, but techniques of inserting delay times before correlation or application of corrections proposed by MAYO [32,35] may be used if necessary. A compensating advantage is that correlation can also be used to process LDV signals.

Design of "multistation" arrangements for simultaneous acquisition of many correlograms is highly dependent on the experiment, but note that the acquisition of, say, 10 sets of data simultaneously does not necessarily reduce experiment time by a factor of 10.

3.3 How should we process our LTV data?

Consider the form of the cross-correlation function of two signals of the form produced by a single particle central LTV transit

$$I_1(t) = I_0 \exp\left(-\frac{2t^2}{\tau_c^2}\right)$$

$$I_2(t) = I_0 \exp\left(-\frac{2(t-t_0)^2}{\tau_c^2}\right) \qquad (3.3.1)$$

where t_0 is the transit time, and τ_c is the single-beam transit time. The cross-correlogram of these signals is

$$G_s(\tau) = \frac{\tau_c\sqrt{\pi}I_0^2}{4} \exp\left\{-\frac{(\tau-t_0)^2}{\tau_c^2}\right\} \qquad (3.3.2)$$

which has its peak at the transit time t_0 and is velocity biased by the τ_c term. In practice, the intensity due to an arbitrary single particle transit has the form

$$I = \frac{kI_0}{r(z)} \exp\left\{-\frac{2(x^2 + y^2)}{r^2(z)}\right\} \qquad (3.3.3)$$

and the form of the turbulent correlogram is

$$G(\tau) = \iiiint\!\!\int_{-\infty}^{+\infty} p(u,v,w,y,z)\, G_s(\tau)\, du dv dw dy dz \qquad (3.3.4)$$

which is very complicated and requires approximations to be made.

To remove z dependence, most approaches assume cylindrical beams of uniform intensity cross-section, the problems of this assumption have already been discussed. A lack of flow reversals is also assumed, limiting turbulence validity to $\simeq 20\%$. Further assumptions are that $v \ll u$ and u is Gaussian distributed. The correlogram form is now

$$G(\tau) = \frac{k_1}{(r_0^2 + 2\sigma^2\tau^2)^{\frac{1}{2}}} \exp\left\{-\frac{(u\tau - D)^2}{(r_0^2 + 2\sigma^2\tau^2)}\right\} \qquad (3.3.5)$$

an example being shown in Figure 24. It is unbiased to first approximation, exhibits skewness and has a peak position generally different from the peak position given by the mean velocity. This form is indistinguishable from that of a linear velocity gradient across the sampling volume [36], i.e. the shape is not uniquely related to the velocity distribution.

The usual experimental procedure is to acquire correlograms at a variety of angles of inclination of the beam plane to the mean flow direction. Based on the Gaussian assumptions discussed, the approach of ROSS [37] is first to determine mean flow angle by fitting a Gaussian curve through the distribution of correlogram peak amplitudes, mean velocity and turbulence estimates being calculated by fitting the Gaussian correlogram form to the data acquired nearest the mean flow direction. This approach cannot extract

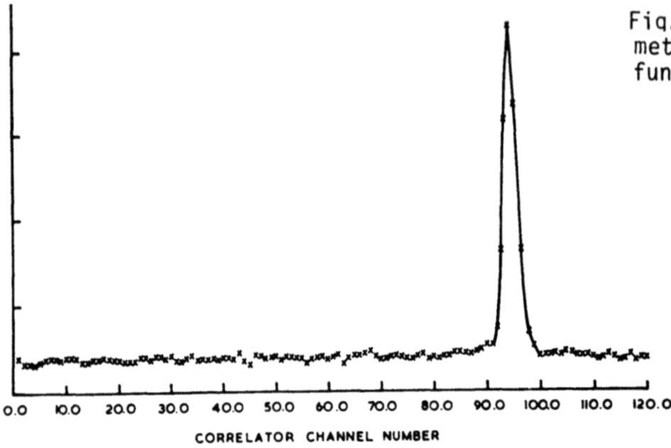

Fig.24. Laser transit velocimeter cross-correlation function

information about the form of the velocity distribution or shear stress details. SCHODL's approach [28,39] is more general and involves integrating the correlograms in different ways to find mean velocity, turbulence intensity and shear stresses, etc.

To date, however, no general procedure for processing data acquired from arbitrary u,v distributions has been published, though some researchers have made further progress by transformation of many correlograms to velocity distributions in an attempt to extract information on the velocity distribution structure as well as the moments. LTV data reduction has received much less attention than that of LDV and there is much to be done.

Another problem, discussed in this volume by GILL et al., is that of normalisation of correlograms, ideally requiring noise and particle concentration levels to remain constant throughout the entire period of acquisition of a set of correlograms.

Finally, note that spatially random (Poisson) seeding is essential for level base lines in LTV correlograms (on timescales short compared to turbulence timescales only). Non-Poisson seeding effects destroy the usual form of the correlogram and render turbulence estimates of no use, but such effects yield new flow information from LTV, vortex shedding frequency [38]. This is discussed further in this volume by BROWN and RICHARDS.

3.4 Can we estimate the accuracy of an LTV experiment?

Just as for LDV (section 2.1) we can simulate or multidimensionally integrate at great computational expense, neglecting particle slip and beam spacing errors. The distance between the two LTV laser beams can be measured to good accuracy, say < 1/10 beam diameter, which is usually a small fraction of one percent of the beam spacing. Growth of the beam waist through the axial observation length can be important, eg 10 μm grows to 16 μm in 200 μm path length; this is neglected in most analyses by assuming cylindrical beams, but low turbulence accuracy is affected. Mixed seeding concentration effects alter accuracy as in LDV, though this has yet to be quantified.

For low turbulence situations, mean velocity estimates to between 1% and 1/10% have been claimed, and flow angles may be found to $1/5^{0}$ or better, assuming that the experiment is statistically stationary. Accuracies at higher turbulence intensities, $\gtrsim$ 15%, can be poor, due mainly to poor data reduction models; two-component LDV is competitive in such circumstances for flow angle estimation, if usable.

Finally, note that focussing on to and parallelism of the laser beams is of paramount importance, as very small errors in both these parameters simultaneously can cause large systematic errors.

4. The Future

In this review, we have concentrated on the problems of perfection and ignored some remarkable achievements, but what of the future?

Many problems to be researched have been illustrated already but other goals exist also.

Currently available phototubes "saturate" at mean count rates of $\simeq$ 10 MHz. To reduce pulse pile-up problems we wish this value to be increased to greater than 100 MHz. Longer record length, faster and cheaper correlators are needed, 1 nanosecond sample times would be invaluable, so too would be a method of direct all-digital estimation of the velocity distribution without the requirement of transformation of a correlogram.

Electronic developments have always been at the heart of progress in laser velocimetry and new circuit architectures such as Uncommitted Logic Arrays, Systolic arrays and very high speed integrated circuitry (VHSIC) will probably have a role to play. Research and development programs having these aims in mind exist at RSRE Malvern and elsewhere, and we hope to realise at least some of the aims in the future.

Acknowledgements

I am indebted to Dr. E.R.PIKE FRS for countless enjoyable and stimulating discussions over a period of some years, and for critical advice concerning the present text. I thank Dr. J.G.McWHIRTER of RSRE, Malvern, Dr. J.B.ABBISS and Mr. P. R.SHARPE of RAE, Farnborough for valuable discussions and making available their data analysis computer software. I thank also Dr. L.J.S.BRADBURY of Surrey University, Dr. P. H.RICHARDS of CEGB Marchwood Labs and Mrs. Mary GILL of Cranfield Institute of Technology for enthusiastic and clarifying conversations.

References

1. H.Z. Cummins, E.R. Pike (eds): Photon Correlation Spectroscopy and Velocimetry (Plenum Press, New York 1977).

2. J.B. Abbiss, T.W. Chubb, E.R. Pike: Opt. Laser Tech, 6, 249 (1974).

3. F. Durst, A. Melling, J.H. Whitelaw: Principles and practice of Laser Doppler Anemometry, (Academic Press, London 1976).

4. L.H. Tanner: Opt. Laser Tech, 5, 108 (1973).

5. H.Z. Cummins, E.R. Pike (eds): Photon Correlation and Light Beating Spectroscopy, (Plenum Press, New York 1974).

6. W.T. Mayo: Proc. Minnesota Symp. on Laser Anemometry, 455, (University of Minnesota 1975).

7. A.E. Smart, W.T. Mayo: Physica Scripta, 19, 426 (1979).

8. J.B. Abbiss: Physica Scripta, 19, 388 (1979).

9. D.P. Robinson: Opt. Laser Tech, 13, 85 (1981).

10. A.D. Birch, D.R. Brown, J.R. Thomas: J Phys D, 8, 438 (1975).

11. J.G. McWhirter, E.R. Pike: Physica Scripta, 19, 417 (1979).

12. J.G. McWhirter: Optica Acta, 27, 83 (1980).

13. J.G. McWhirter: Optica Acta, 28, 1453 (1981).

14. T. Cochrane, J.C. Earnshaw: J Phys D, 11, 1509 (1978).

15. L.J.S. Bradbury: Physica Scripta, 19, 360 (1979).

16. P.R. Sharpe: Physica Scripta, 19, 411 (1979).

17. R.G.W. Brown et al: Physica Scripta, 19, 365 (1979).

18. R.G.W. Brown, M.E. Gill: to be published.

19. C.J. Moore, A.E. Smart: J. Phys E, 9, 977 (1976).

20. R.G.W. Brown, E.R. Pike: Proc. 4th Int. Conf. on Photon Correlation Techniques in Fluid Mechanics, paper 2, (Stanford University 1980).

21. C. Bendjaballah: Opt. Comm, 34, 2, 164 (1980).

22. W.J.S. Hirst: Royal Aircraft Establishment, Report IRP 211 (1973).

23. A.D. Birch, M.G. Dodson: Optica Acta, 27, 1, 3 (1980).

24. C.J. Oliver: J. Phys D, 13, 1145 (1980).

25. J.B. Roberts, J. Downie, M. Gaster: J. Phys E, 13, 977 (1980).

26. P. Buchave, W.K. George, J.L. Lumley: Ann.Rev. Fluid Mech., 11, 443 (1979).

27. E. Jakeman, C.J. Oliver, E.R. Pike: Advances in Physics, 24, 3, 349 (1975).

28. R. Schodl: Trans, ASME. J. Fluid Eng, 102, 412 (1980).

29. A.E. Smart et al: Proc. ASME Conference, New Orleans, 149 (1980).

30. R.G.W. Brown, E.R. Pike: Opt. Laser Tech, 10, 319 (1978).

31. D.A. Gedcke, W.J. McDonald: Nucl.Instr.Meth, 58, 253 (1968).

32. W.T. Mayo: Optica Acta, 27, 53 (1980).

33. J.B. Abbiss, R.G.W. Brown, E.R. Pike, P.R. Sharpe: Proc. 4th Int.Conf. Photon Correlation Techniques in Fluid Mechanics, paper 9 (Stanford University 1980).

34. C.F. Haxell, R.J. Parker: Rolls Royce Ltd, Aero Division report RR(OH) 717 (1979).

35. W.T. Mayo, A.E. Smart: Proc. ISL Conf. on Long Range and Short Range Velocimetry, (Institute Saint-Louis, France 1980).

36. L. Lading: AGARD Conference proceedings No 193, paper 23 (1976).

37. M.M. Ross: Optica Acta, 27, 511 (1980).

38. P.H. Richards, R.G.W. Brown: CEGB Marchwood UK, Report No TPRD/M/1230/N82 (1982) and accepted for publication in J. Phys.D.

39. R. Schodl: DFVLR Köln, Thesis DLR-FB 77-65 (1977).

4. Examples of the Data Reduction of Photon Correlation Laser Doppler Anemometer Autocorrelation Functions Using Fourier Transforms

P.N. Inman and L.J.S. Bradbury

Department of Mechanical Engineering, University of Surrey, Guildford, Surrey, UK

Criteria are established for the Fourier cosine transformation data-reduction technique to be a valid procedure in the analysis of photon correlation data. It is shown that if there are many fringes in the focal volume, and more importantly, if the parameter β ($=2\pi^2(\sigma/U\ r/s)^2$) , linking the number of fringes to the turbulence intensity, is large, precise knowledge of the beam profile characteristics is not required. Practical problems associated with the implementation of the Fourier transformation technique are discussed. Measurements have been made in a turbulent jet mixing layer over a range of local intensities from 4% to 100%. Photon correlation data analysed by Fourier transform are compared with hot-wire and pulsed-wire data.

1. Introduction

There are a number of techniques [1,2] for analysing the autocorrelogram obtained from a photon correlation laser Doppler anemometer. These techniques often make use of the expression due to ABBISS [1] for the autocorrelation of the photon-resolved Doppler difference signal. This signal arises from particle transits through a probe volume formed by the intersection of two coherent laser beams having assumed Gaussian intensity distributions. To extract the autocorrelation from a measured autocorrelogram it is necessary to estimate all of the characteristics of the beam profile model in the cross-over region. The velocity results so obtained are sensitive to the errors in the accuracy of the beam profile representation.

In a practical experiment there are several causes for the measured beam profile to be non-Gaussian. The prime assumption is that the laser beam has an accurately Gaussian intensity distribution. Imperfections in the optics and their arrangement are sources of error, particularly if the probe volume is inaccurately imaged. Scattering over large distances by naturally occurring atmospheric particles can modify the profiles of incident beams in long range measurements.

More fundamentally, the turbulent flow autocorrelation model used for data reduction takes no account of the finite extent of the focal volume. By analytic integration of the laminar flow autocorrelation over a viewing slot of finite width, BRADBURY [3] predicted the departure of the measured beam profile from Gaussian form with increasing slot width. The apparent beam radius, defined where the intensity is $1/e^2$ of its maximum value, increases with aperture width. Clearly large errors will result if a Gaussian model is used to analyse such data.

Under certain combinations of fringe spacing, beam radius and local turbulent intensity the magnitude of the beam component of the autocorrelation is almost constant over the record length. In these circumstances it is not necessary to make a detailed estimation of the beam parameters and a simple, physically plausible parabolic beam profile model suffices to remove both the large d.c. component which appears in measured correlograms, and any remnant of the beam profile. The velocity probability distribution is then the Fourier cosine transformation of the normalised autocorrelation.

The aim of this paper is to indicate the relevant parameters and their values for the Fourier cosine transformation analysis techniques to be practically useful. To do this we start by considering an idealised one-dimensional autocorrelation model.

2. Model Investigation of the Accuracy of the Fourier Transformation Technique

The autocorrelation for a single particle transit in laminar flow, velocity u, aligned normal to the real fringe pattern is given by (1).

$$G(\tau) = \frac{1}{u} \exp(-(u\tau/r)^2)\ [1 + \tfrac{1}{2}\, m^2 \cos 2\pi(u + U_f)\tau/s]\ ; \qquad (1)$$

τ is the time lag, r and s are beam radius and fringe spacing respectively and m is the fringe visibility. U_f is an additional velocity due to frequency shifting. The autocorrelation in turbulent flow is effectively modelled by the integral over velocity of the laminar autocorrelation, weighted by the product up(u) which characterises the turbulence velocity field and implicitly allows for velocity biasing in the particle arrival rate

$$G(\tau)_T = \int_{-\infty}^{\infty} up(u)\ G(\tau)\ du \quad . \qquad (2)$$

We now assume a Gaussian velocity field with probability density given by (3), where U and u are mean and instantaneous velocity and σ is the fluctuation intensity

$$p(u) = \frac{1}{\sqrt{2\pi}\,\sigma} \exp\left[\frac{-(u-U)^2}{2\sigma^2}\right] \quad . \qquad (3)$$

The turbulent autocorrelation is then

$$G(\tau)_T = \frac{\exp(-\ (U\tau/r)^2/F\ )}{\sqrt{F}}\ [1+\tfrac{1}{2}m^2\exp(-2\frac{(\pi\sigma\tau/s)^2}{F})\ \cos\ 2\pi(\frac{U}{F} + U_f)\frac{\tau}{s}\ ] \qquad (4)$$

$$F = 1 + 2(\sigma\tau/r)^2 \quad . \qquad (5)$$

For comparison we write down the Fourier cosine transformation of the Gaussian velocity distribution above

$$R(\tau) = \exp(-2(\pi\sigma\tau/s)^2)\ \cos\ 2\pi\ U\tau/s \quad . \qquad (6)$$

The velocity and intensity information to be recovered from (4) by Fourier transform lies in the cosine term and its exponential factor. Comparison of (4) and (6) shows that for the calculated mean velocity, $U/F + U_f$ and intensity, σ/F, to be accurate the factor F must be close to unity, ie

$$2(\sigma\tau/r)^2 \ll 1 \quad . \tag{7}$$

To ensure the decay of the autocorrelation the argument of the relevant exponential must be large, leading the the first indication that many fringes are required

$$\pi^2(r/s)^2 \gg 1 \quad . \tag{8}$$

It is more important however, that the envelope of the autocorrelation has decayed to a small value before the form of the beam profile has become significant. Comparing the relative magnitudes of the arguments of the autocorrelation and beam exponentials gives (9) linking local turbulent intensity and the number of fringes in the beam radius

$$\beta = 2\pi^2(\frac{\sigma}{U}\ \frac{r}{s})^2 \gg 1 \quad . \tag{9}$$

It is the magnitude of the parameter β which indicates the ease and accuracy with which the Fourier cosine transformation technique may be applied. The beam profile and autocorrelation envelope function can now be expressed in terms of β and intensity σ/U

$$B(\frac{U\tau}{r}, \frac{\sigma}{U}) = \frac{\exp\ (-(U\tau/r)^2/F)}{\sqrt{F}} \tag{10}$$

$$E(\frac{U\tau}{r}, \beta, \frac{\sigma}{U}) = \exp\ (-\beta(U\tau/r)^2/F) \tag{11}$$

$$F = 1 + 2\ (\frac{\sigma}{U})^2\ (\frac{U\tau}{r})^2 \quad . \tag{12}$$

The beam profile is shown in Fig. 1. The deviation from the Gaussian form found in laminar flow ($\sigma/U = 0$) is very marked for large values of $U\tau/r$ at the higher intensity levels. Note that additional distortion effects due to

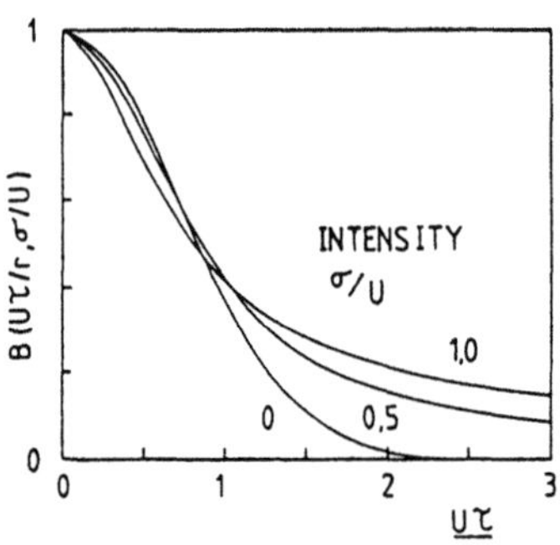

Fig.1. Apparent beam profiles in turbulent flow

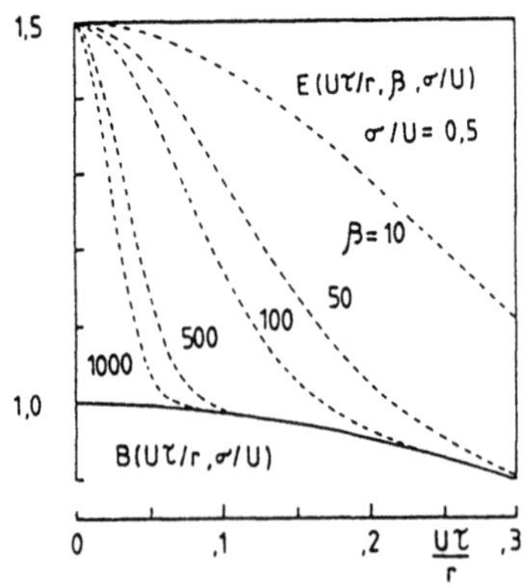

Fig. 2. Autocorrelation envelope E ----, and beam profile B ——, in turbulent flow

the necessarily finite viewing aperture would be expected. It is constructive to examine in Fig. 2 the relationship between the autocorrelation envelope $E(U\tau/r, \beta, \sigma/U)$ and beam profile $B(U\tau/r, \sigma/U)$ at small values of $U\tau/r$ as implied by (7). The explicit dependence upon intensity (σ/U) alone is very weak and is not shown here. It is immediately apparent that the influence of the beam profile becomes increasingly less significant for values of β greater than 500. This implies that, ideally, the product of local intensity and number of fringes within the radius should be greater that 5, requiring, for example, 40 fringes in the focal volume at 25% intensity.

3. Practical Problems Associated with the Application of the Fourier Cosine Technique to Measured Autocorrelograms

The form of the autocorrelogram measured by a correlator having N_c channels operating on a sampling interval T_s is given in (13)

$$G(k) = a_1 \; B(kT_s) \; [1 + \tfrac{1}{2} m^2 \; R(kT_s)] + b_1 \qquad k=1,2...N_c \quad . \tag{13}$$

The constant a_1 depends upon many factors, including the optical efficiency of the experiment and its duration. The background d.c. contribution to the correlogram is represented in b_1. There are a number of practical problems associated with the analysis of measured autocorrelograms.

3.1 Removal of DC Level and Remaining Beam Profile

At large values of β the influence of the beam profile over the record length is small, but may not be totally insignificant. In these cases the beam profile in (13) is approximated by a parabolic model with constants implicitly including the experimental values of a_1 and b_1

$$G(k) \propto (C-D(kT_s)^2)[1 + \tfrac{1}{2}M^2 R(kT_s)] \quad . \tag{14}$$

This model is the simplest form satisfying the symmetry condition and gives a maximum error of ½% when compared with the exponential model over the range $0 < U\tau/r < 0.3$. A second-order dependence of M upon τ is ignored in the approximation of (14) from (13). A least squares fitting procedure is employed to determine the constants C and D.

3.2 Estimation of Initial Autocorrelation Coefficients

The random content of the photomultiplier signal autocorrelates to a delta function at zero time delay making the measurement of the coefficient G(k=0) impossible. In addition, the output of the initial time delay channel G(1), and possibly subsequent channels, is distorted by significant dead-time effects in the photomultiplier tube at small sampling intervals.

Initial estimates for G(0) and G(1) are made using a polynomial approximation to (6), providing there is adequate resolution in the first quarter Doppler cycle. Using these estimates an initial value of mean velocity is calculated by Fourier transformation of the normalised autocorrelation $R(\tau)$. The values of R(0) and R(1) are then corrected by fitting the autocorrelation to a modified form of (6) in the modified form (15)

$$R(\tau) = C \exp(-D\tau^2) \cos 2\pi U\tau/s \quad . \tag{15}$$

Since an accurate estimate is not given by the initial Fourier transformation the intensity implied in the constant D is maintained as a free parameter.

3.3 Selection of Sampling Interval

The Fourier transform of an N lag autocorrelation gives N+1 independent estimates of probability distribution equispaced over a velocity range from 0 to $s/2T_s$, the analogous 'Nyquist' or folding velocity. Although the Fourier cosine transformation is capable of resolving within this range between arbitrary velocity limits, the sampling interval is selected so that the anticipated probability distribution lies within the implied transformation limits with the maximum number of independent estimates in the velocity range of interest.

The selection of sampling interval is also influenced by the requirement for adequate Doppler cycle resolution for the estimation of initial autocorrelation coefficients.

3.4 The Effect of Limited Record Length

At low intensity it is not possible to include the complete correlogram in the limited record length of a correlator. The Fourier transformation of a truncated correlogram is of the form sinX/X with oscillatory wings on either side of a broadened central peak. The apparent intensity $(\sigma/U)_a$ is obtained by taking the weighted moment of this probability density distribution to the first zero crossing on either side of the main peak. ABBISS [1,4] notes that it is possible to establish an empirical relationship linking the apparent and true intensities and the number of cycles in the truncated autocorrelation, N, of the form

$$N(\sigma/u)_a = f[N(\sigma/u)_t] \quad . \qquad (16)$$

A simple procedure, employing this relationship, is used to correct intensity results obtained from truncated data.

3.5 Corrections to High Intensity Turbulence Measurements made Without Frequency Shifting

At turbulence intensities above 30% there is an increasingly significant contribution to the velocity field from negative velocities. Relative shifting of the laser beam frequencies can be used to create a moving real fringe pattern with an effective velocity bias sufficiently large to ensure that the largest real negative velocity is detected as positive. Without frequency shifting a velocity ambiguity exists and the measured probability distribution is folded about the velocity origin. For an assumed Gaussian velocity field it is a simple procedure to predict the error in mean velocity and turbulence intensity calculated from a folded probability density distribution. Skewing of the velocity field to the positive velocity side will tend to reduce the error.

4. Turbulent Mixing Layer Experiment

A series of L.D.A. photon correlation measurements have been made in the mixing layer of an axial turbulent jet. Measurements were taken 2.5 diameters downstream of the nozzle exit, Fig. 3.

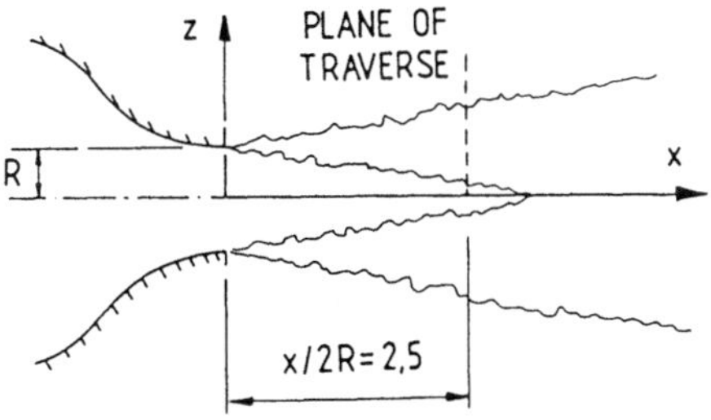

Fig.3. Coordinates of turbulent jet

The jet exit velocity of 6.7 m/s gave a Reynolds number ($2UR/\nu$) of 10^5. The focal volume, of approximately 1.5 mm diameter, was viewed in 90 degree side scatter and contained approximately 60 fringes. Frequency shifting was not available. Comparative measurements were also made using hot-wire and pulsed-wire probes.

Fig.4a shows the mean velocity results, scaled by centre line velocity. Fair agreement is achieved between the different measurement techniques. The intensity results, non-dimensionalised by local mean velocity, are shown in Fig.4b. The increasing scatter for intensities greater than 30 per cent reflects the inherent inaccuracy of hot-wire measurements at high intensites. No corrections to the hot-wire data have been attempted. At high intensities greater confidence is placed on the pulsed-wire results as a consequence of the superior yaw response of this instrument.

The photon correlation results show accuracy comparable with the other techniques over most of the intensity range. For turbulence intensities in excess of 40% the intensity results calculated directly from the folded probability density distributions are shown linked with results adjusted using the Gaussian folding correction of Section 3.5. The corrections to the mean flow results are not discernible at this scale.

Examples of the autocorrelation functions and the resulting calculated probability density distributions are given in Figs. 5 and 6. Near the

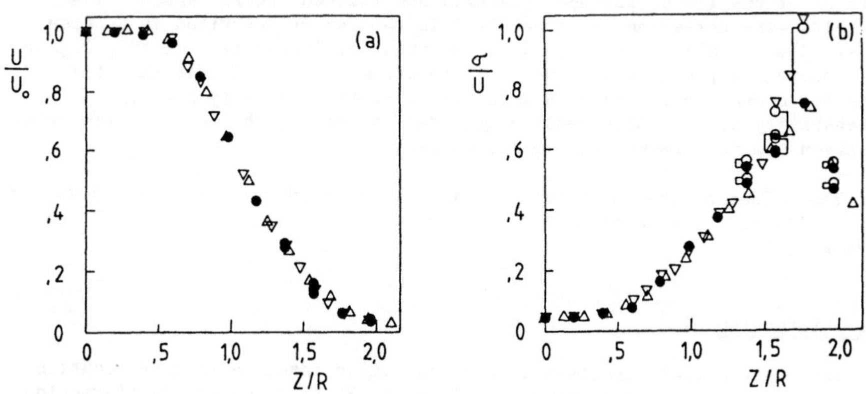

Fig.4. Turbulent jet mean velocity (a) and intensity (b) results
● laser (○ corrected for folding), △ hot-wire anemometer
▽ pulsed-wire anemometer

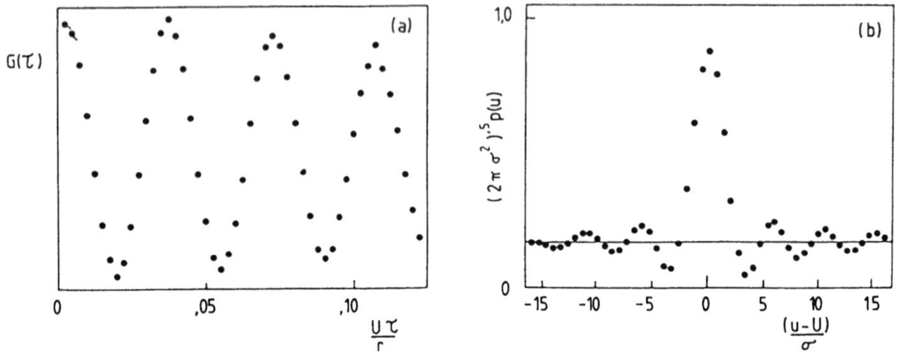

Fig.5. Autocorrelogram (a) and probability density distribution (b)
z/R = 0, σ/U = .044, β = 31. Tagged symbols initial estimates

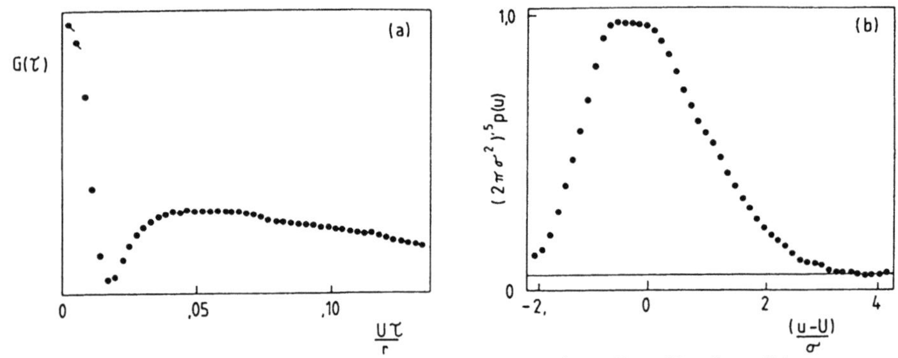

Fig.6. Autocorrelogram (a) and probability density distribution (b)
z/R = 1.38, σ/U = .50, β = 3966. Tagged symbols initial estimates

centre line, Fig. 5a, the raw autocorrelogram is necessarily truncated by virtue of the low turbulence intensity and limited record length. The results were corrected for truncation in the manner described in Section 3.4. The probability density distributions for intensities of 50%, Fig. 6b and above are positively skewed. In consequence it is likely that the results corrected for folding using the assumption of a symmetrical probability density distribution are over-estimates, the true results lying between the incorrected and corrected estimates.

The values of the parameter β defined by a fixed value of r/s and varying intensity σ/U , range from 31 on the centre line to greater than 10^4 near Z/R = 1.75.

5. Discussion

The experimental results indicate that it is feasible to make accurate photon correlation measurements, analysed by Fourier cosine transformation, without precise knowledge of the details of the beam profile if the value of the parameter β is maximised. Provided frequency shifting is available to remove the velocity ambiguity the analysis at high turbulence intensities is easier than at low intensities at which the data truncation problem arises. The use of frequency shifting to increase the apparent intensity has no effect on the value of β which depends upon the true intensity (9).

The attainable value of β is usually limited by the minimum correlator sampling interval and spatial resolution requirements. In high velocity applications moderate fringe spacing is required to give adequate Doppler cycle resolution even with the faster available correlators, and the number of fringes in the beam is restricted by the desired spatial resolution of the focal volume. At low intensities it is beneficial to maximise the number of fringes in the focal volume, thus reducing the effect of the beam profile on the truncated correlogram. It is interesting to note that the truncation problem is not alleviated by using frequency shifting to increase the number of cycles, N, in the autocorrelation record. For a given value of σ, the apparent intensity $(\sigma/U)_a$ is inversely proportional to N. Hence the value of $N(\sigma/U)_a$, the measure of the truncation error, is independent of changes in the apparent mean velocity due to frequency shifting.

The beam profile in a measured correlogram is subject to a large number of unknown factors. If β is maximised the influence of these factors is reduced and accurate results can be obtained by Fourier transformation.

References

1. J B Abbiss: Physica Scripta, 19, 388 (1979)
2. J G McWhirter: Optica Acta, 28, 1453 (1981)
3. L J S Bradbury: Physica Scripta, 19, 360 (1979)
4. J B Abbiss: PhD Thesis, Department of Mechanical Engineering, University of Surrey (1982)

5. A Characterization of the Near Field of a Two-Dimensional Wake by Photon Correlation

G.D. Catalano
Louisiana State University, Baton Rouge, LA 70803, USA

R.E. Walterick
Georgia Institute of Technology, Atlanta, GA, USA

H.E. Wright
Air Force Institute of Technology, Wright Patterson AFB, OH 45433, USA

1. Introduction

The importance of turbulent shear flows such as boundary layers, jets, and wakes in applied problems of engineering has led to considerable effort being focussed on developing and experimentally testing simple scaling laws for these inhomogeneous flows [1]. Recently due in large part to the observations of KLINE and RUNSTALER [2] and BROWN and ROSHKO [3] there has been a concentration on searching for the existence of large scale structures in the turbulent shear flows. Their observations clearly pointed to the highly intermittent nature of the turbulent flow. The violent bursts of activity at local regions of space and time can provide a significant contribution to quantities such as mean velocities, turbulence intensities, and Reynolds stress. The intermittent activity directly contradicts the premises on which simple scaling theories are built.

Some numerical work has been started in the area of relating intermittency to flow structure, and eventually, scaling laws. SIGGIA and AREF [1,4] simulated the plane wake and the mixing layer which are the most common two-dimensional shear flows. The most unusual feature of the simulations was the extreme intermittency of both the plane wake and the mixing layer and the persistence of the intermittency in time. In addition, two non-equivalent length scales were found to exist with one based on the mean velocity profile and the other related to the size and distribution of the vortical structures.

In addition, attention has been focussed on the effects of free stream turbulence on free shear flows. The interaction of a shear flow with a turbulent outer stream is a common and important occurrence in engineering. FINK [5] studied the influence of external turbulence on axisymmetric coaxial jets and found that the turbulence has the effect of enhancing the axial development of the shear flows. SYMES and FINK [6] investigated the effects of external turbulence upon flow past cylinders. The conclusions reached were that the action of the external turbulence produced both higher turbulence intensities and shear stresses within the wake and that the large eddies of a conventional wake are broken down into more intense smaller eddies.

The present ongoing investigation has sought to document in the detail the near field of a two-dimensional wake. The measurements of interest are mean velocities, turbulent intensities, and various length scales. With a carefully defined turbulent ambient stream, it is also possible to concentrate attention on determining the resultant effects on the velocity field. In essence, a boundary condition for the developing wake flow is altered.

Finally, the effects of rotation of the cylinder on the near field wake are examined. HUSSAIN [7] has suggested that in a plane jet and an axisymmetric free shear layer the characteristic average measures are noticeably dependent on initial conditions. Thus, the rotation of the cylinder will allow the possibility of changing one of the large scale initial conditions of the flow field.

Clearly, the need exists for a better understanding of the quantitative effects of the highly intermittent nature of turbulent activity. This knowledge is required if simple or more complex model scaling laws are to be pronounced and received with any credibility. Simple dimensional arguments for mean velocity profiles, or turbulent intensity profiles are no longer adequate, and, in fact, fundamentally incorrect. The initial step is to carefully document a specific turbulent shear flow. Once that has been accomplished for various boundary and initial conditions, new scaling models can then be proposed and evaluated.

2. Equipment and Experimental Procedure

The flow system consists of a circular cylinder with an L/D ratio (i.e., depth to diameter) equal to 0.60, mounted in the test section of a two-dimensional smoke tunnel. The blockage area ratio of the mounted cylinder is equal to 0.12. The tunnel flow is kept at a constant value of 5.9 m/sec with the resulting Reynolds number based on the diameter of the cylinder equal to 50,000.

The production of an homogeneous turbulent flow with a relatively large value for the turbulent intensity (i.e., $u_{rms}/U = 0.05$) is a complex problem. For this experiment, the presence of a stack/injector delivery system for flow visualization made the spatial homogeneity problem even more difficult. Just downstream of the smoke tubes in fact, u_{rms}/U is approximately equal to 0.10. After the installation of a series of grids, the mean velocity variance in the test section is less than ± 1% and the local turbulent intensities are equal to .05 ± .02, both measured at the lateral centerline. For lateral spatial homogeneity, mean velocity profiles and surface pressure profiles are determined. For $X/D < 5$, uniform tunnel flow is present. The pressure survey around the cylinder matches inviscid theory quite well.

Cylinder rotation ($\omega = 500$ rpm) is accomplished by use of a small variable speed motor with rotation rates kept constant to within ± 5 rpm.

The turbulence measurements are made by using a laser Doppler velocimeter in conjunction with a photon correlation processing technique.

The signal from the photomultiplier tube is passed directly to a Malvern correlator. The correlator is used to determine the autocorrelation functions of the turbulent velocity field.

The power spectra as found by LUMLEY et al. [8] is

$$P(w) = \frac{1}{4}[\exp\{-(w + w_0)^2 a^2/2\} + \exp\{-(w - w_0)^2 a^2/2\} \quad , \tag{1}$$

where w_0 is the Doppler frequency, and a is a constant. Taking the Fourier transform yields the autocorrelation function

$$R(t) = \exp(-t^2/2a^2)\cos(w_0 t) \quad . \tag{2}$$

Consider the case when the autocorrelation curve has been distorted by either a lack of number of fringes in the control volume or significant background scattered light. The calculation of the turbulent intensity from an autocorrelation function can be approximated by determining the decrease in the amplitude of the cosine-like wave. One approximation used is the following

$$\frac{u_{rms}}{U} = \frac{1}{\pi}\left(\frac{1}{2}(R - 1) + \frac{1}{2n^2}\right)^{\frac{1}{2}} \tag{3}$$

$$R = \frac{R_2 - R_1}{R_2 - R_3}$$

and R_1, R_2, and R_3 correspond to the voltage(s) at the first minimum, the first maximum, and the second minimum locations, and n is the number of fringes in the control volume. Note that if the autocorrelation curve is skewed or distorted, the absolute value of the voltage R_3 may be less than the absolute value of the voltage, R_1. For $R_3 < R_1$, the value of R is less than unity with (3) resultantly undefined.

Intuitively, this would seemingly indicate a growth in the amplitude of the cosine wave that is not physically realizable for a turbulent auto correlation function. Thus the need exists to eliminate this skewness.

The approach used is to fit a second-order polynomial curve through the first several maxima and another second-order polynomial curve through the first several minima. That is,

$$f_1(t_{max}) = a_1 + a_2 t_{max} + a_3 t_{max}^2 \tag{4}$$

and

$$f_2(t_{min}) = b_1 + b_2 t_{min} + b_3 t_{min}^2 \quad . \tag{5}$$

The constants are then determined from the measured maxima and minima voltages.

The polynomials are next averaged yielding

$$f_3(t_{avg}) = c_1 + c_2 t_{avg} + c_3 t_{avg}^2 \tag{6}$$

where

$$c_1 = \frac{a_1 + b_1}{2}, \quad c_2 = \frac{a_2 + b_2}{2}, \quad c_3 = \frac{a_3 + b_3}{2} \quad . \tag{7}$$

The function $f_3(t_{avg})$ can be considered the new reference datum for the autocorrelation function. By subtracting the voltages corresponding to the appropriate t values, the autocorrelation curve is then centered about the zero reference level. That is

$$R_{new}(t) = R_{old}(t) - f_3(t) \quad . \tag{8}$$

Now the voltage maxima and minima can be calculated and used in (3) to determine the turbulent intensity.

3. Experimental Results

Several of the results documented in the experimental investigation will next be presented. The free stream turbulence intensity, u_{rms}/U_{FS}, is maintained at a value equal to 0.05.

Mean velocity profiles for two downstream locations ($X/2R_o$ = 0, and 5.00) are presented in Figs. 1 and 2. The ratio of the local defect velocity, $U_{FS}-U$, non-dimensionalized by U_{FS} is plotted versus lateral displacement, y, non-dimensionalized by the wake half width, $Y_{½}$. Note that the half width is defined as the lateral location where the velocity is approximately one half the average of the mean velocity at the centerline and the mean velocity at the edge of the wake. Two cases are plotted on each profile corresponding to a fixed case (ω = 0) and a rotating case (ω = 500 rpm).

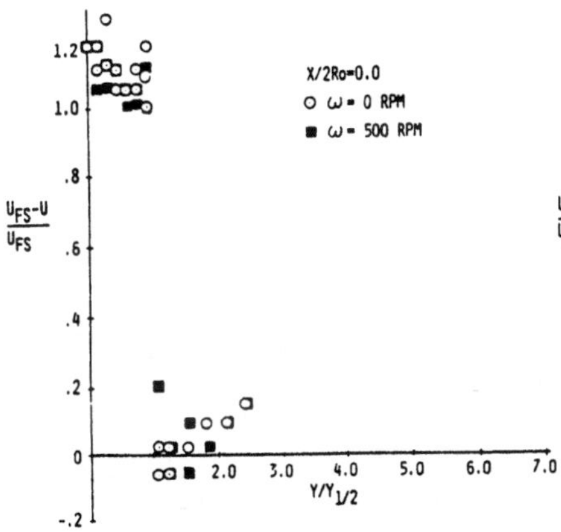

Fig. 1 Wake velocity defect profile, $X/2R_o$=0.0

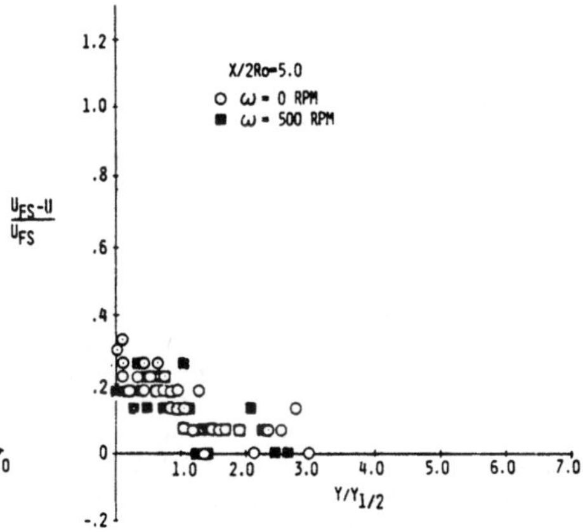

Fig. 2 Wake velocity defect profile, $X/2R_o$=5.0

The existence of the recirculation region is clearly shown for $X/2R_o \leq 0.75$. Note that the maximum absolute velocity occurs for $Y/Y_{½}$ slightly different from zero. Though considerable scatter exists, one effect of rotation appears to be present. The wake width with rotation seems to at first lag behind the non-spinning case but this trend is reversed further downstream. This observation is supported with the data in Fig. 4 which presents the development of the mixing width as a function of downstream location.

A comparison with the investigation by HUSSAIN and RAMJEE [9] (Re = 120) indicates the importance of the Reynolds number on the flow features. At $X/2R_o$ = 4, Hussain measured a minimum mean velocity equal to approximately 55% of free stream. This contrasts with values U/U_{FS} = 0.75 for the present investigation.

The non-dimensionalized wake momentum thickness, $\theta_o/2R_o$, calculated by integrating the momentum equation over the flow cross section is plotted versus non-dimensionalized downstream location, $X/2R_o$ in Fig. 3. The values of θ_o plotted for the rotating case, which are consistently larger for $X/2R_o \geq 0.75$, represent the sum of the momentum thickness above the cylinder centerline (i.e., rotation in the direction of the flow) and below the cylinder centerline (i.e., opposite to the rotation direction). The

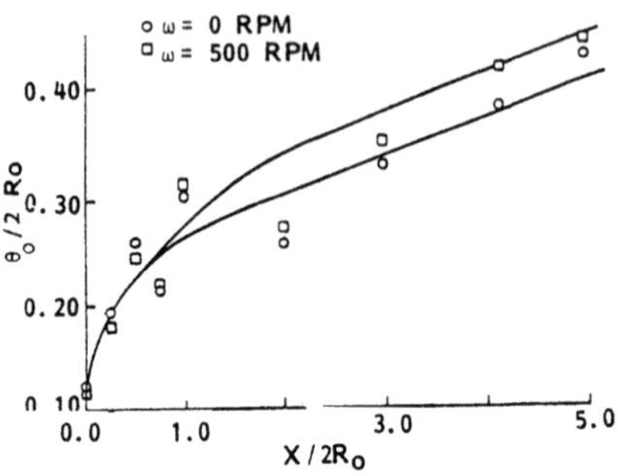

Fig. 3 Momentum thickness growth

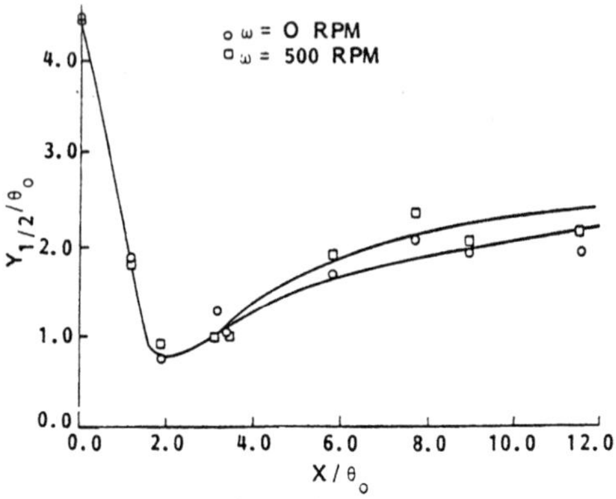

Fig. 4 Mixing width growth

momentum thickness is asymetric for ω = 500 rpm. Consider a kinetic perspective. The rotation creates an asymmetric pressure distribution around the cylinder, which coupled with the uniform flow produces a Magnus-type or lift force. The associated increment in the drag is manifest in the increase in the value of θ_o.

The mixing width, $Y_{1/2}$, non-dimensionlized by θ_o is plotted versus non-dimensional distance downstream, X/θ_o, in Fig. 4. As is the case for the momentum thickness, $Y_{1/2}$, for ω = 500 rpm, is consistently larger after the initial flow development. The mixing width is a possible length scale with which to describe the flow field and is related to the slope of the mean velocity profile. In the recirculation region, the cross stream momentum advection term ($V\, \partial V/\partial y$) is much larger than for the fully developed case. The associated transverse flow strain rate would tend to compress eddies in the cross-stream direction and thus result in a decrease in the magnitude of a characteristic length scale. This is a possible explanation for the initial decrease in $Y_{1/2}$.

The downstream centerline development of the longitudinal turbulence intensities, u_{rms}/U, is presented in Fig. 5. Initially note the rapid

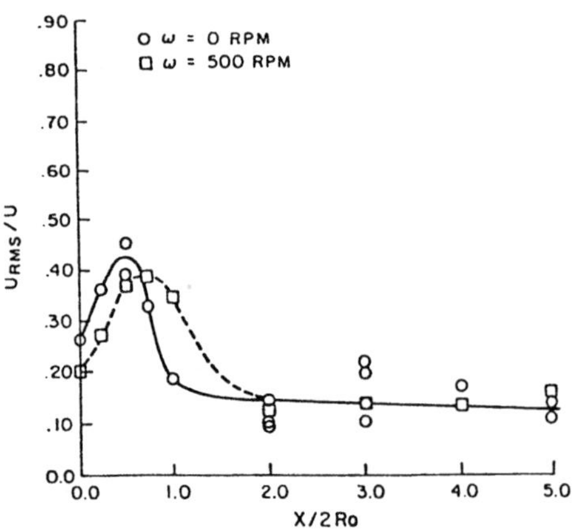

Fig. 5 Turbulent intensities at centerline

increase and tnen decrease in the magnitude of u_{rms}/U as the flow progresses downstream. An explanation for this occurrence is the presence of a recirculation region immediately downstream of the cylinder. The mean velocity near $Y/2R_o = 0$ is zero initially, reaches a non-negligible magnitude in the upstream direction before passing through zero again and eventually becoming positive. Turbulent intensities based on the local mean velocity will clearly behave as shown in Fig. 5. The effect of the free stream turbulence appears as an increase in the intensities at each downstream location as has been observed by HUSSAIN et al. [7]. Rotation of the cylinder appears to delay the initial increase of u_{rms}/U. This is in part due to the larger negative mean velocity at several downstream positions for $X/2R_o < 1.0$. Also, the cross-product terms involving the velocity and vorticity fields strongly affect the Reynolds stress gradients and thus the transfer of energy from the mean to the turbulent flow. The introduction of an asymmetric force distribution and/or a variation in the vorticity field may be reflected in the turbulent kinetic energy at a given flow location.

4. Conclusions

The recirculation region extends approximately one diameter downstream. The momentum thickness increased for the rotating cylinder case indicating a resultant drag increase. The mixing width experienced a compression in the recirculation region behind the cylinder for both the fixed and rotating case. After this initial region of the near field, the value of the mixing width increased, again, at a slightly greater rate for the rotating case. The free stream turbulence appeared to cause a corresponding increase in the values of the turbulence intensities in the wake. The possible resultant effects on the coupled velocity/vorticity fields due to an asymmetric pressure distribution caused by rotation is discussed.

References

1. E. D. Siggia and H. Aref: Nonlinear Dynamics, Annals of New York Academy of Sciences, 357, (1980).

2. S. J. Kline and P. W. Runstalder: J. Applied Mechanics, 26E; 166.

3. G. L. Brown and A. Roshko: J. Fluid Mechanics, 64, 775-816 (1974).

4. H. Aref and E. D. Siggia: J. Fluid Mechanics, 100, 705-737 (1980).

5. L. E. Fink: In Proc. Symposium on Turbulent Shear Flows, 1, Penn. State University, 1977.

6. C. R. Symes and L. E. Fink: In Structure and Mechanisms of Turbulence 1, ed. H. Fiedler (Springer Verlag, Berlin 1977).

7. A. K. M. F. Hussain: In Structure and Mechanisms of Turbulence 1, ed. H. Fiedler (Springer Verlag, Berlin 1977).

8. J. Lumley, W. George, and Y. Kobashi: In Proceedings of Symposium of Turbulence in Liquids, Univ. of Mo. at Rolla, (Plenum Press, New York, 1970).

9. A. K. M. F. Hussain and V. Ramjee: Aeronautical Quarterly, (1976).

6. Experimental Investigation of the Turbulent Boundary Layer Associated with Mixed, Free and Forced Convection Inside a Vertical Cylinder

P.M. Brdlik, Y.P. Semenov, V.A. Belyahkov, and V.P. Alabooshev
Moscow Technological Institute for Forest Production, Moscow, USSR

1. Introduction

As is well known, convection is said to be free or natural if it is caused only by buoyancy due to temperature differences; it is called forced if it is caused only by external forces; and it is called mixed if both causes act simultaneously. Earlier investigations of natural [1, 2] and mixed and forced [3, 4] convection do not give sufficient data for a characterization of the hydrodynamic flow conditions. This is a consequence of the limited accuracy and resolution of traditional flow measuring instruments such as the Pitot tube and the hot-wire anemometer. Laser Doppler Anemometers, on the other hand, should offer sufficient information for a description of the flow patterns.

This paper reports measurements of the boundary layer characteristics due to either free or combined free and forced convection of air inside a vertical hollow cylinder. The data refer to REYNOLDS numbers (for the vertical flow component in x direction) $Re_x = 10^4 \ldots 2 \cdot 10^6$, GRASHOF numbers $Gr_x = 3 \cdot 10^9 \ldots 10^{11}$, a ratio $Gr_x/Re_x^2 = 0.01 \ldots 50$, and to a ratio of length to boundary layer thickness, $L/\delta \lesssim 130$, and under stationary heat input through the cylinder wall. The parameters of the flow medium entering the REYNOLDS and GRASHOF numbers refer to the operating temperature. During the experiment the air temperature was typically 40^0 C, the cylinder wall temperature 100^0 C, and the flow velocity between zero and 6 m/s. For the measurements a Laser Doppler Anemometer with signal processing by the Malvern Photon Correlator was used.

This type of instrument is well suited for the investigation of gas flows owing to its high sensitivity. Nevertheless, there are some problems in applying it to boundary layer flows, but some of these problems arise with any type of Laser Doppler Anemometer.

2. Experimental Investigation

The measurements have been made with experimental equipment described in Ref. [5]. It is basically a vertical-flow low-speed wind tunnel. The dimensions are 0.5 x 0.8 m^2 for the cross section and 3 m for the vertical length. The turbulence at the entry to the test tunnel was not greater than 0.7 %. The canal walls of 3 m length and 2 mm thickness were made of stainless steel. These walls were heated by electric currents. In this way, mixed convection developed along the canal surface.

During the experiment the heat power inputs into the canal walls were noted. In the boundary layer measurements were made of the temperature distribution, the distribution of mean vertical velocity and of vertical turbulent velocity fluctuations.

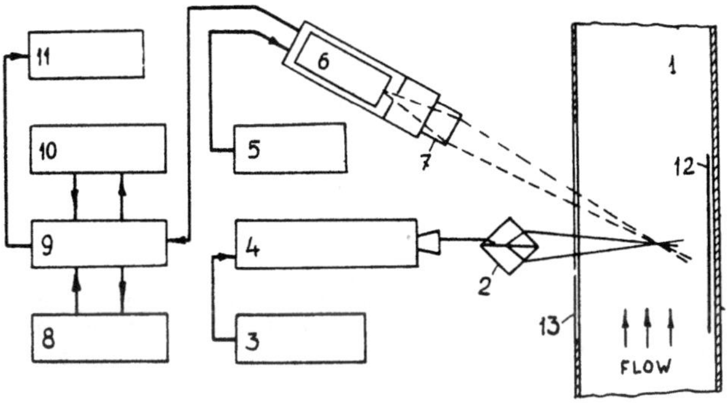

Fig. 1. Experimental setup. (1) Test section of vertical canal; (2) beam-splitter; (3) power supply; (4) laser Spectra Physics model 124 B; (5) high-voltage supply; (6) photomultiplier; (7) lens; (8) data store; (9) correlator; (10) computer; (11) oscilloscope; (12) black screen to avoid flare; (13) glass window in tunnel wall. Coordinate x is in the vertical direction, y in the horizontal perpendicular to the plane of the drawing

The experimental setup is schematically shown in Fig. 1. The experimental conditions made it necessary to use the back-scatter mode of Doppler difference anemometry. A screen covered with black woolen cloth was attached to the opposite canal wall to reduce the effect of flare. The laser and photomultiplier were mounted on a two-coordinate adjustable table. The horizontal travel could be measured to an accuracy of 0.01 mm by a calibrated gauge. The focal distance of the receiving lens was 300 mm and the diameter of the measuring volume 0.4 mm.

As it turned out, the concentration of light scattering particles in the ambient air was sufficient for anemometry of natural convection. In our investigation of mixed convection, on the other hand, we used a closed circuit air flow system, and there the particles were found to disappear gradually. It was therefore necessary to add a small quantity of about 1 μm diameter particles. The suitable amount was determined empirically, with the visible growth rate of the correlogram in the correlator serving as the criterion. For our investigations of pure natural convection where the root-mean-square of vertical velocity fluctuations was more than 30 %, we used the DISA type 55L02 optics unit with Bragg cells in place of the simple beamsplitter. This lead to a more satisfactory operation of the measurement system.

As mentioned above, measurements very close to the side wall proved difficult. We covered the rear wall by a black screen, and for measuring close to the side wall the plane of the optical rays had to be rather parallel to the wall. The particle concentration tends to be lower close to the wall than in the main flow. A further difficulty was the determination of the distance, y, between the wall and the measuring volume. We found the origin of this coordinate by direct visual observation of the coincidence of the measuring volume with the wall surface. This method gives the y coordinate with an error which cannot be larger than the diameter of the measuring volume, i.e. 0.4 mm in our case. In regions very close to the wall, the velocity gradient is large. One would expect that photon correlation gives the correct average velocity over the measuring volume. However, if the particle density closer to the wall tends to be smaller, the measurement will give an average value which is biased towards higher values. This effect and the uncertainty

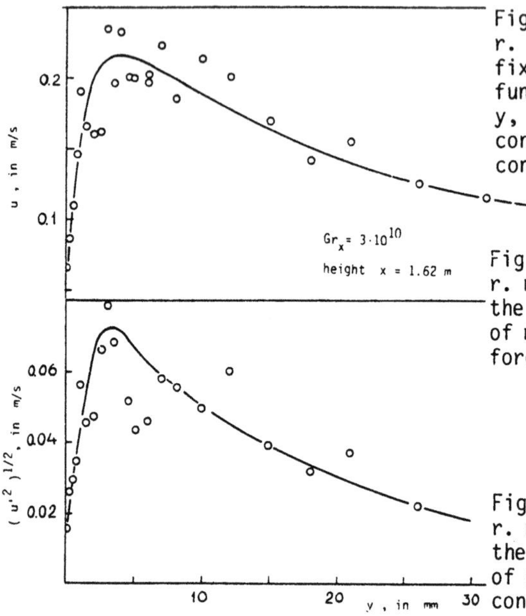

Fig. 2. Mean vertical velocity and r. m. s. velocity fluctuations at fixed height, x = 1.62m, as a function of distance from the wall, y, in the boundary layer under conditions of pure natural convection

Fig. 3. Mean vertical velocity and r. m. s. velocity fluctuations in the boundary layer under conditions of mixed convection dominated by forced convection

Fig. 4. Mean vertical velocity and r. m. s. velocity fluctuations in the boundary layer under conditions of mixed convection dominated by free convection

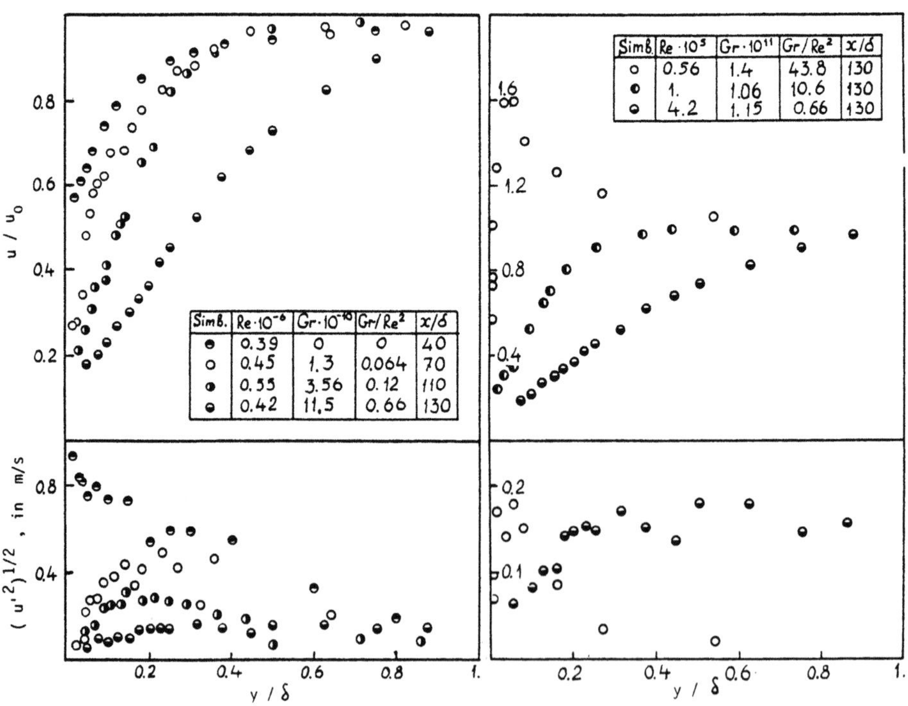

to an impractically low level. This approach could have been successful with a greater concentration of scattering particles in the flow or, perhaps, with greater laser power.

3. Experimental results

Our data on velocity measurements under conditions of pure natural convection are given in Fig. 2. Notice the large scatter of data near the velocity maximum. Every experimental point represents data accumulated in the correlator over a period of 5 minutes. Despite this the scatter is large near the velocity maximum and may signify slow variations of the true velocity. The distribution of the turbulent fluctuations follows a similar pattern as the mean velocity.

Figure 3 shows our results on mixed convection in a regime where forced convection seems to be dominant. Here u_o is the flow velocity in the central part of the canal, and δ is the thickness of the boundary layer. All measurements were taken at a REYNOLDs number close to $0.4 \cdot 10^5$, but with different GRASHOF numbers, i. e. different temperature differences. The mixed convection parameter Gr_x/Re_x^2 varied between 0 and 0.61. For larger GRASHOF numbers a significant widening of the boundary layer is observed which corresponds to a laminarization of the flow in the boundary layer. Accordingly the distribution of velocity fluctuations shows decreasing turbulence with increasing GRASHOF number. The measurement of heat flux, not reported here in detail, showed the expected decrease of heat transfer rate with GRASHOF number.

Measurements in the regime close to natural convection are shown in Fig. 4. The velocity profile gradually passes from that typical for free or natural convection to one typical for mixed convection. But purely natural convection is associated with a turbulent boundary layer, and the superposition of a forced flow leads to flow laminarization and, as a consequence, to a decrease in the efficiency of heat transfer.

4. Conclusions

Our investigations show that Photon Correlation Anemometry is well suited for a study of turbulence caused by natural or mixed natural and forced convection. Yet some improvements are desirable in our view. For example, a smaller measuring volume is desirable in order to get better spatial resolution. Algorithms for data extraction from photon correlation functions are not sensitive enough to identity turbulence levels smaller than 2 % with confidence. It would be desirable to have a way of determining the time scales of turbulent fluctuations.

References

1 R. Cheesewright, Turbulent Natural Convection From a Vertical Plane Surface, Journal of Heat Transfer, Transactions of the ASME, Series C, vol. 90, N1, 1968.

2 G. C. Vliet, C. K. Liv, An Experimental Study of Turbulent Natural Convection Boundary Layers, Journal of Heat Transfer, Transactions of the ASME, Series C, vol. 91, N4, 1969.

3 W. B. Hall, P. H. Price, Mixed Forced and Free Convection from a Vertical Heated Plate to Air, Paper NC 33, Proc. 4th Int., Heat Transfer Conf., Paris, 1970.

4 P. M. Brdlik, Y. P. Semenov and B. N. Kovalev, Heat Transfer Investigation for Mixed Free and Forced Turbulent Convection on Vertical Plate, 2nd Symposium on Turbulent Shear Flows, pp. 935-940, London, 1979.

5 Брдлик П.М.,Семенов Ю.П.,Беляков В.А.,Алабушев В.П., Исследование теплообмена при смешанной конвекции около вертикального цилиндра, Научные труды, Вып.138,М.,МЛТИ, 1982г .

7. Turbulence Measurements in an Ejector Wing Flow Field

G.D. Catalano
Louisiana State University, Baton Rouge, LA 70803, USA

H.E. Wright
Air Force Institute of Technology, Wright Patterson AFB, OH 45433, USA

R.E. Walterick
Georgia Institute of Technology, Atlanta, GA, USA

1. Introduction

Consider the principle of an ejector [1]. In the simplest case, coaxial jets are confined to a duct rather than to a constant pressure atmosphere. For this flow field, the mass flow rate averaged mean axial momentum is not conserved and the static pressure may vary with downstream location. There will be an increase of pressure with increasing downstream position as the jet cores are being consumed by rapid shear layer mixing. In fact, the pressure may also continue to rise in the developing flow zone downstream of the disappearance of the cores. This pressure rise can be considered the source of the pumping effect of the ejector.

Significant and fundamental developments in thrust augmenting ejectors have been accomplished in the last several years. Hypermixing nozzles have been developed with a resultant increase in ejector compactness realized [2]. Mixing and diffusion of flows have been achieved simultaneously with performance advantage. Thrust augmentation ratios on the order of two in an ejector of inlet area ratio 23 have been achieved experimentally [3]. A theoretical methodology which can evaluate the performance of the ejectors subject to a wide range of variation in the thermodynamic parameters of the injected and the entrained fluids has been developed for incompressible and compressible flows for a constant area duct [4]. High lift characteristics of an ejector-flapped wing have been evaluated [5]. A numerical prediction of three-dimensional ejector flows has been proposed [6].

Although the literature on ejectors in general and, particularly, thrust augmenting ejectors, is quite extensive, the turbulence field has been essentially ignored. The information that is available is predominantly concerned with flows in constant area pipes with RAZINSKY and BRIGHTON [7] presenting an extensive set of one-point statistical measurements for varying mean velocity ratios and jet/pipe diameter ratios.

The purpose of this investigation is to conduct an extensive survey of the resultant velocity flow field of a given ejector wing design. The effectiveness of the ejector will be assessed by comparing the flow field with the ejector powered and with the ejector unpowered. The data in this experiment is obtained by use of a laser velocimeter in conjunction with a photon correlation processing technique.

2. Experimental Equipment and Technique

The facility used for this investigation is a two-dimensional smoke tunnel which employs an open return system of flow, capable of subsonic incompressible velocities up to 23 meters/second. All measurements are taken at a nominal freestream velocity of 8 m/sec with a free stream turbulence intensity equal to 0.05 ± .02.

The specific flow field investigated is an ejector wing design conceived by Vought under contract with the Flight Dynamics Laboratory at Wright Patterson AFB (refer to Fig. 1) [8]. A two-dimensional model is constructed and placed in the test section of the wind tunnel. The ejector plenum is supplied from the laboratory compressed air reservoir. Considerable effort was expended in attempting to achieve a uniform exit velocity profile with relatively low values of the turbulent intensities. The aspect ratio of the rectangular nozzle is 4.1:1 and the solidity ratio is 0.327:1. The mean velocity at the nozzle exit plane, U_0, is kept at a constant 16 m/sec.

The velocity measurements are obtained from a photon correlation based laser velocimeter. The photon correlator possesses a resolution time of from 50 nanoseconds to 1 second. Measurements were taken in the single clipped autocorrelation mode and at an infinite sample rate.

Special note should be made of the turbulent intensities measurement technique. Care is taken to minimize the problems of background flare light and photon pileup. The effects of these two phenomena can result in a skewness or a distortion of the photon correlation function from which the mean velocity and local turbulent intensity are calculated. Therefore a numerical technique [9] is employed which results in the alleviation of the skewness problem.

3. Experimental Results and Discussion

In Fig. 1, the location of the mean velocity and turbulent intensity data obtained are shown. Note that in all cases X is measured longitudinally, from the leading edge and Z is measured vertically from the airfoil surfaces.

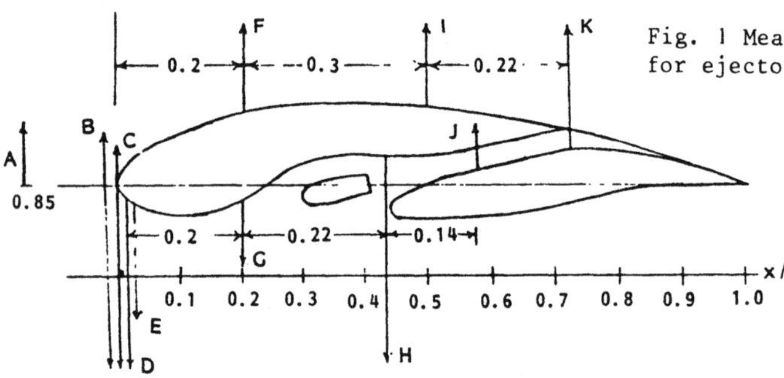

Fig. 1 Measurement location for ejector wing flow

Turbulent intensity is defined here as the ratio between the root mean square of the velocity fluctuations, u_{rms}, non-dimensionalized by the local mean velocity, U.

Fig. 2 shows mean velocity profiles upstream of the ejector wing while the turbulent intensity profiles are presented in Fig. 3. The effect of the ejector shown in the mean velocity profiles is to accelerate the mean flow above the upper surface and to decelerate the mean flow beneath the ejector wing. This effect is quite pronounced immediately upstream of the leading edge. The turbulent intensities are not as well behaved but the

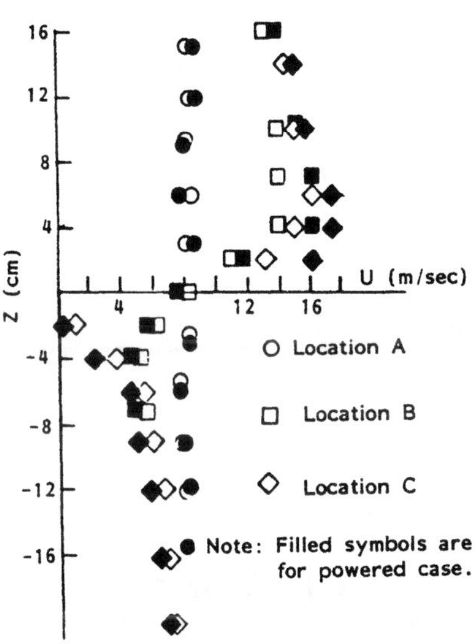

Fig. 2 Mean velocity profiles

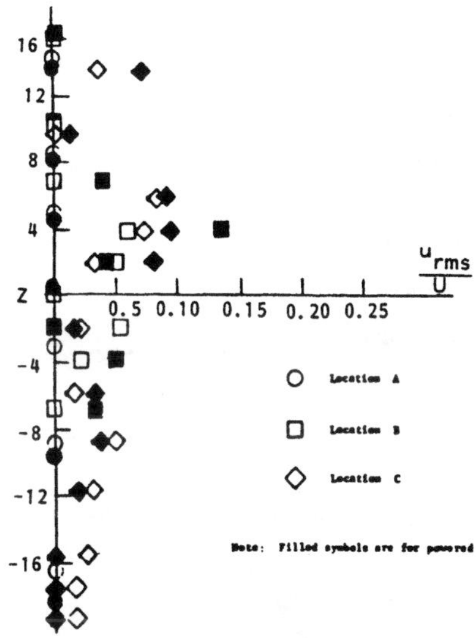

Fig. 3 Turbulent intensitv profiles

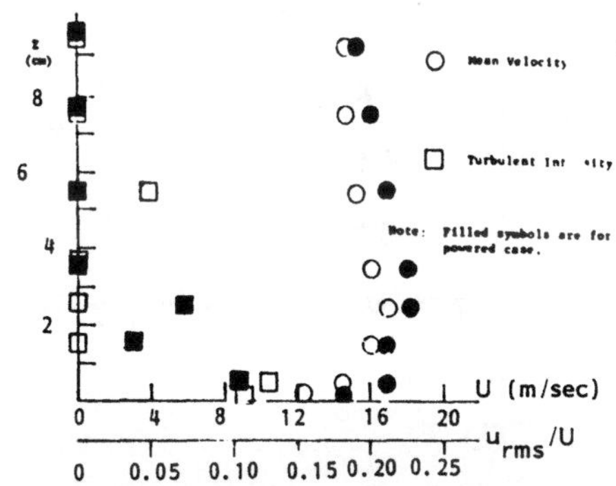

Fig. 4 Mean velocities and turbulenct intensities at X/C = 0.2 (F)

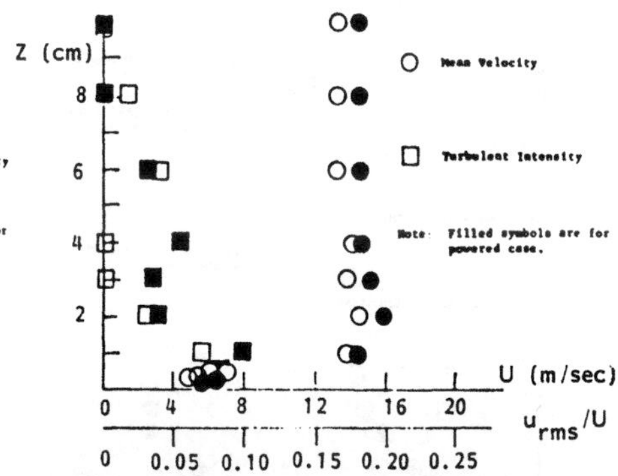

Fig. 5 Mean velocities and turbulent intensities at X/C = 0.5 (I)

value of u_{rms}/U does seem to increase above the upper surface with the ejector powered.

Mean velocity and turbulent intensity profiles are shown for the downstream locations X/C = 0.2, (F), and X/C = 0.5, (I), in Figs. 4 and 5. Both these locations are above the upper surface and the mean flow is consistently faster in the ejector powered case. The turbulent intensity data reinforces an observation made with reference to flow visualization evidence. The value of the turbulent intensities reduce to the free stream value closer to the wing surface with the ejector working. This would again indicate a shift of the potential flow down toward the upper surface.

Consider the data presented for the lower surface locations for X/C =0.02, X/C =0.04, and X/C = 0.2. In Figs. 6 and 7, at each flow location, the mean longitudinal velocity U is less in the powered case. This suggests the movement of the stagnation further along the lower surface as

Fig. 6

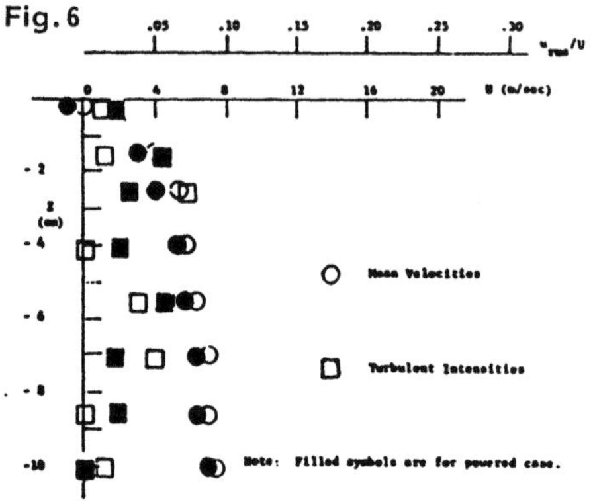

Fig. 7

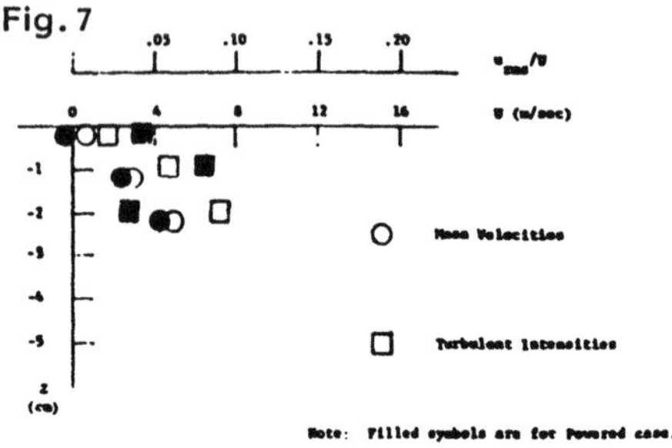

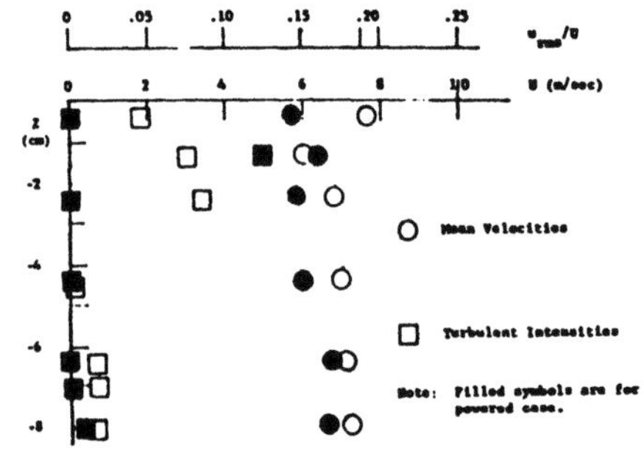

Fig. 8

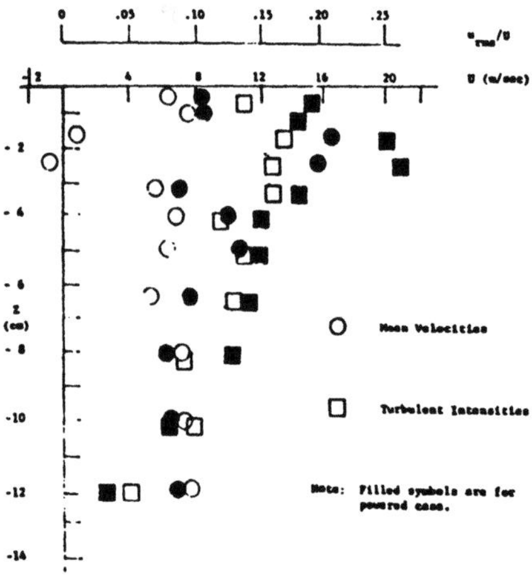

Fig. 9

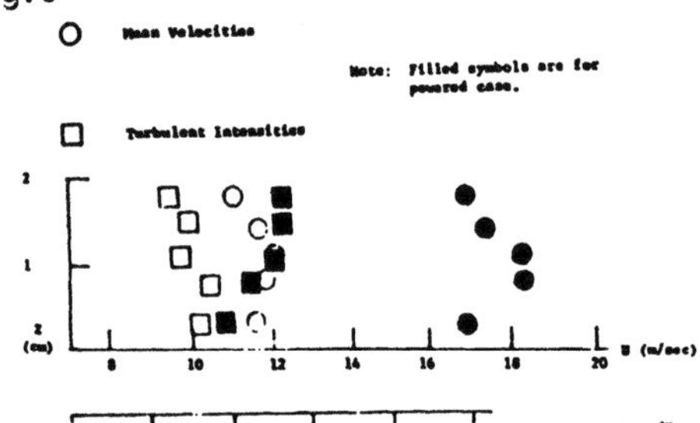

Fig. 10

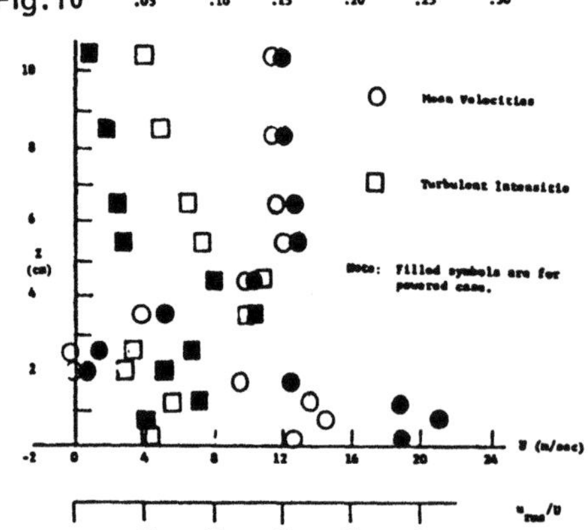

Fig. 11

Fig. 6 Mean velocities and turbulent intensities at X/C = 0.02 (D)

Fig. 7 Mean velocities and turbulent intensities at X/C = 0.04 (E)

Fig. 8 Mean velocities and turbulent intensities at X/C = 0.2 (G)

Fig. 9 Mean velocities and turbulent intensities at X/C = 0.44 (H)

Fig. 10 Mean velocities and turbulent intensities at X/C = 0.58 (J)

Fig. 11 Mean velocities and turbulent intensities at X/C = 0.72 (K) can be argued from Fig. 8. As the stagnation point moves downstream, the apparent angle of attack of the airfoil will increase. With the increase of α, the mean velocity measured in the longitudinal direction will necessarily decrease close to the stagnation point.

The mean and turbulent velocity field downstream of the ejector nozzle is examined in Figs. 9 through 11 for X/C = 0.44, 0.58, and 0.72. Note that the flow for both cases actually accelerates after it enters the constant area mixing duct. Also consider the relatively high turbulent intensities in the confining duct for the ejector powered case. Values of u_{rms}/U equal to 0.25 are measured which is indicative of jet mixing rather than characteristic of duct type flow. Once again for the field above the wing, the velocity returns to the free turbulence condition faster with the jet working. For example, for z = 10 cm in Fig. 11, u_{rms}/U is equal to .05 for the nonpowered case but possesses a value of .01 for the ejector working case.

4. Summary

An experimental investigation of the resultant turbulent flow field about an ejector wing design incorporating a constant area mixing duct is made. Mean velocities and turbulent intensities are calculated from the photon correlation functions. Comparison between the ejector powered and nonpowered cases are made. The following results are obtained.

1) The ejector consistently accelerated the flow field above the wings' upper surface. The influence is felt upstream of the model's leading edge.

2) The stagnation point moved further downstream along the lower surface for the ejector powered case, indicating an apparent increase in the effective angle of attack.

3) The turbulent intensities in the confining duct are of free or coaxial jet magnitude.

4) The streamlines above the wing's upper surface are compressed downwards toward the airfoil,indicating a reduction of the wake region.

References

1. J. P. Johnston: In Topics in Applied Physics, Turbulence ed. P. Bradshaw, 12, (Springer-Verlag, New York, 1978).
2. B. Quinn: J. Aircraft, 10, 8, 481-486 (1973).
3. B. Quinn: ARL 72-0084, (1972).
4. K. S. Nagaraja, D. L. Hammond, and J. E. Graetch: AIAA 73-1184, (1973).
5. H. W. Woolard: J. Aircraft, 12, (1975).
6. D. W. Roberts and G. C. Paynter: Workshop on Thrust Augmenting Ejectors, Ames Research Center, Moffett Field, California, June 28-29, 55-70 (1971).
7. E. Razinsky and J. A. Brighton: "Confined Jet Mixing for Nonseparating Conditions," Trans ASME, D, 93, 333-349 (1971).
8. Ejector Wing Design, Vought Corporation Advanced Technology Center, ATC Report No. R-91100/9CR-44, (1979).
9. G. D. Catalano, R. E. Walterick, and H. E. Wright: AIAA Journal, 19, 3, 403-405 (1981).

8. Long-Range Laser Velocimetry with Particular Emphasis to the Visible Domain*

Lars Danielsson and E.R. Pike**

The Aeronautical Research Institute of Sweden

Summary

A review is made of the field of long-range atmospheric wind velocity measurements by laser anemometry. The emphasis is on visible systems but frequent references are made to the much better developed infra-red systems. Special attention is paid to the type of measurements required for wind power development and, of many possible systems, a critical assessment is made for selection on grounds of spatial resolution, number of velocity components required, safety, efficiency and cost.

1. Introduction

Measurement of wind velocities by means of infra-red laser velocimetry with the heterodyne reference beam technique and coherent detection has been well known for more than a decade [1-4]. In the beginning, it was still something of an open question whether infra-red or visible systems [5-7] should be chosen [8]. The development of lasers, detectors and signal analysis as well as signal-to-noise considerations all decided this in favour of the infra-red systems working in the heterodyne reference-beam mode with coherent detection. This system will have higher laser power, larger coherent detection area, lower Doppler frequencies, smaller quantum energy than a visible system and eye safety. A basic difference between the Doppler-difference system and the reference beam configuration for long-range applications is that the former gives the velocity transverse to the optical transmission while the latter gives the parallel velocity. Thus two reference beam systems at two different angles of incidence would be required to obtain the transverse velocity while two Doppler-difference systems at different angles would be required to give the in-line velocity. An expensive duplication of conventional in-line infra-red systems to obtain transverse velocities would, moreover, not give as high a spatial resolution as a visible Doppler-difference system might be designed to provide.

We should mention here that several attempts have been made to extract cross-wind speed (without direction) by analysis of the transit time of the particle signals [9-11] but with no real practical success. In some situations knowledge of the transverse com-

* Abbreviated version of a paper to appear in J. Phys. E: Sci. Instrum.

** Royal Signals and Radar Establishment, Malvern, England.

ponent only would suffice and in this case we arrive at the difficulty at which this present report is directed. Clearly, if a duplicated system is undesirable or if high spatial resolution is required we are forced back to consider the visible laser alternatives. An obvious question, if low resolution will suffice, is why not use the CO_2 laser in a long-range Doppler-difference mode. However, for these wavelengths detector noise must be dominated by a reference beam [8]. This, incidentally, also serves the purpose of dominating daylight background. While reference beam Doppler-difference systems are possible [12,13] the multiplicity of signal frequencies received makes interpretation difficult, although it may be feasible to alleviate this problem by using multiple frequency shifts.

Although at visible wavelengths detector noise is not a problem the daylight background is. Thus, in a CW visible laser Doppler-difference instrument without reference beam an optical filter with a response sufficiently narrow to render the daylight background negligible would have to be used. It would appear that a Fabry-Perot interferometer of free spectral range 2 nm and finesse about 30 would be adequate. This would present no difficulty to implement being close to moderately priced commercially available models.

An alternative approach to the suppression of daylight background in a visible-wavelength Doppler-difference system would be to use a pulsed laser with gated detection so that for the same mean laser power, for example, the return pulsed signal is larger than would be a CW return during the gate time [14]. It is not clear at the present time, however, that this will be an effective solution since a laser of suitable coherence length and stability must first be developed and then the cost of such a special system must be compared with that of a standard CW laser with etalon suppression of the background. The difficulty of digital photon correlation processing of a larger number of photons in each pulse may also lead to a requirement for analogue processing which would again give rise to counter-productive extra analogue noise problems. We do not, therefore, consider this method further in this text.

A second basic difference between conventional Doppler-difference velocimeters and the reference-beam systems is that the former usually are operated with incoherent detection. At close range the beams can usually be focussed so finely that the condition of only one detected particle in the sample volume can be achieved. At long range, however, there will always be many particles in the scattering volume, even in the cleanest air, and thus either coherent detection must be used or the detector threshold set so high that only the occasional, exceptionally large seed particles are registered. There seems to be no real reason to avoid using a coherent detection design save that the transmitter lenses must then be as large as that of the receiver.

More recently also the laser transit velocimeter has been suggested for long-range transverse velocity measurements [15,16]. However there seems to be no basis for expecting improved results by using this technique over a coherent detection Doppler-difference instrument unless the narrow band filter of the latter system is left out, when the higher rejection of ambient daylight of the LTV would be advantageous. On the contrary, the LTV will not respond unless the angle is set correctly, which at best means a search in angle for a steady, almost laminar flow, and at worst a very low signal rate as the turbulence level of the flow increases.

Much quantitative theory has been done in the field of photon correlation spectroscopy, not only on the basic quantum statistics, but also on errors of measured quantities [17,18]. However, not a great deal is applicable to the problem at hand or indeed to laser velocimetry in general. We have, in this problem, to consider simultaneous fluctuations of three random variables, namely, velocities, number of seeding particles and photon detections. To make things more difficult we have also many possible experimental arrangements both for optics and for electronic processing and data reduction [19]. Thus while, for example, it is known exactly how many scattered photons are required to measure molecular diffusion coefficients by laser scattering, this cannot be calculated for the measurement of velocity in many practical cases as yet although some numerical simulation has been done [20-22].

To summarize, although complete theoretical studies have not yet been done to give authoritative prescriptions for system design and comparison, enough is known to conclude reasonably safely that infra-red systems give superior signal-to-noise performance to visible ones and that if a visible system is to be used, photon correlation LDV is expected to give the best results.

2. Spatial Resolution

The depth in range of the atmosphere which is sampled to provide either transverse or in-line velocity information is determined by the wavelength of the light used, here essentially either 0.5 μm or 10 μm, and by the optical geometry of the transmit-receive (T-R) system. The latter can conveniently be classified for our present purposes under the following three types.

1. The simplest monostatic or single-ended (that is T and R through the same physical aperture) system consisting of a modest aperture high quality refracting component, say, a Germanium lens of ~ 100 mm diam in the infra-red case or, say, a camera lens in the visible case, will be called a <u>lens</u> system. Of course catadioptric components of the same modest apertures will give the same range resolution but will nevertheless be classified here as lens-type systems, Fig.1a.

2. More elaborate single-ended systems requiring higher apertures will be most conveniently realized by the use of catadioptric telescopic optics, of say 300-600 mm diameter. Again, although in principle single refractive components at these high apertures may be possible, we shall call such systems <u>telescopic</u>. We need to distinguish the two cases in which, respectively, the laser beam fills the telescope aperture in a monostatic arrangement and in which two or more parts of the same telescopic objective are used in a multiple monostatic or multistatic system, see Figs.1b and 1c. We call the first "full-aperture telescopic" and the second,

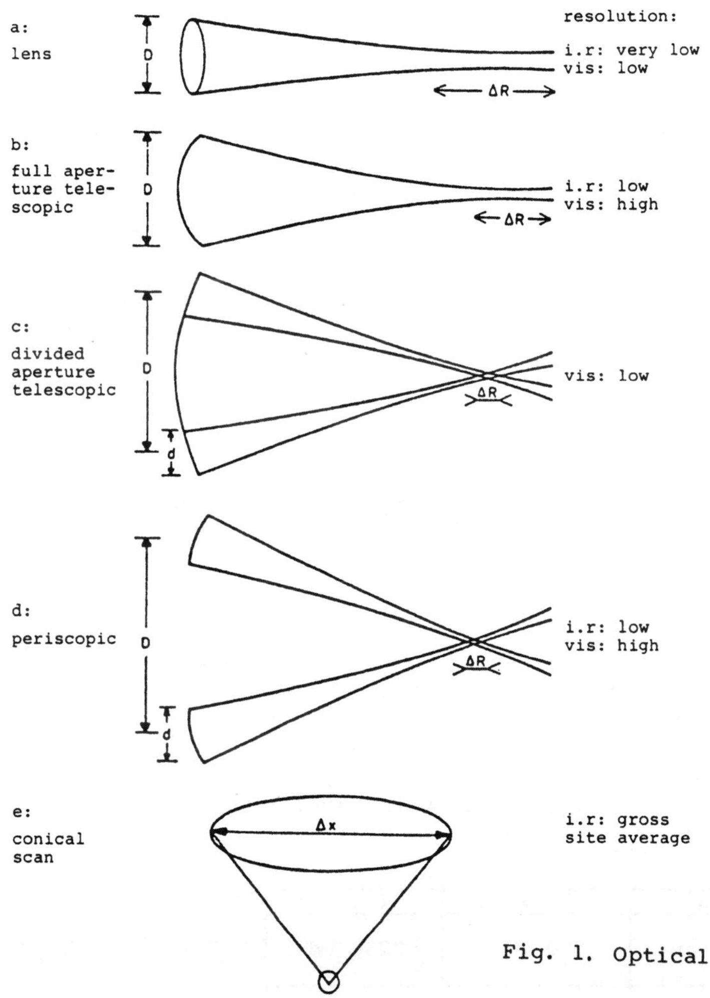

Fig. 1. Optical systems

including the case when the partial aperture consists of separate separate lens-type components, "divided-aperture telescopic".

3. For even larger base lines for transmission and reception separate high-aperture telescopes (multiple monostatic, bistatic or tristatic systems) are sometimes necessary. We will denote such systems as periscopic, Fig.1d.

Fig.1e depicts a further arrangement in which a monostatic system is used with conical scanning. A gross site average of transverse components as well as in-line velocities can be extracted from the output of such a system and we will classify its spatial resolution for our further purposes as "very low".

In the other arrangements we denote by ΔR on the diagrams of Fig.1 the range resolution in each case. For 1a and 1b we have in the full aperture case [23].

$$\Delta R_F = \frac{8R^2\lambda}{\pi D^2} , \quad (8)$$

where R is the range, λ the light wavelength, and D the transmitting aperture diameter. For the divided aperture cases 1c and 1d, by simple geometry,

$$\Delta R_D = \frac{8R^2\lambda}{\pi Dd} , \quad (9)$$

where D denotes the beam separation at transmission, d, the beam diameter at transmission, and where we have used the well-known formula for the beam size, w, at its waist

$$w = \frac{4R\lambda}{\pi d} . \quad (10)$$

If we compare R_F for an infra-red system with R_D for a visible system of the same aperture D we see that

$$\frac{\Delta R_F}{\Delta R_D} = \frac{\lambda_{i-r}}{\lambda_{vis}} \cdot \frac{d}{D} \quad (11)$$

which we can convenienty take, in a typical case of a twenty-fringe visible system, to be of the order of unity.

Table 1 ΔR in meters for various ranges R, for various full apertures D, using 10.6 μm wavelength or various divided apertures for twenty fringe 0.5 μm wavelength geometry.

Case	(1a)	1(b) 1(c)	(1d)
D / R	110 mm	500 mm	2250 mm
50 m	5	0·25	0·01
100 m	20	1	0·05
200 m	80	4	0·2

In Table 1 we show round-figure values of the theoretical range resolution ΔR for cases 1a and 1b according to the formula (8) for $\lambda = 10.6$ μm, as a function of range, for three apertures D, chosen to simplify comparisons, but typical, respectively, of the lens, telescopic and periscopic cases. According to the previous paragraph this will also give the order of magnitude of the range resolution for cases 1c and 1d for visible light using the same total aperture. Cases 1a and 1b can be converted to the visible wavelength by dividing by 20 and, conversely, cases 1c and 1d can be converted to the infra-red wavelength by multiplication by 20.

The spatial resolution of the various systems can be presented in a simplified way shown in Table 2. At R = 200 m and D = 500 mm for example, high relates to $\Delta R = 0.05$ m, low to $\Delta R = 1$ m and very low to $\Delta R = 20$ m.

Table 2. Relative range resolution performance for systems of Figure 1 at visible and infra-red wavelengths

Case / Wavelength	(1a)	(1b)	(1c)	(1d)
Visible	low	high	low	high
Infra-red	very low	low	very low	low

It will be seen immediately that, unfortunately, in no case does the infra-red wavelength offer a high resolution system in these terms. We have not seriously considered two-spot systems as a contender for long-range work after the discussion of the previous section but the range resolution performance of such a system would be the same as case 1b above.

3. Daylight and Detector Noise Background Suppression

In all practical anemometry systems any theoretically ideal performance will be degraded due to the presence of noise. For infra-red systems the detector itself is noisy and it is irradiated with ambient infra-red radiation from the viewing field and the detector surroundings. In the visible region although detectors with very low intrinsic noise (dark current) are now available, there will also be the problem of rejecting the ambient daylight background when present. Two remedies are possible to supress these noise sources, the use of a narrow band optical filter which will restrict incoming ambient radiation and the use of a reference beam which has the effect of dominating all other noise sources. A thorough and comprehensive treatment of the comparative merits if these two methods for visible and infra-red radiometry and radar applications has been given by Jakeman et al. [24]. Application to the present problem, however, requires only a few straightforward statements.

1. Infra-red systems must always use reference beams due to the internal detector noise. It follows that narrow band filters need not be considered in this case.

2. Reference beam systems can be designed with essentially ideal performance, that is to say that the noise sources under consideration may be perfectly suppressed.

3. Reference-beam or space-multiplexed systems for multi-dimensional measurements may introduce problems due to mul-

tiple signal frequencies [12]. These may be avoided either by use of time multiplexing or by using multiple detectors and "uncrossing" slightly the beams at the target. This also gives an extra velocity component. One-dimensional transverse measurements using one detector with crossed beams [3] is not, therefore, considered further.

4. For visible systems a Fabry-Perot interferometer of modest performance will reduce daylight background to negligible levels excluding a direct view of the sun.

4. System Specifications

4.1 Geometrical Specifications

Three geometrical features determine the geometrical part of the system specification, viz, the dimension of the flow to be studied, the particular velocity components to be measured and the necessary degree of geometrical resolution. These factors can be arranged in a "decision flow diagram", see Fig.2, for finding a minimal configuration. At this point we shall become more specific and direct this study to parameters relevant to wind power development.

The advent of wind power generation systems - wind farms - has created a need for a detailed exploration of the lower part of the atmospheric boundary layer, say up to 200 m. Two main objectives should be stressed, namely the mapping of statistical wind data for possible localization of wind farms and the study of the interaction of the individual wind generators in a wind farm.

The first objective - wind data mapping - is characterized by an essentially two-dimensional (horizontal) flow and by moderate requirements for range resolution. The characteristic scale length for the wind vector variation, which is derived from the wind gust statistics, can be as short as 10 m (10-50 m), placing a similar length requirement on ΔR, the range resolution.

The second objective - wind generator interaction - is characterized by a three-dimensional flow field and scale lengths of the order of 0.1 m (0.1-1 m).

It may not be economical to design one single system to fulfill both these objectives. Instead, in designing the measuring system, the requirements on range resolution and dimension of the flow field must first be roughly specified. Using the decision flow diagram of Fig.2 this leads to a minimal system configuration. The high, low and very low resolution may then be determined more precisely using the formula of section 3 above. Thus, for example, to determine the horizontal vector velocity for an atmospheric boundary layer with $\Delta R = 1$ m at $R = 100$ m, exit 6 in the flow diagram would be appropriate.

For generator interference with $\Delta R = 0.1$ m at $R = 100$ m and a three-dimensional velocity the sole possible system configuration would be the periscopic visible at exit 11.

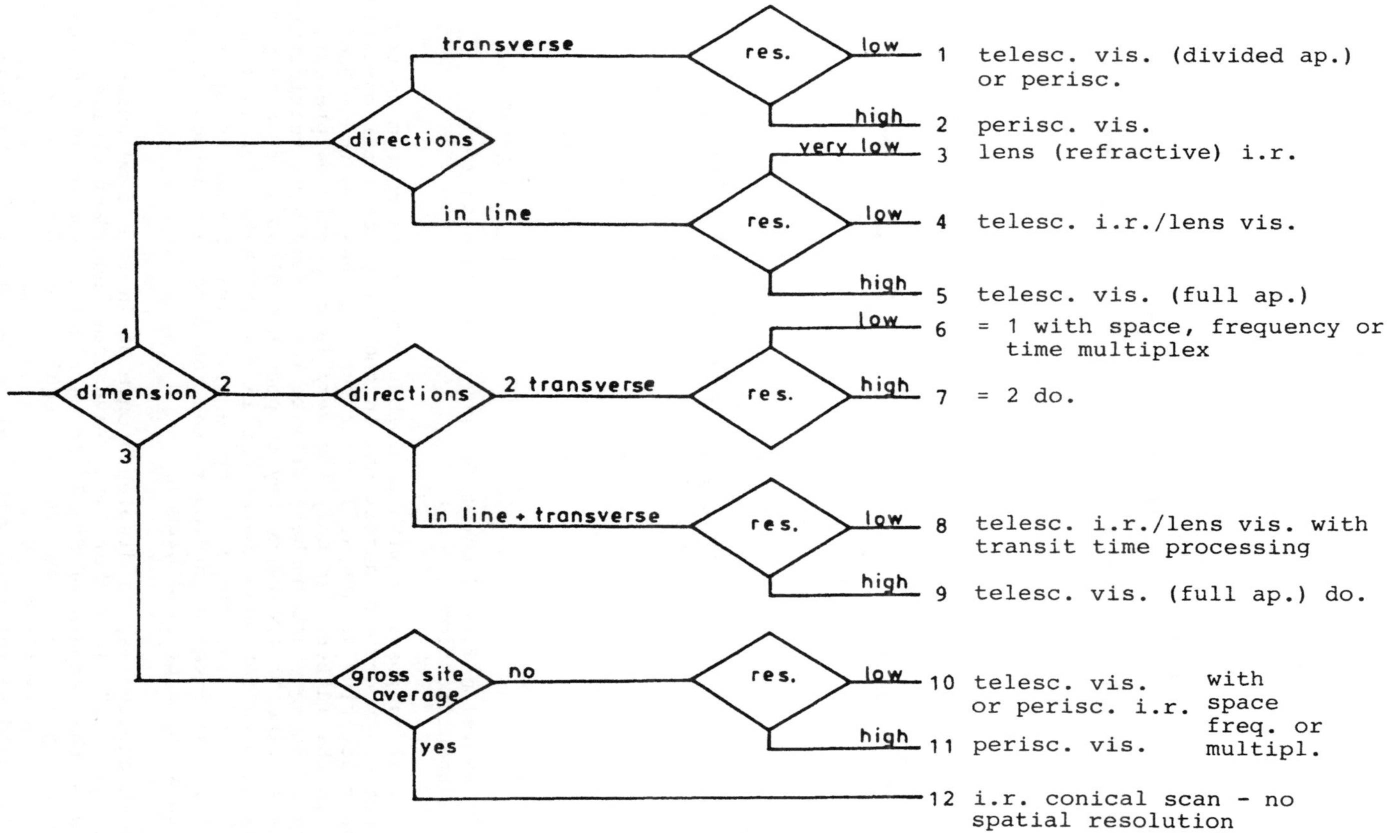

Fig. 2. Decision flow diagram for geometrical specifications

4.2 Physical and Electronic Specifications

The main optical features of a long-range LDV-system were specified by the geometrical considerations of the previous section. There will still remain a number of possible physical and electronic configurations amongst which a choice must be made. Firstly at a number of exits of the Fig.2 system choices are available at infra-red or visible wavelengths. Secondly, although if the infra-red wavelength is chosen a reference beam system is necessary, there remains a choice if the visible wavelength is selected.

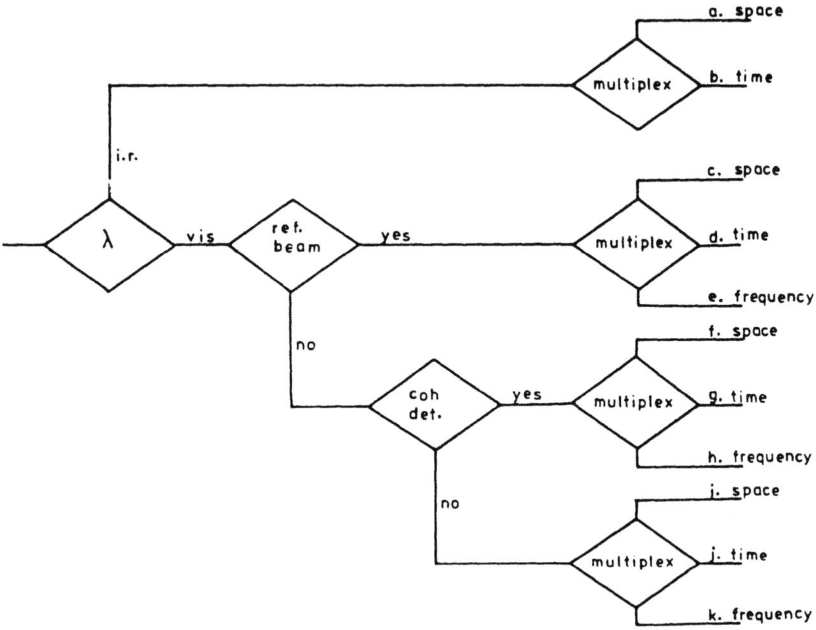

Fig. 3 Decision flow diagram for physical and electronic specification

Thirdly, it is possible with a visible wavelength system to use coherent or incoherent detection. In the former case all particles in the scattering volume will contribute in phase to the signal while in the latter case only the very large particles present, singly or in very low number, will be seen. Fourthly, multiplexing in space, time or frequency may be possible although frequency multiplexing does not seem convenient in the infra-red for lack of a suitable laser. The choice of space multiplexing when using reference beams is taken according to section 3 to mean using multiple detectors and uncrossed beams.

These choices may be displayed once again as a flow chart depicted in Fig.3. This chart will have some redundant choices when following the outputs of Fig.2, in particular, for exits 1-5 the multiplexing alternatives are not required and for 2, 3, 5, 7, 9, 11 and 12 the wavelength choice is already made. (In principle exit 12 could be realized with visible radiation but the lower signal to noise ratio makes this option of little value).

We now consider the options following the geometrical choices for the wind power generation problem discussed in the previous section. At exit 6 all the options of Fig.3 are open in principle. To make a choice we now have to take into account economic and safety factors. We can make the following comments.

1. Solutions (a) and (b) give eye-safe solutions.

2. Solution (b) uses only one detector and processing chain while option (a) uses four. Beam splitting in (a) is replaced by mechanical beam switching in (b). (Beam switched systems have been successfully designed and operated, as in the Marconi Avionics Scanning Laser Anemometer.)

3. Solutions (a) and (b) require periscopic optics while the rest use a telescope. The optical complication of three separate infra-red telescopes with beam switching on an extended base area may be cumbersome compared with a single optical telescope. However, one must bear in mind that this single component, due to the higher finish required, might be equally as costly as the infra-red optics.

4. Solutions (c), (d) and (e) have the disadvantage of requiring high frequency signal processing although they are self aligning.

5. Solutions (f), (g) and (h) seem preferable to (i), (j) and (k) since, at the cost of making the transmission optics of the same aperture as the receiver, necessary in any case for spatial resolution purposes, all particle signals are used.

6. Comment 2 also applies in a modified way to solution (g) in which a crossed beam Doppler-difference system could be mechanically switched between the x and y transverse measurements to save a second detector and processing chain. Since in solutions (f), (g) and (h) Fabry-Perot background suppression is necessary in each processing chain this is a useful advantage.

To summarise, solutions (b) and (g) look economical and interesting, (b) especially so in view of its eye-safe property and (g) because of its compact optics.

At exit 11 the infra-red option is not open and we have only to consider seriously solutions (f), (g) and (h) for the same reasons as given above. Solution (g) would seem to be the most economic.

5. Summary of Published Long-Range Field Experiments

5.1 Methods Using Visible Light

The long-range visible measurements known to us are summarized in Table 3. The earliest experiment on long-range visible laser anemometry seems to be by Hughes et al. [5]. This was a pure test case measuring the velocity and turbulence of an artificial wind seeded with smoke particles. A 100 mW argon-ion laser at 488 nm was used to establish a coherent Doppler-difference measurement in off-axis backscatter at a range of 30 m. The analysis was by means of photon

Table 3. Wind velocity measurements by visible laser LDV/LTV

	RANGE	SCATTERING	OPTICAL SYSTEM	SIGNAL ANALYSIS	LASER
Hughes et al. 1970	30 m	Off axis backsc. smoke seeding day	Coh. Doppl. diff.	Photon corr. integration	A^+ - 488 nm 100 mW
Bourke & Brown 1971	30 m	Forward sc. nat. seed. day	Incoh. Dopp. diff.	Oscilloscope filter bank	HeNe - 633 nm 20 mW
Farmer & Hornkohl 1973	30 m	Off axis backsc. single part. burst day	2-comp. incoh. Doppl. diff. freq. shift	Oscilloscope	A^+ - 514 nm 1000 mW
Bartlett & She 1976	60 m	Off axis backsc. nat. seed. night	Incoh. Doppl. diff.	Photon burst correlation	A^+ - 514 nm 350 mW etalon
Bartlett & She 1977	100 m	Off axis backsc. nat. seed. night	Laser transit	Photon burst correlation	A^+ - 514 nm 350 mW
Lading et al. 1978	70 m	On axis backsc. nat. seeding day	Laser transit lens-system	MCA integration	A^+ - 488 nm 500 mW
Danielsson 1980	60 m	On axis backsc. nat. seed. day	Incoh. Doppl. diff. freq. shift telescopic	Photon corr. integration	A^+ - 514 nm 1400 mW
Durst et al. 1980	105 m	Off axis backsc. nat. seed.	Incoh. Doppl. diff. freq. shift periscopic	Photon corr. integration	A^+ - 514 nm 20 W

correlation. The experiments were performed out-doors on a hot day and neither large scale beam wander, nor small scale wave-front distorsion seemed to distort the measurements severely. The problem of the ambient background light was solved by means of some kind of screen arrangement.

Another artificial test experiment is the one reported by Bourke & Brown in [6]. These authors studied the beam propagation and fringe stability. Velocity measurements were made at a range of 30 m. Two 10 mW HeNe laser beams were used for an incoherent Doppler-difference measurement in forward scatter from the natural aerosol. (The receiving optics were also 30 m away.) Analysis was made by means of oscilloscope recording of the photo current from large particle bursts.

With the beam propagation at a height above ground of 30-60 m and 200 m distance on a warm and sunny day the air turbulence gave rise to a beam wandering amplitude of 10% of the spot size. When fringes were created at a distance of 60 m and 2 m above ground the peak velocity of the fringes due to the air turbulence was estimated to be a few cm/s, i.e. not likely to interfere with any wind measurements of interest. Also this experiment seems to have relied on some kind of black cloth arrangement for the background supression.

In a paper by Farmer and Hornkohl [25], brief mention is made of an experiment which does not seem to have been published in detail. The essential feature is that they employ a two-dimensional Bragg cell for two-component beam splitting and frequency off-set. Thus the transverse vector velocity is measured by means of a two-component incoherent Doppler-difference system. A 2 W argon-ion laser at 514 nm was used and the scattering from the natural aerosol at at a range of 30 m was observed in off-axis backscatter. The photo current from a single detector unit was essentially bandpass filtered and heterodyned into a suitable frequency range before display on a double beam oscilloscope. The Farmer & Hornkohl technique seem to rely very heavily on the big bursts from the rare large particles in the far tail of the particle distribution.

Bartlett and She have performed a series of experiments measuring transverse wind velocities by means of the (incoherent) Doppler difference (LDV) method [26,27] as well as the laser transit (LTV) method [16]. For both methods an off-axis back-scatter optical arrangement was employed with an argon-ion laser at 514 nm and 0.1-0.35 W as the light source. The range was 60-100 m and measurements were done at night and with natural aerosol conditions. Photon correlation was used for the signal analysis. In these works a technique for single particle burst correlation was developed, thus only the very few particles in the large size end of the particle distribution contributed to the velocity information of the signal under the incoherent operating conditions. The velocity correlation for each individual large particle was analysed in a simplified fashion and the ensemble of results for a distribution of such particles was treated statistically. The clip-level of the sampling for the correlation function could then be set high without loss of information and thus the back-ground noise was suppressed. It was concluded that instantaneous atmospheric wind speeds could be measured at a range of up to 100 m by using a low-power, continous wave laser and single burst photon correlation. Air turbulence was not a problem and was not expected to be so for ranges below 1000 m. Laser transit veloci-metry was preferred by these authors for several reasons: simpler, more effective and less sensitive to beam wander (!) due to air turbulence.

The experiment reported by the Risö group [15,28] is the only one using visible light we have found where the atmosphere actually was scanned vertically (and compared with readings from a nearby meteorological tower). Here a time-of-flight laser anemometer (= laser transit velocimeter, LTV) with a fairly complicated optical arrangement was employed, see Fig.4. The two light beams scattered from the focal points were mixed under perpendicular polarizations and subsequently separated into two photomultipliers and the signals were analysed by means of a multi-channel-analyser (MCA). These measurements were performed at daytime using natural seeding and on-axis back-scatter. The light source was an argon-ion laser operated at 0.5 W and 488 nm wavelength. The range was up to 70 m. Here Lading [28] claims superiority of the LTV technique over the LDV for reasons which do not seem quite conclusive. It seems to us that neither Lading nor Bartlett & She have considered realistic measurement situations with large turbulence and large variance of the wind vector.

The experiment by one of us [29] had the objective of demonstrating the feasibility of single-ended Doppler-difference measurements of the wind velocity vectors over ranges of interest for windpower generation, i.e. up to 200 m height. An argon-ion laser operated at up to 1.5 W and 514 nm wavelength was used as light source and a 800 mm f/3.8 Cassegrain-type telescope (400 mm clear aperture) transmitted the laser beams and received the scattered

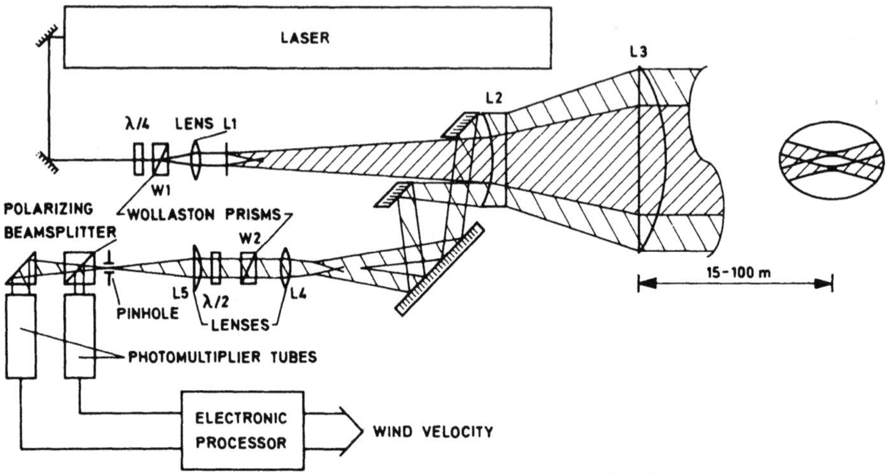

Fig. 4. Experimental arrangement by Lading [15]

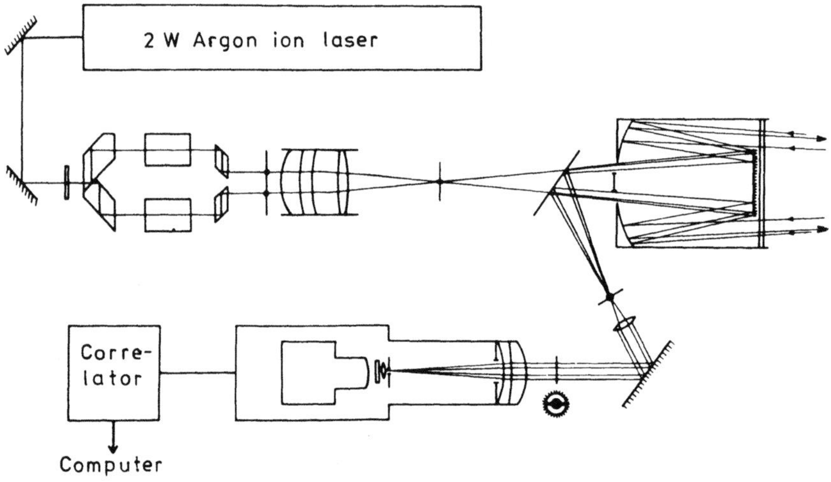

Fig. 5. Experimental arrangement by Danielsson [29]

light (on axis), see Fig.5. The signal analysis was done by means of integrated photon correlation (as distinguished from single burst p.c.) with Fourier analysis. For this procedure individual frequency shifting of the two laser beams was essential due to the very large turbulence (or rather: wind vector variance).

Wind velocity measurements proved possible to perform in daylight under natural aerosol conditions at a distance of about 60 m, however, the background light created considerable difficulties; a suggested improvement here would be to add a Fabry-Perot interferometer of say free spectral range 3 nm and finesse 100 in front of the photo-multiplier in the set-up of Fig.5. As in almost all visible light measurements in the atmosphere, incoherent conditions were met also here, i.e. only the rare very large particles contributed to the signal.

The Karlsruhe group has made a theoretical study of the feasibility of visible LDV measurements over large distances in the atmosphere [30] and an experimental verification of the theoretical results [31,32]. The conclusions were that single-ended Doppler-difference measurements of atmospheric wind speeds are feasible up to a distance of about 500 m with an Argon-ion laser of 20 W. For longer distances pulsed lasers would be necessary. This group has now completed the construction and building of a transportable measuring container including a -5° to +50° levitated optical table for vertical scanning, a 20 W (green line) argon-ion laser, stepping motor operated transmitting and receiving optics, a photon correlator, a data aquisition & analysis system and a graphics terminal display. The system uses incoherent detection and hence will only detect the largest particles. Artificial measurement tests have been performed up to 106 m distance from the receiving optics with a rotating grating for beam-splitting and frequency shift, and natural seeding [32]. The resulting velocity measurements have been compared with sonic measurements and show very good agreement in view of the quite different sampling techniques and control volume size. Studies have been made of the signal-to-noise variaation under varying conditions. Measurements could be recorded at the excellent rate of up to one per second under optimum conditions (one per 8 seconds in the worst cases).

5.2 Methods Using Infra-Red Light

In contrast to the tentative experiments discussed in the previous section, long-range measurements using the CO_2 laser at 10.6 µm wavelength are now routine. Apart from the experiments mentioned above in which, firstly, single-beam transit-time effects were investigated to obtain transverse wind speed and, secondly, in which a reference beam Doppler-difference scheme was constructed, all practical CO_2 laser anemometry systems employ the in-line coherent-detection, reference-beam technique. A suitable atmospheric window exists at this wavelength and lasers of the required power (several watts) and stability (1 part in 10^8) have been developed, more recently as "waveguide" devices which are rugged and compact.

The first experiments were performed in the late sixties as mentioned above and used single-ended telescopic optics with apertures of the order of 15-30 cm giving range resolution according to the formula of equation (8) of 3 - 1.5 m or, in practice, due to beam truncation effects, of about twice these values [33]. The only developments of note other than system engineering since these first experiments have been the use of polarizing optics to improve efficiency [34], the use of pulsed heterodyne systems [35] for very long ranges, the use of waveguide lasers [36] and the development of acousto-optic frequency shifters to resolve the directional ambiguity [37]. Among many developments on the signal processing side the use of surface acoustic wave analysis [38] is worthy of mention although the flexibility of digital signal processing brings it many advocates. As mentioned in the Introduction, the first papers on infra-red anemometry appeared in 1970 [1-4] where already field measurements of aircraft wing-tip vortices were reported [3]. Atmospheric boundary layer measurements were further published in the following years [39]. Quite recently work in this area was begun also at the NOAA Wave Propagation Laboratory [40-42]. An example of a modern "lens type" instrument [33] is shown for example in Fig.6. Related work of interest in the present context is also the study of wake fields of windpower power generators [43].

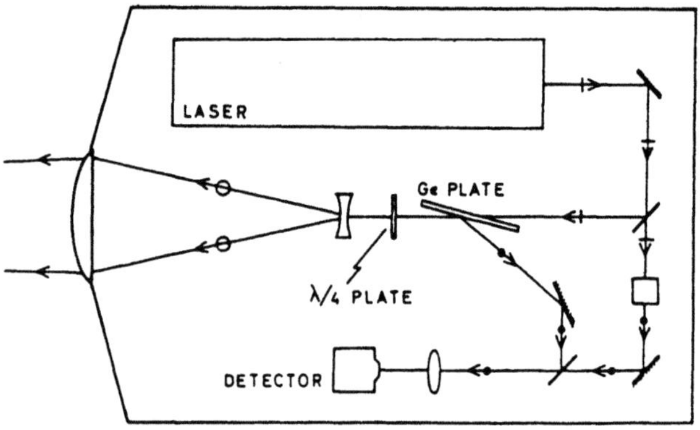

Fig. 6 Typical layout of a monostatic CO_2 laser velocimeter. Polarization techniques with Ge and λ/4 plates promote good efficiency (Vaughan [33])

6. Conclusions

The present report gives a general review of the field of long-range laser anemometry. It seems impossible to design one single ultimate system to fulfill all measurement situations at long range in the atmosphere, even with the limitation to measurements of interest for wind power development. The minimal system with regard to a specific measurement was deduced in section 4 above, but this still gave a choice of wavelength, optical system, multiplexing principle, etc. In addition to these considerations, others, such as safety, economy and flexibility, have to be taken into account. We have not extended our considerations to the very long range case where pulsed laser systems are desirable as this is subject to intensive study elsewhere for global wind measurement satellite systems.

The advantages and disadvantages inherent in the choice of wavelength and optical system are summarized in Table 4. It can be concluded that whenever the requirements permit - firstly those on range resolution, but also those on economy and flexibility - an infra-red system should be chosen. It should be pointed out that for infra-red wavelengths there are well-established, commercial systems available, while for visible wavelength systems, in some cases, further development is required. For 3-dimensional application a combination of an infra-red reference beam system for the in-line component and a 2-component visible Doppler-difference system for the transverse velocity vector might be the optimal solution [32].

Table 4. Evaluation of optical systems

OPT. SYST.	ADVANTAGES	DISADVANTAGES
coherent local oscillator	automatic self-alignment, well established technique, high detection efficiency, little propagation distortion, eye-safe, compact, stable lasers (5 W)	only in-line component, detector cooling, low spatial resolution.
coherent local oscillator visible	good signal-to-noise performance, ambient background unimportant, good spatial resolution, automatic self alignment.	only in-line component, difficult frequency range, frequency shift necessary to cover the velocity range.
coherent Doppler difference visible	transverse (vector) velocity, reasonable frequency range, uses all particles, good spatial resolution.	large total optical aperture, relies on spectral filtering (FP)
incoherent Doppler difference	cheaper optics, transverse (vector) velocity, reasonable frequence range.	uses only the largest particles - low S/N-ratio, otherwise - burst correlation, statistics software relies on spectral filtering (FP), less power density in measuring volume.
laser transit	good background suppression, good spatial resolution.	uses only a few of the largest particles, most sensitive to air-turbulence beam wander no good in turbulent situations, angular search necessary.

The choice of multiplexing principle is of particular interest since the ideas developed here are partly new: time-multiplexing does not seem to have been discussed before for these applications. Space and frequency multiplexing - in our sense of the terms - are well established. It seems to us that considerable economic advantages could be won by time-multiplexing, thus saving the cost of duplication of detectors, beam splitters, Fabry-Perot's, etc.

Frequency shifting has not been discussed - in any detail at least - in this report. It is understood that this option can be realized for most of the systems discussed - and will in reality be necessary in many cases.

References

1. A.J. Hughes, R. Jones, E.R. Pike, J. O'Shaughnessy: CO_2 laser anemometry in the atmosphere. RRE Newsletter 11, 17/1-17/2, 1972(a).

2. A.J. Hughes, J. O'Shaughnessy, E.R. Pike, A. McPherson, C. Spavins, T.H. Clifton: Long range anemometry using a CO_2 laser. Opto-electronics 4, 379-384, 1972(b).

3. R.M. Huffaker, A.V. Jelalian, J.S.L. Thomson: Laser Doppler system for detection aircraft trailing vortices. Proc. IEEE 58, 322-326, 1970.

4. R.M. Huffaker: Laser Doppler detection systems for gas velocity measurement. Appl. Optics 9, 1026-1039, 1970.

5. A.J Hughes, C.J. Oliver, E.R. Pike, J. O'Shaughnessy: Measurement of wind velocity by laser scattering. RRE Newsletter 9, 7/1-7/3, 1970.

6. P.J Bourke, C.G. Brown: Remote measurement of mean wind velocity and turbulence by laser anemometry. Optics and Laser Tech. 4, 23-25, 1971.

7. W.M. Farmer, D.B. Brayton: Analysis of atmospheric laser Doppler velocimeters. Appl. Optics 10, 2319-2324, 1971.

8. A.J. Hughes, E.R. Pike: Remote Measurement of wind speed by laser Doppler systems. Appl. Optics 12, 587-601, 1973.

9. L.Z. Kennedy, J.W. Bilbro: Remote intensity fluctuation measurements with a laser Doppler radar. Appl. Optics 15, 2008, 1976.

10. L.Z. Kennedy, J.W. Bilbro: Remote measurement of the transverse wind component using a laser Doppler velocimeter. Appl. Optics 18, 3010-3013, 1979.

11. J. O'Shaughnessy, W.R.M. Pomeroy: Single beam atmospheric transverse velocity measurment. Opt. and Quant. Elect. 10, 270-272, 1978.

12. E.R. Collins: Some notes on Doppler LDV systems and anemometry. RRE 1973. (Unpubl.)

13. R.L. Schwiesow, R.L. Cupp, M.J. Post, R.F. Calfee: Coherent differential Doppler measurements of transverse velocity at a remote point. Appl. Optics 15, 1145-1150, 1977.

14. H.J. Peifer: Fringe type laser Doppler Anemometry for remote cross wind velocity measurements in the atmosphere. In: Proceedings of the Symposium on Long Range and Short Range Optical Velocity Measurements (H.J Pfeifer, ed.) Rapport ISL R 117/80, 1980, p. VIII/1-VIII/8.

15. L. Lading, A. Skov Jensen, C. Fog, H. Andersen: Time of flight laser anemometer for velocity measurements in the atmosphere. Appl. Optics 17, 1486-1488, 1978.

16. K.G. Bartlett, C.Y. She: Single-particle correlated time-of-flight velocimeter for remote wind speed measurement. Optics Letters 1, 175-177, 1977.

17. H.Z. Cummins, E.R. Pike (eds.): Photon Correlation and Light Beating Spectroscopy. New York and London 1974.

18. H.Z. Cummins, E.R. Pike (eds.): Photon Correlation Spectroscopy and Velocimetry. New York and London 1977.

19. E.R. Pike: Laser Doppler Anemometry. A comparative study of the measurement of motion by light scattering. In: The Engineering Uses of Coherent Optics (E.R. Robertson, ed.) Cambridge 1976, p. 431-457.

20. E.R. Pike: How many signal photons determine a velocity? In: Laser Velocimetry and particle sizing (W.H. Stevenson and H.D. Thompson, eds.). New York 1979, p. 285-289.

21. R.G.W. Brown, E.R. Pike: Seeding levels and size distributions in LDV and LTV. In Photon Correlation Techniques Techniques in Fluid Mechanics (W.T. Mayo Jr. and A.E. Smart, eds.) Stanford 1980, p. 2/1-2/9.

22. C.J. Oliver: Accuracy in laser anemometry. J. Phys. D: Appl. Phys., 13, 1145-59. 1980.

23. C.M. Sonnenschein, F.A. Horrigan: Signal to noise relationships for coaxial backscatter systems that heterodyne backscatter from the atmosphere. Appl. Optics 10, 1600-1604, 1971.

24. E. Jakeman, C.J Oliver, E.R. Pike: Optical homodyne detection. Adv. in Physics 24, 349-405, 1975.

25. W.M. Farmer, J.O. Hornkohl: Two-component, self-aligning laser vector velocimeter. Appl. Optics 12, 2636-2640, 1973.

26. K.G. Bartlett, C.Y. She: Remote measurement of wind speed using a dual beam backscatter laser Doppler velocimeter. Appl. Optics 15, 1980-1983, 1976.

27. C.Y. She: Individual particle scattering and single-burst digital correlation for remote atmospheric wind measurements. Optica Acta 26, 645-657, 1979.

28. L. Lading: The time-of-flight laser anemometer versus the laser Doppler anemometer. In: Laser Velocimetry and Particle Sizing, (W.H. Stevenson and H.D. Thompson, eds.), New York 1979, p. 26-40.

29. L. Danielsson: Laser Doppler velocity measurements over large distances in the atmosphere. In: Proceedings of the symposium on Long Range and short Range Optical Velocity Measurements (H.J. Pfeifer, ed.) Rapport ISL R 117/80, 1980, p. XII/1--XXII/11.

30. F. Durst, B. Howe, G. Richter: Long range LDA wind velocity measurements using visible laser radiation. In: Photon

Correlation Techniques in Fluid Mechanics (W.T. Mayo Jr. and A.E. Smart, eds.) Stanford 1980(a), p 3/1-3/13.

31. F. Durst, B. Howe, G. Richter: LDA-system design for long-range wind velocity measurements. In: Proceedings of the Symposium on Long Range and Short Range Optical Velocity Measurements (H.J. Pfeifer, ed.) Rapport ISL R 117/80, 1980(b), p. XI/1-XI/9.

32. F. Durst, B. Howe., G. Richter: Laser-Doppler-measurements of cross wind velocity. Appl. Optics, to be printed (1982).

33. J.M. Vaughan: Remote wind measurements in the atmosphere using laser Doppler methods. In: Proc. Laser 79, Guilford 1979, p. 461-469.

34. S.J. Alexander, A. Barbour, J. Foord, J. O'Shaughnessy, W.R.M Pomeroy: Long range laser anemometry. In: Proc. Laser 77, Guildford 1977, p. 720-727.

35. J.W. Bilbro, J.M. Vaughan: Wind field measurements in the non-precipitous regions surrounding severe storms by an airborne pulsed Doppler lidar system. Bull Amer. Met. Soc. 59, 1095-1100, 1978.

36. R. Callan, J. Cannell, R, Foord, R. Jones, J.M. Vaughan, D.V. Willetts, A. Woodfield: Laser velocimetry and measurement of true air speed. In: Advances in Transducers for Measurement. Proceedings of IEE Colloquium, London, 12 Dec. 1980.

37. W.R.M. Pomeroy, J.M. Vaughan, D.V. Willetts: Performance study of an acousto-optic frequency shifter in a CO_2 laser velocimeter. J. Phys. E: Sci. Instrum. 13, 982-985, 1980.

38. M. Aldritt, R. Foord, R. Jones, J.M. Vaughan: An integrating SAW spectrum analyser for infra-red Doppler signals. RSRE Res. Rev. 2, 6/1-6/4, 1978.

39. T.R. Lawrence, D.J. Wilson, C.E. Craven, I.P. Jones, R.M. Huffaker, J.A.L. Thomson: A laser velocimeter for remote wind sensing. Rev. Sci. Instr. 43, 512-518, 1972.

40. R.L. Schwiesow, R.E. Cupp: Remote Doppler velocity measurements of atmospheric dust devil vortices. Appl. Optics 15, 1-2, 1976.

41. F.J. Pratte, R.M. Huffaker, T.R. Lawrence, R. Loveland: System considerations of a long range coherent lidar wind sensor with particular emphasis on real weather effects. NOAA Technical Memorandum ERL WPL-42, 1972.

42. S.F. Clifford, T.R. Lawrence, G.R. Ochs, T-i. Wang: Study of a pulsed coherent lidar for crosswind sensing. NOAA Technical Memorandum ERL WPL-48, 1980.

43. D.J. Wilson, A.D. Zalay, M.R. Brashears, C.E. Craven, K.R. Shrider, A.J. Jordan: Full scale wake flow measurements with a mobile laser Doppler velocimeter. J. Aircraft 16, 155-161, 1979.

9. Comparison of LDA Signal Strength in Photon Correlograms Obtained by CW and Pulsed Laser Operation

E. Sommer and H.J. Pfeifer

German-French-Research Institute Saint-Louis (ISL)
F-68301 Saint-Louis, France

Abstract

When photon correlation methods are used for processing signals in laser Doppler anemometry (LDA), the signal strength is different according as to whether long pulse lasers or continuous wave lasers are used. In the present paper experimental results will be reported which quantitatively show the advantages observed in the case of pulsed laser operation. A first explanation will be given.

This paper is part of a range estimation for LDA applications to atmospheric wind velocity measurements.

1. Introduction

In a series of papers |1,2,3,4| first experiments have been reported to determine the cross-wind velocity by LDA methods. The light source used in all these experiments was a cw-Argon laser, whereas the signal processing system employed was a photon correlator. The use of the Argon laser was necessary because at the large distance involved the aerosol particles backscatter only an extremely small amount of the light emitted. Continuous wave Argon lasers provide the highest TEM_{oo}-mode output power at a single wavelength in the visible range.

In field experiments, however, Argon lasers turned out to be poor candidates for wind velocity measurements. In effect, the laser head is of a relatively large size and the necessary water cooling system is quite inconvenient. Moreover, an electrical input power of approximately 20 kW is needed for obtaining an optical output power equalling 2 W. In a paper reporting on the applicability of LDA methods to measure the wind velocity |5|, it has therefore been suggested to replace the high power cw-laser by a dye laser or a frequency doubled Nd:YAG laser with pulses of 50 to 200 µs duration. Pulsed type lasers offer a number of advantages. Thus they have a compact configuration and a relatively straightforward cooling system. The input power is only a fraction of the power needed for cw operation. It is therefore possible to replace relatively bulky measuring devices by portable units. At an assumed pulse duration of 50 µs, a pulse energy of 10 mJ, and a repetition rate of 100/s which is attainable with 3 kW input energy only |6|, the time averaged output power of the laser equals the output power of a 1 W cw laser, i.e. in both cases the same number of photons per second is emitted. For simplicity we assume here identical wavelength for both lasers.

The question, however, arises to what extent the results of range estimations for cw operation may be transferred to pulsed laser operation. In the following, experimental results achieved in laboratory tests will be reported in which both laser types have been simulated.

2. Experimental Arrangement

With the aid of a continuous wave laser a conventional fringe-type laser Doppler anemometer was built up, i.e. via a beam splitter and an image forming system, an optical interference fringe pattern was generated. A small fan blew particles through the interference pattern at low velocities. The particles were natural aerosols present in the laboratory air. At the measuring point the air flow had a turbulence intensity of less than 1 %. Light scattered by the particles was received by a photomultiplier through a collecting lens. In order to obtain photon resolved signals at the photomultiplier output, the laser power was strongly reduced by decreasing the plasma current and by a series of neutral density filters. The photon pulses were fed into a pulse former so that a uniform amplitude of -3V and a uniform pulse duration of 50 ns were achieved. Furthermore the pulse former could be gated such that for simulating pulsed laser operation, it was sufficient to transmit to that gate pulses with corresponding duration and repetition rate.

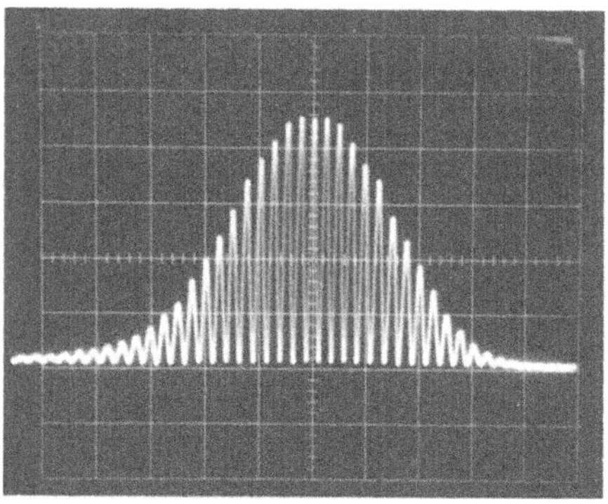

Fig.1. Intensity distribution in the fringe system

The quality of the interference fringe pattern was checked by observing, with full laser power, the passage of individual particles through the probe volume using thereby a second receiving system. Figure 1 illustrates such a record taken by means of a transient recorder. The envelope of the signal is a Gaussian curve according to the TEMoo-mode of the laser. The visibility in the fringe pattern is obviously close to unity.

The fringe spacing was set to 8.6 µm. From Fig.1 the extension of the probe volume defined by the $1/e^2$ points of the intensity perpendicular to the propagation direction of the laser beams has been determined. It turned out to be 205 µm, which corresponds to 24 fringes.

For the experiments described in the present paper, two different flow velocities were chosen, namely 1.38 m/s and 0.86 m/s. This corresponds to signal durations of 149 µs and 238 µs, respectively.

Now in assuming a real experiment conducted in the atmosphere in which at a distance of 100 m, probe volumes of 1 mm in diameter are typically generated, the signal durations mentioned above correspond to wind velocities of 6.7 m/s and 4.2 m/s, respectively. The signal frequencies of 160 kHz and 100 kHz including the Doppler frequency and the Bragg cell frequency shift, are in a range which has already been considered to be optimal in previously conducted investigations |7|.

3. Data Processing

From the correlation values obtained by a Malvern photon correlator of type K7023 the pedestal was removed first by curve-fitting methods. Then the new values were mirrored with respect to zero time shift. This resulted in a total of 192 data points. A succeeding Fourier transform gave an estimation of the power spectrum.

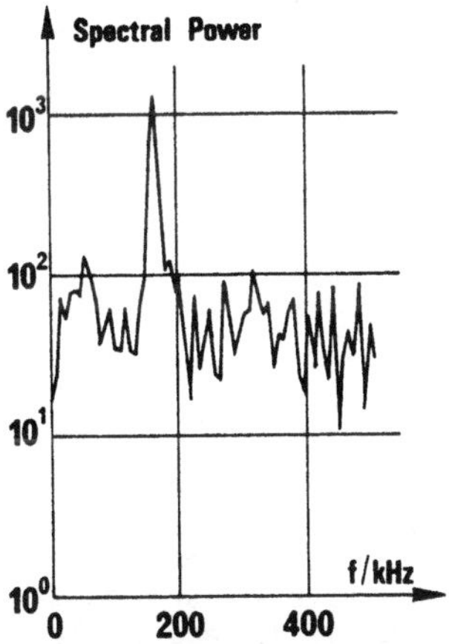

Fig.2. Spectral power of a 100 kHz signal

Figure 2 shows as an example the first 64 Fourier coefficients which were obtained during a run with a flow velocity of 1.38 m/s corresponding to a signal frequency of 160 kHz. For the comparison of the results obtained in cw and in pulsed laser operation, respectively, a signal strength was defined on the basis of the Fourier representation as in Fig.2. To this the values of the maximum Fourier coefficient and its two immediate adjacent values were linearly added. It turned out that this approach provided very reproducible results even if the signal frequency varied slightly from one run to the next.

The tests described here were aimed to give information on the dependence of the signal strength on laser power in the cw and pulsed mode, respectively. For the pulsed laser operation the repetition rate was chosen to 100/s, whereas the duration of the individual pulses was varied between 50 and 500 μs. These values agree with characteristics of real long pulse lasers. As increment time of the correlator 500 ns was set for all the tests, which yields a maximum time delay of 48 μs corresponding to the 96 channels available. The total integration time was fixed at 50 seconds. By this it was guaranteed that statistical variations caused by occasionally appearing large particles are smoothed out.

As a measure of the time-averaged laser power the number of detected photons was taken. This value is printed out along with the correlogram by the microcomputer which performs the calculations. Comparison with a laser power meter proved that this procedure is very accurate and that the variations of the particle concentration could be neglected in the course of these tests.

4. Experimental Results

Figure 3 shows two examples of experimental results obtained in cw and in pulsed laser operation with pulses of 500 μs duration. In both cases the sig-

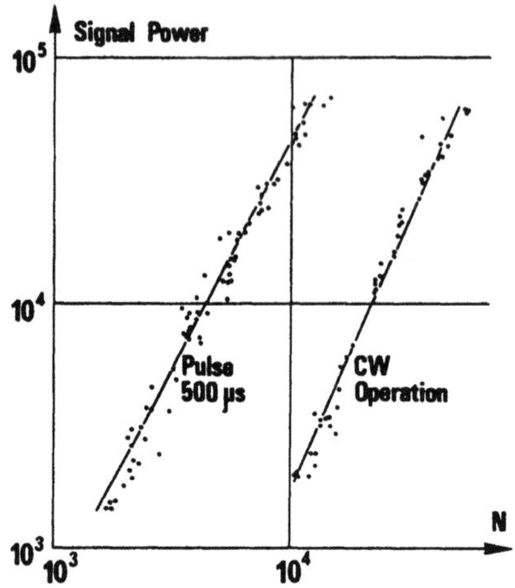

Fig.3. Measuring points of maximum spectral power

nal frequency was 100 kHz. In the ordinate direction the signal strength is plotted in relative units while the abscissa indicates the number of photon pulses detected within a time period of 50 seconds. At a double logarithmic scale the measuring points can be approximated in an adequate way by a fitting straight line. The lower limit of the measuring range is defined by the laser power or in other words, by the number of photon pulses, for which the signal frequency is just still detectable in a reliable way. The upper limit of the measuring range is given by the beginning of saturation in the correlator.

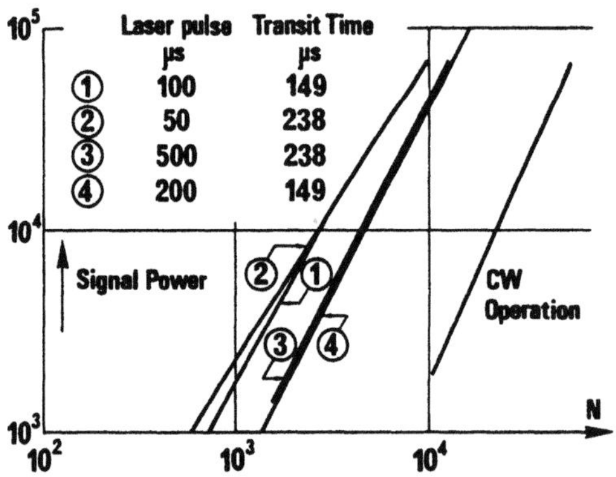

Fig.4. Fitting straight lines for various pulse durations and transit times

In Figure 4 all the experimental results achieved are summarized. For the two straight lines, plotted in the left part of Fig.4, the duration of the laser pulses was shorter than the residence time of the particles in the probe volume. In these cases the pulse durations were 50 µs for a residence time of 238 µs and 100 µs for a residence time of 149 µs, respectively.

For the straight lines plotted in the central part of Fig.4, we have pulse durations of 500 µs and 200 µs for residence times of 238 µs and 149 µs, respectively. The straight line at the right of Fig.4 applies to cw laser operation.

From Fig.4 it can clearly be seen that for obtaining the same signal strength, quite a different amount of time averaged laser power is required. It will obviously be best to have a pulse duration which is shorter than the residence time of the particles in the probe volume as indicated by the straight lines (1) and (2) in Fig.4. In the case of low signal strength the time averaged laser output power needed, for obtaining the same result equals only one tenth of that required in cw laser operation.

In the case of the straight line (2) corresponding to laser pulses of 50 μs duration, the lower limit of the signals which can still be identified in a reliable way, corresponds to a detected number of 538 photons. In the case of cw laser operation at least 11800 photon pulses are needed for identifying a signal with high confidence level.

If the laser pulse duration is somewhat longer than the residence time of the particles, which applies to the straight lines (3) and (4) in Fig.4, the gain factor is equal to 5 as compared to cw operation.

The five straight lines in Fig.4 do not allow to conclude quantitatively the relationship between pulse duration, residence time, signal strength, and time averaged laser power. To this a study of parameters would be necessary including another magnitude, that is the increment time of the correlator. Such a study, however, was beyond the scope of the investigation reported herein.

5. Conclusions

The experiments have shown that for the same time-averaged laser output power, long-pulse laser operation offers substantial advantages over cw laser systems. If the duration of the laser pulses is shorter than the residence time of the particles in the probe volume, the signal strength is by a factor of ten higher than in cw operation. For the number of detected photons which are necessary to define a signal, even a factor of 20 was found. For instance, let us consider a laser with 200 W peak power, with a repetition rate of 100/s, and with a pulse duration of 50 μs. This laser is equivalent to a 10 W cw laser as far as the attainable signal strength is concerned. It is equivalent to a 20 W cw laser when the minimum necessary number of detected photons is considered.

This phenomenon can be explained qualitatively in the following way. All the events involved, i.e. emission of the photons by the laser, scattering process, and detection of the photon pulses are linear processes. Provided there is a sufficiently long time averaging, the number of the detected photon pulses is proportional to the time-averaged laser output power, independently of the mode of laser operation. This has been checked experimentally. The advantage of pulsed laser operation is only obtained by the correlation process. A detected photon will only contribute to the establishment of the definite correlogram if within the longest delay time of the correlator, that is 48 μs in these experiments, a second photon is detected which has been scattered from the same particle. Let us assume a particle scatters a number n of detectable photons. Then we have to distinguish between the following two cases for the number M of products which contribute to the single clipped correlogram. In the first case the residence time T_R of the particles in the probe volume exceeds the pulse duration T_P of the laser. Here a constant α may be defined in the following form:

$$\alpha = \frac{\tau_{max}}{T_P} ,$$

where τ_{max} is the maximum time shift in the correlator.

In the second case the residence time is shorter than the pulse duration. We have:

$$\alpha = \frac{\tau_{max}}{T_R} \quad .$$

This definition includes cw laser operation.

The number M of products due to a single particle is then given by:

$$M = (\alpha - \frac{1}{2}\alpha^2)n^2 - \frac{1}{2}\alpha n \quad .$$

The nonlinear relationship between the number of detected photons and the number of correlation products is the reason for the superiority of pulsed lasers over cw lasers. In effect the number of products is proportional to the signal strength because the Fourier transform is a linear process.

6. Final Remarks

As compared with cw lasers, pulsed lasers show some favourable characteristics such as low input power, compact configuration of both the laser head and the current supply system, and relatively simple cooling system operated in a closed circuit. Furthermore, long-pulse lasers offer the possibility to eliminate electronically background light in a very efficient way. This cannot be achieved in a comparable manner with cw lasers.

The analysis of the experimental results showed that with the same time-averaged laser power, pulsed type lasers allow to attain signal strengths which are ten times higher than those attainable with cw type lasers. For the minimum detectable signal in the correlogram is even 1/20 of the laser power sufficient.

Therefore a pulsed laser emitting, for instance, 50 µs pulses with an energy of 10 mJ and a repetition rate of 100/s, is equivalent to a 10 or 20 W cw laser, respectively.

This offers also promising perspectives considering the load of various optical components. In particular the Bragg cells should be mentioned here which are needed for identifying the sign of the wind velocity. At a high time-averaged laser output power these Bragg cells may generate a thermal lens causing thereby a strong deformation of the emitted radiation.

References

[1]W.M. Farmer and D.B. Brayton, Analysis of Atmospheric Laser Doppler Velocimeters, Appl. Optics, Vol. 10, pp. 2319-2324, 1971.

[2]C.Y. She, Individual Particle Scattering and Single-Burst Digital Correlation for Remote Atmospheric Wind Measurements, Optica Acta, Vol. 26, No. 5, pp. 645-657, 1979.

[3]K.G. Bartlett and C.Y. She, Single-Particle Correlation Techniques for Remote Measurement of Wind Speed: Aerosol Condition and Measurement Rate, J. Opt. Soc. Am., Vol. 69, No. 3, pp. 455-459, 1979.

[4]F. Durst, B. Howe, and G. Richter, LDA-System Design for Long Range Wind

Velocity Measurements, Proc. Symp. on Long Range and Short Range Optical Velocity Measurements, Ed.: H.J. Pfeifer, ISL-Report R 117/80, 1980.
[5]H.J. Pfeifer, Fringe Type Laser Doppler Anemometry for Remote Cross Wind Velocity Measurements in the Atmosphere, Proc. Symp. on Long Range and Short Range Optical Velocity Measurements, Ed.: H.J. Pfeifer, ISL-Report R 117/80, 1980.
[6]A. Hirth and H. Fagot, High Average Power from a Long Pulse Dye Laser, Optics Communications, Vol. 21, No. 2, pp. 318-320, 1977.
[7]E. Sommer and H.J. Pfeifer, A Simulation Model of the Photon Correlation and Application to Multiple Particle Problems in Laser Doppler Anemometry, Proc. 4th Int. Conf. on Photon Correlation Techniques in Fluid Mechanics, Stanford University, Stanford, CA 94305, 1979.

10. Laser Doppler Measurements of Wind Velocities Using Visible Radiation

F. Durst and G. Richter

Institut für Hydromechanik, Universität Karlsruhe, Kaiserstraße 12
D-7500 Karlsruhe 1, Fed. Rep. of Germany

It has been the aim of the authors to augment the existing information on wind field distributions by developing and applying a remote crosswind sensing device based on the conventional dual-beam laser Doppler anemometer operating with naturally occurring aerosols. The measurement distance covered in this work is 20 to 200 meters.

In the present paper, the design considerations for the development of LDA systems applicable for field measurements are described, based on theoretical and experimental investigations carried out at the University of Karlsruhe. This work shows to what extent the concentration, the size distribution, the material, and hence, the light-scattering properties of the naturally occurring aerosols influence the resulting measurement signals.

On the basis of theoretical studies, a prototype measuring system was constructed and mounted inside a transportable container. With this system measurements have been performed on the grounds of the University of Karlsruhe up to distances of 106 m. The results are compared with simultaneous sonic-anemometer measurements obtained at the same location and show good agreement.

1. Introduction

In meteorological studies of the earth's lower atmosphere, detailed information on the vertical distribution of wind velocity is of prime interest. The knowledge of the wind field is becoming of increasing interest in various fields such as the investigation of the flow properties of the planetary boundary layer, monitoring the wind field close to airport runways, or the study of flow around obstacles such as land formations, wind mills, chimneys and cooling towers. Measurements of wind fields in such cases are normally limited to observations made by means of weather masts and/or balloons.

The need for a single-ended remote-sensing instrument led to the present work which basically is a verification study carried out with the aim to determine whether the "fringe mode" anemometry, based on visible laser radiation, is a practical method for the measurement of crosswind components over long distances. Several attempts to measure the crosswind component using the dual-beam method are discussed in literature, for example FARMER and BRAYTON [1], BARTLETT and SHE [2] and DANIELSSON [3], but no practical continuous measurements have yet been published. Thus further research and development work was needed to extend the understanding on this measuring method.

General investigations concerning the influence of atmospheric turbulence on the laser beams propagating through the atmosphere were carried out by BOURKE and BROWN [4], WALL [5], PFEIFER et al. [6], ABELE and RAIDT [7] and others. The result of their work showed that the transmitting laser beams are distorted by local variations of the refractive index of the air, but, restricted to measurement distances shorter than 500m, these distortions do not prevent the crossing and interfering of the two laser beams in the measuring volume.

Based on Mie's formulae for light scattering by small spherical particles, computations were carried out to predict the achievable signal quality and the maximum range of LDA measurement. These theoretical estimations are summarized in the next section. The following sections describe the actual measurement system and some of the experimental results of crosswind measurements.

2. Signal Detectability and Aerosol Properties

LDA measurements over long distances require the application of backscatter systems. Systems of this kind usually result in low scattered and detected light powers due to the long distances involved and also due to the fact that the backscattered light is less than that scattered in forward direction by a factor of 500 to 1000. Depending on the particle size and the particle material, LDA long-range anemometers result in signals resulting from individual photons. Signals of this kind show an average photon rate $\dot{N}_p$ which varies with time as described by the following relationship

$$\dot{N}_p = \frac{P_S}{h\nu} = N_0 \exp\left[-\left(\frac{t-t_i}{\Delta\tau_i}\right)^2\right]\left\{1+\eta_M \cos\left[2\pi\left(f_D+f_{sh}\right)\left(t-t_i\right)\right]\right\} . \quad (1)$$

Here P_S is the scattered light power, η_M the modulation depth of the signal, f_D is the acquired Doppler frequency and f_{sh} the shift frequency. In connection with computing the Doppler frequency from the photon rate variations, the question arises about the minimum photon rate required to ensure the detectability of the Doppler frequency. It was experimentally verified that the Doppler information can be deduced from photon resolved signals, if the maximum photon rate exceeds the Doppler frequency. That means at least one photon per Doppler cycle is required on the average. With less photons than this limiting rate in the center of the signal, LDA signal cycles will be missed and the correct Doppler frequency cannot be determined. This yields the following relationship to ensure detectability of the Doppler frequency:

$$\dot{N}_{p,min} = \frac{P_S}{h\nu} \geq |f_D| \quad . \quad (2)$$

The scattered light power P_S is a function of the laser power inside the measuring control volume, of the measurement distance, of the receiving optics and, first of all, of the properties of the scattering particles, the aerosols. These properties can vary widely and may change in time and space. Nevertheless typical aerosol properties exist and can be used to predict the influence of particle properties on signals.

Aerosols can be classified into three groups:

MARITIME	Oceanic (75%) and continental origin, mean refractive index m = 1.418 - 0.001i
RURAL	Water soluble material (70%) and dust-like components (30%), m = 1.53 - 0.007i
URBAN	65% rural and 35% soot-like components, m = 1.607 - 0.169i

To calculate light scattering properties for the aerosol groups given above, computer programs, based on Mie's scattering theory for spherical particles, were set up and are reported by DURST, MACAGNO and RICHTER [8]. These programs were used to compute for each aerosol group the detectable light power scattered by single particles available inside the measuring control volume. Fig. 1 shows as an example the scattered light power for the optical system described in section 3 for a measurement distance of 100m.

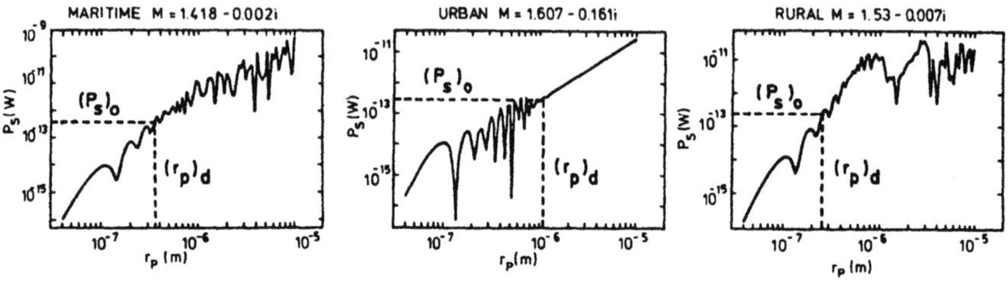

Fig. 1. Scattered light power as a function of aerosol size, measurement distance 100m, laser power 1 Watt

Taking into account the above-derived detection limit for the minimum photon rates necessary for signal processing, the computations resulted in the following conclusions: Single particles with diameters in excess of

a) maritime $d_p \geq 0.7$ μm

b) rural $d_p \geq 0.7$ μm

c) urban $d_p \geq 2.0$ μm

will scatter sufficient light to permit laser Doppler measurements. These results are valid for measurement distance of 100m, laser power 1 Watt and with the optical system described in section 3. Smaller particles will produce an electron pulse rate that does not permit the Doppler frequency to be resolved, larger ones will yield good evaluable signals.

This short survey indicates that the signal quality of real outdoor measurements will strongly depend on the aerosol distribution and concentration and thus on the weather conditions. Nevertheless, with a laser power of 10 Watt available in the control volume, measurements up to distances of 500m should be feasible.

3. Laser-Doppler Anemometer for Wind Velocity Measurements

3.1 Measurement Container

To ensure successful long-time outdoor measurements, a mobile measuring cabin was built up consisting of a standard size container (8 x 8 x 20 feet^3) with extra reinforcement to guarantee mechanical stability. This container is shown in Fig. 2.

A special optical table mounted in the container supports the transmitting and receiving optics and the laser. The table can be tilted by a hydraulic lifting system which also opens part of the roof for clearance of the outgoing beams. This arrangement allows measurements along an arc between -5 and +45 degrees to the horizontal and thus at different heights above the ground.

Fig. 2 Container for LDA wind velocity measurements

In the rear part of the container all units necessary for functioning of the complete system are concentrated in a control center for the laser power, the adjustment of the transmitting and receiving optics, the digital autocorrelator and the computer for the evaluation of the signal information.

3.2 Optical System

A schematic view of the optical system mounted on the top of the base plate of the optical table is given in Fig. 3.

Referring to this figure, mirror M1 takes the beam coming from the laser under the table and guides it to a short focal length, exchangeable lens L1, which focuses the beam onto a rotating diffraction grating. The grating splits the incident beam into two beams and provides frequency shifting proportional to the angular velocity of the grating. The two beams coming from the grating are directed by mirrors M2 and M3 through focusing lenses L2 (f=2250mm). Transmitting mirrors M4 cross the beams at the desired measurement distance. The focal point of the single beams is shifted into the crossing point by variation

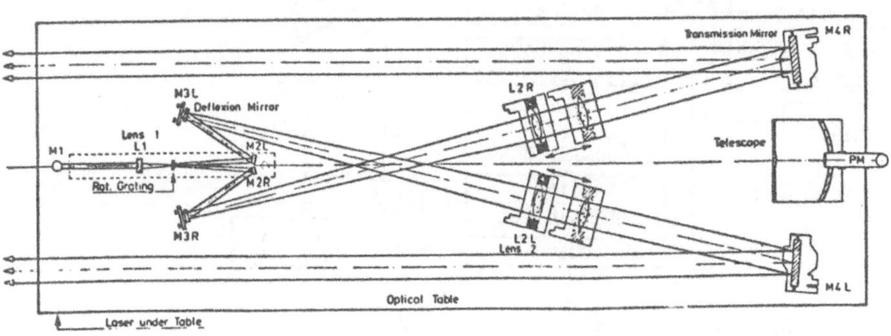

Fig. 3. Transmitting and receiving optics of the LDA container system

of the distance L1 to L2. This transmitting optics guarantees a beam diameter of between 2mm and 5mm inside the measuring control volume at distances from 50m to 500m.

The receiving optics is separated from the transmitting optics and consists of a Schmidt-Cassegrain telescope (C14, 14" diameter of clear aperture, f = 3910mm).

3.3 Signal Processing

Natural aerosols crossing the interference pattern of the measuring control volume scatter light which is modulated according to the intensity distribution inside the measuring volume. This light, detected by the receiving optics, results in electron pulse trains at the output of the photomultiplier. These pulse rates are fed into a digital autocorrelator (Malvern K7023, 50 nsec., 72 channels) which provides an autocorrelogram yielding the signal information to be separated from the noise. Provided that the laser intensity inside the measuring control volume is Gaussian and the turbulence of the flow is smaller than about 15%, then the autocorrelation function $G(\tau)$ is given by ABBISS [9] as

$$G(\tau) = A \exp\left(-\left(\frac{\tau}{T}\right)^2\right) \int_{-\infty}^{\infty} p(u) \left\{1 + \frac{\eta^2}{2} \cos \frac{2\pi u \tau}{\Delta x}\right\} du + B \quad . \qquad (3)$$

Here τ means the correlation lag time, T the mean flight time of the particles through the measuring volume, η_M the modulation depth of the fringe pattern, Δx fringe distance, A amplitude of laser beam profile and B background noise. The signal recovery is processed to determine $p_u(u)$, the probability distribution of the velocity. This is done by Fourier transforming the autocorrelation function in equation (3), which gives directly $p_u(u)$, and thus the required velocity information from which the mean value and turbulence intensity can be deduced.

The main components of the complete electronic system employed by the authors are summarized in Fig. 4. Details can be obtained from ref. [10].

4. Wind Velocity Measurements

With the complete system described in section 3, continuous measurements were made on the grounds of the University of Karlsruhe. Due to the container loca-

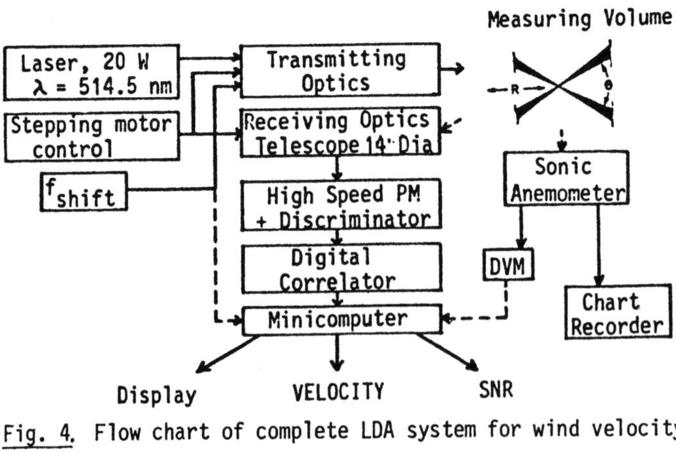

Fig. 4. Flow chart of complete LDA system for wind velocity measurements

tion, the measurement distance was limited to 106 m. A black wooden wall served as a beam stop about 20m behind the control volume to hinder the laser beams from propagating into traffic regions.

For comparison purposes, simultaneously with the LDA measurements, velocity data were taken with a commercial sonic anemometer, which was arranged around the LDA-measuring volume. An example of a time history of the crosswind component measured with both systems in a distance of 106m is demonstrated in Fig. 5.

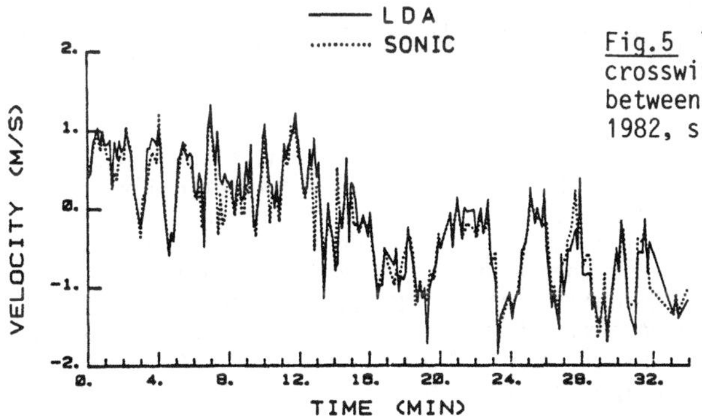

Fig.5 Time history of the crosswind component, comparison between LDA and SONIC data, 18.3. 1982, start 16:00 h

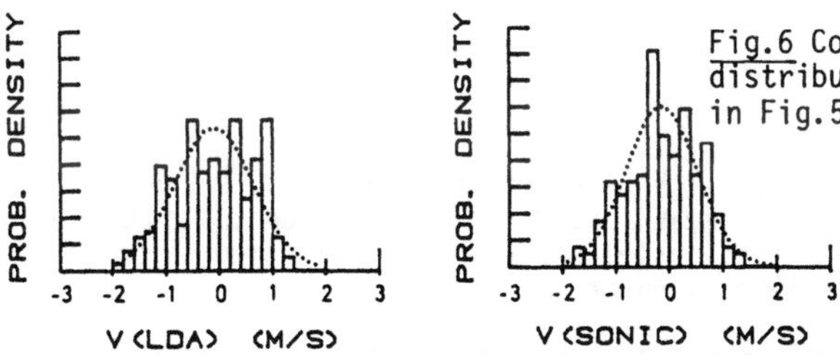

Fig.6 Comparison of the probability distribution of the wind field shown in Fig.5

This plot shows 256 measurement points, each representing the velocity of a time interval of 8 sec. Considering the differences between the measuring principles of the two instruments, the correlation of the results is obviously very good. This good agreement is confirmed by the probability distribution of the velocities given in Fig. 6.

A regression analysis of the cross plot in Fig. 7 yields more quantitative comparison results. The correlation coefficient of the data, indicating how closely the points cluster about the regression line, was found to be 94% indicating a high degree of agreement. The slope of the regression line of Fig. 7 was 1.06, the intercept 0.05 m/sec.

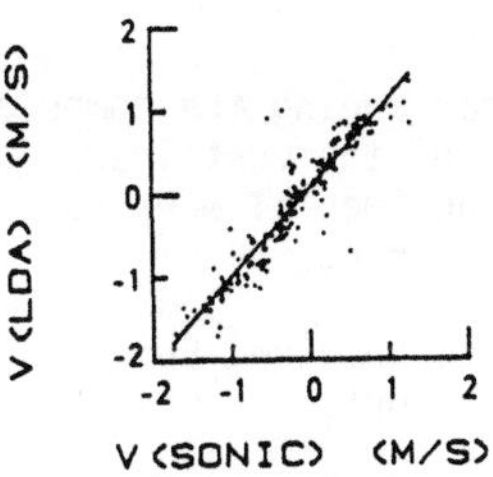

Fig. 7 Cross plot of wind velocities measured with LDA and SONIC anemometers

The analysis of the measurement example presented above showed the following results:

	LDA	SONIC
mean velocity (m/sec)	-.10	-.14
RMS-value (m/sec)	.74	.66

In addition to computing the frequency of LDA signals from the recorded correlograms, information was also obtained on the signal-to-noise ratio (SNR) after the Fourier transform. This SNR is necessary to eliminate signals of no or improper information content, for example if no large particle was crossing the measuring volume during the correlation time interval. 79.4% of all measurement points of the wind field presented in Fig. 5 contributed to the time history.

The signal quality is strongly dependent on the size distribution and the material of the available aerosols. These influences, as theoretically predicted in section 2, were confirmed by the experimental investigations. To quantify the particle size distribution during the LDA-measurement period, simultaneously particle size measurements were performed. The evaluation of these size distributions and the quality of the LDA signals showed, roughly speaking, the following dependencies: LDA measurements of wind velocities are restricted to large scattering particles, that means, the higher the concentration of aerosols larger than 1μm in diameter, the better the signal quality will be.

5. Conclusions

Based on theoretical predictions of light scattering properties, an LDA system was designed for crosswind velocity measurements and was set up and housed in a mobile measurement container. This system was used for long-time comparison measurements of wind fields at 106m distance.

Measurements were performed in winter and summer, at day and night, during rain and sunshine. At all times good velocity information could be obtained.

The signal quality was strongly dependent on the weather conditions and thus the time interval per measurement point had to be adapted to the prevailing conditions.

Though there is still a lot of work to be done to define the dependence of signal quality on weather conditions, the presented investigations proved the feasibility of the LDA method for crosswind component measurements.

6. Acknowledgements

The authors are indebted to B. Howe, who was engaged in rearranging the computer programs for signal evaluation and assisted the setting up of the container. The excellent mechanical work of R. Gerber has made the experimental work possible. Thanks are extended to G. Bartman who typed the manuscript.

This work has been financed by the Bundesamt für Wehrtechnik und Beschaffung in Koblenz. The authors gratefully acknowledge this support.

References

1 W.M. Farmer and D.B. Brayton, Applied Optics, Vol. 10. No. 10 (1971).

2 K.G. Bartlett and C.Y. She, Applied Optics, Vol. 15, No. 8 (1976).

3 L. Danielsson, Proc. 4th Int. Conference on Photon Correlation Techniques, Stanford, USA (1980).

4 P.J. Bourke and C.G. Brown, Optics and Laser Technology, Vol. 3, No. 23 (1971).

5 L.S. Wall, J. of the Opt. Soc. of Amerika, Vol. 64, No. 7 (1974).

6 H.J. Pfeifer, M. König and B. Koch, Rept. R127/78, German-French Research Institute ISL, Saint Louis, France (1978).

7 J. Abele and H. Raidt, Rept. Forschungsinstitut für Optik, FFO 1978/9, Tübingen, FRG (1978).

8 F. Durst, M. Macagno and G. Richter, Rept. SFB80/TM/195, Universität Karlsruhe, FRG (1981).

9 J.B. Abbiss, Physica Scripta, Vol. 19 (1979).

10 G. Richter, Entwicklung und Anwendung eines Laser-Doppler-Anemometers für Windgeschwindigkeitsmessungen, Dissertation, Institut für Hydromechanik, Universität Karlsruhe, FRG (1982).

11. The Use of Laser Velocimetry in the Full-Scale Ship's Environment

C. Norris

The British Ship Research Association, Wallsend, Tyne and Wear, England

Summary

The environmental conditions encountered on board ships pose severe problems for the application of laser velocimetry techniques. Typical problems encountered include severe vibration, noise, high humidity levels, an environment in which contamination of instruments by dust, grease, salt, etc. is difficult to avoid, restricted space and limited technical back-up facilities. The design of a L.D.V. system must take all of these factors into account, the end product being a system which is extremely rugged, relatively unsophisticated and yet capable of measuring flow velocities several metres into the water in which no control of the seeding is practicable.

This paper describes equipment developed at BSRA to meet these requirements and used to carry out, at sea, the measurement of flow velocities just forward of the working propeller. The equipment has a photon correlator as its central data processor and uses a real fringe system in the full back-scatter mode of operation. Measurements up to about 5 metres from the hull were successfully achieved and some specimen results are presented.

Some possible future marine applications of the equipment are considered and these include boundary layer and ship speed measurements.

1. Introduction

The flow around a ship and particularly in the stern region of a ship is very complex and is known to have a direct influence on the performance of the propeller and also on the oscillatory forces emanating from the propeller. These forces are transmitted to the hull and can sometimes cause severe vibration [1]. The shape of the submerged aft-end hull form influences the flow and considerable research on a worldwide basis has been directed towards improving the flow characteristics. BSRA in conjunction with other research groups have carried out full-scale and analytical studies to investigate the phenomena of propeller-induced vibration the results also being compared with scaled models. A major objective of the work was to study the unsteady cavitation of ships' propellers with a view to achieving design methods which would provide means of avoiding serious vibration problems in new ships.

One area of work has been the measurement of the total inflow velocity just forward of the working propeller, whilst complementary measurements of the induced hull pressures and cavitation extent were made. In order to carry out total inflow velocity measurements Laser Doppler Velocimetry was adopted as it had a number of advantages over previous methods.

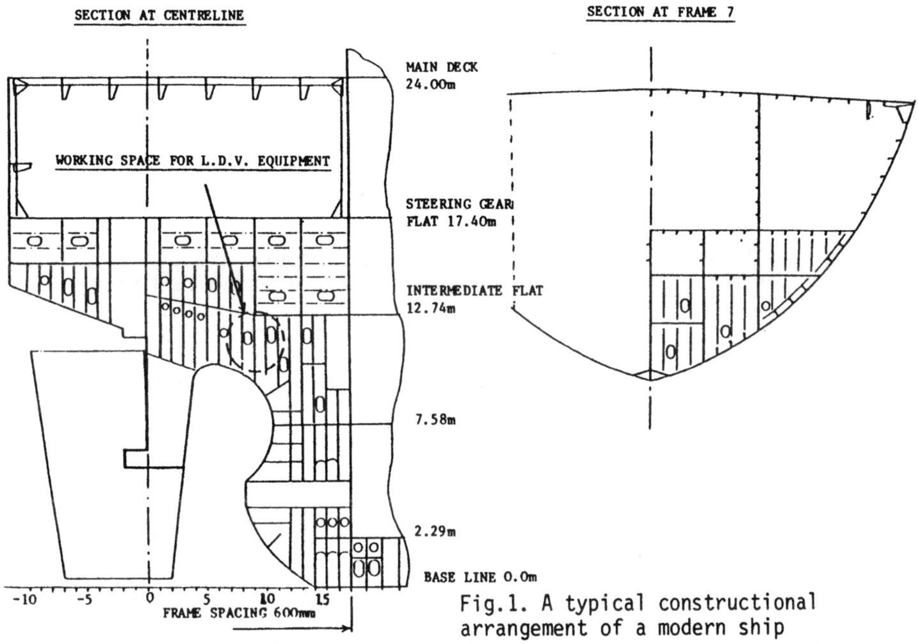

Fig.1. A typical constructional arrangement of a modern ship

Pitot rakes have been used previously but these have a number of obvious limitations and difficulties and at least in one case they have caused damages to the propeller.

A number of establishments have applied LDV to the measurement of flow around ship models [2] and in general it has been possible to use fairly straightforward adaptions of commercially available equipment as the water penetration required is generally of the order of 1 metre and side or forward scatter configurations are often feasible. Some controls of the seeding concentration are possible and the measurement can be carried out under controlled laboratory conditions.

In the full scale, however, the measurement is much more difficult not only because of the extended range required, typically water penetrations of 5m, but also because of the stringent environmental and other logistics problems involved. The aim of this paper is to describe the particular problems encountered and the equipment developed to overcome these problems. Some typical results are also presented.

2. Design Considerations

2.1 Environmental Problems

The constructional arrangement of modern merchant ships, see Fig.1, requires that when carrying out LDV measurements the optical equipment needs to be housed in a ballast water tank. In order to gain access to this space and to minimise structural modifications, which can involve difficult complications with classification society requirements, it is necessary that equipment be compact, typically no wider than 400mm. The aft-peak tank is normally a ballast water tank and is usually subjected to high levels of vibration, typically up to 0.5g and with a wide frequency spectrum from 0.1 to 10000Hz (sea motions to noise frequencies) and this vibration can not be fully isolated if relative movement between the LDV equipment (and thus measuring point) and the ship is to be maintained within tolerable limits. The environment in the aft-peak tank can be particularly uncomfortable with temperatures as high as 35°C and humidity of 100% in tropical areas. Bearing in mind that noise levels of 120dB(A) are normal in these tanks it is clear that the operators will not be able to work to normal technical standards and due account of this must be made when designing the equipment.

The sea state can affect the wake of a ship, and may vary considerably over the period of a day. It is desirable therefore that a full wake scan of about 150 points be carried out in a single work period of about 8 hours. Clearly, it is necessary that the equipment should hold optical alignment for at least a few hours at a time and that optical realignment should be easy and rapid and take no more than about 2-3 minutes.

2.2 Logistics Problems

The planning of a ship trial is a complex problem and it is desirable if possible to use equipment which allows the maximum flexibility in trials arrangements. It is necessary to have equipment which can be moved at short notice to almost anywhere in the world, by air, road or sea. The packaged equipment must be capable of being loaded by crane by unskilled and often unsupervised persons and stand on open deck in storm conditions.

It is desirable that the equipment should be operable by a team of only two or three persons who will almost certainly need to carry out a range of other tasks and therefore be non-specialist in the LDV technique. Technical back-up facilities at sea are virtually non-existent and therefore the operating team must be completely self-reliant. Maintenance, which in the laboratory may be considered routine, is often a major task in a ship because of the general lack of facilities.

2.3 Technical Problems

Because of the geometric arrangement of a ship it is necessary to adopt the full back-scatter mode of operation. The extended range requirements (focal lengths of about 7m in air are required) means that transit velocimetry is not feasible and it is necessary to adopt a real fringe configuration. The range of water speeds typically encountered in a ships wake is from about 2 to 10 metres/second.

Only very limited information is available regarding particle and air bubble sizes and concentrations [3] but from this it seems reasonable that a fringe spacing of 20-30μm is suitable. In order to maintain a scattered frequency of about 100kHz a beam separation of about 100-150mm is required.

The measuring position needs to be moved over a considerable area. Two possible solutions are possible. One is to use a zoom optics for transmission and receiving. This is an expensive solution and results in a complicated arrangement. It was decided to reject this solution as it may result in optical re-alignment problems. A second and more pragmatical solution is to use a fixed focal length transmitting optics and to move the optics within the ship. Beam steering is necessary to overcome structural obstacles within the ship and to move the measuring point over the required scan field.

Because of the colour absorption characteristics of water it is necessary to use blue or green light. The argon-ion laser is suitable; typically a 2 watt laser is required. These lasers however consume what in ship terms is considerable quantities of fresh water at controlled pressures for cooling purposes. The pressure fluctuations in the cooling water supply can be large and can cause difficulties with the laser operation. In tropical conditions it can be difficult to obtain sufficient quantities of cool water. The full back-scatter configuration combined with the fact that the equipment must operate in a range of water clarities (or particles size/concentrations) means that the equipment is likely to produce an inherently poor signal quality particularly as the optical alignment slowly changes. A photon correlator was therefore selected as the most suitable signal processor as it is inherently capable of handling with widest range of signal quality.

Some preliminary analysis of the measured data is necessary during the measurement period but more complete analysis of the data can reasonably be carried out ashore where the more complicated curve fitting routines can be used.

In order that the effect of ship motions on the total inflow velocity can be determined it is useful if estimates of velocity and turbulence can be obtained in as short a period as possible, say 1-5 seconds.

Longer sample periods may however be necessary if a better estimate of the mean wake is required.

3. Experimental Arrangement

3.1 General

To cover the required scan field various optical arrangements have been used. One such arrangement is shown in Fig.2. Variation of water penetration is accomplished by moving the laser trolley transversely across the ship. Various angles of view are selected by moving and rotating

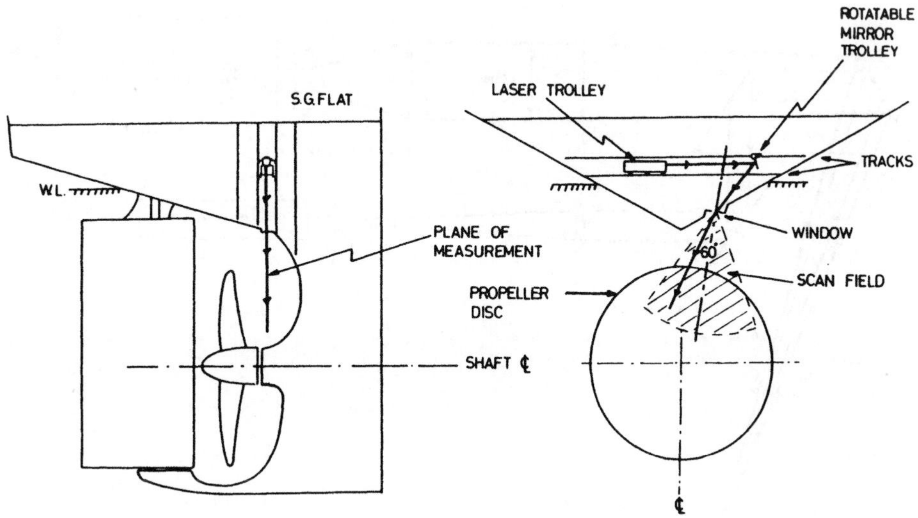

Fig.2. General arrangement of optics

the mirror on the other trolley. The viewing window is inclined to the shell of the ship to achieve the desired scan field. A scan field of approximately ± 30° in water is generally used requiring angles of incidence to the window in air of about ± 40°. Water penetration of about 5 metres have been used in previous studies and this will be increased to 7 metres in future work.

3.2 Transmission and Receiving Optics

The optical arrangement used is shown schematically in Fig.3. An Argon-Ion laser (Spectra-Physics Model 165) is employed capable of producing up to 1W at the wavelength of λ = 514.5nm with a beam diameter 1mm. After beam steering by two coated mirrors (M_1 and M_2), the polarization of the laser beam is adjusted by a half-wave plate so as to yield two equal intensity beams from the beam splitter at 52mm spacing. After reduction of this spacing to approximately 32mm by two prisms the beams are then passed through a beam expander to increase the separation to 135mm and the diameter to approximately 4.2mm. The parallel beams are then passed through a final plano-convex lens, L_3, of pre-selected focal length (typically 3.5 to 7.0 metres). The beams are then passed through holes of approximately 8mm diameter in mirror M_3. After final beam steering using a variety of optical arrangements the beams pass through a specially designed window into the sea. The scattered light is then transmitted by the beam-steering mirrors to mirror M_3 and M_4 and finally to a short focal length lens L_4 onto the pin-hole where it is detected by photomultiplier (Malvern Type RR127). The aperture of lens L_4 can be adjusted and typically a diameter of about 5cm is used.

This optical arrangement results in a measuring volume of about 1mm diameter and 40-70mm length with fringe spacings of about 20-30μm. Whilst this volume may seem large when compared with other work fields, it should be remembered that the aim of the work is to compare the

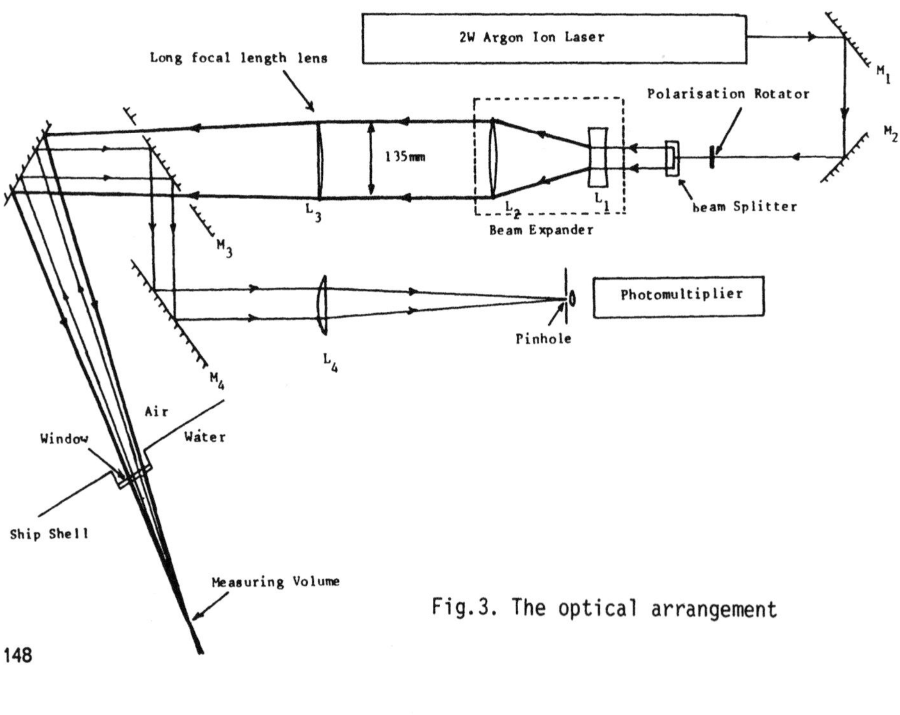

Fig.3. The optical arrangement

results with experimental model data where a typical scaling factor would be 30:1 and thus the volume is sufficiently small for this purpose.

The equipment is normally arranged to measure one flow component only but by simply rotating the beam splitter by 90° the orthogonal component can be measured if desired.

The whole of the optical arrangement is very rugged and can be lifted on a block and tackle and moved into position under sea-going conditions without loss of alignment. In any event, after preliminary setting up, only 6 thumbwheel adjustments are employed to realign the optics occasionally; this operation normally takes only a minute or two to carry out. Two further adjustments exist on the laser which require retuning at approximately 2-hourly intervals.

The various optical components have been carefully selected for operation in the ship environment so as to produce a non-critical design. For example, it has been found that it is possible to rotate the beam splitter several degrees relative to the optical axis without moving the measuring volume significantly.

3.3 Signal Processing Electronics

The signal processing equipment is shown in Fig.4. The photomultiplier signal is analysed using a Malvern correlator Type K7023. The correlograms can be analysed on board using a Malvern Velocity and Turbulence Microprocessor, which uses a simple turning point algorithm data analysis routine, to give first estimates of the velocity and turbulence levels. The correlograms are stored on a digital cassette recorder for more complete analysis ashore using a five parameter curve fitting routine.

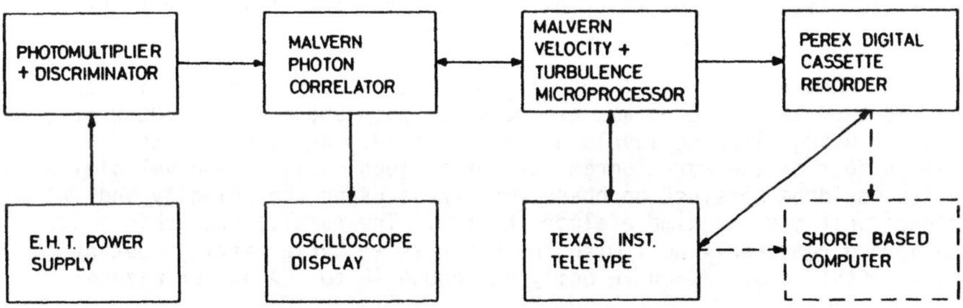

Fig.4. Schematic arrangement of signal processing equipment

4. Measurements

A typical measurement programme would consist of measuring the ships wake on both the port and starboard sides of the hull. The wake would be measured approximately 1 metre forward of the propeller and over as large an area as possible depending on the geometrical constraints of the hull form. Similar measurements would be repeated at various forward ship speeds.

A complementary programme of measurements would be conducted simultaneously and would typically involve the measurement of the propeller rotational speed, main shaft torque, vibration and pressures around the propeller arch. High speed ciné and stereo photography of the propeller cavitation are also carried out to study the growth and collapse mechanism. Recordings of the ship's speed, ship's heading, roll, pitch and rudder angle are usually recorded as well as other environmental parameters such as wind speed and sea state being noted. The hull roughness of the ship is also measured either in dry-dock or afloat using BSRA's hull roughness gauge [4].

The measurement equipment is normally installed on the ship during loading operations, the structural alterations having been carried out during a dry-docking or building period. Setting up would normally take 24-36 hours.

A typical wake survey may involve the measurement of about 150 points on either side of the ship with water penetration of 0.5 to 5 metres and would require about 4-6 hours for each side.

A complete trial, involving about three such surveys at various ship speeds, can usually be accomplished, given suitable weather conditions, in a 7-to 10-day voyage.

5. Results

Successful measurements have been completed on a number of ships trading in various parts of the world including the Mediterranean, Caribbean, English Channel, West Coast of Africa and the North Sea.

Given the quantity of information obtained it is not possible to discuss the results in detail here. Some typical results are however presented to give some indication of the type of data obtained from the LDV work.

Figure 5 shows some typical correlograms obtained from one particular trial. Most of the correlograms are obtained with about 200 to 300mW of laser power using clipping levels of about 1 to 4. As will be noted from Fig.5 most of the correlograms are of a 'good quality' and velocity estimates could be obtained on board from these using the velocity and turbulence unit and the simple algorithm fit. The results from this method agreed remarkably well with the 5-parameter curve-fitting routine generally within about 2% on velocity and about 10 to 15% on turbulence.

The 'moderate' specimens could not be fitted by the microprocessor, the odd point being 'out of position'. The curve-fitting routines could be used with some confidence on this type of data. Some of the correlograms (about 3-4%) were of a 'poor' quality and could only be fitted for mean velocity with any degree of confidence.

Some of the correlograms (about 1-2%), most of which came from an area just forward of the propeller tips, were unintelligible. The reason for these poor correlograms is not understood and requires further investigation. The time required to obtain a "satisfactory" correlogram varies depending on the position in the wake but under ideal conditions takes as little as 1 second at 5m penetration.

The spatial distribution of velocity estimates is reduced to an isowake form of presentation as shown in Fig.6.

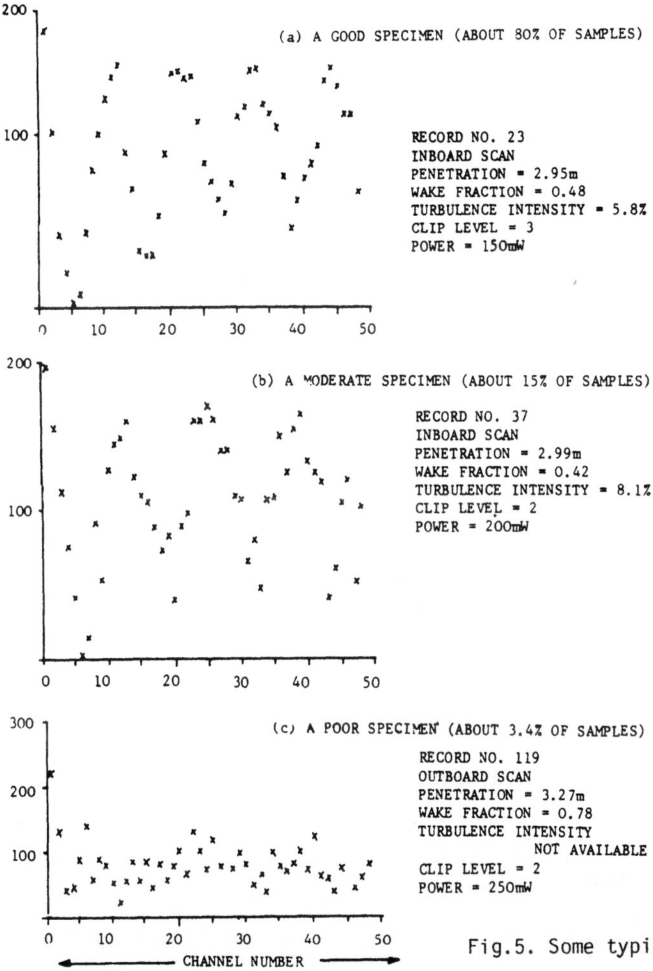

Fig.5. Some typical correlograms

6. Conclusions

The work carried out so far has demonstrated the practicality of installing LDV equipment on ships and measuring water velocities up to 5.0m range. The results obtained have formed a valuable basis on which to assess analytical and modelling techniques. From the experiences gained one may suggest that measurements by means of the techniques used here would be feasible up to 10.0m with fairly small improvements to the equipment. It is also clear that the measurement of a ship's boundary layer would be a fairly simple task for this type of equipment.

Being able to measure at these ranges leads to the suggestion that a highly accurate laser speed log could be constructed to measure the speed through the water, a measurement which is so important for the detailed study of a ship's performance. Such a device could be fitted as a standard or for more research orientated investigations.

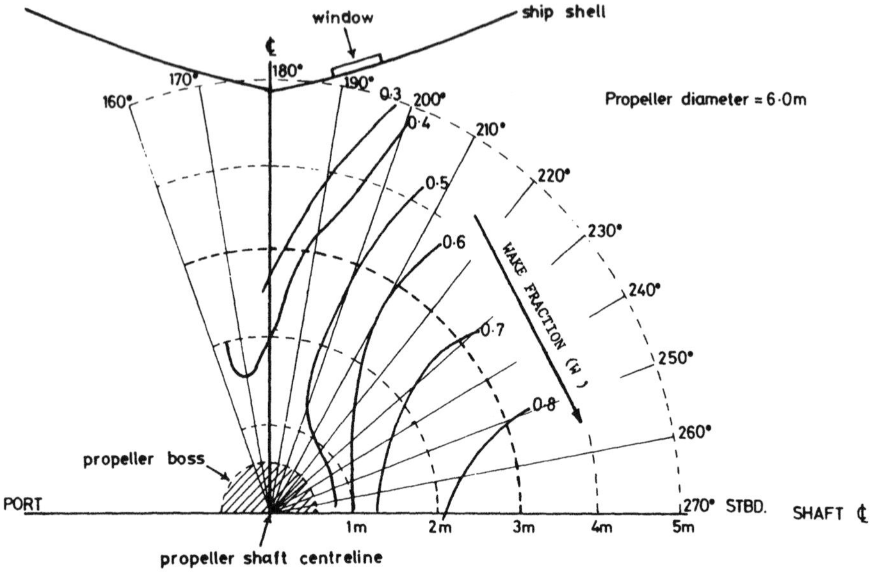

Fig.6. A typical isowake

7. References

1. Fitzsimmons P.A. Symp. Propeller Induced Ship Vibration. R.I.N.A. London 1979.
2. Kirschneck E. and Landau J. Schiff und Hafen, Heft 4, 1981, 33.
3. Weitendorf E.A. et al. 12th Symp. Naval Hydrodynamics, Washington 1978.
4. Byrne D and Ward G. Inst. Mar.Eng.Conf. Priorities for Reducing the Fuel Bill. March 1982.

12. Wake Measurements by Laser Doppler Velocimetry on Full-Scale Ships

J. Kux and H. Stöhrmann

Institut für Schiffbau, Universität Hamburg, Lämmersieth 90
D-2000 Hamburg 60, Fed. Rep. of Germany

Summary

The application of Laser Doppler Velocimetry to the measurement of the axial velocity component in the wake field of full-scale ships by the Institut für Schiffbau is reviewed. Typical results obtained on one voyage are presented.

1. Introduction

This contribution, conceived originally as a discussion to the paper by C. NORRIS [1], could not be presented during the meeting due to the tight schedule. It is included now in the proceedings by kind permission of the editor.

At the Institut für Schiffbau (IfS) the development of a Laser Doppler Velocimetry (LDV) system designed to measure the velocity in the flow field around full-scale ships was started in 1975 (plans dating back to 1973) and its performance was since tested on four different ships. First measurements in a ship boundary layer [2] were presented at the International Symposium on Ship Viscous Resistance in Göteborg 1978 [3]. In the present report only the measurements on the 'Sydney Express' 1980/81 [4] will be reviewed. Similar wake surveys have been obtained on the 'St. Michaelis' in 1981. Results have been presented in Lisbon at the International Symposium on Applications of Laser Doppler-Anemometry to Fluid Mechanics, 1982, [5].

The flow around the ship being as complex as it is, measurements of the velocity in the flow field of various and different vessels will be necessary, since only with such data sets numerical methods and modelling assumptions may be thoroughly tested. The more information acquired in the experiment, the better these tests. The Reynolds numbers for wake flow of model and full-size ships differ by three orders of magnitude. Hence one should not expect that wake flows in both cases are related by simple similarity rules. Instead one is left to finding empirical rules for the prediction of the wake flow of a full-size ship with given hull shape, from data on a similar small-scale model. To derive such rules there is the need for detailed wake surveys at different scales. In addition, wake flow data are required for a realistic prediction of propeller performance.

We fully share the point of view of NORRIS on the importance of such measurements and agree with his remarks on the environmental and logistics difficulties. We also agree with his description of technical problems. Our approach differs from his by the choice of a zoom optics and by incorporation of a Hewlett-Packard HP 1000 minicomputer into our system; it allows data analysis immediately after acquisition on board including curve fitting and isoline plotting.

2. Experimental Arrangement

2.1 General

Figure 1 shows the location of the measuring plane and the propeller, the details of the arrangement of the system optics are shown in Fig. 2. Three windows had been installed in the shell, mounted flush with the outer surface of the hull. Variation of water penetration is achieved by a variation of the focal length of the zoom optics while the different angles of view are selected by tilting the mirror in front of the window. Figure 3 shows a typical pattern of measuring points (some of them omitted in the region near to the windows). To eliminate the influence of the oblique orientation of the windows, prismatic water basins with horizontal top cover plates were installed on top of them.

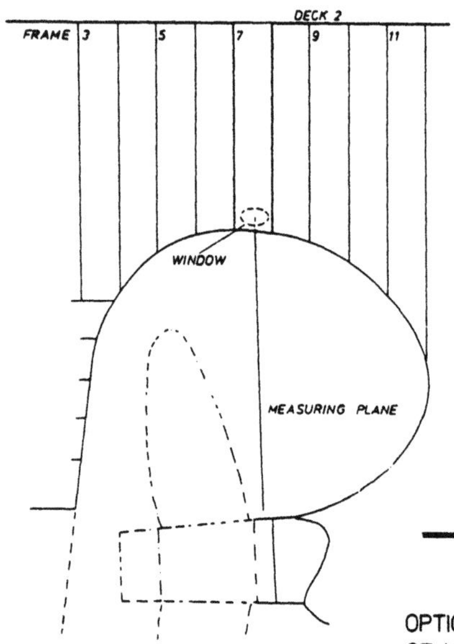

Fig. 1. Section at centerline showing location of measuring plane and frames

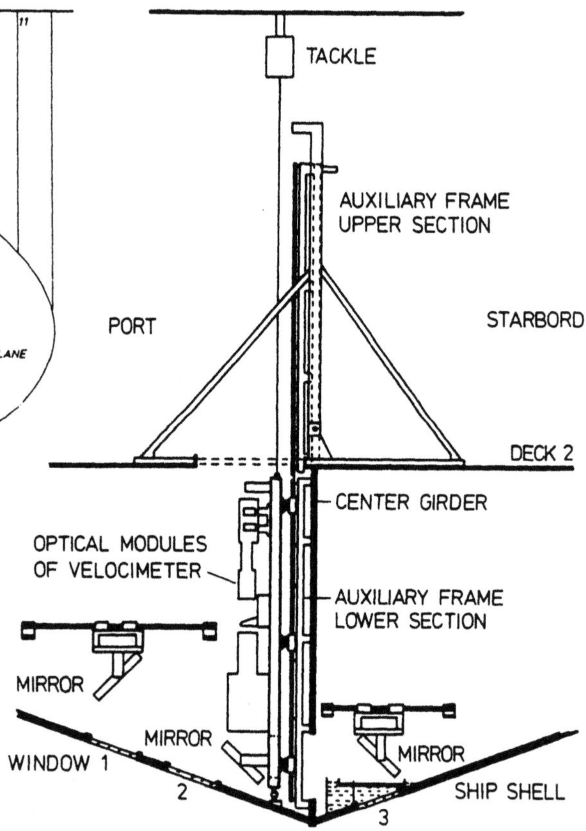

Fig. 2. Section parallel to frame showing arrangement of windows, mirrors and set up of system optics

2.2 Optics

In order to reduce the effects of the high vibration level, the optical system was designed to consist of several integrated moduls and these were

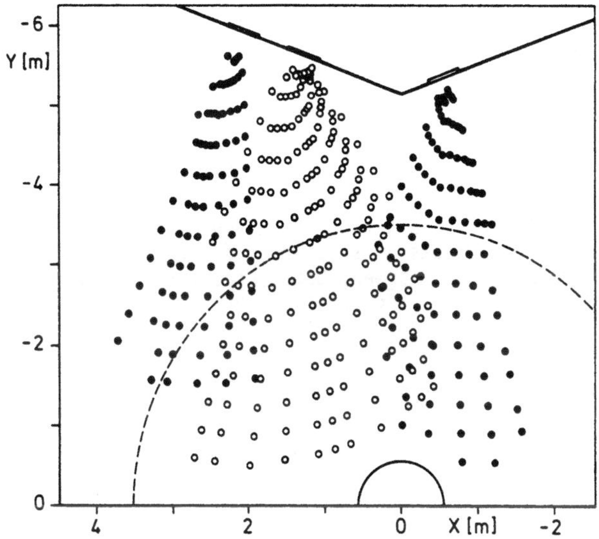

Fig. 3. Typical pattern of measuring points

mounted on a vibration-isolated base plate. The modular approach made it easier to deal with the space restrictions of different ships.As shown in Fig. 2, we used the same laser as NORRIS, however in a vertical position.

The central optical module is a modified standard two-colur device from DISA ELEKTRONIK (55 X 00). A specially designed zoom optics with front lens diameter 300 mm was placed in front of it. It allows a continuous variation of the focal length and has been designed to keep the intersection angle constant under changes of focal length. A measuring volume of similar size as reported by NORRIS is obtained, its diameter is about 6 mm, its length about 100 mm. The device has been conceived as a two-component velocimeter but on the 'Sydney Express' only the axial component of the wake velocity was measured.

2.3 Signal Processing Electronics

On the first journey /2,3/ where the ship boundary layer was investigated, the heart of the signal processing electronics was either a correlator or, alternatively, a transient recorder. For the present journey, a counter-processor (55 L 90a by DISA) was chosen. Signals of a quality that was acceptable for the validation algorithm of the processor were received at satisfactory rates. The counter easily permits time-gating of received Doppler bursts in relation to the angular position of the propeller. After some redesign, a correlator can also be adapted to this mode of operation/6/.

2.4 Computer

To process and store the digital information, an HP-1000 computer had been integrated into the system. This made it possible to obtain processed values during the measurement. The computer is indispensible for getting time resolved velocity data which reveal the periodicity with propeller rotation. It does the appropriate bookkeeping of the accepted Doppler bursts. In addition it performs the bias correction, statistics of the results, etc., it records the ship speed, torque and r.p.m. of the propeller. Its software contains routines to fit curves and to draw isolines. Mini-cassettes - their drives are integral parts of the display terminals - have proven to be well suited for our data storage requirements.

3. Measurements

The measurements on the "Sydney Express" took place on a voyage to Australia between Suez and Fremantle.Small variations in ship speed are unavoidable. The results were grouped according to ship speed. Data rates varied between 10 events per second (close to the window) and about 0.5 events per second (at points of maximum penetration). Note that the data rate depends highly on the margins one prescribes for the validation algorithm of the processor. The evaluation of the data includes a correction for the velocity bias (more particles of higher velocity in a turbulent flow contribute to the mean value than particles with lower velocity).

4. Results

4.1 Averaged Axial Component of Mean Velocity

Fig. 4 is a typical isoline pattern for the axial component at a ship speed of about 10.5 m/s. A separate pattern of isolines was obtained from measurements through each of the three windows. These patterns overlap in regions just above the shaft and about 1.5 m to port. The degree of matching in these regions may be taken as an indication of the accuracy and reliability of the results. Note that the coordinate origin is at the shaft axis.

4.2 Instantaneous Mean Axial Velocity as Related to the Propeller Angular Position

For these measurements, the instantaneous angular position of the propeller was measured by a shaft encoder and, in the computer, this value was attached to every validated velocity value. From these stored data the mean velocity corresponding to particular shaft positions is built up. This measuring procedure is economical since no valid velocity data are discarded. Valid data would be wasted if one would accumulate velocity data for one given shaft position at a time. The variation of mean velocity as a function of shaft angle is given in increments of the angular sampling interval. In the example shown in Fig. 5, a full revolution of the propeller is divided into 62 increments. The abscissa shows the corresponding angles. The data refer to a ship speed of 9.6 m/s and to a position close to the shaft as indicated on the insert to Fig.5. The data reveal an essentially periodic variation of velocity which is caused by the five blades of the propeller.

4.3 Turbulence

A rough pattern of the turbulence distribution is presented in Fig. 6. To be more precise, the plot shows the relative mean square fluctuations of axial velocity. These data are long-time averages and include contributions of genuine turbulence and contributions of time-periodicity that are induced by the propeller.

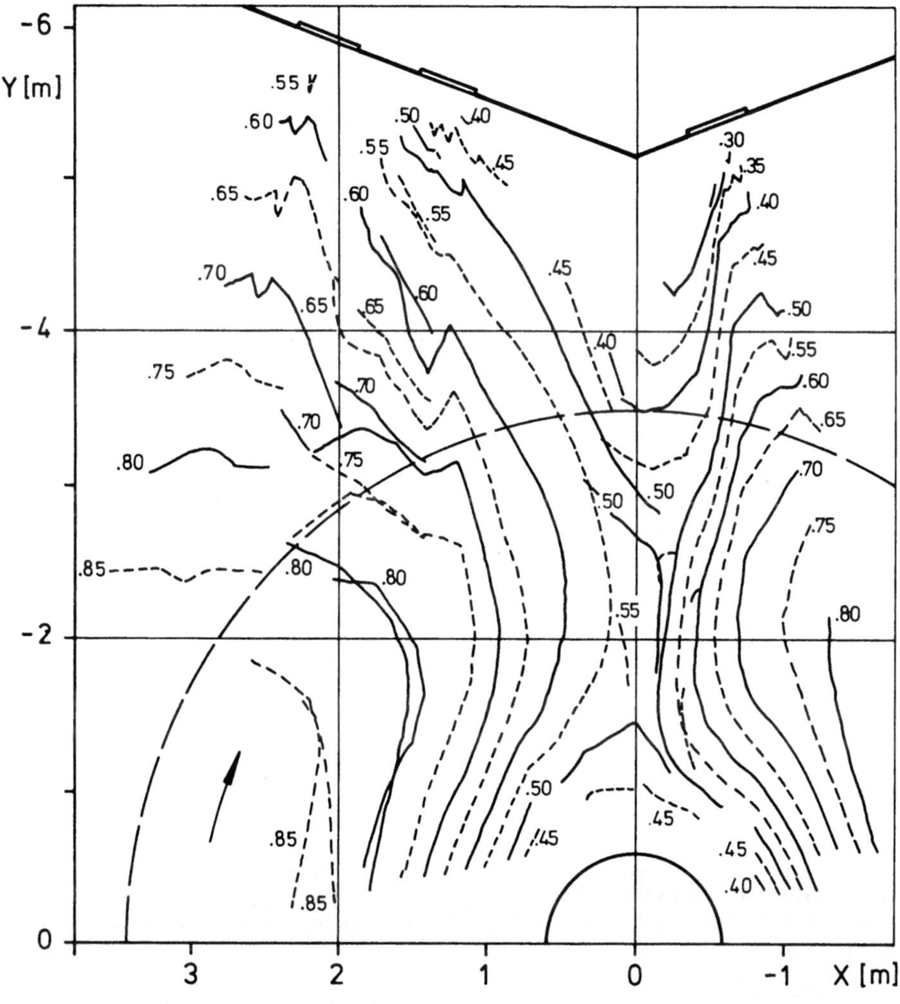

Fig. 4. Isolines of axial velocity component (as fraction of ship speed) as obtained from several runs (ship speed 10.5 m/s)

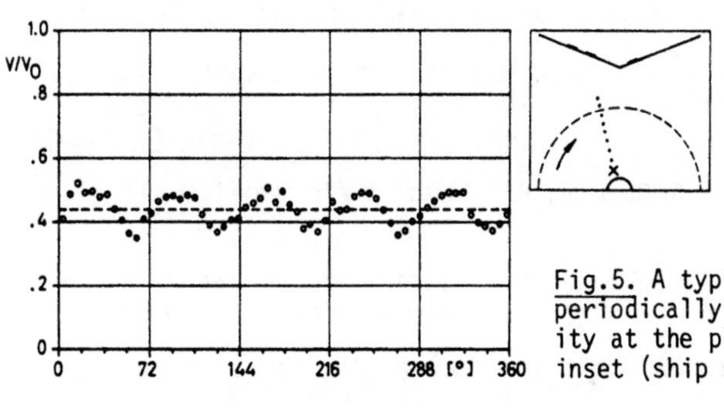

Fig.5. A typical measurement of the periodically instationary mean velocity at the place indicated in the inset (ship speed 9.6 m/s)

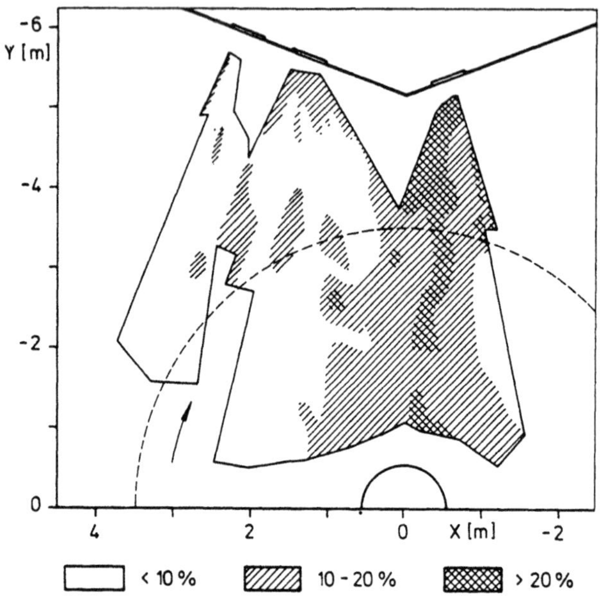

Fig. 6. Distributation of turbulence in the measuring plane (ship speed 10.5 m/s)

5. Conclusions

The results presented in this discussion demonstrate the feasibility of LDV measurements on board of full-size ships. The importance of realistic velocity data for further advances in the art of ship building, particularly of hull and propeller, cannot be overemphasized. Further measurements in the wake, especially in the immediate vicinity of the propeller, and in the boundary layer, especially in the region of three-dimensional separation, are required for a better understanding of the complicated hydrodynamic problems that control the efficiency and economy of ocean-going ships.

6. References

1. Norris, C. 5th International Conference on Photon Correlation Techniques in Fluid Mechanics, Damp 2000, Germany 1982, proceedings, Springer-Verlag.

2. Kux, J., Niemeier, Th. Institut für Schiffbau, Bericht Nr. 361, 1977.

3. Kux, J. International Symposium on Ship Viscous Resistance, Göteborg, 1978, Schweden, proceedings, Liber Distribution, S-16289 Vällingby, Schweden.

4. Kux, J., Niemeier, Th., Stöhrmann, H. Institut für Schiffbau, Bericht Nr. 424, 1982.

5. Kux, J., Stöhrmann, H. International Symposium on Applications of Laser Doppler Anemometry to Fluid Mechanics, Lisbon, 1982, proceedings.

6. Stöhrmann, H. Institut für Schiffbau, Bericht Nr. 380, 1979.

13. Turbulent Engine Flow Measurements Using LDA with Correlator

E. Max and G. Fernström

Applied Physics, 6100 AB VOLVO, S-405 08 Göteborg, Sweden

Abstract

An LDA system for in-cylinder gas velocity measurements in engines is described. The special feature of this instrument is the use of photon correlation processing in measurements with high repetition rate. Velocity values can be taken continously with a repetition rate up to 20 kHz. The data is stored in a mini computer and evaluated afterwards.

In-cylinder measurements in a motored diesel research engine have been done. Accurate single cycle measurements were obtained in consecutive engine cycles. Both cycle-to-cycle variations and turbulence could be observed in the gas flow and separated from each other.

1. Introduction

The performance of a combustion engine is strongly dependent on the air flow behaviour inside the cylinder, see,e.g. [1]. In a direct injected diesel engine, for example, the swirl is important for the fuel mixing with the air. In a petrol engine the gas flow at the spark plug is of special interest [2]. Furthermore the small-scale high frequency variations in the flow, called turbulence, affect the combustion process. Usually, the flow is divided into a time varying mean flow, describing the large-scale flow, and the turbulence [3]. The mean flow is however not exactly repeated every engine cycle and that is often called cycle-to-cycle variation. (Different definitions of these parameters exist in the literature.)

The in-cylinder gas flow is not well understood and also very difficult to predict. Therefore measurements are needed and used for several activities, e.g. to understand the behaviour of the flow, for modelling work, and as a tool in the design work of engines. In all cases mean flows, turbulence, and cycle-to-cycle variations can be of interest. To be able to separate turbulence and cycle-to-cycle variations single cycle measurements with high crank angle resolution have to be made [3].

The measurement technique mostly used now days is the Laser Doppler Anemometer technique (LDA). This is a non-contact optical technique very suitable for measuring complex gas flows, see e.g. [4-5]. In this paper only a short description of the LDA technique itself will be given. Instead a specific LDA system used for incylinder measurements will be described. The special feature of the actual system is the high measurement rate obtained using a digital correlator as signal processor. Examples of measurement results are given in the last part of the paper.

2. The LDA technique

The basic principle of an LDA system for one dimensional measurements is shown in Fig.1. Two laser beams are crossing each other in the measurement volume producing an interference fringe pattern [4]. Small particles, suspended in the gas, are passing through the measurement volume where they scatter the laser light. A typical light burst from a particle passing through the center of the measurement volume is shown in Fig.2.

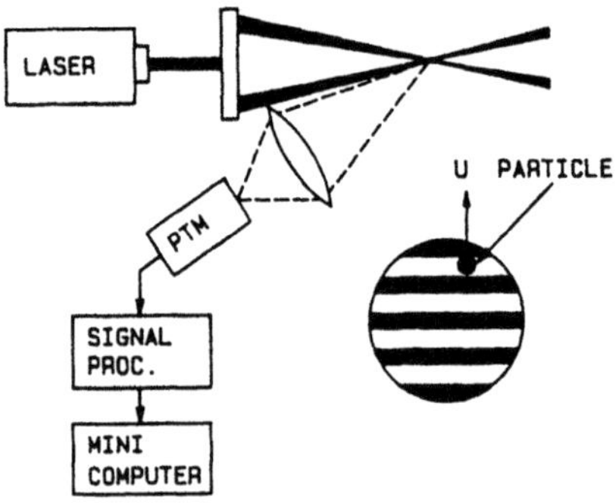

Fig.1. Basic principle of a LDA system

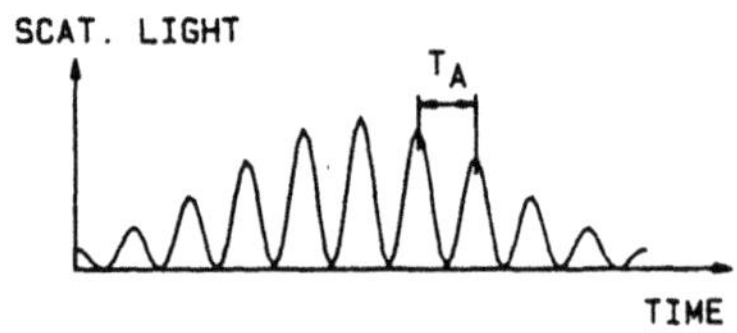

Fig.2. Light burst from a particle

The velocity component U, in the x direction perpendicular to the fringes, is given as

$$U = \frac{s}{T_A} \quad , \tag{1}$$

where s is the fringe spacing and T_A is the period time defined in Fig.2.

To be able to distinguish between positive and negative velocities, one of the two laser beams is frequency shifted Δf [Hz] relative-the other. The fringes will then move in the x direction with a velocity U_0 given by

$$U_0 = s \cdot \Delta f \quad . \tag{2}$$

Thus the particle velocity is evaluated from Fig.2 as

$$U = s(\frac{1}{T_A} - \Delta f) \quad . \tag{3}$$

Some of the scattered light is detected and the detector output is coupled to a signal processor. If the scattered light from a particle is sufficiently strong, the detector output signal can almost have the "perfect" shape as in Fig.2. In that case a counter, actually counting the frequency in the detector signal, works very well. If the signal, for some reason, is very weak and/or disturbed, the signal quality may not be good enough for the counter. Then a digital correlator, able to produce useful data out of less than one hundred detected photons, can be very suitable [6]. Weak detector signals may occur when back-scatter light is detected, small particles are used, or small detector aperture optics have to be used [4].

In an LDA system a mini computer is often used for data acquisition, reduction and control.

3. LDA system

A block diagram containing the major parts of the LDA system is given in Fig.3. A photograph of the optical unit and the engine is shown in Fig.4 and the correlator, interface and computer is shown in Fig.5.

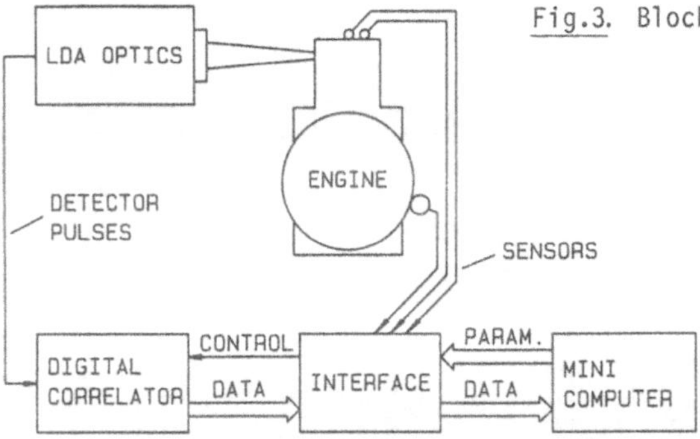

Fig.3. Block diagram of the LDA system

Fig.4. Optical unit and engine

3.1 Optics

In Fig.6 the optical unit (based upon Malvern 4772) is shown. The two beams are frequency shifted 80 MHz resp. 86.25 MHz by two Bragg cells. The fringe spacing is s = 16 µm which gives, according to (2), a fringe velocity U_0 = 100 m/s. The laser power used is 0.3 W at the wavelength 488 nm. Beam diameter at the measurement point is 150 µm.

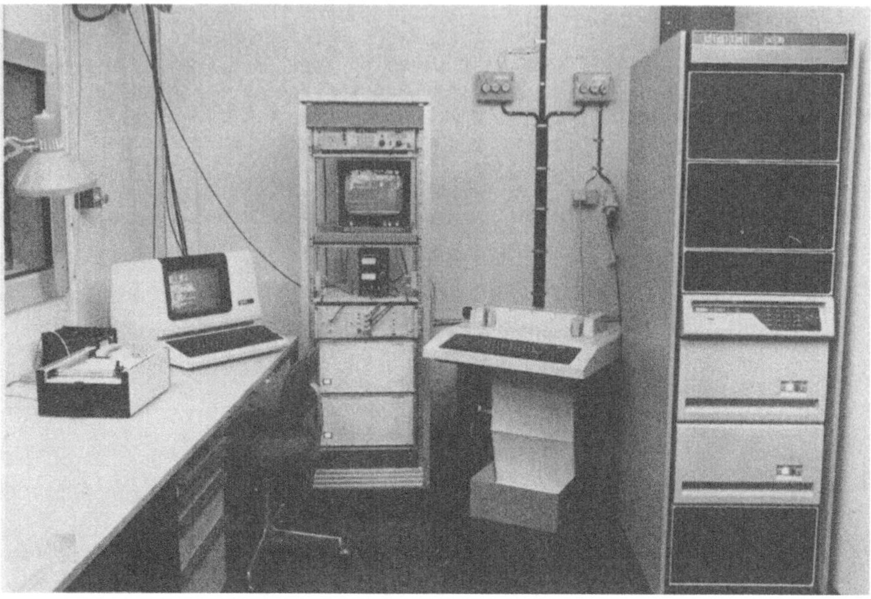

Fig.5. Correlator, interface and mini computer

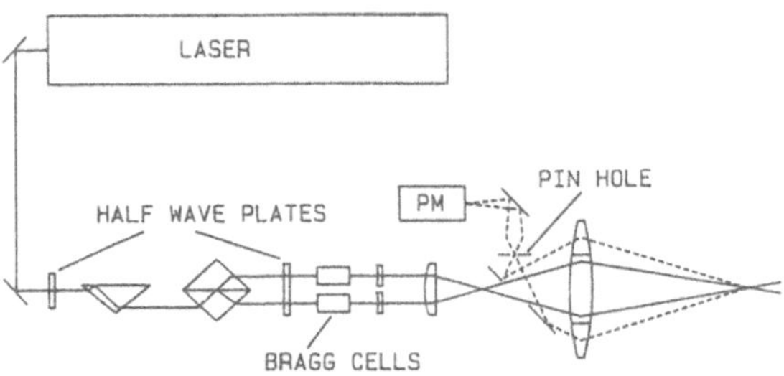

Fig.6. Optical unit

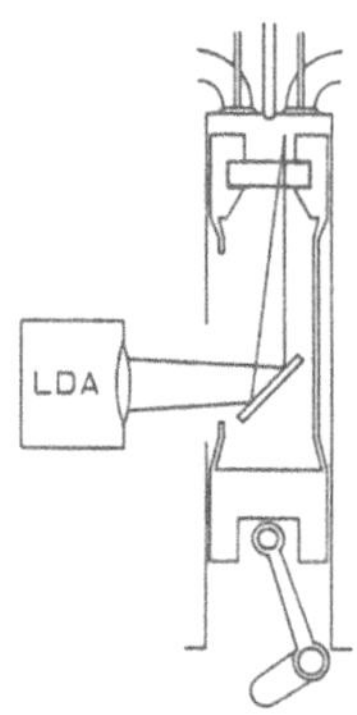

Fig.7. Engine setup

3.2 Engine Setup

The engine used in these measurements is a single cylinder diesel research engine with a cylinder diameter of 85 mm. To get optical access to the cylinder space a double piston arrangement shown in Fig.7 is used. The upper piston is equipped with a quartz window in the bottom and connected to the original lower piston through a sliced spacer. The laser beams are directed into the cylinder space via the mirror located between the two pistons. The mirror is connected to the laser optics and by moving the whole optical unit the space in the piston bowl can be probed.

3.3 Digital Correlator

The signal processor is a 20 ns digital correlator (Malvern K 7026). For a detailed description of a correlator of this kind, se e.g. [7]. A summarized description is as follows. The measurement time T, which are used for measuring one velocity value, is divided into N intervals each with a length of $\triangle T$. (For the correlator used here, $\triangle T$ is a multiple of 20 ns.) If one (or more) photon is detected in the i:th interval ($1 < i < N$) the detector signal $S_i = 1$, else $S_i = 0$. In real time the correlator computes the auto correlation function G_j of S_i as

$$G_j = \sum_{i=j+1}^{N} S_i \cdot S_{i-j}, \qquad j = 1, 2, \dots N_{CH} \quad , \tag{4}$$

where N_{CH} is the number of channels in the correlator ($N_{CH} = 64$ in our case). At the end of the measurement time T, the autocorrelation function G_j is available for read out and a new measurement can immediately be started.

3.4 Interface Electronics and Computer

To syncronize the measurements with the engine crank angle and to store all the correlator data a mini computer (PDP 11/34 incl. two RL02 disc-memory) and special interface electronics are used.

First the computer asks the operator to give all the parameters necessary to define the experiment, e.g. crank angle window resolution for one individual velocity point (0.1^0 - 25^0), crank angle range (0.1^0 - 720^0) in the engine cycle, and the number of cycles. The parameters are then transmitted to the interface electronics where they are buffered.

For each crank angle window (with a duration of T [s]) one auto correlation function, ACF, is produced by the correlator. During the time for one ACF to be produced the ACF from the preceeding crank angle window is read out from the correlator. The ACF is given in 64 points and every point has a dynamic range of one to four 8-bit bytes. The interface reads out the desired part of the correlator store according to the measurement parameters. Usually all the 64 ACF points and the least (or the two least) significant bytes are used. One byte is transfered from the correlator every 1.0 µs.

The interface electronics also include a data acquisition system for sensors, measuring, e.g. cylinder pressure. These data are sampled for every crank angle window and transfered, together with the correlator data, into the computer.

The data transfer can proceed continuously until the computer memory has been filled up which occurs after 200 kbyte of data. The measurements then have to make an interuption of less than one second while the computer memory is dumped to the disc. (The disc can store 10 Mbyte of data.)

3.5 Data Processing

All the ACF data from an experiment is processed afterwards in the mini computer. The method used is a curve-fitting procedure that assumes a constant velocity during the measurement time T. First the measured ACF is high-pass filtered. The result, A_j ($j = 4, 5, \dots 60$), is then compared with a set of 64 theoretical ACF curves $B_{j,k}$ ($j = 4, 5, \dots 60$; $k = 1, 2, \dots 64$) based on velocities U_k (-50 m/s $< U_k <$ 50 m/s). The theoretical ACF is taken from [6] and is an approximation of the high-pass filtered ACF as

$$B_{j,k} = e^{-\left(\frac{U_k \cdot j \cdot \Delta T}{r}\right)^2} \cdot \cos\left(\frac{2\pi(U_k + v_0)}{s} \cdot \Delta T \cdot j\right) \quad . \qquad (6)$$

Curve fitting numbers M_k are calculated as

$$M_k = \frac{\sum_{j=4}^{60} A_j \cdot B_{j,k}}{\sqrt{\sum_{j=4}^{60} A_j^2 \cdot \sum_{j=4}^{60} B_{j,k}^2}} \quad , k = 1, 2, \ldots 64 \quad . \qquad (5)$$

An example of this procedure using data from a real measurement is shown in Fig.8. The peak value M_{peak} of M_k is thereafter accurately located using a closer spaced set of theoretical curves around the peak in M.

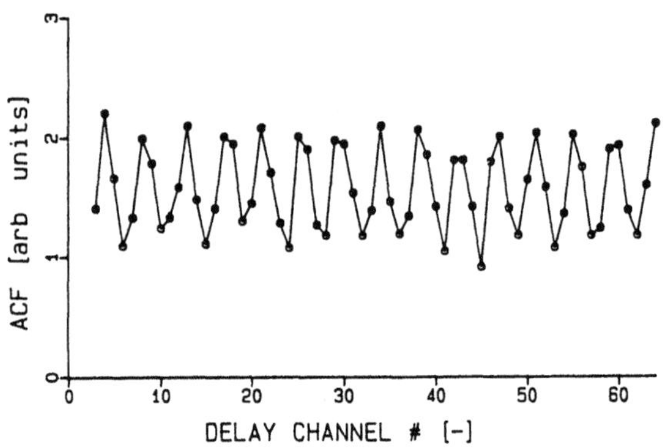

Fig.8a. Example of an auto correlation function

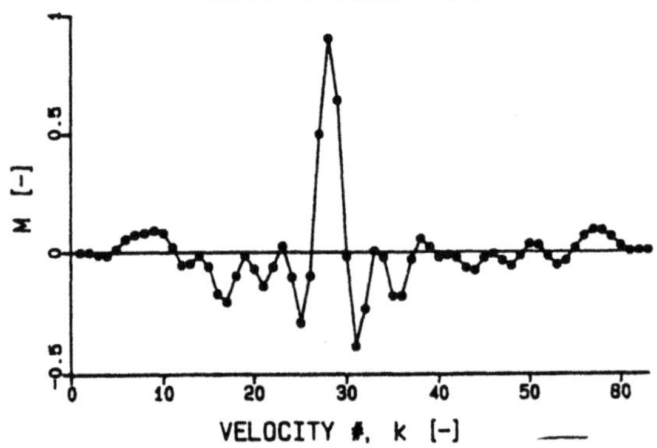

Fig.8b. Curve-fitting numbers M_k resulting from the autocorrelation function in Fig.8a

The peak valve M_{peak} is used to make an estimate of the uncertainty ΔU in the calculated velocity U. An analysis has lead to

$$\Delta U = \frac{1}{2} \cdot \sqrt{\frac{1}{M_{peak}^2} - 1} \quad . \qquad (7)$$

This formula has been checked by making computer simulations of a large number of auto correlation functions giving different $M_{peak}/\Delta U$ results.

4. Measurements

During the in-cylinder measurements, the engine was motored at a speed of 15 rps. The horizontal locations of the measurement points were chosen according to Fig.9. The vertical location was always 15 mm below the cylinder top. For every point the velocity was first measured in the x direction and then in the y direction for 48 cycles each. The crank angle resolution was 2 degrees and the correlator parameter ΔT was 40 ns.

TiO_2 particles were used to seed the air. Powder of 0.2 μm particles was dispersed into the air by a fluidized bed (TSI 2400). The seeded air can not be produced with the required flow rate and was therefore buffered in a large plastic bag before the measurement was started.

The x component of the velocity in point 31 is shown in Fig.10. The mean curve is obtained by evaluating the velocities from the ensemble avergage of all the 48 auto correlation functions. The uncertainty ΔU according to (7) is typically less than 0.3 m/s for the mean curve and less than 0.7 m/s for the single cycle curve. In Fig.11 three consecutive single cycle curves are shown. The high-frequency components (turbulence) differ very much from cycle to cycle with respect to phase and fine structure. Close investigation shows that the low-frequency components also differ, which means that cycle-to-cycle variations exist.

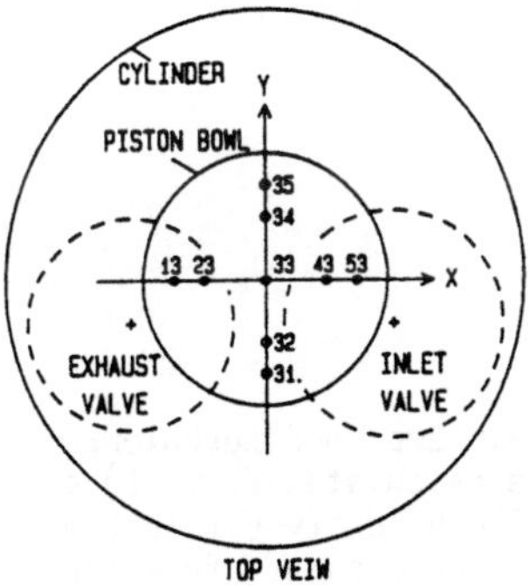

Fig.9. In-cylinder measurement points

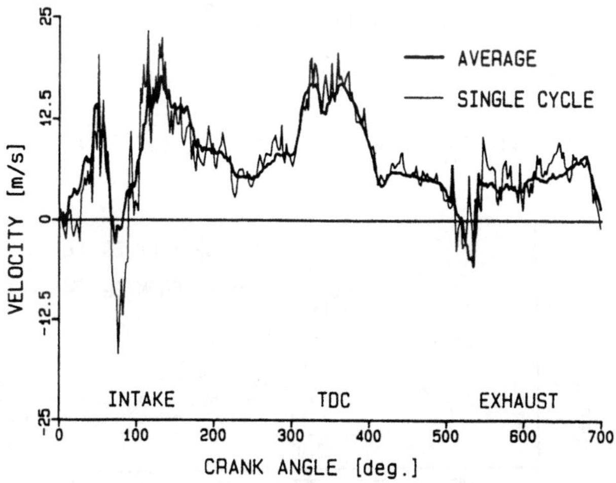

Fig.10. Velocity U (x component) in measurement point 31

The turbulence, $u(\emptyset)$, is calculated as

$$u(\emptyset) = \sqrt{\frac{\sum_{l=1}^{48} \left[U(\emptyset,l) - \overline{U}(\emptyset,l)\right]^2}{48}} \qquad (8)$$

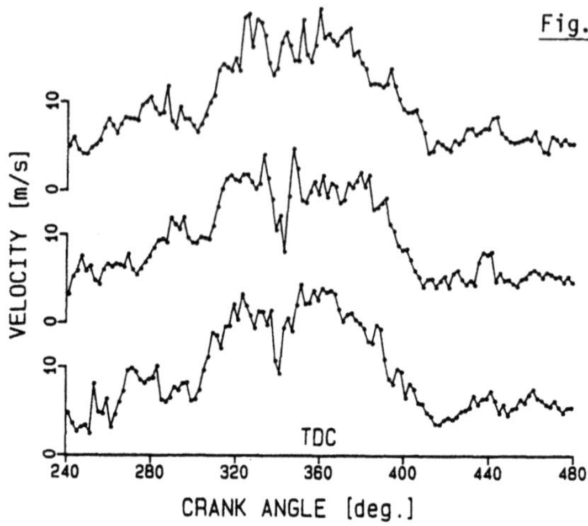

Fig.11. Single cycle velocity curves (x component), point 31

where $U(\emptyset,l)$ is the instantaneous velocity at crank angle $\emptyset$ in cycle l and $\overline{U}(\emptyset,l)$ is representing the mean velocity in cycle l and is obtained by lowpass filtering $U(\emptyset,l)$. If the mean velocity is defined as the ensemble average $\overline{U}_E(\emptyset)$, the turbulence $u_E(\emptyset)$ includes the cycle-to-cycle variations and is given by

$$u_E(\emptyset) = \sqrt{\frac{\sum_{l=1}^{48} \left[U(\emptyset,l) - \overline{U}_E(\emptyset)\right]^2}{48}} \quad . \tag{9}$$

In Fig.12 the turbulence according to (8) and (9) is shown for measurement point 31.

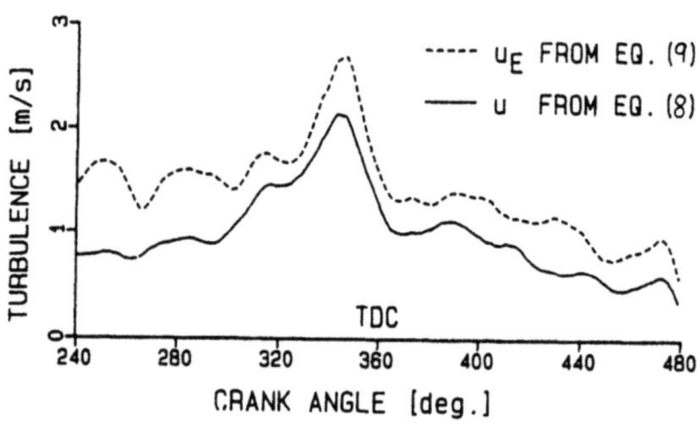

Fig.12. Smoothed turbulence curves calculated from (8) and (9). The velocity data are 48 single cycle curves from measurement point 31, x-component

5. Conclusions

The LDA system using photon correlation processing has shown to work very well for high repetition rate in-cylinder measurements. The maximum repetition rate (20kHz) due to the data transfer is believed to be high enough. To make sure that the obtained velocity values are accurate enough the calculated estimates of the uncertainties are very useful.

The in-cylinder measurements show that accurate single cycle measurement was obtained in consecutive engine cycles. Both cycle-to-cycle variations and turbulence could be observed in the gas flow and separated from each other.

The LDA-system has already proved to be a very useful instrument for flow measurements in engines.

References

1 F. Brandl, I. Reverencic, W. Cartellieri and J. C. Dent, SAE-paper 790040 (1979).

2 J. B. Cole and M.D. Swords, SAE-paper 800043 (1980).

3 R. B. Rask, "Fluid Mech of Combustion System", Presented at the Fluid Eng Conf, Boulder, Colcorado, USA, June 22-23 1981 (ASME, New York, 1981) pp. 11-20.

4 F. Durst, A. Melling and J. H. Whitelaw, "Principles and Practice of Laser-Doppler Anemometry" (Academic Press, New York, 1976).

5 Warren H. Stevenson, Proc. IEEE, 70, No. 6 (1982), pp. 652-658.

6 J. B. Abbiss, T. W. Chubb, E. R. Pike, Optics and Laser Technology (December 1974), pp. 249-261.

7 "Photon Correlation Spectroscopy and Velocimetry", edited by H. Z. Cummins and E. R. Pike (Plenum Press, New York and London, 1977) pp. 246-343.

14. Particle Statistics in Laser Transit Velocimetry

Joachim C. Erdmann

Boeing Commercial Airplane Company, Systems Technology-Instrumentation, Seattle, WA 98124, USA

1. Introduction

It is a general experience that the cross-correlation function obtained by laser transit velocimetry sensitively depends on the turbulence intensity, but relatively little is known about this dependence. In this article we shall first review the known methods to process two-spot transit information and then proceed with an analysis by means of particle statistics. The main reason for the latter approach is simply that only in this way the response of a transit velocimeter in flows of medium and high turbulence intensity can be analyzed. It is hoped that such studies of particle transport in turbulent fluids will stimulate wider acceptance of two-beam transit techniques even at high turbulence. It will be shown that in principle the two-beam configuration not only can be utilized to sample the velocity distribution as in laser Doppler velocimetry, but also has the interesting, so far unused capability of facilitating statistical analysis of turbulence structure and spectral properties.

2. Methods to Process Two-Spot Transit Information

A well-known approach to process the light intensity burst pairs received by a transit velocimeter with 2 detectors is due to SCHODL [1] and TANNER [2]. The time between the peaks of consecutive bursts belonging to the same pair is measured. In a turbulent flow this quantity is a random variable and can be sampled in a histogram which yields both the expectation and the variance of the time of flight. Particle statistics is not considered in this approach except that one has to make the provision that particles follow one another at time intervals larger than the transit time. If this is not the case, the occurrence of overlapping transits causes invalid results. The associated statistics has been analyzed by ABBISS et al. [3] who considered Poisson-distributed arrival rates of the particles.

SCHODL's technique works very well in flows of low turbulence and has the great advantage of yielding a time-of-flight histogram.

A way to better recognize the bursts belonging to the same particle pair even in the presence of a background of unpaired bursts or noise owing to various sources is the cross-correlation technique. An analysis of the cross-correlation function can be founded on the video signal that is observed in the output of either one or two detectors. This waveform as a function of the time can be written down if one knows the optical parameters of the observing instrument and the intensity profile in the focal waists of the two laser beams. In this way LADING constructed analytical expressions of the video signal [4]. He then calculated various correlation functions of interest, considering cases such as spatial velocity gradients and temporal velocity fluctuations. Again, such calculations can be carried out without much reference to particle statistics.

LADING's approach employs an analog signal and its correlation function. DURRANI and GREATED converted the analog signal into a digital photon count rate thus obtaining expressions for the digital cross-correlation function [5]. Their conclusions, as those of the other workers, were based on the assumption that the velocity does not vary during the transit time interval. Furthermore, they included a consideration of particle statistics assuming a Poissonian arrival rate distribution with a parameter that remains constant even in turbulent flow.

At low or medium turbulence intensity broadening of the cross-correlation function is caused significantly also by the single-spot transit time. MAYO has shown how transit time broadening can be reduced by deriving discriminator pulses prior to correlation that coincide with the center of the photon bursts. The cross-correlogram then again becomes a time-of-flight histogram as in SCHODL's approach [6].

All of these analytical approaches have in common that assumptions about the signal are combined so that the signal waveform can be stated *a priori*. Furthermore, it is taken for granted that the path between the two spots is a deterministic, that means predictable straight line. In this way practical results can be obtained that are true at low turbulence, but become questionable at high turbulence intensity.

As an alternate way to analyze transit signals we shall now invoke particle statistics. It will be seen that the results allow a discussion of the transit velocimeter even when the turbulence intensity is high.

3. Probability Theoretical Concepts in Transit Velocimetry

In the following an analysis will be presented for the case when all three velocity components have a common stationary normal distribution. It is also assumed that this distribution is the same in both probe volumes. To obtain a satisfactory description under the assumptions above, one must not neglect the fact that the velocity in the two beam probes is a correlated time-dependent vector pair. The mathematical treatment then requires path integral methods of the form developed by PAPADOPOULOS [7]. These techniques were first applied by CHANDRASEKHAR in a treatment of Brownian motion [8], and since then have become central in much more general form through the use of Green's functions or *propagators* [9]. The propagator concept can also be usefully employed in a description of the transit of seed particles embedded in a turbulent fluid.

One must also properly address the *motion modes* of seed particles in the transit path. It is often assumed that a particle on its way from point A to point B in a turbulent fluid carries out a random walk superimposed on a convective component so that the motion can be described as diffusion. For this to be true, however, A and B must be separated by a distance that is large compared with at least the small turbulence scale. The random walk model yields a stochastic propagator that possesses the Markov property. This description, however, is rarely applicable to a transit velocimeter in which the distance between the two beams is very small. The particle motion is then essentially deterministic and leads to a non-Markovian propagator. Its time dependence is different from that in the diffusion case. In general, transitions from deterministic to stochastic propagation must be included, and the relevant calculations are given in Secs. 4 - 6. The results may become subject of future experimental interest. In fact, the configuration of a transit velocimeter lends itself well to the determination of the propagator. In

Secs. 7 - 9 the close relation of the propagator to the transit correlation function is analyzed. Systematic measurements of its time dependence may in some cases admit of an interpretation of the turbulence scale.

Another analytical tool used in this note is *conditioning* by the initial velocity. This means that results are first obtained for a fixed given value of the turbulent velocity, properly allowing for the decorrelation following some initial epoch. In other words, a particle entering probe volume A at an initial epoch τ with velocity $\vec{v}(\tau)$ will in general have a different velocity at epoch $\tau+s$ in probe volume B after a transit time s. The method of conditioning allows one to precisely account for the difference *on the average*. This occurs by randomization of the result obtained for fixed velocity, weighting it with the proper velocity distribution. In this way one can then derive the finer features of the transit correlation function including biasing of the mean sampled velocity. Combination of conditioning with particle arrival statistics has been used recently to treat the statistical bias in laser Doppler velocimetry [10]. A similar approach is taken in Sec. 11.

4. Stochastic and Deterministic Motion

Consider two small regions V_1 and V_2 within a turbulent fluid. Seed particles carried by the fluid through volume V_1 occasionally also pass through volume V_2 at some later time. For a probabilistic model we evidently require the function $g(\vec{r}_2,t_2|\vec{r}_1,t_1) \equiv g(2|1)$ which is the density function of the conditional probability that a particle will pass through a differential neighborhood of $\vec{r}_2$ at epoch t_2 given that it was in the vicinity of $\vec{r}_1$ at epoch t_1. We shall see later that $g(2|1)$ may be appropriately called the *propagator* of the particle density. Adhering to this nomenclature, we first discuss briefly a fundamental difference in the probabilistic meaning of the propagator as applied to two different models of particle motion.

If V_1 and V_2 are separated by a large distance, a particle on a path connecting V_1 and V_2 will experience many interaction events of turbulent eddies, and its momentum will be frequently changed in random fashion. A random walk model appears most appropriate in this case. There are, in fact, many different paths possible that can carry the particle from V_1 to V_2 , and the actual path taken is a matter of chance.

A large body of literature has been dealing with this problem. Of particular interest is CHANDRASEKHAR's treatment of Brownian motion [8], PAPADOPOULOS' article on functional integral calculus [7], and a recent work by SCHUSS discussing many applications [11]. The diffusive propagator is given by

$$g(\vec{r}_2,s|\vec{r}_1,0) = (4\pi Ds)^{-3/2} \exp\{-(\vec{r}_2-\vec{r}_1-\vec{v}_0 s)^2/(4Ds)\} \tag{1}$$

which is the normal density of $\vec{r}_2-\vec{r}_1$ with expectation $\vec{v}_0 s$ and variance $2Ds$, $\vec{v}_0$ being the mean velocity vector and D the diffusion coefficient.

Characteristic for the solution (1) is its independence of the initial epoch t_1 and the linear dependence of the variance on the transit time s. Since normal distributions are closed under convolutions, this propagator generates a convolution semi-group in which the transit times are additive [12]. Thus there holds the convolution property

$$g(\vec{r}_2,s_1+s_2|\vec{r}_1) = \int g(\vec{r}_2,s_1|\vec{r})g(\vec{r},s_2|\vec{r}_1)d\vec{r} \tag{2}$$

where the integration extends over the whole space. That (2) has indeed the form of a convolution can be seen by substituting (1). Eq.(2) is identical with the Chapman-Kolmogorov identity expressing the Markov property of the propagator [13].

The situation is quite different if the two volumes V_1 and V_2 are separated by only a small distance. Then a turbulent eddy very often covers both volumes simultaneously. For a given eddy an embedded particle moves essentially in a deterministic way which means there is only one path possible. In fact, by Hamilton's principle a particle moves along a path of minimum *action* [14]. A probabilistic treatment is still necessary, however, since the actual paths taken by consecutive particles observed in large quantities over a long time are forming an ensemble depending on the flow parameters. Again we obtain a propagator $g(2|1)$, but it should be understood that its nature is quite different from the one given by (2). It will be calculated in the next section.

The conditional probability density $g(2|1)$ has been defined for a single particle with the tacit assumption that all particles propagate independently. Suppose now that $n(\vec{r}_1,t_1)$ is the density of seed particles in volume V_1 at epoch t_1. Then the density in volume V_2 at epoch t_2 is evidently given by the integral

$$n(\vec{r}_2,t_2) = \int_{V_1} g(\vec{r}_2,t_2|\vec{r}_1,t_1)n(\vec{r}_1,t_1)d\vec{r}_1 \qquad (3)$$

where the integration extends over the volume V_1. While the kernel or Green's function $g(2|1)$ is stochastic only in the diffusion case (meaning that it satisfies (2)), Eq.(3) is generally valid. It describes the propagation of the particle density from V_1 to V_2 in the time interval $\{t_1,t_2\}$, and thus $g(2|1)$ is called the propagator.

5. Calculation of the Propagator

Consider now a homogeneous and isotropic velocity field $\vec{v}(t)$. The isotropy condition is imposed mostly to avoid multiple subscripts and tensorial expressions. General expressions can probably be derived, at least when the coordinate axes coincide with the principal axes of the Reynolds stress tensor. Non-homogeneity is likely to present a more serious problem, and solutions including dependence on the space coordinates appear possible only in special cases.

The coordinate vector $\vec{r}(t+s) = \vec{r}_2$ of a particle at epoch $t+s$ relative to its position $\vec{r}(t) = r_1$ at epoch t can be written as

$$\vec{r}_2 - \vec{r}_1 = \Delta\vec{r} = \int_t^{t+s} \vec{v}(\tau)d\tau \ . \qquad (4)$$

The conditional density $g(\vec{r}_2,t+s|\vec{r}_1,t)$ must now be calculated as the *functional* ensemble average of the deterministic density $\delta\{\vec{r}(t+s)-\vec{r}(t)-\Delta\vec{r}\}$ with respect to the velocity distribution:

$$g(2|1) = \int \delta\{\vec{r}(t+s) - \vec{r}(t) - \vec{v}_0 s - \int_t^{t+s} \vec{\vartheta}(\tau)d\tau\}\, D[\vec{\vartheta}] \ . \qquad (5)$$

This means that for a given velocity vector $\vec{v}$ with mean value $\vec{v}_0$ and turbulent component $\vec{\vartheta}$ only one path is possible such that the argument of the δ function vanishes at all times. But because the vectors $\vec{v}$ as seen by consecutive particles form an ensemble, the superposition of many δ functions generates a "broadened" propagator as s increases. The *functional differential* $D[\vec{\vartheta}]$ is given by

$$\mathcal{D}[\vec{v}] = f_t^{t+s}[\vec{v}(\tau)] \prod_{t<\tau<t+s} d\vec{v}(\tau) \tag{6}$$

in PAPADOPOULOS' notation [7]. Eq.(6) involves the symbolic notation $f_t^{t+s}[\vec{v}]$ for the infinitely dimensional joint probability density function of *all* possible velocity values that can occur in the time interval $\{t,t+s\}$, and $\Pi d\vec{v}(\tau)$ is the product of infinitely many vector differentials. In other words, the random velocity record in $\{t,t+s\}$ is regarded as a function of infinitely many variables. These variables, however, are not independent, and we must take care to properly account for the velocity correlation. In this respect our treatment differs from functional integration applied to Brownian motion under a stochastic force without correlation [7].

In practice, the functional integral (5) is carried out as follows: Rewrite (5) after representing the δ function by its Fourier transform:

$$g(2|1) = (2\pi)^{-3} \int \exp\{i\vec{k}\cdot[\vec{r}(t+s)-\vec{r}(t)-\vec{v}_0 s]\}d\vec{k} \times \int \exp\{-i\vec{k}\cdot \int_t^{t+s} \vec{v}(\tau)d\tau\}\mathcal{D}[v] \quad . \tag{7}$$

Then calculate the functional integral

$$I(\vec{k}) = \int \exp\{-i\vec{k}\cdot \int_t^{t+s} \vec{v}(\tau)d\tau\}\mathcal{D}[\vec{v}] \tag{8}$$

by partitioning the time interval $\{t,t+s\}$ into N subintervals

$$\{t,t+h\},\ \{t+h,t+2h\},\ \ldots,\ \{t+(N-1)h,t+s\}, \qquad h = s/N \ ,$$

writing $\vec{v}_j = \vec{v}(t+jh)$, $j=0,1,\ldots,N-1$ and approximating $I(\vec{k})$ by

$$I_N(\vec{k}) = \int \exp\{-ih\vec{k}\cdot \sum_{j=0}^{N-1} \vec{v}_j\}f_N(\vec{v}_0,\vec{v}_1,\ldots,\vec{v}_{N-1})d\vec{v}_0 d\vec{v}_1\ldots d\vec{v}_{N-1} \quad , \tag{9}$$

where f_N is the joint density of the N vectors $\vec{v}_0, \vec{v}_1, \ldots, \vec{v}_{N-1}$. $I(\vec{k})$ is obtained by letting $N \to \infty$.

Obviously $I_N(\vec{k})$ is the Fourier transform of f_N which means it is the characteristic function of the vector sum $\Sigma\, \vec{v}_j$, $j=0, 1, \ldots, N-1$. The density f_N is normal in $3N$ dimensions. Because of the assumed isotropy of the velocity field the covariance matrix can be reduced to an $N\times N$ matrix with elements

$$c_{jk} = \langle\vec{v}_j\cdot\vec{v}_k\rangle \quad , \quad j,k = 0, 1, \ldots, N-1 \tag{10}$$

formed by scalar multiplication. Thus, by the calculus of characteristic functions Φ in higher dimensions [15] we obtain

$$I_N(\vec{k}) = \langle\exp\{-ih\vec{k}\cdot(\vec{v}_0+\vec{v}_1+\ldots+\vec{v}_{N-1})\}\rangle = \Phi = \exp\{-\tfrac{1}{2}\sigma_N^2 k^2 h^2\} \tag{11}$$

with

$$\sigma_N^2 = \sum_{j,k=0}^{N-1} c_{jk} \quad . \tag{12}$$

Also, because of the assumed stationarity of the velocity distribution there follows

$$c_{jk} = \langle\hat{v}^2\rangle\, \rho(|k-j|h) = 3\sigma^2\rho(|k-j|h) \tag{13}$$

where σ^2 is the variance of a single velocity component and $\rho(\tau)$ the velocity correlation coefficient. Eq.(12) now becomes

$$\sigma_N^2 = N + 2(N-1)\rho(h) + 2(N-2)\rho(2h) + \ldots + 2\rho[(N-1)h] \ . \tag{14}$$

If we now let $N \to \infty$ and $h \to 0$ such that $Nh = s$ stays constant and substitute

$$h \to d\tau \quad , \quad \nu s/N \to \tau \quad , \quad \nu = 0, 1, \ldots, N-1 \tag{15}$$

we see that

$$\lim h^2 \sum_{j,k=0}^{N-1} \rho(|k-j|h) = 2 \int_0^s \tau\rho(s-\tau)d\tau = 2 \int_0^s (s-\tau)\rho(\tau)d\tau \ . \tag{16}$$

Hence, and from (11) through (13),

$$I(\vec{k}) = \lim_{\substack{N\to\infty \\ h\to 0}} I_N(\vec{k}) = \exp\{-3\sigma^2k^2 \int_0^s (s-\tau)\rho(\tau)d\tau\} \ . \tag{17}$$

Substituting the last equation in (7) we obtain after Fourier inversion

$$g(\vec{r}_2,t+s|\vec{r}_1,t) = [2\pi F(s)]^{-3/2} \exp\{-(\vec{r}_2-\vec{r}_1-\vec{v}_0 s)^2/(2F(s))\} \tag{18}$$

with

$$F(s) = 2\sigma^2 \int_0^s (s-\tau)\rho(\tau)d\tau \ . \tag{19}$$

Thus the propagator is a normal density of $\vec{r}_2-\vec{r}_1$ with expectation $\vec{v}_0 s$ and variance $F(s)$ which depends on the time non-linearly as found from (19). If we define the *correlation integral*

$$\gamma(s) = \int_0^s \rho(\tau)d\tau \tag{20}$$

and the *correlation time*

$$\Theta_c = \gamma(\infty) \quad , \tag{21}$$

we can write (19) after partial integration in the form

$$F(s) = 2\sigma^2[s\gamma(s) - s\gamma(s) + \int_0^s \gamma(\tau)d\tau] = 2\sigma^2 \int_0^s \gamma(\tau)d\tau \ . \tag{22}$$

Two special cases are of interest: a) $s \ll \Theta_c$. This means the transit time s is much smaller than the velocity correlation time Θ_c. Then $\rho(\tau) \to 1$ for $\tau < s$, $\gamma(\tau) \to \tau$ and

$$F(s) \to \sigma^2 s^2 \ . \tag{23}$$

Thus, for short transit times the variance F is proportional to the square of the transit time.

b) $s \gg \Theta_c$. Then we find from (21) and (22)

$$F(s) \to 2\sigma^2\Theta_c s \ . \tag{24}$$

But this means that we recover the result given by (1) with the diffusion constant

$$D = \sigma^2\Theta_c \ . \tag{25}$$

The variance F is then proportional to the transit time.

It is interesting to see that although we started out with a single-path deterministic propagator, randomization subject to the condition $s \gg \Theta_c$ renders the propagator stochastic. This is evidently a consequence of the ergodic principle: The random walk model in which the transit path is chosen from an ensemble of *simultaneously* probable paths is equivalent to the model of the average path in the ensemble of *consecutive* deterministic paths, provided the transit times are large compared with Θ_c.

6. Conditioning by the Initial Velocity

The transit probabilities calculated in the preceding section were derived for arbitrary initial velocities $\vec{v}(t)$ such that *all* possible velocity records of length s were included. For a variety of problems, most notably the calculation of the statistical bias (see Sec. 11), it is useful, however, to be more selective and to ask for the propagator subject to the condition that all velocity records in the ensemble commence with the same vector $\vec{v}(t)$. We expect that for increasing values of s the influence of the initial velocity decreases and that for large values of s the propagator should again be of the stochastic type (1).

The calculation follows exactly the same steps as in Sec. 5, but with the conditional functional differential

$$D[\vec{v}\,|\,\vec{v}(t)] = f_{t}^{t+s}\,[\vec{v}(\tau)\,|\,\vec{v}(t)] \prod_{t<\tau<t+s} d\vec{v}(\tau) \quad . \tag{26}$$

Again, the interval $\{t,t+s\}$ is partitioned into N subintervals, and instead of (9) we now obtain

$$I_N[\vec{k}\,|\,\vec{\vartheta}(t)] = \int \exp\{-ih\vec{k}\cdot \sum_{j=o}^{N-1} \vec{\vartheta}_j\} f_{N-1}(\vec{\vartheta}_1,\vec{\vartheta}_2,\ldots,\vec{\vartheta}_{N-1}\,|\,\vec{\vartheta}_0)\, d\vec{\vartheta}_1 d\vec{\vartheta}_2 \ldots d\vec{\vartheta}_{N-1} \;. \tag{27}$$

Here the conditional multi-dimensional normal density is [16]

$$f_{N-1}(\vec{\vartheta}_1,\vec{\vartheta}_2,\ldots,\vec{\vartheta}_{N-1}\,|\,\vec{\vartheta}_0) = \frac{|M|^{-\frac{1}{2}}}{(2\pi)^{3N/2}} \exp\{-\tfrac{1}{2}\sum_{j,k=o}^{N-1} q_{jk}\vec{\vartheta}_j\cdot\vec{\vartheta}_k\} \Big/ \frac{e^{-\frac{1}{2}\hat{v}_0^2/\sigma^2}}{(2\pi\sigma)^{3/2}} \tag{28}$$

where $|M|$ is the determinant of the matrix

$$M = Q^{-1} \quad , \quad Q = (q_{jk}) \quad . \tag{29}$$

To calculate $f_{N-1}(\vec{\vartheta}_1,\vec{\vartheta}_2,\ldots|\vec{\vartheta}_0)$ we introduce the substitutions

$$\begin{aligned} \vec{y}_0 &= \vec{\vartheta}_0 \\ \vec{y}_1 &= \vec{\vartheta}_1 - \rho(h)\vec{\vartheta}_0 \\ \vec{y}_2 &= \vec{\vartheta}_2 - \rho(2h)\vec{\vartheta}_0 \\ &\ldots \\ \vec{y}_{N-1} &= \vec{\vartheta}_{N-1} - \rho[(N-1)h]\vec{\vartheta}_0 \;. \end{aligned} \tag{30}$$

The associated isotropic covariance matrix is of rank N and has the form

$$M^{+} = (1/3)(\langle\vec{y}_j\cdot\vec{y}_k\rangle) = (\sigma^2[\rho(|k-j|h)-\rho(kh)\rho(jh)]),\; j,k = 0,\,1,\,\ldots,\,N-1. \tag{31}$$

This matrix should be compared with the original matrix M, also of rank N, but given by

$$M = (c_{jk})/3, \quad j,k = 0, 1, \ldots, N-1 \quad . \tag{32}$$

With c_{jk} given by (13) we can evidently write the right side of (31) as

$$\sigma^2\{\rho(|k-j|h)-\rho(kh)\rho(jh)\} = (1/3)\{c_{jk}-\rho(jh)c_{oj}-\rho(kh)c_{ko}+\rho(kh)\rho(jh)c_{oo}\}. \tag{33}$$

This, however, is the well-known rule for the reduction of determinants: In the original matrix M we subtract from the row with index $j > 0$ the row with index 0 after multiplying each element by $\rho(jh)$. This operation we carry out for each row $j > 0$ and then also for each column with index $k > 0$ from which we subtract the column $k = 0$ after multiplication with $\rho(kh)$. It is then easily seen that

$$|M^+| = |M| = \left(\begin{array}{c|ccc} c_{00}/3 & 0 & 0 & \ldots \\ \hline 0 & & & \\ 0 & & M^* & \\ \vdots & & & \end{array}\right) \tag{34}$$

where

$$M^* = (1/3)(\langle\vec{y}_j\cdot\vec{y}_k\rangle) \quad , \quad j,k = 1, 2, \ldots, N-1 \quad . \tag{35}$$

Thus the original $N\times N$ matrix M can be reduced to the product of $c_{00}/3$ with the $(N-1)\times(N-1)$ matrix M^*. This factorization is just what is needed in (28) to carry out the division by the normal density of $\vec{v}_0$. The result after substitution of (30) in (35) is then

$$f_{N-1}(\vec{v}_1,\vec{v}_2,\ldots,\vec{v}_{N-1}|\vec{v}_0) = \frac{|M^*|^{-\frac{1}{2}}}{(2\pi)^{3(N-1)/2}} \exp\{-\tfrac{1}{2}\sum_{j,k=1}^{N-1} q^*_{jk}[\vec{v}_j-\rho(jh)\vec{v}_0]\cdot [\vec{v}_k-\rho(kh)\vec{v}_0]\}. \tag{36}$$

with

$$q^*_{jk} = M^*(k,j)/|M^*| \quad , \quad M^*(k,j) = (-1)^{k+j}|M^*_{kj}|. \tag{37}$$

$M^*(k,j)$ is the cofactor of the element $\langle\vec{y}_k\cdot\vec{y}_j\rangle$ in the matrix M^* with subdeterminant $|M^*_{kj}|$.

We now obtain from (27) and (30)

$$\begin{aligned} I_N[\vec{k}|\vec{\vartheta}(t)] &= \langle\exp\{-ih\vec{k}\cdot(\vec{\vartheta}_0+\vec{v}_1+\ldots+\vec{\vartheta}_{N-1})\}|\vec{\vartheta}_0\rangle \\ &= \exp\{-ih\vec{k}\cdot\vec{\vartheta}\sum_{j=o}^{N-1}\rho(jh)\}\langle\exp\{-ih\vec{k}\cdot(\vec{y}_1+\vec{y}_2+\ldots+\vec{y}_{N-1})\}|\vec{v}_0\rangle \\ &= \exp\{-ih\vec{k}\cdot\vec{v}(t)\sum_{j=o}^{N-1}\rho(jh)\}\Phi(-h\vec{k},\ldots,-h\vec{k}) \end{aligned} \tag{38}$$

where Φ denotes the characteristic function of the vector sum $\sum_{j=1}^{N-1}\vec{y}_j$. Instead of (11) we now find

$$I_N[\vec{k}|\vec{v}(t)] = \exp\{-ih\vec{k}\cdot\vec{v}(t)\sum_{j=o}^{N-1}\rho(jh)\}\exp\{-\tfrac{1}{2}\sigma^2_{N-1}k^2h^2\} \tag{39}$$

with

$$\sigma^2_{N-1} = \sum_{j,k=1}^{N-1} \langle \vec{y}_j \cdot \vec{y}_k \rangle = 3\sigma^2 \sum_{j,k=o}^{N-1} \rho(|k-j|h) - \rho(kh)\rho(jh) \quad . \tag{40}$$

It is then readily found that

$$\lim_{\substack{N\to\infty \\ h\to o}} h \sum_{j=o}^{N-1} \rho(jh) = \int_o^s \rho(\tau)d\tau \tag{41}$$

and

$$\lim h^2 \sum_{j,k=o}^{N-1} \rho(jh)\rho(kh) = [\int_o^s \rho(\tau)d\tau]^2 \quad . \tag{42}$$

Furthermore, Eq.(16) is also valid in the present case. Hence,

$$I[\vec{k}|\vec{\tilde{v}}(t)] = \lim_{\substack{N\to\infty \\ h\to o}} I_N[\vec{k}|\vec{\tilde{v}}(t)]$$

$$= \exp\{-i\vec{k}\cdot\vec{\tilde{v}}(t)\int_o^s\rho(\tau)d\tau - 3\sigma^2k^2[\int_o^s(s-\tau)\rho(\tau)d\tau - \tfrac{1}{2}[\int_o^s\rho(\tau)d\tau]^2]\} \quad . \tag{43}$$

After Fourier inversion we obtain the result that replaces (18):

$$g[\vec{r}_2,t+s|\vec{r}_1,\vec{v}(t),t] = (2\pi)^{-3}\int\exp\{i\vec{k}\cdot[\vec{r}_2-\vec{r}_1-\vec{v}_0s-\vec{\tilde{v}}(t)\int_o^s\rho(\tau)d\tau]\}$$

$$\times \quad \exp\{-3k^2F^*(s)/2\}d\vec{k}$$

$$= [2\pi F^*(s)]^{-3/2} \exp\{-\frac{[\vec{r}_2-\vec{r}_1-\vec{v}_0s-\vec{\tilde{v}}(t)\gamma(s)]^2}{2F^*(s)}\} \tag{44}$$

where $\gamma(s)$ is again the correlation integral (20) and

$$F^*(s) = 2\sigma^2 \int_o^s (s-\tau)\rho(\tau)d\tau - \sigma^2[\int_o^s \rho(\tau)d\tau]^2 = 2\sigma^2 \int_o^s \gamma(\tau)d\tau - \sigma^2\gamma^2(s) \quad . \tag{45}$$

Notice that in the first of the Eqs.(44) use can be made of the conditional expectation of $\vec{v}$ at epoch t+s given $\vec{v}$ at epoch t for normal distributions [17]:

$$\langle\vec{v}(t+s)|\vec{v}(t)\rangle = \vec{v}_0 + \vec{\tilde{v}}(t)\rho(s) \quad . \tag{46}$$

The integral

$$\vec{w}(s|t) = \int_o^s \langle\vec{v}(t+\tau)|\vec{v}(t)\rangle d\tau \tag{47}$$

denotes the average displacement of a particle in the time interval {t,t+s} [18]. Thus, substituting (47) in (44) we finally obtain

$$g[\vec{r}_2,t+s|\vec{r}_1,\vec{v}(t),t] = [2\pi F^*(s)]^{-3/2}\exp\{-\frac{[\vec{r}_2-\vec{r}_1-\vec{w}(s|t)]^2}{2F^*(s)}\} \quad . \tag{48}$$

As in Sec. 5 we consider the special cases of very short and very long transit times: a) $s \ll \Theta_c$. Then $\gamma(s) \to s$ and

$$F^*(s) \to 0 \quad . \tag{49}$$

Eq.(44) then becomes

$$g[2|1,\vec{v}(t)] \rightarrow \delta[\vec{r}_2-\vec{r}_1-\vec{v}(t)s] \quad , \tag{50}$$

and this means the propagator is fully deterministic in this case.

b) $s \gg \Theta_c$. Now there holds $\gamma(s) \rightarrow \Theta_c$ and

$$F^*(s) = 2\sigma^2\Theta_c s - \sigma^2\Theta_c^2 \quad . \tag{51}$$

Neglecting Θ_c as compared with s we see that again the stochastic propagator (1) is recovered with the diffusion constant $D = \sigma^2\Theta_c$. The influence of the initial velocity value is eventually lost as s increases. In this limit the propagator again possesses the Markov property.

More generally we can compare the conditional and unconditional propagators (48) and (18) by the time dependence of the expectation and variance of the displacement $\vec{r}_2-\vec{r}_1$:

	Unconditional	*Conditional*
Expectation of $\vec{r}_2-\vec{r}_1$ *:*	$\vec{v}_0 s$	$\vec{v}_0 s + \vec{\vartheta}(t)\gamma(s)$
Variance of $\vec{r}_2-\vec{r}_1$ *:*	$2\sigma^2\int_0^s \gamma(\tau)d\tau$	$2\sigma^2\int_0^s \gamma(\tau)d\tau - \sigma^2\gamma^2(s)$.

7. Detection Probabilities in Transit Velocimetry

The measurement configuration of a two-spot laser velocimeter utilizes two parallel, almost cylindrical probe volumes V_1 and V_2 of circular cross sections, generated by two parallel focussed laser beams of intensities I_1 and I_2. We denote by a_1,a_2 the radii of the two cylinders and by d the distance between their center lines (see Fig.1). The two beams are aligned in the turbulent flow such that the plane spanned by their axes forms only a small angle ε with the mean velocity vector $\vec{v}_0$. Ideally there holds $\varepsilon \rightarrow 0$. A particle that has first entered probe volume V_1 and has been detected there at *some time* during the interval ΔT_1 required for transit of V_1 will possibly also traverse probe volume V_2 and, if it does, will possibly be detected.

For a digital correlator of the type employed in laser velocimetry, time is a discrete variable measured in units of magnitude T, the so-called *sample increment*. Detection of a particle, if it happens, occurs within a particular sample increment, say the *m*th. We then say the detection event "occurs at epoch mT".

The probability that a particle first detected at epoch mT in volume V_1 will also be detected during its transit of V_2 at some later epoch $m'T$ is $q(m,m')$. This function can be expressed as the product

$$q(m,m') = q^*(m,m')\chi \quad . \tag{52}$$

Here q^* is the probability that a particle first detected at epoch mT in V_1 will be *observable* in volume V_2 at epoch $m'T$. Furthermore, χ is the probability of a detection given that the particle has arrived in V_2.

The probability χ depends on the photon detection process in great detail. The laser beam profile, the number of photons per particle, the quantum efficiency and other instrumental parameters are most important. This probability is best determined experimentally (see Sec. 10). In the present calculation the Gaussian beam profile with unity maximum is accounted for in the ex-

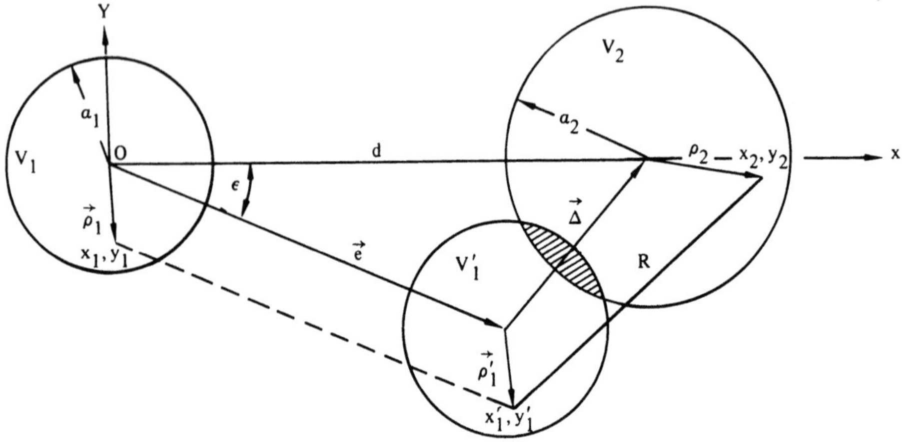

Fig.1. Configuration of two probe volumes V_1,V_2 and the displaced particle cloud in V_1'. A particle initially in probe volume V_1 is found in V_1' after displacement by the vector $\vec{e}$ (see also Appendix)

pression for q*. Thus χ depends on the maximum beam intensity in V_2 and on the detector sensitivity in ways we shall not investigate in this article.

The probability $q^*(m,m')$, on the other hand, can be calculated given the propagator derived in Se. 5 or 6. We shall restrict ourselves to the calculation in terms of the conditional propagator given the fluctuation value $\tilde{\vec{v}}(mT)$ during the mth sample increment. We shall assume that T is sufficiently small so that $\tilde{\vec{v}}$ can be regarded as constant during T. This condition can be very well satisfied with modern correlators which are now designed with large numbers of channels [19].

A particle at the time of its detection in V_1 occupies its position within V_1 with the density W_1/V_1 where W_1 is the laser beam profile function (see Appendix) in probe volume V_1, assumed to be Gaussian. This means that a particle in V_1 at the time of its detection is more likely to be in the center of the probe volume than in the marginal region. Integrating

$$W_1 V_1^{-1} g[\vec{r}_2, m'T|\vec{r}_1, \vec{v}(mT), mT]$$

with respect to $\vec{r}_1$ over V_1 with g given by (48) we obtain the conditional probability density of a particle passing through the vicinity of $\vec{r}_2$ at epoch $m'T$ given $v(mT)$ and given that it was in V_1 with density $W_1 V_1^{-1}$ at epoch mT. During the subsequent epochs $(m+1)T$, ..., $(m+k)T$, ... the particle moves randomly with an average velocity given by (46) and can be found in a region V_1'. Thus, translating all possible initial particle locations by the vector

$$e(m,m') = \vec{v}_0[(m'-m)T] + \tilde{\vec{v}}(mT)\gamma[(m'-m)T] \tag{53}$$

which is the integral of the average velocity and thus the average displacement that we first encountered in (44), we obtain the replica V_1' of V_1 as shown in Fig.1.

Let W_1' be the beam profile function in the displaced volume V_1' and let W_2 be the profile in V_2. The probability of detecting the particle at epoch $m'T$ is then the integral

$$q(m,m') = (\alpha\chi/V_1)\int\int W_1' W_2 g[\vec{r}_2, m'T|\vec{r}_1, \vec{v}(mT), mT] d\vec{r}_1' d\vec{r}_2 \quad . \tag{54}$$

Here α denotes a normalization constant discussed in the Appendix. Contributions to (54) arise from the *intersection* or overlap volume of V_1 and V_2. In the proper sense this is the entire space, but mostly the cylinder with the shaded area shown in Fig.1 as cross section.

The integral (54) is calculated in the Appendix with the result

$$q(m,m') = \frac{\chi\alpha a_2^2}{2F^*+a_1^2+a_2^2}\exp\{-\frac{\Delta^2}{2F^*+a_1^2+a_2^2}\} \; ; \; \alpha = \begin{cases}(a_1^2+a_2^2)/a_2^2, & a_1 \leqq a_2 \\ (a_1^2+a_2^2)/a_1^2, & a_1 > a_2\end{cases} \tag{55}$$

where $\Delta(m,m')$ is the distance between the centers of V_1' and V_2, shown in Fig.1. As m' increases, the overlap between V_1' and V_2 reaches a maximum and then decreases again. If the flow angle ε is zero, maximum overlap occurs when $\Delta = 0$. Then (55) can also be presented as

$$q(m,m') = \frac{\chi\alpha a_2^2}{2F^*+a_1^2+a_2^2}\exp\{-\frac{[d-sv_0-\hat{v}_x\gamma(s)]^2}{2F^*+a_1^2+a_2^2}\} \, . \tag{56}$$

F^* is the variance of the displacement and is given by (45). From (55) one concludes that the largest values of q are obtained for vanishing turbulence or for short transit times, because in both cases there holds $F^* \to 0$. In the special case $a_1 = a_2 = a$ the maximum detection probability must be 1. This condition is used to determine the normalization constant α. For an additional discussion of the normalization requirement see the Appendix.

8. The Dual-Beam Transit Correlation Function

Let $N[v(mT)]$ be the random number of particles counted in V_1 during the mth sample increment. $v(mT)$ is the absolute amount of the velocity in V_1 at epoch mT. Furthermore, let M be the number of particles counted in V_2 during the m'th increment. For M we can write

$$M = J + K \tag{57}$$

where J is the number of particles that have not been detected in V_1 while K is the number of particles that were detected in V_1. J and N are independent while K depends on N and follows the binomial distribution

$$b(K{=}k;N{=}n) = \binom{n}{k}q^k p^{n-k} \quad ; \quad p = 1 - q \, . \tag{58}$$

q is the probability that a particle detected in V_1 at epoch mT is again detected in V_2 at epoch $m'T$ and is clearly identical with $q(m,m')$ given by (55) and (56). The random variable N follows the Poisson distribution

$$p_n[T|v(mT)] = \frac{[A_1 v(mT)T]^n}{n!} e^{-A_1 v(mT)T} \tag{59}$$

which depends on the seed rate constant A_1 and on the absolute value of the velocity at epoch mT. The constant A_1 equals the number of seed particles per unit length of streamline through probe volume V_1. The velocity includes the turbulent component. Notice that (59) is a simplification of a more general distribution derived previously [18]. Thus, if in accordance with (47) for $T \to 0$ we introduce

$$w(T,m'|mT) = \int_0^T \langle v(m'T+\xi)|v(mT)\rangle d\xi \to \langle v(m'T)|v(mT)\rangle T \tag{60}$$

the distribution of the random variable M becomes

$$p_m[T,m'|v(mT)] = \frac{[A_2 w(T,m'|mT)]^m}{m!} \, e^{-A_2 w} \quad . \tag{61}$$

A_2 is in general different from A_1 because of different size of V_2 and V_1, different intensities I_1, I_2, and different sensitivities of the detectors. A_1 and A_2 are treated here as constants, although at high particle velocities they are in general also velocity dependent.

With (57) there follows now

$$<MN|v(mT)> = <JN|v(mT)> + <KN|v(mT)> = <J|v> <N|v> + <<KN|N,v>> \, . \tag{62}$$

For the expectations one finds

$$<J|v> = A_2 w - A_1 v(mT) T q(m,m') \tag{63}$$

(this follows from (57) and (58)),

$$<N|v> = A_1 v(mT) T \quad , \tag{64}$$

and

$$<<KN|N,v>> = q<N|v>(1+<N|v>) \tag{65}$$

(because $<KN|N,v> = qN^2$ and $<N^2|v> = <N|v> + <N|v>^2$).

We then obtain from (62)

$$<MN|v(mT)> = A_1 A_2 w(T,m'|mT) v(mT) T + q(m,m') A_1 v(mT) T \quad . \tag{66}$$

The dual-beam transit correlation function follows from the last equation by taking its expectation with respect to the distribution of $v(mT)$ which is a generalized Rice distribution [20]. This calculation requires finding the two correlation functions $<w(T,m'|mT)v(mT)>_v$ and $<q(m,m')v(mT)>_v$.

Accounting for (60) one finds for $T \to 0$

$$<w(T,m'|mT)v(mT)>_v = <v(mT)v(m'T)>T \tag{67}$$

which is essentially the correlation function of the absolute amount of the velocity. Since by assumption only stationary random processes are involved, the right side of (67) is expected to depend only on the difference $m' - m$.

The calculation of $<qv>_v$ is more complicated because it involves the components of the velocity as well as its absolute amount. For low and medium turbulence and small flow angles ε, however, we can put $v \sim v_x$. Then a normal distribution replaces the Rice distribution [20]. Furthermore, instead of (53) we can write for the absolute amount of $\vec{e}$

$$e(m,m') = v_0(m'-m)T + \hat{v}_x(mT)\gamma[(m'-m)T] \quad . \tag{68}$$

Since ε is assumed to be quite small, we can also write

$$\Delta^2 \to (d-e)^2 \, . \tag{69}$$

Thus, from (55) there follows

$$<q(m,m')v(mT)> = \frac{\chi\alpha a_1^2}{(2F^*+a_1^2+a_2^2)\sigma\sqrt{2\pi}} \int_{-\infty}^{+\infty} \exp\{-\frac{(d-e)^2}{2F^*+a_1^2+a_2^2} - \frac{(v_x-v_0)^2}{2\sigma^2}\} v_x dv_x$$

or, since according to (22) and (45) with $j = m' - m$

$$F^*(jT) = 2\sigma^2 \int_0^{jT} \gamma(\tau)d\tau - \sigma^2\gamma^2(jT) = F(jT) - \sigma^2\gamma^2(jT)$$

this becomes

$$<q(m,m')v(mT)>_v = \frac{\chi\alpha a_2^2}{(2F^*+a_1^2+a_2^2)\sigma\sqrt{2\pi}} \int_{-\infty}^{+\infty} (v_0+\hat{v}_x)\exp\{-\frac{\hat{v}_x^2}{2\sigma^2} - \frac{[\hat{v}_x\gamma(jT)+v_0 jT-d]^2}{2F+a_1^2+a_2^2-2\sigma^2\gamma^2(jT)}\} dv_x \quad . \qquad (70)$$

Here the exponential function in the integrand can be brought into the form

$$\exp\{-\frac{(d-v_0 jT)^2}{2F+a_1^2+a_2^2}\} \exp\{-[\hat{v}_x - \frac{\gamma(jT)(d-v_0 jT)}{(2F+a_1^2+a_2^2)/2\sigma^2}]^2[(2F+a_1^2+a_2^2)/2\sigma^2]/[2F^*+a_1^2+a_2^2]\} \ .$$

Thus, after some rearrangement (70) becomes

$$<q(m,m')v(mT)>_v = \frac{\chi\alpha a_2^2}{\sqrt{2F^*+a_1^2+a_2^2}\ \sqrt{2F+a_1^2+a_2^2}} \{v_0 + \frac{2\sigma^2\gamma(jT)(d-v_0 jT)}{2F+a_1^2+a_2^2}\} \times \exp\{-\frac{(d-v_0 jT)^2}{2F+a_1^2+a_2^2}\} \ . \qquad (71)$$

For vanishing turbulence this simplifies to

$$<q(m,m')v(mT)>_v = \frac{\chi\alpha a_2^2 v_0}{a_1^2+a_2^2} \exp\{-\frac{(d-v_0 jT)^2}{a_1^2+a_2^2}\} \quad , \ j = m'-m \qquad (72)$$

in accordance with (56). In this case the expectation $<qv>$ has a maximum for $j_0 = d/v_0T$. In general, however, turbulence shifts the maximum away from j_0. The deviations will be investigated in Sec. 11 in conjunction with the statistical bias.

Eq.(71) has been derived under the assumption that $v \sim v_x$ is a good approximation. This is indeed the case for low to medium turbulence intensity which is also the range in which the dual-beam transit velocimeter is most often applied. Then there holds

$$<v(mT)v(mT+jT)> = v_0^2 + \sigma^2\rho(jT) \quad . \qquad (73)$$

The ratio of the mean number of particle detections per unit length of streamline in probe volumes V_1 and V_2 is

$$A_1/A_2 = \kappa(a_1^2/a_2^2) \quad , \qquad (74)$$

where κ depends on the ratios of the beam intensities and the detector sensitivities. For balanced conditions these ratios are 1 and thus $\kappa = 1$. On the other hand, if intensities and sensitivities have been adjusted such that $A_1 = A_2$, then there holds $\kappa = a_2^2/a_1^2$.

If we now summarize the result of the randomization of (66), substituting (71), (73), and (74), the dual-beam transit correlation function is seen to assume its general form given by

$$<N(mT)M(mT+jT)> = A_1A_2[v_0+\sigma^2\rho(jT)]T^2 + $$

$$+ \frac{\alpha\chi A_1 a_2^2 T}{\sqrt{2F^*+a_1^2+a_2^2}\ \sqrt{2F+a_1^2+a_2^2}} [v_0 + \frac{2\sigma^2\gamma(jT)(d-v_0 jT)}{2F+a_1^2+a_2^2}] \exp\{-\frac{(d-v_0 jT)^2}{2F+a_1^2+a_2^2}\} \quad . \qquad (75)$$

Here we can introduce the expectations of the number of detections in a sample increment of length T

$$\nu_1 = A_1 v_0 T$$
$$\nu_2 = A_2 v_0 T \tag{76}$$

and the *effective variance* of the displacement in the time jT

$$S^2(jT) = F(jT) + \tfrac{1}{2}(a_1^2+a_2^2) \quad . \tag{77}$$

Eq. (75) can then also be expressed as

$$\langle N(mT)M(mT+jT)\rangle = \nu_1\nu_2[1 + \eta^2\rho(jT)] +$$
$$+ \frac{\alpha\chi\nu_1 a_2^2}{\sqrt{2}S(jT)\sqrt{2F^*(jT)+a_1^2+a_2^2}}\left[1 + \frac{\sigma^2\gamma(jT)(d-v_0jT)}{v_0S^2(jT)}\right]\exp\left\{-\frac{(d-v_0jT)^2}{2S^2(jT)}\right\} \tag{78}$$

where

$$\eta = \sigma/v_0 \tag{79}$$

is the turbulence intensity.

The first term in the transit correlation function (78) is the *base* term proportional to $\nu_1\nu_2$ and dependent on the turbulence intensity η and the velocity correlation coefficient ρ. For vanishing turbulence the base term is simply given by $\nu_1\nu_2$. The second term in (78) is the correlation peak term proportional only to the particle rate ν_1. The shape of the correlation peak depends sensitively on the turbulence intensity η, the beam radii a_1 and a_2, and on the velocity correlation time Θ_c.

For analytical purposes it is more convenient to express (78) in terms of dimensionless quantities. Introducing

$$\begin{aligned} f &= 2F/(a_1^2+a_2^2) \quad , \\ f^* &= 2F^*/(a_1^2+a_2^2) \quad , \\ \lambda &= d/(v_0\Theta_c) \quad , \\ \mu^2 &= 2d^2/(a_1^2+a_2^2) \quad , \\ s^2 &= S^2/(v_0^2\Theta_c^2) = \eta^2 F/(\sigma^2\Theta_c^2) + \lambda^2/\mu^2 \quad , \\ x &= T/\Theta_c \quad , \\ \Gamma(x) &= \gamma(x\Theta_c)/\Theta_c \end{aligned} \tag{80}$$

we replace (78) by

$$\langle N(mT)M(mT+jT)\rangle/(\nu_1\nu_2) = 1 + \eta^2\rho(jT) +$$
$$+ \frac{\alpha\chi a_2^2/(a_1^2+a_2^2)}{\nu_2\sqrt{(f+1)(f^*+1)}}\left[1 + \frac{\eta^2\Gamma(x)(\lambda-x)}{s^2}\right]\exp\left\{-\frac{(\lambda-x)^2}{2s^2}\right\} \quad . \tag{81}$$

It is found that this function depends characteristically on λ, the ratio of particle transit time to the velocity correlation time. Examples are shown in Fig.2 for deterministic (left half, $\lambda = 0.1$) and stochastic (right half, $\lambda = 10$) propagation. The plots shown for different values of the turbulence intensity η vs. the reduced time coordinate x in the deterministic case peak for values of x close to 0.1, but it is seen that the maximum shifts to somewhat lower x values as η increases. This feature as well as the asymmetry

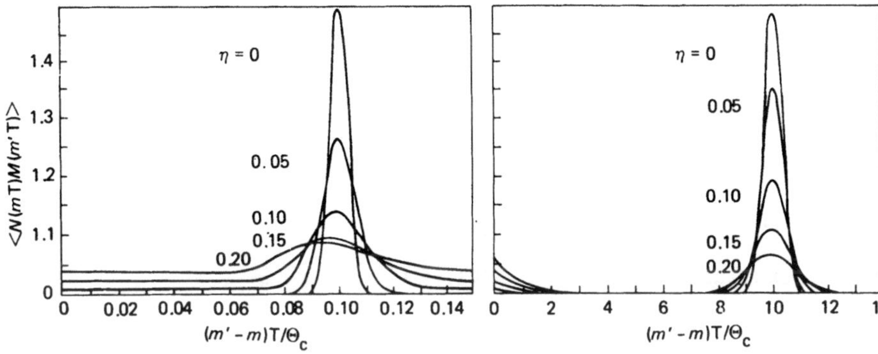

Fig.2. Particle transit cross-correlation function vs. reduced transit time for several values of the turbulence intensity η
Parameters: $a_1 = a_2 = a$; $\mu = 30$; $\nu_1 = 1$; $\nu_2 = 1$; $\chi\alpha = 1$
Left half: Almost deterministic propagation, $\lambda = 0.1$
Right half: Almost stochastic propagation, $\lambda = 10$

of the correlation peak will be further investigated in Sec. 11. It is almost negligible when propagation is stochastic as in the examples shown in the right half of Fig.2. But now one notices the rate correlation decay of the base term at low values of x.

9. Effect of Single Clipping

Correlators operate by averaging large numbers of products, and the multiplication time of each product becomes a serious problem. During the multiplication the incoming signal can not be utilized. Since the multiplication time depends on the number of bits to be processed and not on the sample increment T, the signal utilization efficiency drops sharply when T is small and decreases. The first commercial correlators were of this type. The problem can be much alleviated, however, once it has been realized that one need not process the full number of bits in order to obtain a useful correlation function. Of the various possible schemes two have emerged as particularly useful, the single-clipping and the double-clipping method. We shall restrict ourselves to the discussion of the effect of single clipping which is the more useful scheme of the two.

In single clipping one of the factors in the product is carried with all of its bits. The other factor is hard-limited at a selectable level in the following way: If the clip level is i, then the clipped count N_i associated with a count N is defined as

$$N_i = \begin{cases} 0 \quad , & N \leqq i \quad , \\ 1 \quad , & N > i \quad . \end{cases} \tag{82}$$

The electronics required to process single-clipped multiplication is reduced to a shift register and a bank of AND gates. The multiplication in all storage channels of the correlator is thus a parallel process involving the AND gates simultaneously.

The single-clipped cross-correlation function for the clip level i is obtained from (62) replacing N by N_i:

$$\langle MN_i | v(mT)\rangle = \langle J|v\rangle\langle N_i|v\rangle + \langle\langle KN_i | N_i, v\rangle\rangle \quad . \tag{83}$$

With the distribution (59) one calculates

$$\langle N_i | v\rangle = \sum_{n=i+1}^{\infty} p_n[T|v(mT)] \quad , \tag{84}$$

$$\langle\langle KN_i | N_i, v\rangle\rangle = q\langle N_i \dot{N} | v\rangle = q \sum_{n=i+1}^{\infty} n p_n[T|v(mT)] \quad . \tag{85}$$

$\langle J|v\rangle$ is given as before by (63). Thus,

$$\langle MN_i | v(mT)\rangle = A_2 w(T, m'|mT) \sum_{n=i+1}^{\infty} p_n \; +$$

$$+ \; q(m,m')\{ \sum_{n=i+1}^{\infty} n p_n - A_1 v(mT) T \sum_{n=i+1}^{\infty} p_n \} \; . \tag{86}$$

Of particular interest is the clip level $i = 0$. Since $\sum_{n=0}^{\infty} p_n = 1$ and $\sum_{n=1}^{\infty} np_n = A_1 v(mT)T$, we obtain from (86) with (60)

$$\langle MN_0 | v(mT)\rangle = A_2\langle v(m'T)|v(mT)\rangle T[1 - \exp\{-A_1 v(mT)T\}] \; +$$
$$+ \; q(m,m') A_1 v(mT) T \exp\{-A_1 v(mT) T\} \quad . \tag{87}$$

The observable single-clipped cross-correlation function $\langle MN_0\rangle$ can now be calculated by forming the expectation of (87) with respect to the distribution of $v(mT)$. Notice that

$$g[T|v(mT)] = A_1 v(mT) \exp\{-A_1 v(mT) T\} \tag{88}$$

is the density function of the probability $G[T|v(mT)]$ that at least one particle arrives in probe volume V_1 in the mth sample increment of length T [20].

The calculation of the randomized cross-correlation function requires the expectations $\langle v(m'T)\rangle_v = v_0$, $\langle\langle v(m'T)|v(mT)\rangle \exp\{-A_1 v(mT)T\}\rangle_v$, and $\langle q(m,m')\, g[T|v(mT)]\rangle_v$. We shall restrict ourselves to those cases where the approximation $v \sim v_x$ is justified, i.e. for small ε and for $\sigma \ll v_0$ so that the normal distribution of v_x can be used. Then the conditional expectation of the velocity is given by (46). Hence, with $j = m - m'$

$$\langle\langle v(m'T)|v(mT)\rangle \exp\{-A_1 v(mT)T\}\rangle = \frac{1}{\sigma\sqrt{2\pi}} v_0 e^{-A_1 v_0 T} \int_{-\infty}^{+\infty} \exp\{-A_1 \hat{v}_x T - \frac{\hat{v}_x^2}{2\sigma^2}\}\, d\hat{v}_x \; +$$

$$+ \; \frac{1}{\sigma\sqrt{2\pi}} \rho(jT) e^{-A_1 v_0 T} \int_{-\infty}^{+\infty} \hat{v}_x \exp\{-A_1 \hat{v}_x T - \frac{\hat{v}_x^2}{2\sigma^2}\}\, d\hat{v}_x$$

$$= \frac{1}{\sigma\sqrt{2\pi}} \exp\{-A_1 v_0 T + \tfrac{1}{2} A_1^2 T^2 \sigma^2\}[v_0 \int_{-\infty}^{+\infty} \exp\{-\frac{(\hat{v}_x + A_1 T\sigma^2)^2}{2\sigma^2}\}\, d\hat{v}_x \; +$$

$$+ \; \rho(jT) \int_{-\infty}^{+\infty} \hat{v}_x \exp\{-\frac{(\hat{v}_x + A_1 T\sigma^2)^2}{2\sigma^2}\}\, d\hat{v}_x]$$

$$= [v_0 - A_1 \rho(jT)\sigma^2 T]\, e^{-A_1 v_0 T}\, e^{\frac{1}{2} A_1^2 T^2 \sigma^2} \quad . \tag{89}$$

Also, from (71) and (77) we calculate

$$<q(m,m')v(mT)e^{-A_1 v(mT)T}>_v = \frac{\chi\alpha a_2^2}{\sigma\sqrt{2\pi(2F^*+a_1^2+a_2^2)}} \exp\{-\frac{(d-v_0 jT)^2}{2S^2(jT)}\} \times$$

$$\times \exp\{-A_1 v_0 T - \frac{\sigma^2\gamma(jT)(d-v_0 jT)A_1 T}{S^2(jT)} + \frac{A_1 T^2\sigma^2(2F^*+a_1^2+a_2^2)}{4S^2(jT)}\} \times$$

$$\times \int_{-\infty}^{+\infty} (v_0+\hat{v}_x) \exp\{-\frac{\sigma^2[\hat{v}_x S^2/\sigma^2-\gamma(jT)(d-v_0 jT)+\tfrac{1}{2}A_1 T(2F^*+a_1^2+a_2^2)]^2}{S^2(jT)(2F^*+a_1^2+a_2^2)}\} d\hat{v}_x$$

$$= \frac{\chi\alpha a_2^2}{\sqrt{2}S(jT)\sqrt{2F^*+a_1^2+a_2^2}} [v_0 + \frac{\sigma^2\gamma(jT)(d-v_0 jT)}{S^2(jT)} - \frac{A_1 T\sigma^2(2F^*+a_1^2+a_2^2)}{2S^2(jT)}] \times$$

$$\times \exp\{-\frac{(d-v_0 jT)^2}{2F^*+a_1^2+a_2^2}\} \exp\{-A_1 v_0 T + \frac{\sigma^2[A_1 T(2F^*+a_1^2+a_2^2)-2\gamma(jT)(d-v_0 jT)]^2}{4S^2(jT)(2F^*+a_1^2+a_2^2)} -$$

$$- [\sigma^2\gamma^2(jT)(d-v_0 jT)^2]/[S^2(jT)(2F^*+a_1^2+a_2^2)]\} \quad . \tag{90}$$

For vanishing turbulence one has $F^*=0$, $2S^2= a_1^2+a_2^2$ so that the last equation becomes

$$<q(m,m')v(mT)e^{-A_1 v(mT)T}>_v = \frac{\chi\alpha a_2^2 v_0}{a_1^2+a_2^2} \exp\{-A_1 v_0 T - \frac{(d-v_0 jT)^2}{a_1^2+a_2^2}\} \tag{91}$$

in close correspondence with (72). For low detection rates or when $A_1 v_0 T \ll 1$, (91) becomes equal to (72).

Returning now to (87) and substituting (73), (76), (79), (89), and (90), we find for the randomized single-clipped cross-correlation function

$$<N_0(mT)M(mT+jT)> = \nu_2 - \nu_2[1 - \nu_1\eta^2\rho(jT)]e^{-\nu_1+\frac{1}{2}\eta^2\nu_1^2} +$$

$$+ \frac{\chi\alpha a_2^2\nu_1 e^{-\nu_1}}{\sqrt{2}S(jT)\sqrt{2F^*+a_1^2+a_2^2}} [1 + \frac{\eta^2\gamma(jT)v_0(d-v_0 jT)}{S^2(jT)} - \frac{\nu_1\eta^2(2F^*+a_1^2+a_2^2)}{2S^2(jT)}] \times$$

$$\times \exp\{-\frac{(d-v_0 jT)^2}{2S^2(jT)}\} \exp\{\frac{\eta^2\nu_1^2(2F^*+a_1^2+a_2^2)}{4S^2(jT)} - \frac{\eta^2\nu_1\gamma(jT)v_0(d-v_0 jT)}{S^2(jT)}\} \quad . \tag{92}$$

For $\nu_1 \ll 1$, i.e., for very low detection rates, (92) becomes equal to (78), as one can verify by expansion of the exponential function. This result is, of course, expected because at low particle rates the values of N do not exceed 1.

The effect of single clipping at higher detection rates can best be assessed for the case of vanishing turbulence. With $\eta = 0$ Eq.(92) becomes

$$<N_0(mT)M(mT+jT)> = \nu_2(1 - e^{-\nu_1}) + \frac{\chi\alpha a_2^2\nu_1 e^{-\nu_1}}{a_1^2+a_2^2} \exp\{-\tfrac{1}{2}\mu^2(1-z)^2\} \tag{93}$$

where μ has been defined by (80) and where

$$z = jv_0 T/d \quad . \tag{94}$$

As the particle rate ν_1 increases, the ratio of peak term to base term decreases. From (93) one finds that this ratio is proportional to $e^{-\nu_1}/(1-e^{-\nu_1})$.

10. Integral of the Cross-Correlation Peak

A useful quantity that can be obtained experimentally is the time integral of the peak-generating term $A_1 T\langle q(m,m')v(mT)\rangle_v$ given by (70) in the unclipped case or $A_1 T\langle q(m,m')v(mT)\exp\{-A_1 v(mT)\}\rangle_v$ given by (90) for the clipped case. These expectations represent the expected number of particles that have been detected at epoch mT in probe volume V_1 and are also detected in probe volume V_2 at a later epoch $m'T$. Their time integrals Q, Q_0 are then evidently equal to the expected *total* number of detections in V_2 after prior detection in V_1. Thus we can call Q/T or Q_0/T the *total number of correlated detections* in a sample increment T, unclipped or clipped, respectively. This number can never exceed ν_1.

We restrict ourselves to the case of vanishing flow angle ε so that Δ in Fig.1 is given by

$$\Delta = d - v_0\tau - \hat{v}_x\gamma(\tau) \quad , \tag{95}$$

where τ is the time variable assuming discrete values jT in the correlation display of a digital correlator. If, furthermore, τ is small compared with the velocity correlation time Θ_c, then by (49) $F^* = 0$ and $\gamma(\tau) = \tau$ so that $\Delta = d - v_x\tau \sim d - v\tau$. Substituting q from (56) and applying (76) we find

$$Q = A_1 T \int_{-\infty}^{+\infty} \langle qv\rangle_v d\tau = \frac{\chi\alpha\nu_1 a_2^2\sqrt{\pi}}{v_0\sqrt{a_1^2+a_2^2}} \tag{96}$$

and

$$Q_0 = A_1 T \int_{-\infty}^{+\infty} \langle qve^{-A_1 vT}\rangle_v d\tau = \frac{\chi\alpha\nu_1 a_2^2\sqrt{\pi}}{v_0\sqrt{a_1^2+a_2^2}} \langle e^{-\nu_1 v/v_0}\rangle \quad . \tag{97}$$

This means that a deterministic propagator ($\tau \ll \Theta_c$) leads to a time integral of the correlation peak that is independent of the turbulence for unclipped correlation. The ratio $Q_0/Q = \langle\exp\{-\nu_1 v/v_0\}\rangle$ decreases as the number of detections increases. In fact, at higher particles rates clip levels higher than 0 must be selected.

At small detection rates the expressions (96) and (97) are practically equal. One then obtains one equation for the determination of $\chi\alpha$, a_1, and a_2 from a measured correlogram. The other two equations are both provided by (93),valid at zero turbulence. From the correlogram one finds the value of the maximum of the peak term and the variance of the Gaussian. These values must be equal to $\chi\alpha\nu_1 a_2^2/(a_1^2+a_2^2)$ and $(a_1^2+a_2^2)/d^2$. Thus, if d, v_0, and ν_1 are known, a_1, a_2, and the product $\chi\alpha$ can be calculated.

11. Mean Velocity and Statistical Bias in Transit Velocimetry

In its conditional form (66) the transit cross-correlation function depends on the initial velocity $v(mT)$. In particular, the peak-generating term $q(m,m')A_1 v(mT)T$ is proportional to the velocity. It is evident that this results from particle transport in the flow. In fact, the quantity $A_1 v(mT) = \dot{N}(mT)$ represents the particle *current* through probe volume V_1 at epoch mT. Thus, when the flow is turbulent, more particles are commencing their transit from V_1 to V_2 during a high-velocity fluctuation than on the average, and less particles arrive when the velocity is lower than the mean. The same situation exists in laser Doppler velocimetry and is known to lead to statistical bias of the velocity distribution [21]. We must also expect statistical bias in transit velocimetry.

In the following calculation we define the *bias parameter* β by the equation

$$\beta = (u_0 - v_0)/v_0 \quad , \tag{98}$$

where v_0 is the expectation of the true velocity distribution in the flow, while u_0 is the mean velocity as obtained from the cross-correlation function. To calculate u_0, we can apply two different methods. In the first, u_0 is calculated from the relation

$$u_0 = d/(jT)_m \quad , \tag{99}$$

where $(jT)_m$ is the (possibly interpolated) value of the lag time for the *maximum* of the correlation peak. In the second method we calculate u_0 from the relation

$$u_0 = d/\langle jT\rangle \quad , \tag{100}$$

where $\langle jT\rangle$ is the *first moment* of the lag time in the peak term, or

$$\langle jT\rangle = Q^{-1}\int_{-\infty}^{+\infty} \tau\langle qv\rangle_v d\tau \quad , \qquad Q = \int_{-\infty}^{+\infty} \langle qv\rangle_v d\tau \quad . \tag{101}$$

The integral Q has been calculated for deterministic propagation and unclipped correlation in Sec. 10. This is the simplest case because we can put $F^*=0$ in the expression (55) for q. Restricting ourselves to this case (including zero flow angle ε), we find with

$$a^2 = \tfrac{1}{2}(a_1^2 + a_2^2)/v^2 \tag{102}$$

$$\langle jT\rangle = Q^{-1}\frac{\chi\alpha a_2^2}{\sqrt{a_1^2+a_2^2}}\left\langle \frac{1}{a\sqrt{2\pi}}\int_{-\infty}^{+\infty} \exp\{\frac{(d/v-\tau)^2}{2a^2}\}\tau d\tau\right\rangle_v = \langle d/v\rangle_v \tag{103}$$

as one should expect. For low values of the turbulence we can expand

$$1/v = v_0^{-1}(1+\hat{v}_x^2/v_0^2+\hat{v}_y^2/v_0^2)^{-\frac{1}{2}} \sim [1-\tfrac{1}{2}(\hat{v}_x^2+\hat{v}_y^2)]/v_0 \quad ,$$

hence, with $\langle v_x\rangle = \langle v_y\rangle = \sigma^2 \quad , \quad \sigma^2/v_0^2 = \eta^2$

$$\langle d/v\rangle = (d/v_0)(1-\eta^2) \quad . \tag{104}$$

Then, because $1/\langle d/v\rangle \sim (v_0/d)(1+\eta^2)$ for small values of η, there follows from (98) and (100)

$$\beta = \eta^2 \tag{105}$$

exactly as in the statistical bias of LDV data [10,22]. The three modes of processing velocity data as discussed in [10,22], *free-running* (the present case), *controlled*, and *one-shot* also apply in the same way to transit velocimeters, at least for deterministic propagation. An experimental setup to implement a controlled processor with a Malvern Correlator is reported in [22].

To calculate $(jT)_m$, the position of the peak maximum, we return to the unclipped correlation function (78). Again we assume $T \ll \Theta_c$ so that $\rho = 1$, $\gamma = jT$, $F^* = 0$, $S^2(jT) = \sigma^2(jT) + \tfrac{1}{2}(a_1^2+a_2^2)$. Thus, particle propagation is assumed to be deterministic. Differentiating the peak term with respect to jT and putting the result equal to zero, we obtain an equation for $(jT)_m=\tau_m$ in the form

$$-\tau_m\eta^2+t[1+\eta^2-3\tau_m\frac{\eta^5v_0^3}{2\sqrt{2}S^3}] + t^2\eta^2v_0^2\frac{1+\tau_m}{2S^2} + t^3\tau^2\frac{\eta^4v_0^4}{4S^2} = 0 \tag{106}$$

where

$$t = d/v_0 - \tau_m \tag{107}$$

is the deviation of τ_m from the peak position d/v_0 for $\eta = 0$. Then, if the turbulence intensity is small, the deviation t is also small compared with τ_m so that (106) for $\eta << 1$ reduces to

$$-\tau_m \eta^2 + t(1+\eta^2) = 0$$

or

$$\tau_m = (jT)_m = \frac{d}{v_0} \frac{1+\eta^2}{1+2\eta^2} \sim \frac{d}{v_0}(1-\eta^2-2\eta^4) \quad . \tag{108}$$

It is thus seen that turbulence causes the peak maximum to shift towards lower values of τ_m. From (99) we then obtain

$$u_0 = v_0(1+\eta^2+2\eta^4+\ldots)$$

and from (98) in the limit $\eta \to 0$

$$\beta = \eta^2$$

which is identical with (105). Thus, for very low turbulence the maximum of the transit correlation function shifts together with the biased expectation so that the biased correlation function remains unskewed. At somewhat higher turbulence levels the maximum shifts by a larger amount than the mean value does, and so the correlation function assumes an asymmetrical shape. This asymmetry is caused mostly by the non-linear time dependence of the effective variance S^2 of the particle displacement (see (78), (77), and (19)).

Acknowledgments

The research reported herein was funded by The Boeing Commercial Airplane Company as part of a larger effort in laser-optic instrumentation. I am grateful to Messrs. D.J. Bernitt, L.J. Carpenter, and G.M. Bowes for their support and their interest in this work. I also like to thank Drs. D.C. Soreide and C.D. Tropea for stimulating discussions, particularly of the statistical bias.

Appendix: Integral of the Propagator

We assume that the two probe volumes V_1 and V_2 of a two-spot laser velocimeter are infinitely long cylinders oriented in the z direction of a Cartesian coordinate system. Fig.1 shows the beam configuration as cross sections in the x,y plane. The separation of the centers of the circular beam cross sections is d. The light intensity profile functions of the two beams expressed in terms of the *spot coordinate vectors* $\vec{\rho}_1$ and $\vec{\rho}_2$ are the Gaussians

$$W_{1,2}(\rho_{1,2}) = \exp\{-\rho_{1,2}^2/a_{1,2}^2\} \tag{A1}$$

which define a_1 and a_2 as the effective beam radii.

A particle at the point $\vec{r}_1 = \{x_1,y_1,z_1\}$ at epoch mT drifts to the point $\vec{r}_2' = \{x_2,y_2,z_2'\}$ at epoch $m'T$. The point $\vec{r}_1'$ lies in the replica volume V_1' obtained by translation of V_1 by the vector

$$\vec{e}(j) = \vec{v}_0 jT + \vec{\tilde{v}}(mT)\gamma(jT) \quad , \quad j = m' - m \ . \tag{A2}$$

This vector is the integral of the conditional velocity expectation $\vec{v}_0+\vec{\tilde{v}}(mT) \times\rho(jT)$ in the time interval $\{mT,m'T\}$.

The purpose of this section is to calculate the probability that a particle first detected in probe volume V_1 at epoch mT will be *observable* in probe volume V_2 at epoch $m'T$. This probability was introduced in (52). It is given by the integral

$$q^*(m,m') = V^{-1}\alpha\iint W_1'W_2 g[\vec{r}_2,m'T|\vec{r}_1,\vec{v}(mT),mT]d\vec{r}_1 d\vec{r}_2 \tag{A3}$$

of the propagator

$$g[\vec{r}_2,m'T|\vec{r}_1,\vec{v}(mT),mT] = \frac{1}{[2\pi F^*(jT)]^{3/2}}\exp\{-\frac{[\vec{r}_2-\vec{r}_1-\vec{e}(j)]}{2F^*(jT)}\} \tag{A4}$$

where the variance F^* of the displacement $\vec{r}_2-\vec{r}_1$ has been discussed in Sec.6. W_1' is a function of the form (A1) centered on the displaced beam axis in V_1'. It is evident that contributions to the integral (A3) can stem only from the *intersection* of V_1 and V_2. In the proper sense this is the entire space, but it is common to regard the cylinders with radii a_1, a_2 as the practical boundaries of the beams. This definition, however, requires introduction of the normalization constant α that will be determined from the boundary (certainty) conditions of our problem at the end of this section.

The propagator is the product of the Gaussian densities

$$(2\pi F^*)^{-1}\exp\{-\frac{(x_2-x_1')^2+(y_2-y_1')^2}{2F^*}\}$$

and

$$(2\pi F^*)^{-\frac{1}{2}}\exp\{-\frac{(z_2-z_1')^2}{2F^*}\} \quad .$$

The last factor is a function of z_1' and z_2 alone. Thus the integrations with respect to z_1' and z_2 can be carried out immediately by first integrating with respect to z_2-z_1', then with respect to z_1'. The volume V_1 is the product of the cross-section area πa_1^2 times the length $L \to \infty$. The integration with L^{-1} as normalization constant clearly yields unity so that (A3) after substitution of (A1) becomes

$$q^* = (2\pi^2 F^* a_1^2)^{-1}\alpha\iint e^{-\rho_1'^2/a_1^2}\, e^{-\rho_2^2/a_2^2}\, e^{-\frac{1}{2}R^2/F^*}\, d\vec{\rho}_1' d\vec{\rho}_2 \tag{A5}$$

where (see Fig.1 which illustrates the geometry)

$$R^2 = (\vec{\Delta} + \vec{\rho}_2 - \vec{\rho}_1')^2 \quad . \tag{A6}$$

Furthermore,

$$\vec{\Delta} = \{d-e_x, -e_y\} \tag{A7}$$

is the vector from the center of the cross section of V_1' to that of V_2.

We now introduce the variable transformation with Jacobian +1

$$\vec{\xi} = \vec{\rho}_1' - \vec{\rho}_2 \quad , \qquad \vec{\xi}' = \tfrac{1}{2}(\vec{\rho}_1' + \vec{\rho}_2) \tag{A8}$$

so that

$$\vec{\rho}_1' = \vec{\xi}' + \tfrac{1}{2}\vec{\xi} \quad , \qquad \vec{\rho}_2 = \vec{\xi}' - \tfrac{1}{2}\vec{\xi} \quad . \tag{A9}$$

Then (A6) becomes

$$R^2 = \xi^2 + \Delta^2 - 2\xi\Delta\cos\Phi \quad , \tag{A10}$$

where Φ is the angle between $\vec{\Delta}$ and $\vec{\xi}$. Thus, R depends only on ξ, not on ξ', and (A5) transforms into

$$q^* = (2\pi^2 F^* a_1^2)^{-1}\alpha \int\int e^{-\frac{1}{2}R^2/F^*} E(\vec{\xi})\,d\vec{\xi} \tag{A11}$$

with

$$E(\vec{\xi}) = \int \exp\{-\frac{(\xi'+\frac{1}{2}\xi)^2}{a_1^2}\}\exp\{-\frac{(\xi'-\frac{1}{2}\xi)^2}{a_2^2}\}d\vec{\xi}' \tag{A12}$$

being the overlap integral of the intensity profile functions with radii a_1 and a_2 centered on two points separated by the distance ξ. This integral is easily calculated with the result

$$E(\vec{\xi}) = \exp\{-\tfrac{1}{4}\xi^2(a_1^{-2}+a_2^{-2})\}\int_{-\infty}^{+\infty}\int_{-\infty}^{+\infty}\exp\{-(\xi_x'^2+\xi_y'^2)(a_1^{-2}+a_2^{-2})\}$$

$$\times \exp\{(\xi_x\xi_x'+\xi_y\xi_y')(a_1^{-2}-a_2^{-2})\}d\xi_x' d\xi_y'$$

$$= \exp\{-\tfrac{1}{4}\xi^2(a_1^{-2}+a_2^{-2})\}\frac{\pi}{a_1^{-2}+a_2^{-2}}\exp\{\tfrac{1}{4}\xi^2\frac{(a_1^{-2}-a_2^{-2})^2}{a_1^{-2}+a_2^{-2}}\}$$

or

$$E(\vec{\xi}) = \frac{a_1^2 a_2^2\pi}{a_1^2+a_2^2}\exp\{-\frac{\xi^2}{a_1^2+a_2^2}\} \quad . \tag{A13}$$

It is seen that the overlap integral depends only on the absolute value of ξ. Then (A11) after substitution of (A10) becomes

$$q^* = \frac{\alpha}{2\pi^2 F^* a_1^2}\int_0^{\infty}\int_0^{2\pi}\xi E(\xi)\exp\{-\frac{\xi^2+\Delta^2-2\xi\Delta\cos\Phi}{2F^*}\}\,d\xi d\Phi$$

$$= \frac{\alpha}{\pi F^* a_1^2}\int_0^{\infty}\xi E(\xi)\exp\{-\frac{\xi^2+\Delta^2}{2F^*}\}\,I_0(\Delta\xi/F^*)d\xi \tag{A14}$$

where I_0 is the modified Bessel function of order zero. With (A13) and the substitution

$$h^2 = \frac{2F^*+a_1^2+a_2^2}{2F^*(a_1^2+a_2^2)} \tag{A15}$$

Eq.(A14) becomes

$$q^* = \frac{\alpha a_2^2}{(a_1^2+a_2^2)F^*}\,e^{-\frac{1}{2}\Delta^2/F^*}\int_0^{\infty} e^{-h^2\xi^2}\xi I_0(\Delta\xi/F^*)d\xi \quad . \tag{A16}$$

But [23]

$$\int_0^{\infty} e^{-h^2\xi^2}\xi J_0(\Delta\xi/F^*)d\xi = (2h^2)^{-1}\exp\{-\frac{\Delta^2}{4F^{*2}h^2}\} \tag{A17}$$

where J_0 is the Bessel function of order zero. Thus, since $I_0(z) = J_0(zi)$, (A16), (A17), and (A15) together yield

$$q^* = \frac{\alpha a_2^2}{2F^*+a_1^2+a_2^2} \exp\{- \frac{\Delta^2}{2F^*+a_1^2+a_2^2}\} \quad . \tag{A18}$$

For $\Delta = 0$ there must hold in the absence of turbulence ($F^* = 0$)

$$q^* = \begin{cases} 1 \quad , & a_1 \leqq a_2 \quad , \\ a_2^2/a_1^2 & a_1 > a_2 \quad . \end{cases} \tag{A19}$$

For the normalization constant α in (A18) one then finds

$$\alpha = \begin{cases} (a_1^2+a_2^2)/a_2^2 \quad , & a_1 \leqq a_2 \quad , \\ (a_1^2+a_2^2)/a_1^2 \quad , & a_1 > a_2 \quad . \end{cases} \tag{A20}$$

References

1 R. Schodl: In *Laser Optical Measurement Methods for Aero Engine Research and Development*, AGARD Lecture Series No. 90 (AGARD, Paris 1977)
2 L.H. Tanner: Optics and Laser Technology, June 1973, 108-110
3 J.B. Abbiss, R.G.W. Brown, E.R. Pike, P.R. Sharpe: Proc. of 4th Internat. Conf. Photon Correlation Techniques in Fluid Mechanics, ed. by W.T. Mayo, Jr., A.E. Smart, Stanford University, August 1980
4 L. Lading: In *Laser Velocimetry and Particle Sizing*, ed. by H.D. Thompson and W.H. Stevenson (Hemisphere Publ. Corp., Washington, D.C. 1979)
5 T.S. Durrani, C.A. Greated: Appl. Optics 14, 778 (1975)
6 W.T. Mayo, Jr.: 3rd Internat. Conf. Photon Correlation Techniques in Fluid Mechanics, Cambridge, U.K., March 21-23 1979
7 G.J. Papadopoulos: In *Path Integrals and Their Applications in Quantum, Statistical, and Solid State Physics*, ed. by G.J. Papadopoulos (Plenum Press, New York 1978)
8 S. Chandrasekhar: Revs. Mod. Phys. 15, 1 (1943), also in *Selected Papers on Noise and Stochastic Processes*, ed. by N. Wax (Dover Publications, New York 1954)
9 P. Ramond: *Field Theory, A Modern Primer* (W.A. Benjamin, Inc., Reading 1981)
10 J.C. Erdmann, C. Tropea: Seventh Biennial Symposium on Turbulence, University of Missouri-Rolla, Sept. 21-23, 1981
11 Z. Schuss: *Theory and Applications of Stochastic Differential Equations* (John Wiley & Sons, New York 1980)
12 K. Yosida: *Functional Analysis* (Springer-Verlag, Berlin 1978)
13 W. Feller: *An Introduction to Probability Theory and Its Applications, Vol. II* (John Wiley & Sons, New York 1971), p. 98
14 W. Yourgrau, S. Mandelstam: *Variational Principles in Dynamics and Quantum Theory* (Dover Publications, New York 1968)
15 Ref. [13], p. 521
16 J.S. Bendat: *Principles and Applications of Random Noise Theory* (John Wiley & Sons, New York 1958), p. 118
17 Ref. [13], p. 72
18 J.C. Erdmann: J. Fluid Mech. (submitted)
19 W.T. Mayo, Jr., R.P. Woodward, A. Bolano, A.E. Smart: In Ref.[3]
20 J.C. Erdmann, C. Tropea: Sonderforschungsbereich 80, University of Karlsruhe, Report SFB80/ET/198, August 1981
21 D.K. McLaughlin, W.G. Tiederman: Phys. Fluids 16, 2082 (1973)
22 J.C. Erdmann, C.D. Tropea: International Symposium on Applications of Laser-Doppler Anemometry to Fluid Mechanics, Lisbon, Portugal, July 5-7, 1982.
23 M. Abramowitz, I.A. Stegun: *Handbook of Mathematical Functions* (National Bureau of Standards, Washington 1974), p. 486

15. Base Flow Measurements in a Transonic Cascade Using a Laser Transit Anemometer

R.G.W. Brown

Royal Signals and Radar Establishment, St. Andrews Road
Malvern, Worcestershire, UK

P.H. Richards

Marchwood Engineering Laboratories, Marchwood, Southampton, England

1. Introduction

In recent years the development of the laser transit anemometer has provided significant advantages for measurements in certain types of flows. These include laminar flows, flow through shock fronts and in particular flows very close to boundaries but do not include flows in which there is any appreciable level of turbulence. For transonic flows the laser transit system is particulary useful because it can give a comparatively rapid measure of the mean velocity to an accuracy of better than 2-3%. Also the flow angle can be measured to better than 0.2 degrees provided the flow is reasonably steady and appropriate care is exercised.

This paper describes one interesting aspect of a programme of measurements using a laser transit anemometer to map the complete flow field in the trailing edge region of a blade forming part of a cascade in a transonic wind tunnel. The blade profile was chosen to simulate the conditions in the last stage of the low pressure cylinder in a large steam turbine and the measurements to be described were made in the base flow region,i.e. very close to the blade trailing edge.

2. Base Flow

The base flow region arises from the convergence of the two blade surface boundary layers at some point downstream creating a dead flow region at the trailing edge. The mechanism by which the wake is shed from the trailing edge is the von Karman vortex street. The wake and the associated trailing edge blockage give rise to losses and if the wake is unsteady then the drag and perhaps the shocks will also be unsteady. To calculate the flow field in this region CLEMENTS' [1] inviscid method for two-dimensional vortex shedding was used and a typical result for the normalised velocity contours is given in Fig.1. It shows a discrete array of vortices being shed downstream with dimensions similar to that of the trailing edge thickness. The Strouhal number derived from the calculation was 0.22 giving a vortex shedding frequency of around 130 kHz for a velocity of 480 m/s and a trailing edge thickness of 0.8 mm. This frequency may appear to be high but compares favourably with the results reported by BRYANSTON-CROSS and CAMUS [2] and by HEINEMANN and BUTEFISCH [3] where both groups found frequencies well above 100 kHz for Mach numbers in the range 1.2 to 1.4.

3. Laser Transit Anemometer

The laser transit anemometer used was the Malvern 4772 laser system and has been described in detail elsewhere [4]. Essentially two beams with equal

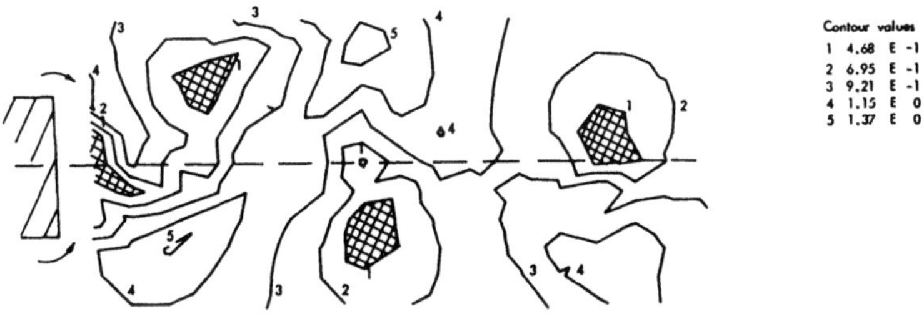

Fig.1 Contours of V_{ABS}/U_0 Showing Vortex Shedding

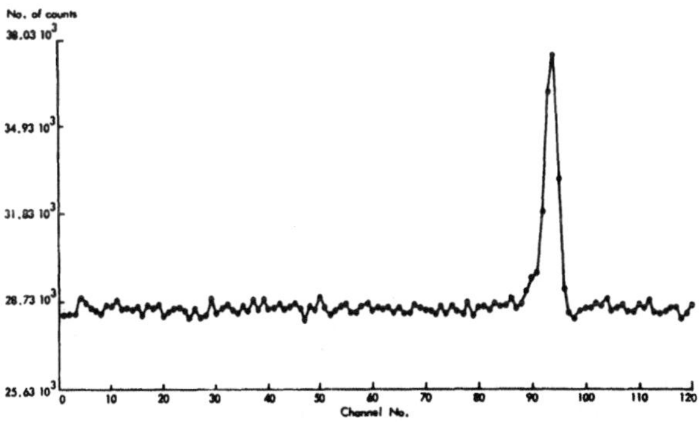

Fig.2 Cross-Correlogram in Laminar Flow (τ = 10 ns)

intensity and orthogonal polarisation were focussed by a large aperture triplet lens to give two spots 15 μm diameter separated by 450 μm. The spots were observed by two photomultipliers, one for each spot. The analogue outputs of the photomultipliers were converted to a time series of pulses by constant fraction discriminators and were then analysed by a 10 nanosecond Malvern K7026 digital photon correlator operating in the cross-correlation mode. A typical cross-correlogram (Fig.2) had the form of a very narrow peak at a delay time corresponding to the particle transit time between the two laser spots [5]. The width of the peak was a measure of the spread in transit times. The flow angle was easily derived from a number of measurements at different angular settings of the laser spots and by plotting the height of the peaks as a function of spot angle. In many cases this gave an accuracy of better than ±0.05 degrees for the flow angle.

In regions close to the shock waves where the particle slip was not uniform and in the wake where the spread in velocities was high, the correlograms were much broader [5]. The accuracy of the measurements consequently decreased but even for the worst cases reported the accuracy was still around 3-4% for mean velocity and 0.5 degrees for the flow angle.

Measurements close to the blade trailing edge and along the centre line of the wake for at least two chord lengths downstream all gave an oscillatory cross-correlogram similar to that shown in Fig.3. The spatial extent of the oscillating feature was only 1 - 1.5 mm and on either side of it the results reverted to the broad peak correlogram typical of the wake. This oscillatory component clearly represented a highly periodic event which we believe to be

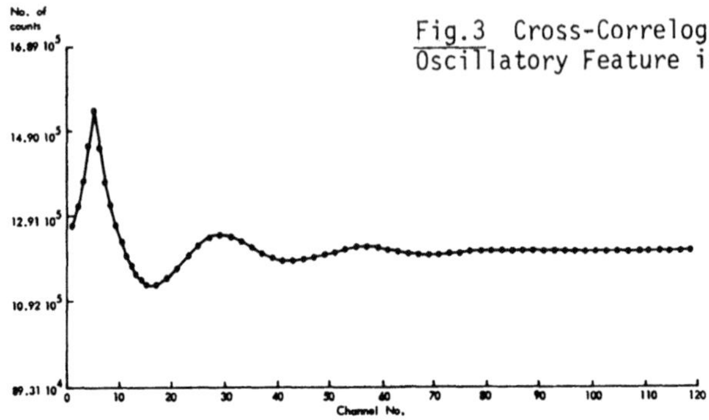

Fig.3 Cross-Correlogram Showing an Oscillatory Feature in the Flow (τ = 200 ns)

the vortices being shed from the blade trailing edge. No such correlogrammes have, to our knowledge, been reported before and they will be discussed in more detail in the next section.

4. Results and Discussion

The results shown in Fig.4 indicate that the base flow region extended downstream for about 2.5 mm and that it was roughly triangular in shape. The flow angle change on the suction face side of the wake shows a fairly sharp change of angle, 4.5 degrees, at the root of the trailing edge shock. From tables of the Prandtl-Meyer function this should correspond to a velocity change from 480 m/s to 430 m/s which was precisely the velocity measured 2 mm downstream of the shock. (This 2 mm roughly corresponds to the slip of a 0.5 μm → 1 μm particle passing through such a step change in velocity). The angle changes on the pressure face side of the wake were much more gradual giving a large compression fan and a weak trailing edge shock compared with the suction face shock.

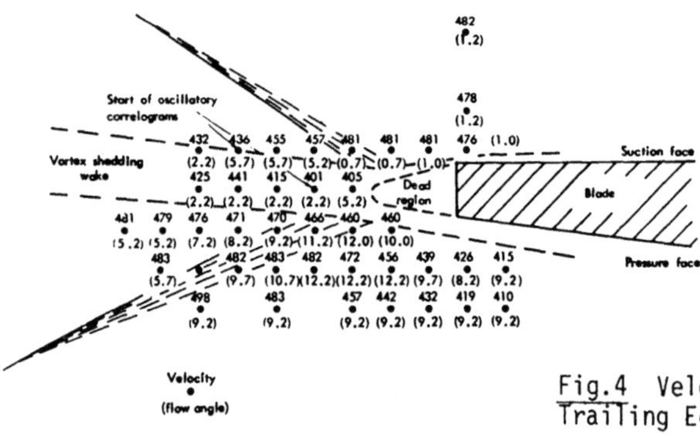

Fig.4 Velocities and Flow Angles in Trailing Edge Region

The first measurements of the vortex street were obtained 4 mm from the trailing edge and the shedding frequency was 160 kHz at Mach number 1.4. A Strouhal number S = t.f/u calculation based on a velocity (u) of 500 m/s and a trailing thickness (t) of 0.8 mm shows that the shedding frequency (f) would be 125 kHz for S = 0.2 and 185 kHz for S = 0.3. The Strouhal number is known to be a function of Mach number [6] and the range 0.2 to 0.3 is typical for transonic flows. Recent measurements by BRYANSTON-CROSS and CAMUS [2] show that above a Mach number of 1.0 the shedding frequency is a rapidly increasing linear function of Mach number with 110 kHz reported at Mach number 1.25.

Extrapolation of their results to Mach number 1.4 would give a frequency of around 150 kHz, well in the region reported here.

The oscillatory cross-correlation function can be derived by assuming a form for the intensity of the light scattered from the two laser spots. The presence of the vortices will clearly modify this intensity since no seeding particles will be able to penetrate the vortices and hence no light will be scattered during their passages across the measurement region. If it is assumed that the probability that a vortex will be traversing the spots at any time is given by a periodic function, then the cross-correlation function could have the form

$$R_{1,2} = I_0^2 \int_{-\infty}^{\infty} [\exp \frac{-2t^2}{\tau_c^2} + K \cos \omega t][\exp - \frac{2(t - t_0 + \tau)^2}{\tau_c^2} + K \cos \omega(t - t_0 + \tau)]dt$$

where I_0 is the incident laser intensity, τ_c is the coherence time $(r_0/\bar{v})$, t_0 is the time for the particle to traverse the two spots, τ is the correlator delay time and K is a constant modifying the strength of the oscillation. It can be shown that this reduces to

$$R_{1,2} = \frac{I_0 \sqrt{\pi}\, \tau_c}{4} |\exp - \frac{(\tau - t_0)^2}{\tau_c^2} + 8 K \exp \frac{-\omega^2 \tau_c^2}{8} \cos \omega(\tau - t_0)|$$

This shows that the peak will occur when $t_0 = \tau$, as expected, and since for the present measurements

$$\exp \frac{-\omega^2 \tau_c^2}{8} \approx 1$$

the peak will be superimposed on a slowly decaying cosine wave. Hence this approximate model gives the essential features measured in Fig.2. It is more likely to be useful at the high frequencies when the time for a vortex to pass a given point corresponds roughly to the time between vortices. At much lower frequencies the cross-correlation function should take the form of a series of Gaussian pulses and any decay would be the result of fluctuations in the time between consecutive vortices.

It is by definition that the above discussion assumes the presence of seed particles between the vortices and it is not clear how the particles would behave in such a flow. Only natural seeding was used in the wind tunnel but as in any device driven by a large compressor, the seeding level due to leaks could be high. However, our experience with count rates and varying the correlator clipping levels indicated that while the whole of the flow was seeded, it was not overseeded and the size of the seeding particles was most likely sub-micron. The particles were obviously being entrained into the wake as the velocity measurements showed but from the flow angle results it was clear that there was a very large fluctuation in flow direction at any point as a function of time [6]. However, the peak was still well defined and used to give an estimate of the mean velocity. Any attempt to extract any further information such as turbulence level from such a correlogram would be completely unjustified.

5. Conclusions

The Laser Transit Anemometer has been used for measurements of the velocity, flow angle and vortex shedding frequency in the base flow region close to the trailing edge of a blade in a transonic turbine cascade. The excellent spatial resolution obtainable with the anemometer was well suited to the challenge of this type of measurement and the results obtained agreed with those predicted.

6. References

1. R. R. Clements: J. Fluid Mech. 57, 321 (1973)
2. P. J. Bryanston-Cross, J. J. Camus: ASME paper 82-GT-132 (1982)
3. H. J. Heinemann, K. A. Butefisch: AGARD CP-277 (1978)
4. R. G. W. Brown, E. R. Pike: Optics and Laser Technology 317, Dec. (1978)
5. P. H. Richards, R. G. W. Brown: CEGB Report TPRD/M/1230/N82 (1982)
6. F. Motallebi, J. F. Norbury: J. Fluid Mech. 110, 273 (1981)

16. Measurements in Turbomachines Using Two-Spot Anemometry

Mary E. Gill, C.P. Forster, and R.L. Elder

School of Mechanical Engineering, Cranfield Institute of Technology, Bedford, England

Abstract

A Two-Spot laser anemometer and correlator have been used for flow measurements upstream of, and within the blade passages of a small turbocharger impeller. Velocity and turbulence values have been extracted from the correlator data using an analysis program based on the theory of ROSS[2]. The operation of this analysis is considered from the viewpoint of the user.

Areas of doubt still exist in the application of this method of analysis, especially as the need to compare correlograms taken at different angular positions necessitates some form of normalisation of the data. The importance of normalisation factors and their effect on the accuracy of the result has been investigated. The conclusions of this study are discussed.

Finally, some results of the measurements within the turbocharger are presented.

1. Introduction

The problem of defining the complex flow existing within small centrifugal compressors, which typically find applications in small gas turbines and turbochargers, is well known. At the same time, it is accepted that improved knowledge of these flow processes will lead directly to the design of more efficient compressors and associated savings in fuel.

The need for improved techniques for measuring the flow in the compressor was recognised both by the turbocharger and gas turbine manufacturing industries and this has led to the development of complex laser anemometry facilities capable of probing the complicated flows involved.

During the past few years at Cranfield a laser Doppler system has been used to measure the flow exhausting from the impeller and its development through both vaneless and vaned diffusers placed downstream of the impeller. In this region of the compressor, flow velocities up to 400 m/s and 15% turbulence are encountered in passage widths of the order of 7 mm wide and at blade passing frequencies of 24 KHz. Together these factors make this one of the most difficult applications of the laser Doppler system. With experience gained, however, and the use of the Fourier Transform data processing technique, many useful measurements have been made on a fairly routine basis and some of these results have been presented, ELDER et al.[1].

The success of this initial work outside the impeller led to increased interest in the flow field within the impeller passages, Fig.1. The proximity of solid surfaces necessarily encountered combined with the restric-

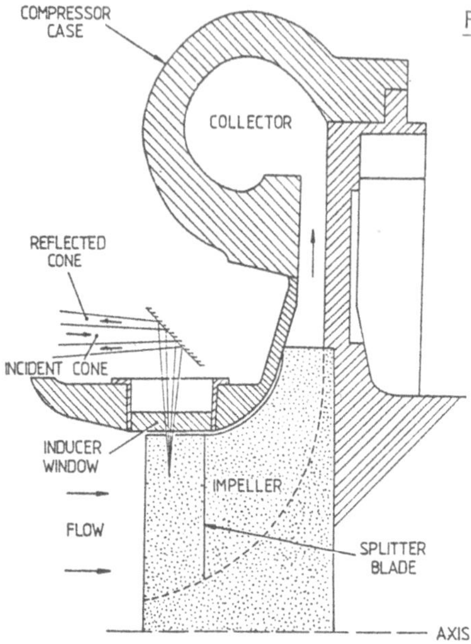

Fig.1. Profile of the compressor showing the impeller and optical access

tion of a backscatter view suggested the applicability of the Transit or Two-Spot anemometer, with its reduced probe length and high light intensity in the probe volume.

The transit anemometer has now been used to take measurements upstream of, and in the inducer region of the impeller, analysis of the data being achieved by a computer program based on the analysis of ROSS[2]. This application, and particularly the accuracy of the analysis procedure is discussed in this paper.

2. Experimental Arrangement

The system used for these measurements was a Malvern Instruments Two Spot/Fringe Anemometer Type 4772, operating in its two spot mode. A 15 mw Helium-Neon laser was employed and a commercial seeding unit was used to produce seeding particles of corn oil of mean diameter 0.5μ. Although the two-spot anemometer will operate successfully using only natural seed in the atmosphere, it was felt that by using artificial seeding, it was possible to ensure that the size of the particles being detected was sufficiently small to be following the air flow of interest. A beam spacing of 450μ and a sample time of 200 ns was selected.

The outputs from the two photomultiplier tubes were passed through similarly adjusted constant fraction discriminators, the output of these providing the input to a Malvern digital correlator type 7026. The computed cross-correlation function is transferred to a DEC mini-computer for analysis and storage.

3. Analysis of Data

The analysis of the data output from the correlator is achieved using the analysis of ROSS [2]. This was discussed by ELDER et al.[1] and it suffices here to say that the analysis relies on an isotropic turbulence model which enables the measured correlation function to be related to the probability distribution of the flow velocity in the direction of measurement (1). This equation can also be expressed in terms of the measured time delay, τ, (2). Integration of (1) over velocity, resulting in (3), represents the probability of any particle passing through both spots at a given orientation.

$$p(U,\alpha) = \frac{U}{\pi\sigma^2} \exp - \left\{\frac{\overline{U}^2 \sin^2(\alpha-\alpha_0)}{\sigma^2}\right\} \exp - \left\{\frac{(U - \overline{U}\cos(\alpha-\alpha_0))^2}{\sigma^2}\right\} \quad (1)$$

$$p(\tau,\alpha) = \frac{ks^2}{\tau^3}\exp - \left\{\frac{(1/\tau - (\overline{U}/s)\cos(\alpha-\alpha_0))^2}{(\sigma/\overline{U})^2_\tau \quad (\overline{U}/s)^2}\right\} \quad (2)$$

$$\int p(U,\alpha)dU = p\alpha\ (\alpha) = K \exp - \left\{\frac{(\alpha-\alpha_0)^2}{\sigma^2/\overline{U}^2}\right\} . \quad (3)$$

Although it is possible to build up a full probability (U,α) distribution for the flow field, in general, all that is required for flow analysis is the mean flow velocity (speed and direction) and a value for turbulence intensity. For this reason, only these values are extracted from the measurements using (2) and (3).

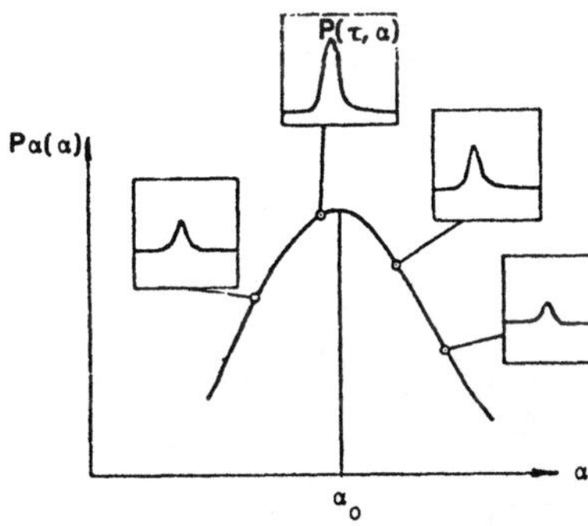

Fig.2. Calculation of mean flow angle from correlation functions taken at different angular orientations

The technique may be summarised as follows, Fig.2. Correlation functions are recorded for various angular orientations of the two laser beams. For each correlogram a count number representing the number of counts under the peak is calculated, and the result displayed on the computer VDU. This number represents the degree of alignment with the flow at the particular orientation of the two spots and its variation with angle is given by (3). These integrated counts are displayed as a function of orientation. When a maximum is obtained, the mean flow direction α_0 can be calculated from a fit of (3) to the data. The mean velocity and turbulence are then determined by fitting (2) to the correlation function taken closest to the mean flow angle.

These computations are carried out online, so that the operator has a check on whether sufficient data has been collected to complete the computations, and also to determine at which orientation subsequent measurement should be taken. The raw data can be stored on 'Floppy disc' for subsequent analysis off-line, but in general, when the quality of the data is considered to be adequate, only the resultant values of velocity, flow angle and turbulence are recorded.

A particular problem that becomes apparent in using this analysis procedure is the necessity of comparing successive correlograms taken at different angles and necessarily at different times. In order to carry out this comparison some form of normalisation is necessary but becomes complicated because, during the time over which this data is accumulated, variations in laser power, particle seeding density, and background flare may occur. The figure of interest in determining mean flow direction is the proportion of particles detected in the measurement volume that have a trajectory passing through both laser beams, thereby giving a contribution to the peak of the correlation function. This figure reaches a maximum when the plane of the beams is aligned with the mean flow direction. In turbulent flow, there are always particles present that have velocities out of alignment with the plane of the beams, and these give rise to uncorrelated signals contributing to the background level of the function. Such events are indistinguishable from random noise counts arising from background flare. Hence, normalisation by the ideal factor, that of the number of particles detected in the measurement volume, is not possible, and other forms of normalisation have been proposed. The following section discusses some of these forms and the associated errors estimated to be introduced by the uncertainty of the normalisation procedure.

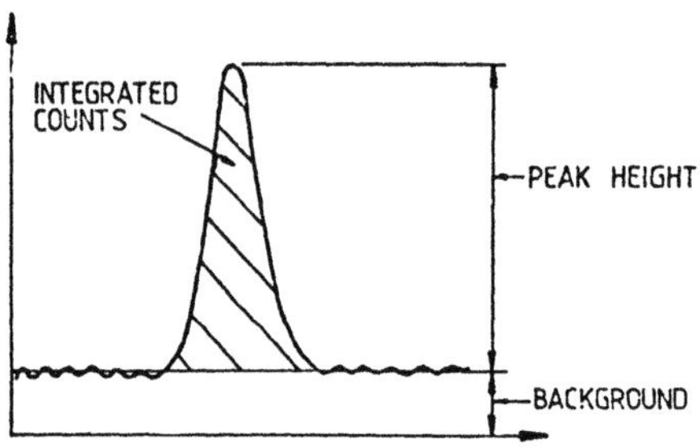

Fig.3. Effect of environmental parameters on the shape of the correlation function

PARAMETER INCREASED	VARIABLES AFFECTED	DUE TO	MAGNITUDE OF EFFECT
LASER POWER	BACKGROUND PEAK HEIGHT	INCREASED SCATTERING " "	BOTH INCREASE PROPORTIONATELY
SEEDINGRATE	PEAK HEIGHT BACKGROUND	INCREASED PARTICLE SCATTERING PARTICLES OUT OF ALIGNMENT WITH BEAMS	DEPENDENT ON ORIENTATION TO FLOW
BACKGROUND FLARE (NOISE)	BACKGROUND	INCREASED NOISE	SECONDARY EFFECT
PMT THRESHOLD	BACKGROUND PEAK HEIGHT	SENSITIVITY & NOISE REJECTION	GREATEST CONTRAST OBTAINED AT OPTIMUM THRESHOLD LEVEL

4. Normalisation Considerations

The various environmental parameters which may affect the correlation function are reviewed in Fig.3 together with the major influence of these parameters. It can be seen that variation of any of these (laser power, seeding rate and background flare) might affect the precise form of the correlation function. As any change in signal rate can be caused by a change in any of these parameters, quantification of these effects and accounting for their presence cannot be achieved. For this reason, some parametric studies have been carried out in an attempt to define the uncertainties involved.

Operating at a real measurement station in a running compressor (so that any effects that occur in real cases might be included) repeated measurements were taken over a period of time at constant compressor operating conditions. Firstly the measurement was repeated at nominally constant conditions of laser power, seeding rate, PMT threshold level and measuring time, and an uncertainty level could be placed on the measured variables of the system. Secondly, the measurements were repeated with varying seeding rates and PMT threshold levels and finally with high background noise levels due to the accumulation of seeding oil in the system. The first of these tests only is discussed here.

Over the duration of the experiment the scatter in the peak height (number of counts in the fullest register) of the correlation function and the (non-normalised) integrated counts was 20%. (The non-normalised integrated counts represent the number of counts under the correlation function having subtracted the background level). The monitor channel contents showed a similar 20% variation. This might be attributed to drift in any of the components of the system, or of the operating condition of the test compressor itself. Whatever the cause of the variation, it would be expected to be present at any time when measurements were being taken.

The integrated counts (area under the peak of each correlogram) were normalised by various factors, as shown in Table 1. The scatter over the experiment could be reduced from 20% for non-normalised counts to 14-16% with normalisation, as shown in column A of the table. Little distinction between most of the commonly used normalisation techniques was found. Column B in the table shows the probable variation that would occur during the time normally required to take a measurement, and is appreciably less, typically 6%.

Table 1 Observed uncertainties in integrated counts after normalisation

FACTOR	MAXIMUM ERROR IN NORMALISED COUNTS (%)	LIKELY ERROR OVER NORMAL OPERATION (%)
AU + BU	14	6
AC + BC	16	6
AU	25	21
$(AU \times BU)^{1/2}$	14	5
$(AC \times BC)^{1/2}$	16	7
AU × AC	35	29
BACKGROUND LEVEL	17	13

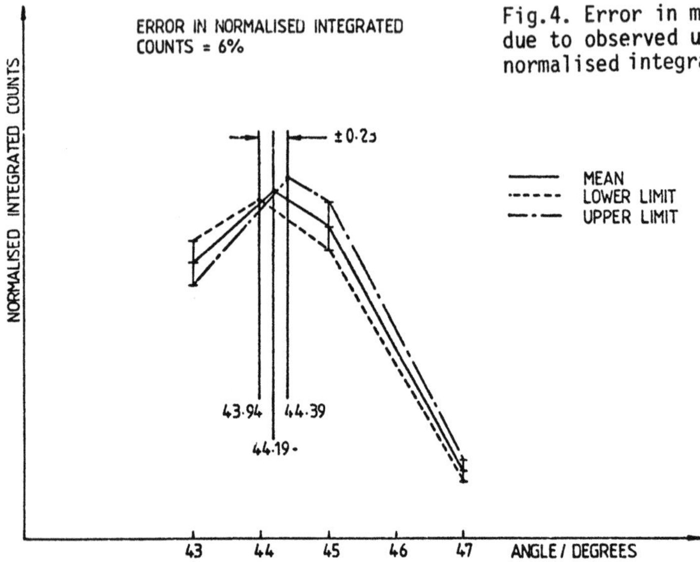

Fig.4. Error in mean flow angle due to observed uncertainty in normalised integrated counts

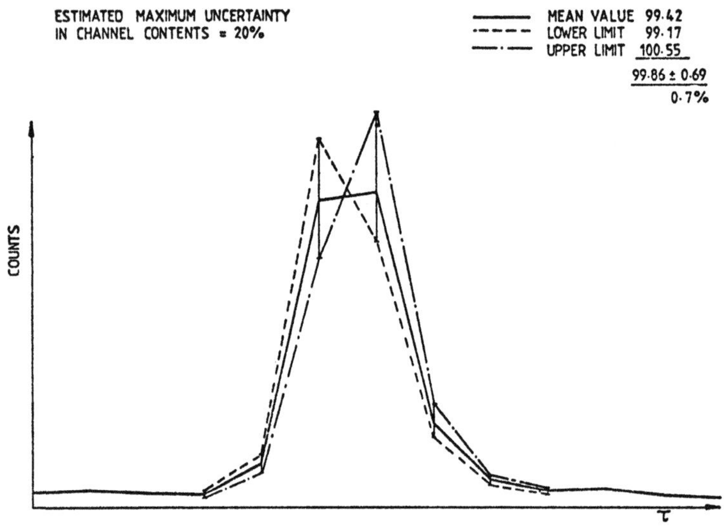

Fig.5. Error in mean velocity due to observed uncertainty in channel contents

It might be concluded from the above that although normalisation (in the format described) tends to produce the desired effect of reducing the scatter neither is the procedure exact nor its precise format crucial.

The effect of these experimental variations on the accuracy of the final result were then considered. By taking the extreme cases, upper and lower error limits were obtained. For the case of normal operation, the observed error of 6% in normalised counts resulted in an uncertainty of mean flow angle estimation of about $\pm 0.2^{o}$, Fig.4. The maximum error in integrated counts of 16% that was observed over a longer period of time resulted in an estimated angle accuracy of ± 1°. The uncertainty in mean velocity estimation is seen to be less than 1%, Fig.5. The accuracies quoted here agree quite well with those claimed elsewhere, SCHODL [3].

The small uncertainty in estimation of flow angle from the apparently very uncertain normalised integrated counts parameter can be explained by consideration of (3), where K is defined as a constant. Variations in K occur between measurements due to the factors discussed here. This variation in K is much smaller than the variation due to the exponential term and the resultant error in the definition of the mean flow angle is correspondingly small. At higher turbulence levels, greater problems would be encountered as the rate of change of the exponential term is less, thereby permitting variations in K to become of more significance.

5. Results

An example of the measurements that have been obtained using the Two-Spot laser anemometer is given in Fig.6. The direction and magnitude of the flow vectors have been measured at positions across the blade passage, between the leading edge of the full blade and that of the splitter blade. Only mid-stream line vectors are shown here but both turning of the flow to take up the shape of the rotating blade passage and splitting of the flow by the splitter blade are clearly defined.

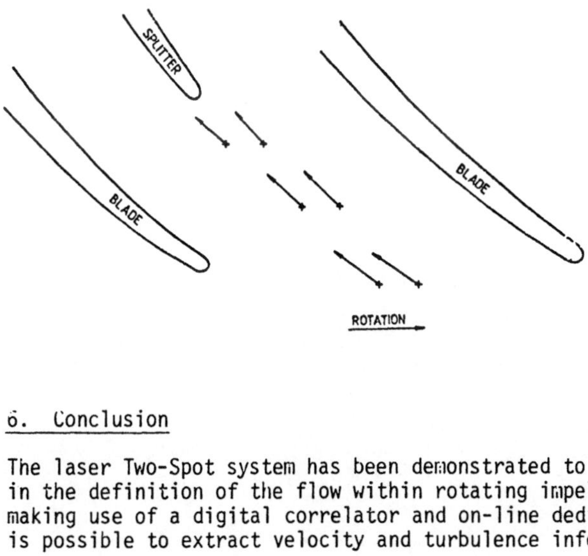

Fig.6. Example of measurements taken within the blade passage of the impeller

6. Conclusion

The laser Two-Spot system has been demonstrated to be an invaluable tool in the definition of the flow within rotating impeller blade passages. By making use of a digital correlator and on-line dedicated mini-computer it is possible to extract velocity and turbulence information from the flow.

For the case considered, where the flow has a turbulence of 5-8% and is also unsteady with time, the error bounds have been found to be well within the accuracy to which the flow might be expected to be defined.

Acknowledgements

The authors would like to thank Holset Engineering Ltd., Huddersfield, for their permission to include an example of the results obtained under their sponsorship.

References

1. Elder R.L., Forster C. and Gill M.E. 1980 Proc. 4th Int. Conf. Photon Correlation Techniques in Fluid Mechanics. 25-27 Aug. 1980

2. Ross M.M. 1980 Optica Acta 27 511-528 April 1980

3. Schodl R. 1980 J. Fluids Eng. 102 412-419 Dec 1980

17. A Scaling Law for Performance Evaluation of a Laser Time-of-Flight Velocimeter

R.F. Kelley, L.S. Hsu, H. Shimizu, and C.Y. She

Department of Physics, Colorado State University, Fort Collins, CO 80523, USA

1. Introduction

During the past decade, a variety of laser techniques have been proposed and experimentally demonstrated for atmospheric wind speed measurements [1] with a varied degree of success and practical usefulness. In most cases, an investigator proposed a technique of laser velocimetry and/or reported a successful atmospheric wind-speed measurement at a convenient range without discussing the time required for such a measurement or the dependence of the measurement rate on laser power and aerosol condition. Such an investigation, in our view, is incomplete; the lack of theoretical understanding makes the evaluation of system performance obscure and the prediction of the potential usefulness of the laser velocimeter impossible.

The need for a figure of merit for performance evaluation in terms of measurement rate taking into account the available aerosol conditions has been recognized in our earlier work [2,3]. We restrict ourselves to a class of laser time-of-flight velocimeter (LTV) using laser backscattering from individual aerosol particles and digital photon-burst correlation for data analysis. With such sensitive photon-counting methods, field wind-speed measurements have been demonstrated for atmospheric ranges around 100 m using a cw laser with approximately 1 W of power [3-6]. For this class of velocimetry, we have recently succeeded in formulating [7] a scaling law predicting the atmospheric wind-speed measurement rate F, as a function of laser power P_o, and measurement range R, under a given aerosol condition. By including several practical considerations, such as the Gaussian laser beam profile, an inverse power law for aerosol particle-size distribution, $dN/dr = C_o r^{-\beta}$, where r is the particle radius, a simple backscatter-differential cross section, $K_o r^2$, and a theoretically calculated minimum detectable total number of photon-burst counts (per transit), n_s, under given experimental conditions, our most recent version of the scaling law has been derived theoretically and tested experimentally under limited laboratory conditions [8].

The scaling law, (8) of [8], can be rewritten as

$$\frac{F}{\ell_{\parallel} v\rho} = \frac{\sqrt{\pi}\, C_o}{(\beta-1)^{3/2}}\, r_s^{(1-\beta)} \tag{1}$$

where

$$r_s^{-2} = \gamma/n_s \tag{1a}$$

$$\gamma = \left(\sqrt{\frac{2}{\pi}}\frac{K_o A}{h\nu}\right)\frac{1}{v}\left(\frac{\eta P_o}{\rho R^2}\right) \tag{1b}$$

$$n_s = \sqrt{\frac{\pi}{2}}\,(\kappa+1)\,\frac{\rho}{v\tau_o}\,\exp\left[\frac{1}{2}\left(\frac{v\tau_o}{\rho}\right)^2 (m-1)^2\right] \tag{1c}$$

and ρ, $\ell_{\parallel}$, v, A, $h\nu$, η, κ, m, and τ_o are, respectively, the e^{-2} beam radius in the viewing volume, the dimension of the viewing volume along the laser beam, wind speed, cross-sectional area of the receiving telescope, photon energy, the efficiency of the detection system, the clip-count level per sample time, number of sample times that the signal exceeds the clip level in a measurement, and the time per digital sample. γ/n_s is identified as r_s^{-2} where r_s is the minimum detectable particle radius under given experimental conditions. The quantity $F/(\ell_{\parallel} v\rho)$ in (1) can be interpreted as the number of particles per unit volume having radius greater than r_s, a quantity familiar to aerosol scientists [9]. When experimental conditions are such that a small range of r_s is encountered, the aerosol parameters C_o and β are constant and (1) may be used to determine these parameters experimentally. This was done in [8] and together with the measured photon-count distribution, our scaling theory for performance evaluation of a laser time-of-flight velocimeter (LTV) was verified in a self-consistent manner.

However, when a large range of r_s is encountered, C_o and β may not be constant. In general, the measured value of β will increase as the aerosol radius increases. To obtain a wide range of r_s in our correlation experiments, the laser power P_o and beam radius ρ must be varied sufficiently. In this manner, the validity of (1) can be verified for a much wider range of r_s. In this paper, we report the results of such a laboratory experiment.

Preliminary field experiments for determining the scaling law (1) using data obtained from measurement ranges between 25 m and 150 m have been carried out during the month of November 1981 at Fort Collins, Colorado. The results of this set of field experiments is reported and a prediction of wind-speed measurements at a longer range of 500 m is made.

2. Experimental Results

Although two laser beams are normally used in an LTV measurement [3], we consider for simplicity the use of a single Gaussian-profiled laser beam. This is adequate for the establishment of the scaling law since, in a typical two-beam LTV setup, most individual particles traverse both laser beams [8]. As a result, a single-beam arrangement identical to that shown in [8] is used.

Following the same procedure outlined in [8] with a laser power P_o ranging from 0.3 mW to 0.4 W and beam radius ρ ranging from 40 to 900 μm by using different focussing lenses with R = 1.7 m and v = 5.2 m/sec, we have conducted extensive laboratory experiments covering r_s values from 0.1 to 4.0 μm. The experimental result when plotted as $F/(\ell_{\parallel} v\rho)$ vs r_s is shown in Fig.1. The resemblance of Fig.1 to the typical cumulative number distribution of aerosols [9] is quite obvious. The β value changes from 3.2 to 6.1 as r_s increases in general agreement with the expected room aerosol distribution. Our previous result of $\beta \simeq 4.3$ occurs at $r_s \simeq 1.5$ μm as a special case.

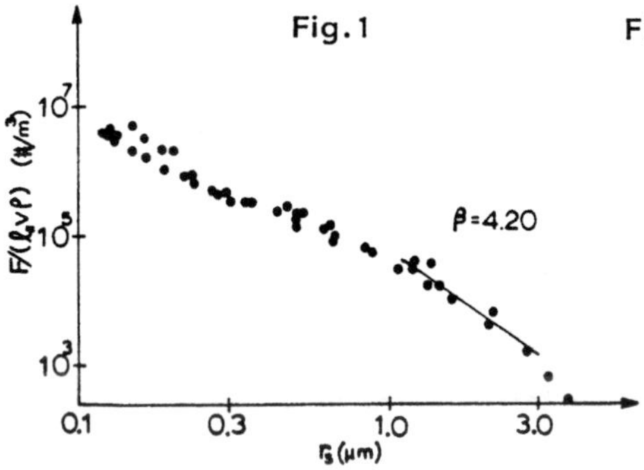

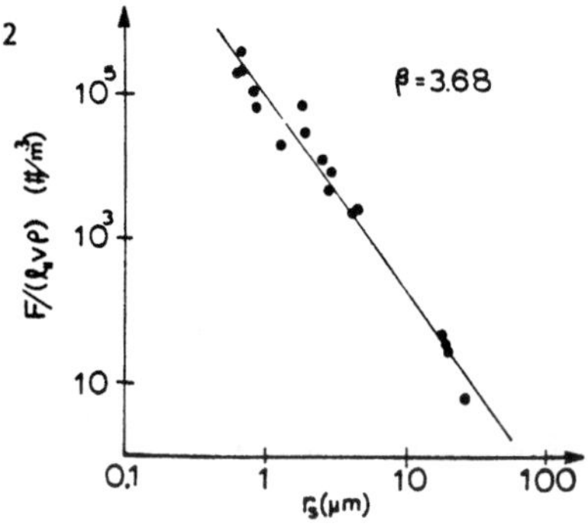

Fig.1. Scaling law of the laboratory experiments. The LTV measurement rate F divided by $(\ell_{\parallel} v\rho)$ is plotted as a function of minimum detectable radius r_s which depends on experimental parameters such as laser power P_0 and detection efficiency η. A wide range of r_s (0.1-4 μm) is covered and the slope of this plot yields varied values for the power of the aerosol-size distribution, $\beta \simeq 3.2$-6.1

Fig.2. Scaling law of the field experiments at Fort Collins, Colorado during the first half of November 1981. The experimental measurement rate F divided by $(\ell_{\parallel} v\rho)$ is plotted against the minimum detectable radius r_s which depends on experimental parameters such as laser power P_0 and measurement range R. The measurement range was varied to 152 m. LTV performance at a longer range such as R = 500 m may be predicted from this graph

Preliminary field experiments for the verification of the scaling law have been conducted during November 1981. A transmitting telescope of 5 cm diameter was used for measurement ranges from 25 m to 152 m. Due to a limiting aperture in the transmitting telescope, the laser power delivered to the viewing volume ranged from 7 mW to 96 mW. The detection telescope was a Celestron 8 with $A = 0.032\ m^2$. The efficiency of the detection system was measured to be around 6.5×10^{-3}. The beam radius in the viewing volume was calculated by a simple diffraction formula because the transmitted beam was circular and nearly uniform. A plot of $F/(\ell_{\parallel} v\rho)$ vs r_s for these field experiments is given in Fig.2 with a least-square fitted β value of 3.68. A smaller β value for field data as compared to the laboratory data is, as expected, the result of less settling in the field. Using the least-square fitted curve, the scaling law for field experiments under the experimental conditions of November 1981 is

$$\frac{F}{\ell_{\parallel} v\rho} = 6.93 \times 10^{-12}\ (r_s)^{-2.68} \quad . \qquad (2)$$

3. Prediction and Conclusions

Using the experimentally determined scaling law, i.e., (2), we can estimate the typical measurement rate F for ranges up to 150 m and predict the measurement rate at a longer range such as R = 500 m. For example, at $r_s = 19.3$ μm, corresponding to R = 152 m, $\rho = 2.09 \times 10^{-3}$ m, $\ell_{\parallel} = 1.41$ m, and v = 0.93 m/sec, the average measurement rate is $F = 8.15 \times 10^{-2}\ sec^{-1}$ (or 12 sec per measurement).

Using (1) and the measurement rate at R = 152 m to predict the measurement rate at R = 500 m, there is difficulty in estimating the value of n_s. As outlined in [8], n_s is calculated form the measured value of $(d\bar{n}/dt)\tau_o$ where $d\bar{n}/dt$ is the average count rate. Since both ρ and τ_o are proportional to R with constant v, we can keep n_s constant by keeping $(d\bar{n}/dt)\tau_o$, thus κ and m of (1c), constant. Since $\ell_{\parallel} \alpha R^2$, to keep the value of $(d\bar{n}/dt)\tau_o$ constant for a longer measurement range, we can, according to (7) of [8], use a smaller laser power by letting $P_o \alpha 1/R$. Using a smaller power for calculating measurement rate at longer ranges, we would arrive at a conservative prediction while simplifying the calculation. By letting $P_o \alpha 1/R$, γ of (1b) will vary as R^{-4} and the predicted measurement rate, F(R = 500 m) can be calculated from the following ratio:

$$\frac{F(R = 500 \text{ m})}{F(R = 152 \text{ m})} = \left(\frac{500}{152}\right)^3 \left[\frac{1}{\left(\frac{500}{152}\right)^4}\right]^{2.68/2} = 6.03 \times 10^{-2} \quad .$$

For F(R = 152 m) = 8.15×10^{-2} sec^{-1}, we arrive at the predicted measurement rate of 4.9×10^{-3} sec^{-1} for R = 500 m. This prediction of approximately 3.4 min for every wind-speed measurement at R = 500 m is not unreasonable in view of the fact that our prediction was quite conservative by using a lower laser power than is available. The use of lower power for our prediction tends to compensate for the effect of turbulence not included. The effect of turbulence which is not known exactly at this point may reduce the signal arrival rate. We are currently in the process of conducting simultaneous C_n^2 measurements and performing field experiments with a transmitting telescope constructed with a surplus 18 cm diameter telescope objective and an eyepiece. The laser beam can transmit through this telescope without being apertured. As a result, more than five times the laser power can be delivered to the scattering volume as before. This increased laser power should improve the predicted LTV performance further.

With this new transmitting telescope, we have tried some wind-speed measurements at 500 m range with one laser beam at night. Our preliminary result indicates a measurement of one every 2 min with P_o = 0.3 W. Further experiments using the new transmitting telescope with simultaneous C_n^2 measurements, and wind-speed measurements at R = 500 m using two laser beams, are being conducted.

Acknowledgement

The support of the U.S. Army Research Office under contract DAAG29-80-C-0038 is gratefully acknowledged.

References

1. J. W. Bilbro, "Atmospheric laser Doppler velocimetry: an overview," Opt. Eng. 19, 533-542 (1980).
2. K. G. Bartlett and C. Y. She, "Remote measurement of wind speed using a dual backscatter laser Doppler velocimeter," Appl. Opt. 15, 1980-1983 (1976).
3. K. G. Bartlett and C. Y. She, "Single-particle correlated time-of-flight velocimeter for remote wind-speed measurement," Opt. Lett. 1, 175-177 (1977).
4. L. Lading, A. S. Jensen, C. Fog, and H. Andersen, "Time-of-flight laser anemometer for velocity measurements in the atmosphere," Appl. Opt. 17, 1486-1488 (1978).

5. F. Durst, B. Howe, and G. Richter, "Long range LDA wind velocity measurements using visible laser radiation," in Proceedings of the Fourth International Conference on Photon Cirrelation Techniques in Fluid Mechanics, W. T. Mayo, Jr. and A. E. Smart, eds. (Joint Institute for Aeronautics and Acoustics, Stanford, Calif., 1980), Paper 3.
6. L. Danielsson, "Experiments with laser-fringe velocity measurements over large distances in the atmosphere," in Proceedings of the Fourth International Conference on Photon Correlation Techniques in Fluid Mechanics, W. T. Mayo, Jr. and A. E. Smart, eds. (Joint Institute for Aeronautics and Acoustics, Stanford, Calif., 1980), paper 4.
7. K. G. Bartlett and C. Y. She, "Single-particle correlation techniques for remote measurement of wind speed: aerosol condition and measurement rate," J. Opt. Soc. Am. 69, 455-459 (1979).
8. C. Y. She and R. F. Kelley, "Scaling law and photon-count distribution of a laser time-of-flight velocimeter," J. Opt. Soc. Am. 72, 365-371 (1982).
9. S. Twomey, Atmospheric Aerosols, (Elsevier, Amsterdam, 1977), p. 9.

18. Flow Measurements with a 5 ns Correlator

A.E. Smart and W.T. Mayo, Jr.

Spectron Development Laboratories, Inc., 3303 Harbor Boulevard, Suite G-3, Costa Mesa, CA 92626, USA

Abstract

Spectron Development Laboratories, Inc., has implemented a 5 ns interval correlator for photon detection, ultrasonic acoustics, and other applications where existing digital correlators have been too slow or had too few stores. The time multiplexable 'CORRELEX' is operable stand-alone and has interface capability for high-speed serial and parallel computer communication. A brief system description is included.

This paper reviews three series of laser flow measurements with CORRELEX processors. Two refer to laser fringe and transit anemometer measurements with a small test jet in the laboratory for the U. S. Air Force Arnold Engineering Development Center. The third is a series of boundary layer measurements taken on a 'bump model' in the NASA Ames 6' x 6' Transonic Wind Tunnel. Graphic computer printouts of selected data are shown and a brief discussion of precision and limitations is included.

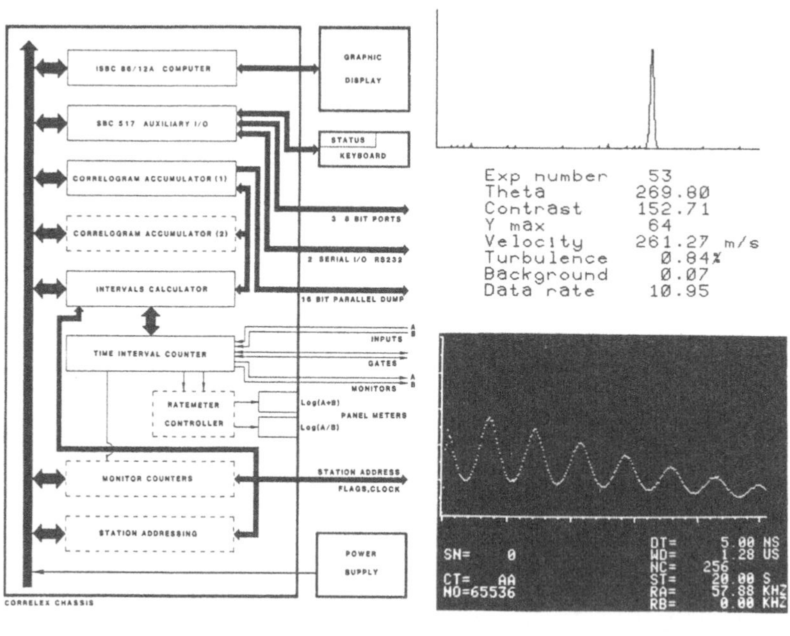

CORRELEX diagram and example data from laser transit and fringe anemometers

19. Velocity, Turbulence, and Shear Stress from Laser Transit Anemometer Correlograms

W.T. Mayo, Jr. and A.E. Smart

Spectron Development Laboratories, Inc., 3303 Harbor Boulevard, Suite G-3
Costa Mesa, CA 92626, USA

Abstract

Several authors have published analysis schemes for both velocity magnitude and direction from laser transit anemometer data obtained from a digital correlator. The present paper extends and clarifies these ideas into a computational scheme which includes the extraction of turbulence and shear stress from a general two-dimensional model. The operation of this scheme is demonstrated upon real LTA data from laboratory and NASA Wind Tunnel sources. The support of the U. S. Air Force Arnold Engineering Development is gratefully acknowledged.

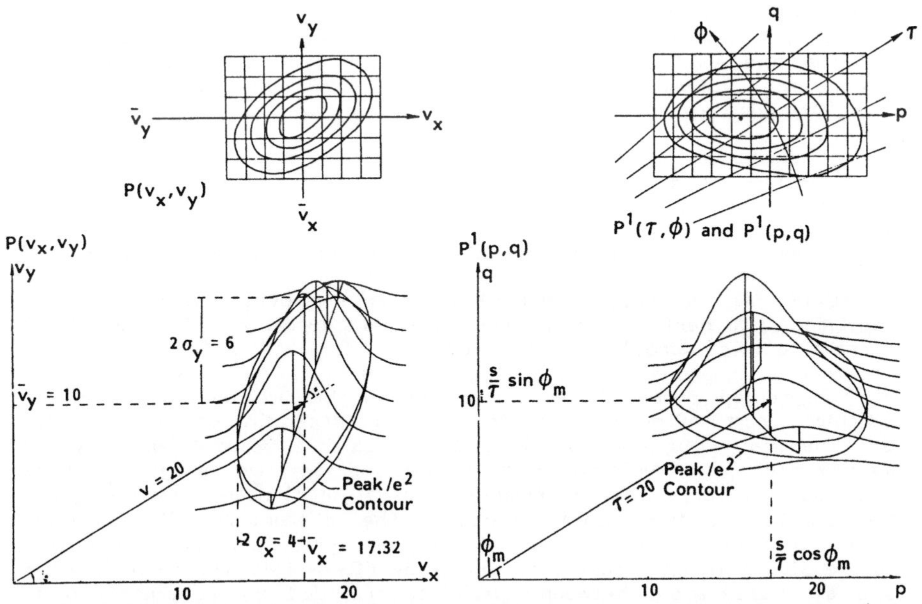

PDF and contour in velocity space $P(v_x,v_y)$ — PDF and contour in measurement space $P^1(p,q)$

Relationship of velocity and measurement spaces for a laser transit anemometer

20. The Laser Gradient Anemometer

Steen Hanson

Risø National Laboratory, Electronics Department, DK-4000 Roskilde, Denmark

Abstract

An optical method for measuring spatial velocity gradients is proposed. The method is based on the enhancement of the coherent beating between two Doppler-shifted light fields from two moving particles in the volume where the measurement takes place. The introduction of various optical filters in the detector plane determines the relative separation between the particles whose scattered fields add coherently on the detector.

1. Introduction

The investigation of various flows often calls for determining the instantaneous velocity gradient, important as the driving force in inducing turbulence. Different detection schemes have been proposed. MISHINA and ASAKURA |1| divide the laser beam and illuminate two measurement volumes with a distance of approximately 20 microns and uses the Doppler-shifted light from these volumes to give the velocity gradient.

FERMIGER et al. |2| describe a method for investigating velocity gradients and turbulent diffusion. A pulsed laser "writes" a phase grating based on the instantaneous flow structure, and a second laser beam "reads" the grating and its altered flow to indicate the change.

FULLER |3| describe a method based on homodyne detection of light seattered from particles in a suspension. The correlation function of the current contains turbulence information.

The method proposed here is based on the Doppler shift of light scattered from moving particles in an amount depending on their velocity and the scattering K vector. In ordinary Laser Doppler Anemometry the major part of the modulated signal stems from single particle contributions, or rather heterodyning of light scattered form a given particle by the two incomming beams. The coherent contribution, i.e. mixing of light scattered from two particles, gives a minor contribution to the signal in most cases, as shown by DRAIN |4|. The restriction in spatial displacement between particles, provided the coherent contribution should still be present, has been analysed by HANSON |5| and the relative strength of the coherent versus the incoherent part of the signal from an LDA as a function of the numerical aperture has been analysed by SHI-KANG WANG |6| who showed that additional information can be extracted from the signals.

The basic concept for this method is the introduction of a spatial filter in front of the receiving aperture which acts as an optical processor to eliminate all coherent two-particle contributions, except those from particles with the correct mutual separation as defined by the filter.

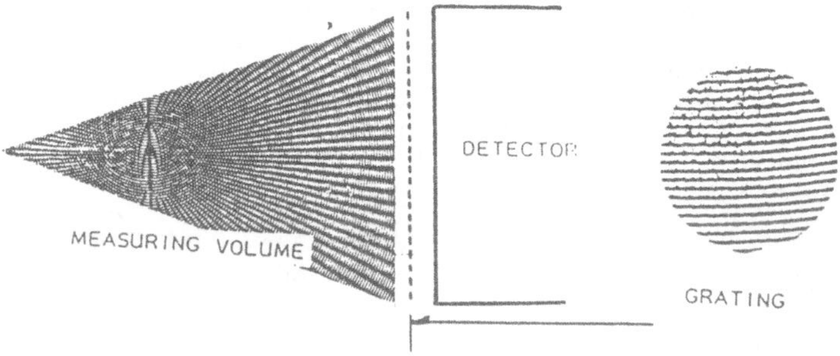

Fig. 1. Interference pattern produced by two spherical waves from two transversally separated particles. A grating is introduced to remove out-of-phase portions of the beat signal

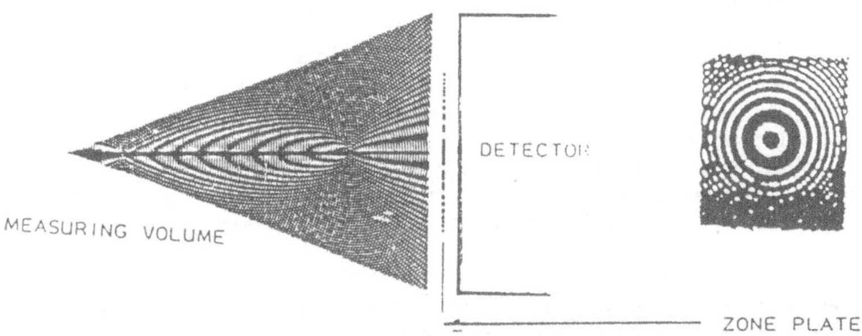

Fig. 2. Interference pattern produced by two spherical waves from two longitudinally separated particles. A zone plate is inserted to remove out-of-phase portions of the beat signal

The principle can be explained by the use of the Moire' figures introduced in |5|. If a single beam is scattered from two particles, two spherical waves emanate and interfere on the detector, as shown in Fig. 1. If the detector were small compared to the distance between the interference lines in Fig. 1, the beat frequency would be present. The output of a detector that is large is a superposition of beat frequencies with alternating phase across the detector, causing the beat frequency to vanish. If a mask is introduced which blocks out all the out-of-phase part of the detector surface, the beat frequency will reappear at the detector output, but only if the relative separation between particles is matched to the mask. The mask which defines a relative separation perpendicular to the optical axis turns out to be an ordinary grating. As shown in Fig. 2, the mask which defines a relative separation along the optical axis is a zone plate. Depending on the numerical aperture of the aperture/mask, a small deviation from the straight fringes at the grating will enhance the signal.

Such a mask is made most conveniently by superimposing two Gaussian beams originating from different points and recording the pattern on a holographic plate. The development could either give a transmission filter or, preferably, a phase filter.

The effect of the grating in a semi-classical photon description is to introduce an ambiguity, as seen from the detector point of view, in determining which of the two scattering particles is the origin of the received photon. This ambiguity is present only when the particle separation matches the grating space and is revealed by the detector as a current which contains information about both scattering particles, i.e. the beat frequency.

2. Fourier Optical Model for the Laser Gradient Anemometer in the transversal mode

The analysis of the LGA is most conveniently performed by use of the Fourier optical model as employed by LADING |7| for the laser Doppler anemometer.

We will treat a two-dimensional model as shown in Fig. 3. Three planes are considered:

- The measuring volume illuminated with a Gaussian beam $\tilde{G}_1$ of width σ_1 and power P. The position of particle no. i is (x_i, y_i) and the scattering k vector is (k_x, k_y). All particles are supposed to have equal scattering cross-sections.
- Receiving aperture with a Gaussian pupil function $\tilde{G}_2$ of width σ_2 followed by the absorption grating with a transmission function H(x).
- Detector plane preceded by a pinhole $\tilde{G}_3$ of Gaussian shape of width σ_3.

The choice of Gaussian-shaped apertures simplifies the mathematical expressions without changing the basic result.

The mathematical expressions for the various optical elements are described using the notation explained in Appendix A:

Field emitted from the measuring volume:

$$U_1 = \sum_{i=1}^{N} \tilde{G}\{x_i, y_i, 0, 0, (P/\pi\sigma^2)^{1/2} \exp\left(-\frac{x_i^2 + y_i^2}{2\sigma_1^2}\right) \cdot \exp(i(k_x x_i + k_y y_i))\} \quad .$$

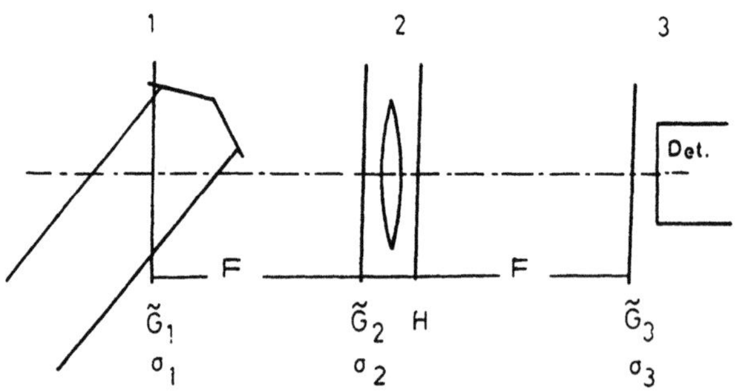

Fig. 3. Fourier optical model: 1) measuring volume, 2) receiving aperture with grating, and 3) detector with pinhole

Detector aperture:

$$\tilde{G}_2\{0,0,\sigma_2^2,\sigma_2^2,1\}\ ,\ H(x)=(1-\alpha/2)+\alpha/2\cdot\cos(2\pi g x)$$

where g is the grating constant in lines pr. meter.

Pinhole function: $\tilde{G}_3\{0,0,\sigma_3^2,\sigma_3^2,1\}$.

The field incident on the detector can now be written

$$U_3=\sum_{i=1}^{N}(F(H\cdot\tilde{G}_2)*\tilde{G}_{1i})\tilde{G}_3 \qquad (1)$$

where F is the Fourier transform and ✱ means convolution. By using standard formulas for Fourier transformation, multiplication, integration, and convolution of Gaussian functions we get the expression for the field in the detector plane from particle no. i. This consists essentially of the diffraction-limited picture of the particle itself called $\tilde{G}_{oi}$ and the two diffracted images of the same particle, $\tilde{G}_{+i}$ and $\tilde{G}_{-i}$. The detector current can be approximated as

$$i(t)\approx\iint_{-\infty}^{\infty}\Big(\sum_i^N|\tilde{G}_{0i}|^2+|\tilde{G}_{+i}|^2+|\tilde{G}_{-i}|^2+2\sum_{ij}^{N}{}'(\tilde{G}_{0i}\tilde{G}_{-j}+\tilde{G}_{0i}\tilde{G}_{+j}+\tilde{G}_{+i}\tilde{G}_{-j})\cos(k_x\Delta x_{ij}+k_y\,y_{ij})\,dx\,dy \quad . \qquad (2)$$

The interpretation of the term is conveniently explained in the case of two particles in the measuring volume as shown in Fig. 3. The three squared single-particle contributions in (2) give a pedestal present whenever a particle is present. The three two-particle contributions control the existence of the time-dependent Doppler term of interest.

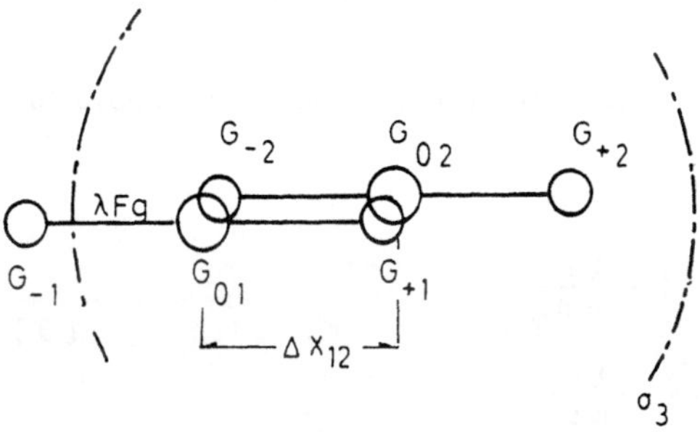

Fig. 4. Field distribution in detector plane arising from two scattering particles in the measuring volume

If the image of one particle overlaps the diffracted image of another particle, i.e.

$$\iint(\tilde{G}_{0i}\tilde{G}_{+j}+\tilde{G}_{0j}\tilde{G}_{-i})dx\,dy \neq 0$$

the information is present. If two diffracted images of two particles overlap, i.e.

$$\iint\tilde{G}_{+i}\ \tilde{G}_{-j}\ dx\,dy \neq 0$$

a weak contribution may arise which will not be considered here.

Going back to (2) we essentially obtain the following expression for the current, provided that two particles cross the measuring volume:

$$i(t) \approx \exp[-(\bar{x}^2 + \bar{y}^2)(\frac{1}{\sigma_1^2} + \frac{2}{\sigma_3^2})] \cdot$$

$$[1 + M_\alpha \, M_g \, M_{\Delta x} \, M_{\bar{x}} \cos(k_x \Delta x_{12} + k_y \Delta y_{12})] \quad (3)$$

where $\bar{x}(t) = 1/2\,(x_1(t) + x_2(t))$

$$\Delta x_{12}(t) = x_1(t) - x_2(t)$$

and it is assumed that

$$(\pi \sigma_2 \sigma_3)^2 \gg (\lambda F)^2 \quad .$$

The complete expression for the current is given in Appendix B.

The physical significance of the modulation terms is:

- M_α: term depending on the modulation depth of the grating.
- M_g: term depending on the relative separation between the real and diffracted images and the width of the pinhole.
- $M_{\Delta x}$: term depending on the distance between the two crossing particles $M_{\Delta x} = 1$ for $\Delta x = \lambda Fg$ and $\Delta y = 0$. The term is time dependent because of the time dependence of Δx and Δy.
- $M_{\bar{x}}$: term depending on the average position of the two particles. Although this term may become greater than unity, its influence is negligible and the term is set equal to unity.
- $\cos(k_x \Delta x + k_y \Delta y)$ is the phase term which gives the modulation frequency

  $$\omega_{Doppler} = \bar{K} \cdot (\bar{V}_1 - \bar{V}_2)$$

  with the constraint

$$|\lambda FG - \Delta x_{12}| < \frac{\lambda F}{\pi \sigma_2} \quad (3)$$

$$|\Delta y_{12}| < \frac{\lambda F}{\pi \sigma_2} \quad .$$

The passage of two particles will result in a current pulse as shown in Fig. 5. The following conditions must be fulfilled in order to have an acceptable Doppler signal from two particles where the conditions in (3) are satisfied:

$$M_g = 1 \Rightarrow \sigma_3 > \lambda FG \quad (4)$$

$$\tau_{burst} \geq \tau_{transit} \Rightarrow \frac{|\bar{r}_0 \cdot \nabla \bar{v}|}{|\bar{v}|} \leq \frac{\lambda F}{\sigma_2} \sqrt{\frac{\sigma_3^2 + 2\sigma_1^2}{\sigma_1^2 \sigma_3^2}} \quad (5)$$

where the last condition is based on the assumption that the instantaneous flow field can be written

$$\bar{v}(\bar{r}) = \bar{v}(0) + \bar{r} \cdot \bar{W} \quad (6)$$

and $\bar{r}_0$ is the vector associated with the grating, here $\bar{r}_0 = (\lambda Fg, 0)$. If the flow field can be further simplified:

$$v(x,y) = (0, v_0 + \frac{dv}{dx} x)$$

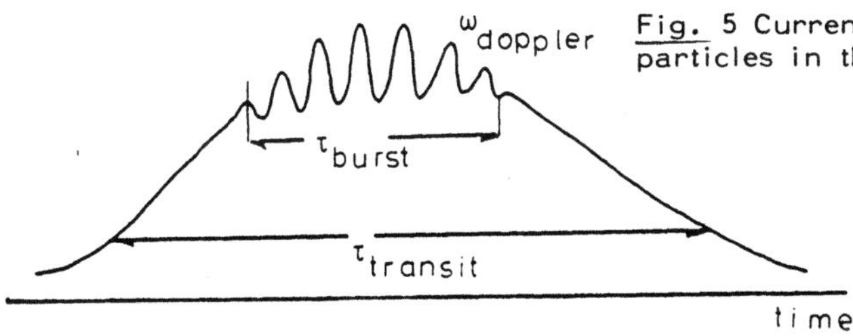

Fig. 5 Current pulse from two particles in the measuring volume.

the condition in (5) will read $\left(\frac{dv}{v\,dx}\right) \leq \left(g\,\sigma_2 \cdot \sqrt{\frac{\sigma_1^2\,\sigma_3^2}{2\sigma_1^2 + \sigma_3^2}}\right)^{-1}$,

where $g\,\sigma_2$ is the number of fringes on the grating and the square root is approximately equal to the smaller of the apertures σ_1 and σ_3. If the relative velocity gradient is too large, the two particles entering the volume at which measurement takes place with the right distance given by (3), will be too far from each other as they leave the volume, as defined either by the Gaussian beam or the pinhole.

Inserting the expansion (6) for the instantaneous flow field gives the measured Doppler frequency

$$\omega_{Doppler} = \bar{r}_0 \cdot \nabla\bar{v} \cdot \bar{K}$$

where $\bar{r}_0$ is determined by the grating, the $\nabla\bar{v}$ tensor is given by the flow field, and $\bar{K}$ is the scattering wave vector defined by the incoming and outgoing wave vectors. It should be noted that $\bar{r}_o$ is proportional to the wavelength of the laser beam, and $\bar{K}$ inversely proportional to the wavelength. Thus, the measured frequency is wavelength independent. Though the requirement for temporal coherence of the illumination source may be alleviated, the demand for spatial coherence of the light striking the two particles still exists.

For convenience, the calculation shown here is made with an absorption grating to avoid higher-order diffraction. Loosely speaking, the effect of the grating is to superimpose two particle images on the detector. This could just as well have been done with a phase grating, a reflection grating, a Wollaston or Rochon prism, or even a Bragg cell.

3. Doppler-broadening contributions

The number of Doppler oscillations is given by

$$N_{osc} = \frac{\lambda F}{2\pi\sigma_2} \cdot \frac{\bar{r}_0 \cdot \nabla\bar{v} \cdot \bar{K}}{|\bar{r}_0 \cdot \nabla\bar{v}|} \qquad (7)$$

where it is assumed that $\tau_{transit} > \tau_{burst}$.
This gives the relative broadening

$$(\Delta\omega_D / \omega_D)_b = 1 / N_{osc} \qquad (8)$$

corresponding to transit time broadening in the differential Doppler set-up.

The second broadening effect, which shall be addressed here, arises from the tolerance in separation between particle pairs as expressed in

(3) which gives rise to a broadening due to a superposition of velocity gradients

$$(\Delta\omega_D/\omega_D)_{ambiguity} = (g\sigma_2)^{-1} \qquad (9)$$

which is the reciprocal value of the number of fringes on the grating.

4. An optical model for the Laser Gradient Anemometer in the axial mode

A stringent mathematical treatment of an LGA system with a zone plate as filtering element (c f. Fig. 2) will not be given. The working priciple will be explained with reference to Fig. 6 where the rotationally symmetric zone plate is placed in front of the detector of radius R_o at a distance F from the measuring volume. Two particles i and j on the optical axis with an axial separation ΔZ_{ij} are considered.

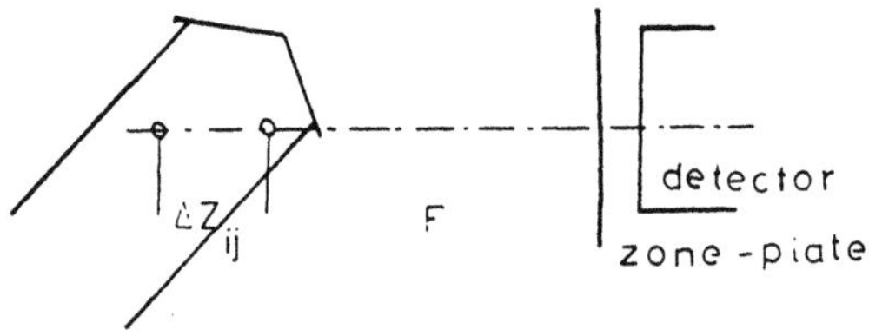

Fig. 6. Optical model for the LGA in axial mode

The two-particle contribution to the detector current is modulated with the Doppler frequency

$$\omega_{Doppler} = \bar{K}(\bar{v}_i - \bar{v}_j)$$

provided that the integral

$$I = \iint^{Aperture} H_z(\bar{r})\, G(\bar{r}-\bar{r}_j)\, G^*(\bar{r}-\bar{r}_i)\, d^2r \neq 0 \qquad (10)$$

where $H_z(\bar{r})$ is the filter and $G(\bar{r}-\bar{r}_i)$ is the Green's function expansion of the spherical waves emanating from the particles:

$$G(\bar{r}-\bar{r}_j) = \exp\left[ik|\bar{r}-\bar{r}_j|\right] \quad .$$

With the stated assumptions and the Fresnel approximation, (10) can be rewritten

$$I \approx \int_0^{R_o} H_z(\varkappa)\, \exp\left[ik\varkappa \frac{\Delta z_{ij}}{2F^2}\right] d\varkappa$$

where $\varkappa = x^2 + y^2$.

A selection rule for ΔZ_{ij} can be obtained in the same way as in the transversal case by using, for instance,

$$H_z(\varkappa) = \tfrac{1}{2}(1 + \cos\gamma\varkappa),$$

or written in the original coordinates,

$$H_z(r) = \tfrac{1}{2}(1 + \cos\gamma r^2) \quad .$$

The selection rule for detecting a Doppler signal becomes

$$\Delta z_{ij} = \lambda \gamma F^2/\pi \quad .$$

As for the transversal case previously treated, a phase grating may prove to be preferable.

5. Optical set-up

The measurement with an LGA system calls only for one incoming beam and one detector system. As the basic set-up is simple, some effort could be spent in gaining further information. For instance, a division of the laser beam into two perpendicularly polarised incident beams may define two scattering wavevectors either of which can be chosen by inserting a polarising filter in front of the detector. A change or rotation of the grating may alter the direction in which the gradient is measured. In the same way the two strong lines in an argon laser may be used to define two scattering wave vectors and thereby provide two velocity components. An array of gratings may be used to select different gradient directions, or in some cases a programmable spatial light modulator may even be adequate.

6. Summary

A simple set-up for measuring the ∇v tensor has been proposed. The constraint for interparticle distances to give a Doppler signal has been demonstrated and the quality of a Doppler burst analysed. The bursts may be processed by an electronic spectrum analyser or preferably by an LDA counter, where each measurement gives the instantaneous gradient.

There are questions which should be addressed: As the signals depend on two-particle contributions, a trade-off must be found between having many coincidences together with an increased incoherent background level or having fewer but better signals. If a correlation between mean velocity and velocity gradients exists some sort of de-biasing may be necesary depending on the data reduction to be performed. The modulation depth of the LGA bursts is lower than that of the LDA bursts from single particles, thus demanding increased signal strength to give the same variance on the parameter (frequency) to be estimated.

REFERENCES

1. Mishina and Asakura, Japan J. Appl. Phys., 15, No 10, pg. 2001-2002 (1976).

2. M. Fermiger et al., Appl. Phys. Lett., 36 (5) pg. 361-363 (1980).

3. G. G. Fuller et al., J. Fluid Mech., 100 (3) pg. 555-575 (1980).

4. L. E. Drain, J. Phys. D: Appl. Phys. 5, pg. 481495 (1972).

5. S. Hanson, Opto-Electronics 6, pg. 263-269 (1974).

6. Shi-Kang Wang, Risø National Laboratory report R-453 (1981).

7. L. Lading, Opto-Electronics, 4, pg. 385-398 (1972).

APPENDIX A

The Gaussian function

$$G(x,y)=A_1 \exp\left[-\left(\frac{(x-x_1)^2}{\sigma_{1x}^2}+\frac{(y-y_1)^2}{\sigma_{1y}^2}\right)\right]$$

will be written:

$$G\{x_1, y_1, \sigma_{1x}^2, \sigma_{1y}^2, A_1\} \quad .$$

APPENDIX B

The detector current from two particles separated by a distance close to the coherent condition selected by the grating, i.e. $\Delta x_{12} \approx \lambda FG$ and $\Delta y_{12} \approx 0$

$$i(t) \approx \iint_{-\infty}^{+\infty} \sum_{i=1}^{2} \left(|\tilde{G}_{0i}|^2 + |\tilde{G}_{+i}|^2 + |\tilde{G}_{-i}|^2\right) + 2 \sum_{ij=1}^{2}{}' \cos(k_x \Delta x_{ij} + k_y \Delta y_{ij}) \cdot$$
$$\mathrm{Re}\left[\tilde{G}_{0i}\,\tilde{G}_{-j} + \tilde{G}_{0i}\,\tilde{G}_{+j}\right] \, dx \, dy$$

where $\Delta X_{ij} = X_i - X_j$ and $\Delta Y_{ij} = Y_i - Y_j$.

Evaluation of the terms based on the assumption

$(\pi \sigma_1 \sigma_2)^2 \gg (\lambda F)^2$

gives (3) where

$$M_\alpha = \frac{(1-\alpha/2)(\alpha/2)}{(1-\alpha/2)^2 + \alpha^2/8}$$

$$M_g = \exp\left[-\left(\frac{\lambda FG}{\sigma_3}\right)^2\right]$$

$$M_{\Delta x} = \exp\left[-\left((\Delta x_{12} - \lambda FG)^2 + \Delta y_{12}^2\right)\left(\frac{\pi \sigma_2}{\lambda F}\right)^2\right]$$

$$M_{\bar{x}} = \cosh\left(\frac{2\bar{x}\lambda FG}{\sigma_3^2}\right) \quad \text{where} \quad \bar{x} = (x_1 + x_2)/2 \quad .$$

21. The Application of Cross-Correlation Measurements to the Determination of Velocity Gradients in Fluid Flow

C. Keveloh and W. Staude

Universität Bremen, D-2800 Bremen, Fed. Rep. of Germany

The application of dynamic light scattering to flow measurements has proven a valuable tool in order to determine cartesian velocity components of a flow seeded with light scattering particles. Main advantages compared to other methods are that, on the one hand, there is no influence of the measuring device on the flow, and on the other hand there is no gauging necessary because only geometrical factors enter the final result.

It is obvious that in the scattered light field information is contained also on more complicated properties of the flow. As FULLER et al. [1] have shown the homodyne spectrum can be used to obtain information on the velocity gradient, because the differently fast moving particles in the scattering volume lead to a characteristic decline of the intensity autocorrelation function. But this method has the disadvantage that the evaluation hinges on the exact knowledge of the light intensity profile in the scattering region, and does not allow to assign the measured value to certain tensor components of grad $\vec{v}(\vec{r})$.

The method we have proposed [2,3] lacks these deficiencies; in fact it bears a definite resemblance to the mentioned methods of velocity measurements: As in those cases one can perform single-and double-beam measurements, and the parameters necessary are essentially the same. The physical basis of this method is the phenomenon of velocity gradient caused speckle motion.

We will show below that a speckle describes a definite path; direction and velocity of this motion is determined by the scattering geometry and the velocity gradient tensor. Placing two light detectors in a speckle path and calculating the cross-correlation function of the detector signals then leads to a peak in this function whence certain tensor components can be determined.

Having outlined the theory in detail elsewhere [3] we will give here an alternative explanation of the basic idea of the method. Let us, for that purpose, consider a typical scattering experiment. The incoming plane and coherent wave is described by the wave vector $\vec{k}_i$, a particular Fourier component of the quasi-elastically scattered wave by $\vec{k}_a$ or by the scattering vector $\vec{q} = \vec{k}_i - \vec{k}_a$. The intensity of this wave is now determined by the $\vec{q}$ component of the Fourier transform $\tilde{n}(\vec{q})$ of the particle density $n(\vec{r})$:

$$(1) \qquad I(\vec{k}_a) \propto |\tilde{n}(\vec{q})|^2$$

If, as we assume, the particles are statistically independent, then so are different Fourier components of the density. That means that the scattered light intensity is statistically independent for different directions. If, however, intensities for different times are considered this statement stays true only if the scattering system is spatially homogeneous [2].

Let us now study the scattering on the inhomogeneous system of a flow with non-vanishing velocity gradient:

$$(2) \qquad \vec{v}(\vec{r}) = \vec{v}_o + \Gamma\vec{r} \quad .$$

Here Γ is a constant tensor that is traceless due to the assumed incompressibility of the fluid.

Neglecting Brownian movement, (2) is the equation of motion of the scattering particles, which is solved by:

$$\vec{r}(t) = \exp(\Gamma \cdot t)\cdot\vec{r}(0) + \int_0^t \exp(\Gamma t')dt'\cdot\vec{v}_o \equiv R(t)\cdot\vec{r}(0) + S(t)\cdot\vec{v}_o \quad .$$

We see that there is a linear mapping of the initial positions $\vec{r}(0)$ onto those at the time t. Because of this linearity and the regularity of $R(t)$ for all t we can deduce an important property of the Fourier transform of the time-dependent particle density. For that purpose we note that the time evolution of the density is given by

$$n(\vec{r},t) = n\big(R(-t)\cdot\vec{r} + S(-t)\cdot\vec{v}_o, 0\big) \quad .$$

Fourier transformation yields:

$$\tilde{n}(\vec{q},t) = (2\pi)^{-3/2}\int n(\vec{r},t)e^{i\vec{q}\cdot\vec{r}}d^3r = (2\pi)^{-3/2}\int e^{i\vec{q}\cdot\vec{r}}n(R(-t)\cdot\vec{r}+S(-t)\cdot\vec{v}_o,0)d^3r$$

On performing a change of variables in the right-most integral according to

$$\vec{\xi} = R(-t)\cdot\vec{r} + S(-t)\cdot\vec{v}_o$$

we obtain

$$(3) \quad \tilde{n}(\vec{q},t) = (2\pi)^{-3/2}\int e^{i\vec{q}\cdot R(t)\vec{\xi} - i\vec{q}\cdot S(t)\cdot\vec{v}_o}\, n(\vec{\xi},0)d^3\xi = e^{-i\vec{q}\cdot S(t)\cdot\vec{v}_o} \otimes \\ \otimes\, \tilde{n}(R^T(t)\vec{q},0)$$

where we have used the relations

$$\begin{aligned} & R^{-1}(t) = R(-t) \\ (4) \quad & R(t)S(-t) = -S(t) \\ & \det(R(t)) = 1 \quad . \end{aligned}$$

Equation (3) exhibits exactly the property that is of importance here, namely that the density component $\tilde{n}(\vec{q},t)$ is, apart from a velocity-dependent phase factor, given by the component $n(R^T(t)\vec{q},0)$ at time zero. According to equation (1) we finally have the result that the light intensity scattered into direction $\vec{k}_a = \vec{k}_i - \vec{q}$ at time t is equal to that scattered into $\vec{k}'_a = \vec{k}_i - R^T(t)\cdot\vec{q}$ at time 0.

The time-dependent function $\vec{q}(t)$, given by

$$\vec{q}(t) = R^{-T}(t)\cdot\vec{q}_o$$

describes a speckle path in q space for a speckle that started at $\vec{q}_o$. The last of the equations (4) establishes the fact that every q in q space belongs to one and only one speckle path.

In order to utilize this speckle motion for the experimental determination of velocity gradients we have to make sure that it can actually be observed. This is apparently only the case if this motion is confined to the Ewald sphere, because all k vectors must have the same length. Choosing the two scattering vectors close together we then have the condition that the speckle motion must be perpendicular to $\vec{k}_a$. Expansion of $R^{-T}(t)$ for small times then leads to

$$\vec{k}_a\cdot\Gamma\vec{k}_a = \vec{k}_i\cdot\Gamma\vec{k}_a$$

as the condition for observability. For the general case where $\Gamma\vec{k}_a$ is neither zero nor collinear to $\vec{k}_a$ one easily finds that this condition is fulfilled for all $\vec{k}_i$ that lie on a cone around $\Gamma\vec{k}_a$ and through $\vec{k}_a$. This property enables the determination of all of the eight tensor components of Γ by choosing three linearly independent $\vec{q}_{oi}$ and measuring the direction and velocity of the respective speckle motion.

For the case of a flow in a rectangular tube, as we have studied experimentally, the velocity gradient tensor has - in a properly chosen coordinate system - only one component. We have chosen the scattering geometry such that $\Gamma\vec{k}_a = 0$, which can apparently easily be done. This assures observable speckle motion. The figure shows the experimental results.

Now it is obvious that Brownian motion and the finite scattering volume renders things more complicated than just discussed. Both these effects lead to a fading of the speckles in time, the finite scattering volume in addition to a finite speckle width. On measuring the speckle motion one therefore obtains a broadened peak in the cross-correlation function with a maximum that is in general shifted towards shorter times. This shift becomes obviously greater the faster the particles move out of the scattering volume as measured on a time scale given by $\|\Gamma\|^{-1}$. It is exactly this effect that leads to the deviations of the values determined by our method from the actual values if one goes towards the middle of the flow tube, i.e. to higher velocities and smaller gradients. Increasing the scattering volume in the direction of $\vec{v}_o$ brings this deviation down, although the influence of Brownian motion has still to be taken care of.

There is another reason for speckle motion as for instance PUSEY [4] has pointed out. If the scattering region is not il-

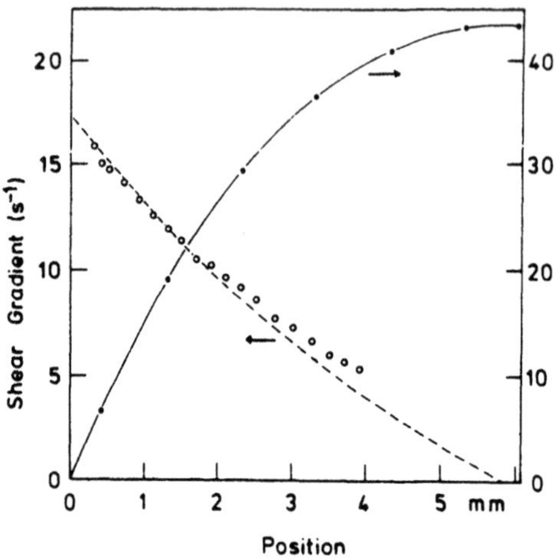

Experimental results of a particular run of measurements on a rectangular flow tube. The circles refer to velocity gradients determined by our method, the dots to velocities determined by the heterodyne technique. The solid line is a least mean square fit of the theoretical velocity curve through the measured values, and the dashed line is the spatial derivative of this curve

luminated by a plane but convergent or divergent wave a speckle motion collinear to the velocity occurs. We have not yet studied this effect experimentally, but in the measurements cited above speckle motion has been found to be perpendicular to the velocity, and is clearly not caused by this effect.

As mentioned above there is an alternative method where two crossing laser beams, but only one detector, are used. In that case the two beams should show no mutual coherence or should be orthogonally polarized in order to avoid interference with the ordinary crossed-beam laser Doppler signal. The measured autocorrelation function here contains, apart from a peak centred at zero time, a shifted peak. This can be explained as follows. The two laser beams produce two identical, but displaced speckle patterns. Due to the velocity gradient these patterns move. If the beam directions are properly chosen one pattern will be at some time t identical to the other one at time 0. This clearly causes the peak at time t in the correlation function.

We hope to have shown that the proposed method seems to be a promising one for the experimental determination of velocity gradients even though problems remain that still have to be solved.

References

1 G.G. Fuller, J.M. Rallison, R.L. Schmidt, and L.G. Leal; J. Fluid Mechanics 100 (1980) 555
2 F. Böner and W. Staude, Optics Communications 40 (1982) 407
3 C. Keveloh and W. Staude, submitted to Appl. Optics
4 P.N. Pusey, J. Phys. D9 (1976) 1399

Part II

Correlation Techniques: Methodical Advances and Applications to Signal Types Other than Photons

22. Rate Correlation and Data Preprocessing with Digital Correlators and Structurators

Klaus Schätzel

Institut für Angewandte Physik der Universität Kiel, Olshausenstraße 40-60
D-2300 Kiel 1, Fed. Rep. of Germany

Abstract

Various applications of fast digital correlators to optical measurement techniques can be described by a generalized rate correlation approach. Examples are the well-known photon correlation techniques, the velocity correlation measurements based upon threshold crossing events, and the correlation of digitized optical phase differences as produced by random phase screens.

On the basis of the general approach, digital preprocessing in order to reduce the bit format of the input data can be discussed for single- and multi-bit correlators. This discussion is easily extended to include digital structurators as well. The most universal preprocessing scheme which avoids distortion of correlograms and structure functions is random preset scaling, if necessary combined with subtraction of a constant background.

Introduction

Fast digital correlators as pioneered by PIKE'S group at Malvern have mostly been used for the analysis of single photon detection pulse signals. This "photon correlation technique" has important applications in studies of diffusing particles and laser velocimetry [1, 2].

During the late 1970's ERDMANN and GELLERT showed another way of processing optical signals with a digital correlator [3-5]. They used a Schmitt trigger to produce pulses from threshold crossing events of light scattered by particles crossing a single probe beam. The "rate correlation function" of these pulses is proportional to a velocity correlation function and thus directly yields spectral information about the flow under inspection which is difficult to access with photon correlation velocimetry.

The combination of the basic rate correlation idea with a standard two-beam laser Doppler velocimeter improves the technique and allows the measurement of auto- and cross-correlation functions of individual Cartesian velocity components [6-8].

A third use of digital correlators was in the real-time analysis of optical path length fluctuations in a random phase screen [9-14]. Two quadrature signals are hardclipped and fed to a digital phase counter. This counter removes the 2π ambiguity and provides a digital output suitable for the correlator. As a final result we obtain the temporal auto-correlation of the optical phase fluctuations, a "phase correlation function".

This paper attempts to give a general approach to all three types of applications - and possibly others not mentioned yet. Based upon this approach we will address the problem of data preprocessing as required by the finite input format of hardware correlators, particularly the still popular single-bit models. Stimulated by the phase measurements mentioned above, we will add a discussion on the alternative use of photon structure functions instead of correlations.

1. Generalized Rate Correlation Approach

In order to cover various correlation techniques of optical signals we consider some continuous stochastic time process $x(t)$, where $x(t)$ stands for physical quantities like optical intensity, flow velocity, or optical phase difference. Ensemble averages over this process $x(t)$ will be denoted by $< \ldots >$.

All three techniques outlined in the introduction, i. e. photon correlation, rate correlation, and phase correlation, can be characterized by a common feature. The underlying continuous physical quantity $x(t)$ is not directly available for measurements. Instead the measurements are based upon a related discrete stochastic process r_i, typically a number of pulses counted within a sample time interval i of duration T centered about t_i. For simplicity we will assume the zero dead-time situation where t_i equals iT.

Particularly for photon correlation and rate correlation, the relation between the continuous process $x(t)$ and the discrete process r_i is a stochastic one. For each individual realization of $x(t)$ there exists an ensemble of pulse count processes r_i. In order to avoid confusion we will use bars to denote averages over this ensemble, i. e. averages over r_i for a given $x(t)$.

Digital correlation of the discrete pulse counts r_i produces a correlogram which shall be used to obtain spectral information about the continuous process $x(t)$. We will now state two conditions for the pulse count process that are sufficient to guarantee a simple linear relationship between the digital correlation of r_i and the continuous temporal correlation function of $x(t)$.

The first condition is the postulate of linearity between $x(t)$ and the mean pulse count rate given by the r_i as

$$\lim_{T \to 0} (\overline{r_i}/T) = A + B\, x(t_i) \tag{1}$$

For processes $x(t)$ which do not vary significantly within a finite sample time interval T, the limit in (1) may be omitted. If the constant A vanishes, e. g. by redefinition of $x(t)$, and $C = BT$ replaces the slope B, (1) can be further simplified to

$$\overline{r_i} = C\, x(t_i) \ . \tag{2}$$

The second condition is a limited correlation length of the discrete process r_i. For some finite number l corresponding to a lag time $\tau_l = lT$, the following must be true for any given $x(t)$:

$$\overline{r_i\, r_{i+k}} = \overline{r_i}\; \overline{r_{i+k}} \quad \text{for all } k > 1 \; . \tag{3}$$

In order to be of practical use, the "forbidden lag time region" below τ_1 must be small compared to characteristic time scales in $x(t)$.

For $k > 1$, both conditions (2) and (3) can be combined to yield

$$\overline{r_i\, r_{i+k}} = C^2\, x(iT)\; x(iT + kT) \; . \tag{4}$$

Taking the ensemble average over $x(t)$ as well, we obtain

$$< \overline{r_i\, r_{i+k}} > = C^2 < x(iT)\; x(iT + kT) > \; . \tag{5}$$

If both stochastic processes are stationary, (5) does not depend on i. We can set $i = 0$ on the right-hand side of (5) and for practical purposes, the ensemble averages on the left-hand side of (5) can both be replaced by a time average:

$$\lim_{M\to\infty} \frac{1}{M} \sum_{i=1}^{M} r_i\, r_{i+k} = C^2 < x(0)\; x(kT) > \; . \tag{6}$$

A suitable estimator for the temporal correlation function of $x(t)$ is obtained by use of a finite number of samples M in (6).

Thus (6) describes the basic rate correlation idea. In spite of the (perhaps large) fluctuations of the instantaneous pulse count rate about its mean, which is proportional to $x(t)$ as stated in (2), the digital correlation of the pulse counts remains proportional to the correlation of the signal $x(t)$, if the lag time exceeds the "pulse correlation time" $\tau_1 = lT$ defined in (3).

The reason for this convenient property is the linearity of all averaging procedures which allows them to be used in any arbitrary sequence. Thus in a practical correlation experiment both, the fluctuations of $x(t)$ and of the pulse counts r_i average out simultaneously. Since there is no additional correlation between the r_i for a given $x(t)$ at the lag times considered, the rate correlogram obtained after averaging over many sample time intervals completely corresponds to the correlation of $x(t)$ alone.

1.1 Photon Correlation

Just to illustrate the wide applicability of the generalized rate correlation approach, we briefly consider the most familiar example of photon correlation.

Let $I(t)$ be an optical intensity integrated over the sensitive aperture of a photo multiplier tube. A photon counting detector then produces a pulse count sequence n_i which is Poisson distributed about a mean value proportional to $I(t)$ if temporal and spatial coherence are sufficient.

Thus our first condition (2) is satisfied immediately. The statistical independence of individual photon detections implies the validity of our second condition (3), in fact with zero correlation length $l = o$. For all $k > o$ we obtain the well-known equivalence of photon correlation functions and intensity correlations

$$\lim_{M\to\infty} \frac{1}{M} \sum_{i=1}^{M} n_i \, n_{i+k} = C^2 < I(O) \; I(kT) > . \tag{7}$$

The zero pulse correlation length limits the forbidden lag time region to the zero channel in the photon correlogram. The finite dead time of real photo multiplier tubes, however, produces some anticorrelation between adjacent samples at high count rates, which tends to increase the forbidden region.

Another dead-time effect is a violation of the linearity condition (2) at high intensities. A complicated distortion of the photon correlation function is the necessary consequence.

The second problem associated with real photo multipliers, a finite dark count rate, does not violate (7) essentially. The more general linearity condition (1) replaces (2). This leads to two additive terms in (7) which both do not depend upon the lag time kT, i. e. shift the base-line only.

1.2 Rate Correlation

In rate correlation as introduced by ERDMANN and GELLERT [3], an optical intensity $I(t)$ is detected by some analog device, e. g. a photo diode. For simplicity we will assume strong signals which allow us to neglect shot noise. The crossing events of $I(t)$ with a threshold I_S are monitored and a pulse is generated for each crossing event, yielding the discrete pulse count process r_i.

The scattering process which produces the signal $I(t)$ is assumed to be time dependent through one velocity $v(t)$ only. This assures a linear relationship between the mean rate of threshold crossings, i. e. our pulse counts, and this velocity, if the full bandwidth of the signal $I(t)$ is present at the threshold comparator. Hence, our first condition (2) is satisfied.

Uniform seeding with noninteracting particles implies statistical independence of light signals produced by nonoverlapping sets of scattering particles. Thus our second condition (3) is satisfied with a forbidden lag time region up to

$$\tau_1 = \underset{t}{\mathrm{Min}} \, \{s/v(t)\} \; , \tag{8}$$

where s is the linear dimension of the scattering volume in the direction of the instantaneous velocity. τ_1 can become quite large if low velocities occur, e. g. at very high turbulence levels. It should be noted, however, that the temporal resolution limit given by (8) exactly corresponds to the finite spatial resolution defined by the size of the measurement volume.

For lag times $kT > \tau_1$ and a stationary velocity process, (6) implies a correspondence of digital rate correlation and velocity correlation

$$\lim_{M\to\infty} \frac{1}{M} \sum_{i=1}^{M} r_i \, r_{i+k} = C^2 < v(0)\ v(kT) > . \qquad (9)$$

Instead of photon correlation velocimetry, where velocity distribution information has to be unfolded from intensity correlation functions, in rate correlation immediate spectral velocity information is obtained from rate correlograms.

1.2.1 Single-Beam Rate Correlation

The original rate correlation technique pioneered by ERDMANN and GELLERT [3, 4] uses a single cylindrical beam to define a scattering volume. This setup is sensitive to a radial velocity component

$$v(t) = (v_x^2 + v_y^2)^{1/2} , \qquad (10)$$

if the z direction is that of the beam. For low seeding levels typical in gas flows, threshold crossings can be identified with the arrival of single particles in the beam [5]. For uniform seeding, (9) is valid for all lag times exceeding the longest particle transit time through the beam.

If seeding is nonuniform, but $v(t)$ remains constant, the same experiment yields the spatial correlation of the scattering particle density where time lags τ correspond to spatial lags $v\tau$ [3].

In cases where both velocity and local particle density fluctuate simultaneously, the two influences can generally not be isolated.

A particular advantage of single beam rate correlation for high speed applications is the fact that the required signal processing bandwidth is much lower than that for laser Doppler techniques with comparable spatial resolution.

1.2.2 Cross-Beam Rate Correlation

The inconvenient radial velocity component (10) measured in single-beam rate correlation can be replaced by a single Cartesian velocity component if cross-beam rate correlation is being used [6, 7].

A typical setup includes a symmetric laser Doppler velocimeter with Bragg cells to provide a frequency shift to overcome directional ambiguity. The scattered light has to be intense enough to produce a reasonable signal to noise ratio in the detector signal, say better than 20 dB.

A sharp cut-off high pass filter is used to remove the low-frequency pedestal from the detected Doppler bursts. The resulting signal is an approximately symmetrical sinusoidal wave with stochastic changes of amplitude and phase due to the scattering particle statistics. An optional low-pass filter can be used to reduce high-frequency noise due to the detection or the electronic processing.

If the cut-off frequencies of both filters are kept well out of the range of the Doppler frequency, changes in the flow velocity v_x perpendicular to the fringe system compress or expand the bandpass-filtered

Doppler signal proportional to v_x without any further changes in the signal.

A zero-crossing detector with enough hysteresis to prevent accidental triggering by shot noise produces a pulse train from the filtered Doppler signal. The mean pulse rate is a linear function of v_x. This satisfies our first condition (1).

The second condition (3) is again tied to the longest particle transit time which defines the forbidden small lag time region τ_1. It should be mentioned that intensity correlation for lag times smaller than τ_1 is the very basis of photon correlation velocimetry. Rate correlation on the other hand excludes this region. This argument already shows that comparatively slow digital correlators are sufficient for most rate correlation measurements. There remains, however, a large bandwidth requirement for the analog front end.

Since both conditions (1) and (3) have been satisfied, the general rate correlation approach is applicable to cross-beam rate correlation. If we ignore frequency shift for simplicity, as can always be done by the consideration of a velocity relative to the moving fringe pattern, the final relation reads

$$\lim_{M\to\infty} \frac{1}{M} \sum_{i=1}^{M} r_i\, r_{i+k} = c^2 < v_x(0)\ v_x(kT) > \text{ for } kT > \tau_1 \ . \qquad (11)$$

As in single-beam rate correlation, careful attention has to be paid to seeding conditions. For low seed levels, say less than one particle per measurement volume, we obtain separated single particle bursts most of the time. The pulse count rate is essentially proportional to both v_x and particle concentration in this limit [6]. Thus seeding must be uniform in order to avoid distortion of the rate correlogram. Furthermore, an absolute knowledge of the seeding level or a calibration measurement are necessary, if c must be determined for correlation measurement in absolute velocity units.

Fortunately, both problems can be avoided at higher seeding levels, i. e. mean particle densities large compared to one per measurement volume. In this situation there exists an almost continuous Doppler signal. It has been shown [6] that in the Gaussian limit the zero-crossing rate of this signal is hardly affected by particle statistics at all. For most practical applications, r_i becomes twice the Doppler frequency, if crossings in both directions are detected. Consequently, there is no influence of moderate fluctuations in the mean scattering particle density on the correlogram and the constant c is determined by the fringe separation alone.

In applications with slow velocity fluctuations, a phase-locked loop tracker [15] can be used to provide a continuous pulse sequence even if finite drop-outs of the detector signal occur [8]. This technique provides an extremely high signal to noise ratio, but the finite response time of the tracker may increase the forbidden lag time region.

Both variants of cross-beam rate correlation have been successfully applied to turbulent channel flows [6-8] and - extensively - to circular Couette flows, where time periodic modes of the Taylor vortices [6-8, 16-19, 31] and the transition to turbulence [19, 20] were under study.

1.2.3 Speckle Rate Correlation

The moving scatterer observed in rate correlation need not consist of particles in a flow, but can be almost any source of diffuse light scattering. Illuminated with coherent light, such an object will produce a speckle pattern. This pattern changes in time at a rate proportional to the velocity of the scatterer. Although absolute threshold crossing rates of such a speckle signal depend on laser power, scattering geometry, detector aperture, and threshold, the proportionality of mean crossing rate and velocity holds for any fixed set of the parameters mentioned [21].

For lag times larger than speckle coherence, again the rate correlogram corresponds to the velocity correlation function as stated in (9).

1.3 Phase Correlation

The research of turbulent random phase screens requires a detailed statistical knowledge about optical path length fluctuations produced by the screen. For deep phase screens, interferometers with bidirectional fringe counting [9, 13] or phase counters operating on hard-clipped quadrature signals generated by polarization coding [10-13] can be used to obtain real-time digital phase information.

On-line processing by fast digital probability analyzers and correlators yields desired statistical information [9-14]. However, this application of digital correlators differs from photon and rate correlation in that no stochastic but a deterministic relation exists between the digital pulse count rate r_i and the continuous phase difference $\Delta\phi$ under study.

Quantization noise created by digitizing $\Delta\phi$ is responsible for a small violation of our linearity condition (1) or (2). This quantization noise also posesses a finite correlation time which can be estimated using the temporal density $1/\tau_s$ of "jumps" in the digitized signal r_i.

For reasonably wide distributions of $\Delta\phi$, i. e. for standard deviations of $\Delta\phi$ in excess of the quantization steps, the effect of quantization noise in the correlogram can be approximated by an additional short-lived contribution [13]

$$\frac{1}{12} \exp\left(-6\tau/\tau_s\right) \qquad (12)$$

in units of r_i, where quantization steps are equal to 1.

For lag times τ much larger than τ_s, (12) can be neglected and the effect of deterministic quantization - although it violates our first condition - looks very much like the forbidden lag time restriction well known in rate correlation.

Missing large scale stationarity typical for atmospheric optics and many laboratory scale turbulent phase objects [13] restricts the usefulness and reproducibility of correlation measurements for such systems. Based on turbulence theory by KOLMOGOROFF, TATARSKI [22] introduced the concept of structure functions to atmospheric optics, which are defined for a wider class of stochastic processes than correlation functions and still provide essentially the same practical information [23-25].

The successful use of structure functions in random phase screen work [12-14] and the construction of a fast digital structurator by our group stimulated an extension of this paper to include structure functions as well as correlations. This extension will be given in section 3.

2. Preprocessing

For many applications of fast digital correlators, particularly in rate correlation and phase correlation, the number of pulses counted in a single sample time interval τ exceeds the limited input capacity of the correlator - typically 1 or 4 bits. Hence, some preprocessor has to be used to reduce the count rate [26].

We will first formulate the general requirements of such a processor and then discuss various concepts, i. e. clipping, random clipping, scaling und subtraction. This discussion will be given for single-bit correlators as well as for multi-bit models.

2.1 General Requirements

Let r_i again denote the original pulse counts and $\tilde{r}_i$ the reduced signal obtained by preprocessing of r_i. The two conditions postulated for the relation between r_i and the underlying continuous process $x(t)$ in section 1 must hold for the reduced signal as well.

The linearity condition is satisfied, if the ensemble averages of r_i and $\tilde{r}_i$ for any given $x(t)$ are related in a linear fashion

$$\overline{\tilde{r}_i} = D + E\,\overline{r_i}\ . \tag{13}$$

D and E are constants, possibly $D = 0$, and the bars denote ensemble averages over the pulse count statistics.

The second condition regards a pulse count correlation between delayed and direct signals. An ideal preprocessor should not introduce any additional correlation of this type into the reduced signal.

This is generally less of a problem, if the processor is to be used in the delayed channel only. The second condition then reads

$$\overline{\tilde{r}_i\, r_{i+k}} = \overline{\tilde{r}_i}\;\overline{r_{i+k}} \qquad \text{for all } k > 1\ . \tag{14}$$

A more restrictive formulation is necessary if both channels require a preprocessed signal:

$$\overline{\tilde{r}_i\, \tilde{r}_{i+k}} = \overline{\tilde{r}_i}\;\overline{\tilde{r}_{i+k}} \qquad \text{for all } k > 1\ . \tag{15}$$

In addition to the stated conditions, there is one more property highly desirable for a good preprocessor. The additional noise introduced by the reduction process should be as small as possible in order to avoid an unnecessary increase of total measurement times or a decrease of accuracy in the correlogram.

2.2 Single-Bit Preprocessing

2.2.1 Clipping

The wide use of single-bit correlators is largely due to the application of clipping [26-28] in the delayed channel, known as single clipping.

This technique generally violates the linearity condition (13). However, for the important special case of photon correlation of Gaussian light, averaging over the intensity statistics restores the validity of this condition and allows the measurement of undistorted intensity correlation functions.

The same is true if weak intensity fluctuations are masked by strong photon noise, as often encountered in reference beam spectroscopy - a phenomenon for which ADRIAN [29] coined the term self-scaling.

Obviously, clipping does not violate our second condition (14) or (15). The procedure is performed independently for each sample.

An attractive property of clipping is the fact that no additional randomness is introduced into $\tilde{r}_i$. Hence, a properly chosen clip level allows single bit processing at largely the same signal to noise ratio as full correlation in many applications of correlation spectroscopy.

Double clipping, i. e. the use of a clipping gate in the direct as well as in the delayed channel has not found many applications due to severe distortion problems, although it allows an extreme simplification of the correlation procedure.

2.2.2 Random Clipping

A stochastic variation of the clip level over the whole range of input counts leaves the linearity condition (13) valid [26]. It also satisfies the second condition (14) or (15), if statistically independent clip levels are being used.

Thus we obtain correct correlograms in both cases, when random clipping is used in one or both correlator input channels.

It should be noted though, that strong quantization noise is introduced by random clipping and long total measurement times may be required to compensate for this extra noise. An estimate of this noise is given by

$$\operatorname{Var} \tilde{r}_i = m^{-2} \operatorname{Var} r_i + \frac{\overline{r_i}}{m} \left(1 - \frac{\overline{r_i}}{m}\right) , \qquad (16)$$

where m is the range of clip levels [26].

2.2.3 Scaling

Most commercial correlators, single- as well as multi-bit types, use a scaling counter [26, 30] instead of a random clip gate. A scaling counter can be interpreted as a pseudo random clip device where the instantaneous clip level is given by the difference of the counter range m and its contents at the beginning of a particular sample time interval, i. e. the sum of all previous counts modulo m.

Thus the technique relies on the random nature of the input counts and caution has to be used for the processing of more deterministic signals as they can well occur in other than photon correlation applications. Particularly for signals with small fluctuations on a large background, scaling produces an artificial periodicity in $\tilde{r}_i$ which may lead to erroneous correlograms if total measurements are not sufficient to average out this periodicity [26].

Of course scaling severely violates the second condition (15), if the scaled signal is to be used in both correlator input channels and no correct correlograms can be expected.

Since scaling, when applicable, simulates a true random clipping, quantization noise considerations again lead to (16).

2.2.4 Subtraction

Problems with single scaling are often connected with a large constant background contribution in the signal r_i. Subtraction [8, 26] of this background reduces the problem and greatly extends the applicability of scaling particularly in cross-beam rate correlation [8].

Being a deterministic, linear procedure, subtraction satisfies both our conditions and hence may be used safely in one or both correlator input channels. No randomness or quantization noise is added to the signal.

As a welcome side effect, subtraction improves the signal to noise ratio of the correlogram by reduction of its base line and hence the noise associated with its finite time estimation.

Subtraction has been applied with great success to rate correlation and phase correlation measurements [8-11] and should certainly be used whenever a constant background is present in the input signal.

2.3 Multi-Bit Processing

Scaling and subtraction can be used with multi-bit correlators as well as single bit types and most of the arguments in sections 2.2.3 and 2.2.4 apply to both cases. The randomness requirements on the input count signal can be relieved for the use of scaling counters with multi-bit correlators, though, due to the decrease in scale level.

However, there remains a slight pulse correlation problem if the scaling counter is to be used in both correlator input channels. In order to avoid this problem, the scaling counter has to be preset at the beginning of each sample time interval.

2.3.1 Fixed Preset Scaling

If the counter is preset to some fixed state, e. g. to zero, we obtain a multi-bit generalization of clipping. Again the linearity condition (13) is violated and distorted correlograms must be expected.

With increasing bit number, the deviations from linearity in (13) decrease quickly and correlograms become less distorted. The effect is essentially identical to that of deterministic quantization discussed in section 1.3. Experience with a software 8-bit device indicates negligible distortions at this input range [13].

2.3.2 Random Preset Scaling

If all distortions, which can still be significant in 4-bit correlators, are to be avoided, the preset state of the scaling counter must be chosen randomly over the whole range of counter states. This random preset scaling is the straightforward generalization of random clipping and the only pre-processing scheme, which satisfies both conditions postulated in section 2.1 even if used in both input channels of either single- or multi-bit correlators.

The additional quantization noise created by random preset scaling for multi-bit correlators can be approximated by [26]

$$\mathrm{Var}\,\tilde{r}_i = m^{-2}\,\mathrm{Var}\,r_i + \frac{1}{6} - \frac{1}{6}m^{-2}\,, \tag{17}$$

where m again denotes the scale level.

In photon correlation applications, the quantization noise in (17) can safely be neglected for 4-bit correlators, if count rates r_i up to the order 10^2 per sample time are considered [26].

A random preset scaler based upon a shift register type pseudo random number generator was successfully used with Malvern single- and multi-bit correlators by us in rate correlation measurements on circular Couette flow. A combined subtractor/random preset scaler will be built into the 4-bit structurator/correlator which is presently under construction in our laboratory.

3. Digital Structure Functions

For all three techniques described in section 1, photon correlation, rate correlation, and phase correlation, the digital correlation function can be replaced by a digital structure function

$$D_M(k) = \frac{1}{M}\sum_{i=1}^{M}(r_i - r_{i+k})^2\,. \tag{18}$$

This function is related to a correlogram by

$$D_M(k) = \frac{1}{M}\sum_{i=1}^{M}(r_i^2 + r_{i+k}^2) - \frac{2}{M}\sum_{i=1}^{M} r_i\, r_{i+k}\,. \tag{19}$$

Apart from a factor of -2, both functions differ by an estimate of the second moment of r_i only. This estimate is derived equally from the delayed and direct data and has a slight dependence on the lag k.

Despite of their close relationship and the fact that both functions contain essentially identical spectral information for $M \to \infty$, the noise performance of structure and correlation functions shows distinct differences.

As it is well known from analog structure functions, they can tolerate much higher low-frequency noise and drifts than correlation functions [22, 23]. The difference in (18) largely reduces low-frequency contributions to the low lag time channels of the structure function. As a consequence, the structure function yields a reliable estimate of short time scale features even if averaging over slower signal components is still incomplete and a correlogram would still show severe distortions, e. g. a skewed base line.

On the other hand, the correlogram copes better with the fast fluctuations of the pulse counts r_i about their expectations given by the underlying continuous process $x(t)$ as considered in section 1 [24, 25].

3.1 Photon Structure Functions

Due to its lower sensitivity to Poisson photon detection noise, the correlogram generally provides the lower estimator standard deviation as compared to the corresponding structure function, if weak photon count signals are processed.

For a fair comparison, the base line has to be subtracted from both functions as suggested by OLIVER and PIKE [24]For this purpose, $\bar{n}^2$ and $2\bar{n}$ are subtracted from correlation and structure function estimators, respectively. The mean count rate $\bar{n}$ is usually estimated as a monitor channel.

For direct detection of Gaussian-Lorentzian light, as well as in the presence of a reference beam acting as local oscillator, the standard deviation of structure functions can be computed to exceed that of equivalent correlograms by a factor between $\sqrt{3}$ and $\sqrt{5}$ in the low photon count rate limit. This factor decreases with increasing mean count rate or if short total measurement times must be used [24, 25].

At high photon count rates, signal noise due to the scattering process dominates the Poisson detection noise and a characteristic drop of the estimator variance for low lag time values shows up in the structure function [25].This behavior is not observed in correlation functions and it again reflects the lower sensitivity of structure functions to low frequency disturbances in the signals.

For mean count rates of the order of one photon detection per sample time interval, the structure function already shows a smaller estimator standard deviation than the correlogram at lag times below the coherence time in direct detection of Gaussian-Lorentzian light [25].

The same occurs in the presence of a reference beam, but at a higher mean count rate depending on the ratio of the photon noise produced by the strong local oscillator to signal fluctuations in the weaker cross-term of reference and scattered light amplitudes [25].

The marked lag time dependence of noise on structure functions suggests their usefulness, if fast scattering components are to be determined in the presence of slower component - a situation typically encountered in highly polydisperse systems of diffusing particles. For a 1:10 ratio of coherence times and equal scattering amplitudes of two discrete Gaussian-Lorentzian contributions, estimator standard deviations of the order $\sqrt{10}$ below those of correlograms have been calculated for photon structure functions [25], using sample times appropriate to determine the fast component.

The advantages of structure functions become even greater, if a short total measurement time implies incomplete averaging over the slowest signal components. This behavior was demonstrated by Monte Carlo stimulations [24] as well as calculated for a simple example [25].

These considerations although done for Gaussian-Lorentzian light typical for diffusion experiments, suggest two situations in Doppler velocimetry as well, where a structurator could outperform a digital correlator. One is long-range velocimetry where atmospheric turbulence causes strong low-frequency modulation of the Doppler signal: The other is burst correlation, where a very short total measurement time is required in order to obtain a reasonable temporal resolution.

Experiments with a 100 ns photon structurator are being prepared in both fields, diffusion study and Doppler velocimetry.

3.2 Rate Structure Functions

Estimator standard deviation calculations can easily be translated from photon correlation to rate correlation by noting the similar Poisson character of photon detection and particle arrival statistics. Thus advantages for correlators are to be expected for low seeding levels whereas structurators gain at high seeding levels, i. e. more than one particle arrival per sample time.

One important special case has to be mentioned, however. In cross-beam rate correlation, the threshold crossing rate becomes almost independent of particle statistics at high seed levels [6] and the same is true whenever a phase locked loop tracker can be used to "clean up" the Doppler signal prior to digital processing [8].

The effect is antibunching of the threshold pulses resulting in an almost negligible amount of noise due to particle statistics in the pulse sequence. Thus the very high count rate limit rarely encountered in photon correlation or "structuration" is the typical case in many cross-beam correlation or "structuration" experiments.

Based on this consideration, significant advantages can be expected for the use of structure instead of correlation functions. This agrees with our experience obtained with a 10 ms/1024 channel/8 bit software structurator [13] on circular Couette flow [19].

3.3 Phase Structure Functions

Since the statistical analysis of random phase screens by direct phase measurements stimulated our work on software and hardware structurators, it can be no surprise to find the concept of structure functions most suitable in this application.

In fact, due to the deterministic analog to digital conversion process, there is very little high-frequency noise present on the digital phase signal. On the other hand we find strong slow drifts in most experiments due to thermal expansion of the interferometer as well as to large-scale turbulence in the phase object under study [12, 13]. Furthermore, turbulent time scales are often as slow as 100 ms, such that incomplete averaging becomes a problem, if exceedingly long total measurement times are to be avoided as it is dictated by long-term stability problems of the setup.

All characteristics of the experiment strongly favor the use of structure functions and drastic improvements in the reproducibility of results were achieved by the use of the software structurator already mentioned in section 3.2 instead of a fast digital correlator [12-14].

3.4 Preprocessing for Digital Structurators

In contrast to digital correlators, a structurator cannot be built using the familiar (1 x n) or (4 x n)-bit designs. The subtraction required in the structure function (18) can only be done for numbers of equal scale and only after all pulses contributing to the delayed and the direct counts have been received. This leads to an (n x n)-bit fully parallel design where the necessary preprocessing has to be done for both input channels of the structurator.

Although a (1 x 1)-bit structurator could easily be designed, the bad noise performance of structurators for low count rates led us to prefer a (4 x 4)-bit approach. Such a device can be used for count rates up to 100 per sample time interval with quantization noise levels still well below the Poisson variance, if proper preprocessing is being used [26].

This preprocessor should not be a single free running scaling counter due to additional correlation problems associated with double scaling as discussed in section 2.2.3. The zero preset approach can be used for devices with high input capacity - like the 8-bit software structurator [13] - only, in order to avoid distortions.

The best possible choice for a 4-bit structurator is a random preset scaler as described in section 2.3.2. This device avoids both distortions and additional correlation effects.

The complete preprocessor includes a subtractor as well. Although structure functions (18) do an implicit subtraction of any background and a subtractor would not be necessary for a full structurator, the limited input range of any hardware device can be used much better after a format reduction of the input data by subtraction.

As a side effect it should be noted that fully parallel 4-bit processing in preprocessor and structurator does allow a 4-bit input even at the shortest selectable sample time of 100 ns. This allows the instrument to be used with pulse count rates in excess of 100 MHz, a significant improvement over the present 10 MHz limit of multi-bit correlators.

Conclusions

The separate consideration of an underlying continuous process $x(t)$ and a discrete pulse count sequence r_i determined by $x(t_i)$ allows the formulation of two criteria - mean value linearity and finite correlation length of the discrete process r_i for a given $x(t)$ - which must be satisfied in order to obtain a correspondence of the temporal correlation function of $x(t)$ with the digital correlation of the r_i. This generalized rate correlation approach can be applied to photon correlation, single-and cross-beam rate correlation, and - with slight modifications - to the correlation of digitized optical phase signals. Other possible applications may well exist, in and outside the field of optical flow measurements.

The finite input format of hardware correlators often calls for some preprocessing of input count rates. Distortions due to clipping and additional correlation possibly introduced by scaling can both be overcome by random clipping or its multi-bit generalization random preset scaling. Subtraction allows to make better use of signals on a constant background as are often encountered in cross-beam rate correlation and phase correlation measurements.

All applications involving low-frequency noise problems or short total measurement times suggest the use of digital structure functions in the place of correlograms. Polydispersity, long range and burst Doppler velocimetry are examples in photon counting. Cross-beam rate correlation at high seeding levels and phase measurements on turbulent random phase screens can almost always be improved by the use of a structurator instead of the correlator.

Acknowledgement

The author thanks E. O. Schulz-DuBois, G. Pfister, and E. R. Pike for discussions and cooperation.

References

1 H. Z. Cummins, E. R. Pike (eds.): Photon Correlation and Light Beating Spectroscopy, Plenum Press, New York (1974)

2 H. Z. Cummins, E. R. Pike (eds.): Photon Correlation Spectroscopy and Velocimetry, Plenum Press, New York (1977)

3 J. C. Erdmann, R. I. Gellert: Recurrence rate correlation in scattered light intensity, J. Opt. Soc. Am. 68, 787-795 (1978)

4 J. C. Erdmann: Application of recurrence rate techniques to turbulence analysis, Physica Scripta 19, 396-401 (1979)

5 J. C. Erdmann: Statistics of random flow analyzed by optical techniques, Opt. Acta 27, 31-44 (1980)

6 K. Schätzel: Ratenkorrelation - ein neues Meßverfahren zur zeitlichen Analyse stationärer Strömungen, Diplomarbeit, Kiel (1978)

7 R. Vehrenkamp, K. Schätzel, G. Pfister, E. O. Schulz-DuBois: Direct measurement of velocity correlation functions using the Erdmann-Gellert rate correlation technique, J. Phys. E 12, 119-125 (1979)

8 K. Schätzel: Advances in cross-beam rate correlation, Opt. Acta 27, 45-52 (1980)

9 K. Schätzel: Interferometric analysis of deep random phase screens, Opt. Lett. 5, 389-391 (1980)

10 K. Schätzel: Two-point statistics of turbulent deep random phase screens, Proc. 4th Conf. on Photon Correlation Techniques in Fluid Mechanics, eds. W. T. Mayo, A. E. Smart, Stanford University, 23-1 (1980)

11 K. Schätzel, E. O. Schulz-DuBois, R. Vehrenkamp: Optical correlation techniques in fluid mechanics, Opt. and Laser Techn. (April 1981), 91-95

12 J. Merz: Lichtstreuung am Phasenschirm, Diplomarbeit, Kiel (1981)

13 K. Schätzel: Fluktuationen der optischen Weglänge in turbulenten Phasenobjekten, Dissertation, Kiel (1982)

14 K. Schätzel, G. Parry: Realtime Moiré measurement of phase gradient, subm. to Opt. Acta (1982)

15 B. S. Fedders, A. Köneke: An LDA system yielding long-term averages and velocity distribution, J. Phys. E 12, 766-769 (1979)

16 G. Pfister: On the structure of non-stationary Taylor vortices, Proc. 4th Conf. on Photon Correlation Techniques in Fluid Mechanics, eds. W. T. Mayo, A. E. Smart, Stanford University (1980)

17 A. Brandstäter, U. Gerdts, A. Lorenzen, G. Pfister: The structure and dynamics of non-stationary Taylor-vortex flow, in "Nonlinear Phenomena at Phasetransitions and Instabilities", ed. T. Riste, Plenum Press, New York, 303 (1981)

18 A. Lorenzen: Messungen zum Existenzbereich einiger Taylormoden, Diplomarbeit, Kiel (1982)

19 G. Pfister: Hydrodynamische Instabilitäten und der Übergang zur Turbulenz in der rotationssymmetrischen Couette-Strömung, Habilitationsschrift, Kiel (1981)

20 T. Mullin, T. Brooke Benjamin, K. Schätzel, E. R. Pike: New aspects of unsteady Couette flow, Phys. Lett. 83A, 333-336 (1981)

21 N. Takai, T. Iwai, T. Asakura: Real-time velocity measurement for a diffuse object using zero-crossings of laser speckle, J. Opt. Soc. Am. 70, 450-455 (1980)

22 V. I. Tatarski: Wave Propagation in a Random Medium, Mc Graw-Hill, New York (1981)

23 E. O. Schulz-DuBois, I. Rehberg: Structure function in lieu of correlation function, Appl. Phys. 24, 323-329 (1981)

24 C. J. Oliver, E. R. Pike: Statistical accuracy in the photon counting structure function of fluctuating light fields, to be publ. in Opt. Acta (1982)

25 K. Schätzel: Noise in photon correlation and photon structure functions, subm. to Opt. Acta (1982)

26 K. Schätzel: Signal preprocessing for digital correlators, Appl. Phys. 22, 251-256 (1980)

27 E. Jakeman, E. R. Pike: Spectrum of clipped photon-counting fluctuations of Gaussian light, J. Phys. A 2, 411-412 (1969)

28 E. Jakeman: Theory of optical spectroscopy by digital autocorrelation of photon-counting fluctuations, J. Phys. A 3, 201-215 (1970)

29 R. J. Adrian: Self-scaling effects in clipped photon correlations, this volume

30 E. Jakeman, C. J. Oliver, E. R. Pike, P. N. Pusey: Correlation of scaled photon counting fluctuations, J. Phys. A 5, L93-96 (1972)

31 A. Lorenzen, G. Pfister, T. Mullin: End effects on the transition to time-dependent motion in the Taylor experiment, subm. to Phys. Fluids (1982)

23. Statistical Stability of Clipped Photon Correlations in the Self-Scaling Limit

Ronald J. Adrian

Department of Theoretical and Applied Mechanics, University of Illinois, Urbana, IL 61801, USA

1. Introduction

High-speed photon correlators commonly use digital clipping techniques to convert the photon rate time series into a series of one-bit numbers that can be multiplied rapidly, thereby improving the product rate (number of multiplication products performed per second) and the overall throughput rate of the correlator. Letting

$$n_i = \text{number of photomultiplier anode pulses in time interval } (t_i,\ t_i + \Delta\tau), \tag{1}$$

the photon rate correlation $\langle n_i n_{(i+p)} \rangle$ with time delay $p\Delta\tau$ is obtained ideally by quantizing each value of n_i with enough resolution to make errors in the products of the samples negligible. Clipping processes effectively quantize n_i with one-bit resolution. In clipping without scaling the quantization occurs with respect to a set threshold level n_{th}, and the clipped value of n_i is given by

$$n_{i1} = \begin{cases} 1, & \text{if } n_i > n_{th} \\ 0, & \text{if } n_i \leq n_{th} \end{cases} . \tag{2}$$

The single-clipped correlation is $\langle n_i n_{(i+p)1} \rangle$, and the double-clipped correlation is $\langle n_{i1} n_{(i+p)1} \rangle$.

In general, clipping with respect to a fixed threshold distorts the resulting correlation. Scaling is a technique that removes the clipping distortion by replacing the fixed threshold with a randomly varying threshold so that the clipping effectively samples the photon count rate at many different quantization levels. The random clipping level for the ith time interval k_i assumes values $0,1,\ldots,K-1$ with equal probability, independent of the value of n_i or the value of k_j or n_j in any other time interval. The scaled and clipped value of n_i is

$$n_{isl} = \begin{cases} 1, & \text{if } n_i > k_i \\ 0, & \text{if } n_i \leq k_i \end{cases} . \tag{3}$$

It is known [1] that the single-clipped and scaled correlation $\langle n_i n_{(i+p)sl} \rangle$ is an undistorted version of $\langle n_i n_{(i+p)} \rangle$, and that the double-clipped, double-scaled correlation $\langle n_{isl} n_{(i+p)sl} \rangle$ is also undistorted [2], provided in each case that the probability of n_i exceeding the maximum scaling level

K-1 is negligible. (As noted in Ref. [3], the term "scaling" originally referred to a particular method of accomplishing random clipping level selection in hardware [4]. In Refs. [2], [3] and [5], and in the present paper, scaling will denote the ideal random clipping level process defined above.)

In a recent paper [5], it has been shown that the random shot noise fluctuations inherent in the photon count rate n_i can, under certain conditions, automatically scale the signals so that clipping yields negligibly distorted correlations. This effect, called "self-scaling", occurs whenever the fluctuations in the count rate that are associated with modulation of the light flux are sufficiently small compared to the fluctuations associated with the Poisson statistics of the emission process. An important case in which this condition is met is a varying light flux signal combined with a substantial mean light flux, i.e. dynamic light scattering spectroscopy with flare from a window, laser Doppler velocimetry with low visibility signals and high background light levels or substantial laser flare, or laser two-spot velocimetry under similar conditions.

The purpose of the present paper is to evaluate the rates at which double-clipped and single-clipped correlations converge to their mean values under self-scaling conditions, and to compare these rates to the convergence rates for unclipped correlation and for clipped and scaled correlation. This information will be used to assess the utility of self-scaling in photon correlation experiments and to define the conditions in which its use is desirable. On the basis that clipping removes statistical information, it has been speculated previously [2] that clipped correlograms will exhibit more variance, e.g. less stability, than unclipped corelations. We will see, interestingly, that this is not always true.

2. Count Rate Model

As in [5] we shall model the pulse train from the photomultiplier anode as a Poisson process whose statistics are governed during the ith time interval by the emission parameter

$$\varepsilon_i = \int_{t_i}^{t_i + \Delta\tau} \frac{\eta_q J(t)}{h\upsilon_0} \, dt \qquad (4)$$

wherein η_q is the quantum efficiency of the detector, h is Planck's constant, υ_0 is the light frequency, J is the light flux received by the detector, and $J/h\upsilon_0 = \dot{\varepsilon}$ is the emission rate parameter, corresponding to the mean number of emissions per second, given that the light flux is J. It is convenient to isolate the randomness in the count rate n_i that is due to randomness in J(t) from that which is due to the inherently random nature of the emission process by using conditional statistics in which J(t) and hence ε_i are fixed. Thus,

$$P(n_i | \varepsilon_i) = \frac{e^{-\varepsilon_i} \varepsilon_i^{n_i}}{n_i!} \qquad (5)$$

is the conditional Poisson probability distribution of n_i given that the emission parameter is ε_i, and $\langle n_i | \varepsilon_i \rangle$ is the mean count rate given that the emission parameter is ε_i. The conditional averages of n_i and n_i^2 are

$$\langle n_i | \varepsilon_i \rangle = \varepsilon_i , \tag{6}$$

$$\langle n_i^2 | \varepsilon_i \rangle = \varepsilon_i^2 + \varepsilon_i , \tag{7}$$

respectively, for a Poisson process.

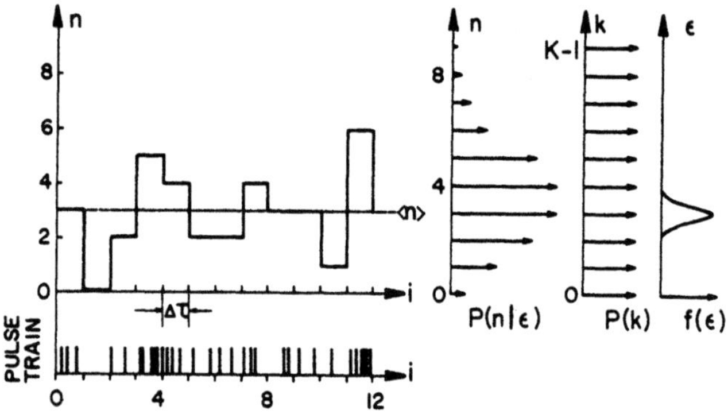

Fig. 1. Pulse counting process and probability distributions associated with scaling and self-scaling

Fig. 1 illustrates the conversion of the anode pulse train into a time series by counting the number in $\Delta\tau$ time intervals, and the principles of scaling and self-scaling. The scaling level probability distribution must be uniform over a range that exceeds the largest likely values of n_i. For the example, in Fig. 1 with $\langle n \rangle = 3$, it is clear by reference to the Poisson probability distribution $P(n|\varepsilon)$ that a value of K-1 greater than or equal to 9 is sufficient to assure that n_i will only rarely exceed the maximum scaling level. Self-scaling occurs if the shot noise fluctuations in the count rate serve the same purposes as a scaling level would. Specifically,

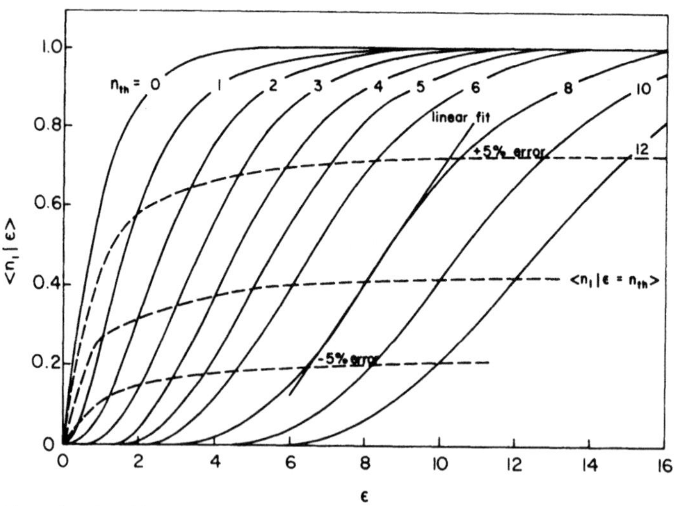

Fig. 2. Conditionally averaged value of the count rate after clipping, as a function of the emission rate

the noise fluctuations should have uniform probability over the range of fluctuations, so that they have the effect of randomly and uniformly varying a clipping level when they are added to the signal. As illustrated in Fig. 1, the probability $P(n|\varepsilon)$ is nearly uniform over the range of the indicated ε-distribution. At first sight, the present explanation of self-scaling appears to be fundamentally different from the explanation offered in [5] wherein linearity of the conditional mean clipped count rate, $\langle n_1|\varepsilon\rangle$ was offered as the basis for the self-scaling effect, c.f. Fig.2. However, they are merely different ways of looking at the same mechanism, because uniformity of $P(n|\varepsilon)$ implies linearity of $\langle n_1|\varepsilon\rangle$. The present explanation exposes the relationship between scaling and self-scaling more fully.

3. Variance of Double-Clipped Correlation Estimates

3.1 Double-Clipped/Double-Scaled

The estimator for the double-clipped/double-scaled correlation is

$$\frac{1}{N}\sum_{i=1}^{N} n_{isl}\, n_{(i+p)sl} \tag{8}$$

with mean value $\langle n_{isl}n_{(i+p)sl}\rangle$. This mean value was related to the unclipped correlation in Ref. 2 by expressing the mean as an average of the conditional average $\langle n_{isl}n_{(i+p)sl}|n_i,n_{(i+p)}\rangle$, noting that this conditional average equaled $\langle n_{isl}|n_i\rangle\langle n_{(i+p)sl}|n_{(i+p)}\rangle$ because independence of k_i and $k_{(i+p)}$ for $p \neq 0$ implied conditional independence of n_{isl} and $n_{(i+p)sl}$ given n_i and $n_{(i+p)}$, and using the relation

$$\langle n_{isl}|n_i\rangle = 1 \cdot \text{Prob}\,\{n_i-k_i > 0|n_i\} \tag{9a}$$

$$= n_i/K\ . \tag{9b}$$

(Equation (9a) appeared incorrectly in Ref. [2]). The result was

$$\langle n_{isl}n_{(i+p)sl}\rangle = \frac{\langle n_i n_{(i+p)}\rangle}{K^2}\ , \quad p \neq 0 \quad . \tag{10}$$

The variance of the estimator in (8) is found as follows:

$$e_{slsl} = \langle[N^{-1}\sum_{i=1}^{N} n_{isl}n_{(i+p)sl} - K^{-2}\langle n_i n_{(i+p)}\rangle]^2\rangle \tag{11a}$$

$$= N^{-2}\sum_{i=1}^{N}\sum_{j=1}^{N}[\langle\langle n_{isl}n_{(i+p)sl}n_{jsl}n_{(j+p)sl}|n_i,n_j,n_{(i+p)},n_{(j+p)}\rangle\rangle$$

$$- K^{-4}\langle n_i n_{(i+p)}\rangle\langle n_j n_{(j+p)}\rangle] \tag{11b}$$

$$= N^{-2}\sum_{i=1}^{N}\sum_{j=1}^{N}[\langle\langle n_{isl}|n_i\rangle\langle n_{(i+p)sl}|n_{(i+p)}\rangle\langle n_{jsl}|n_j\rangle\langle n_{(j+p)sl}|n_{(j+p)}\rangle\rangle$$

$$- K^{-4}\langle n_i n_{(i+p)}\rangle\langle n_j n_{(j+p)}\rangle] \tag{11c}$$

$$+ N^{-2} \sum_{i=1}^{N} [\langle\langle n_{isl}^{2} | n_i\rangle \langle n_{(i+p)sl}^{2} | n_{(i+p)}\rangle\rangle$$

$$-\langle\langle n_{isl} | n_i\rangle^2 \langle n_{(i+p)sl} | n_{(i+p)}\rangle^2\rangle]$$

$$+ 2N^{-2} \sum_{i=1}^{N-p} [\langle\langle n_{isl} n_{(i+p)sl}^{2} n_{(i+2p)sl} | n_i, n_{(i+p)}, n_{(i+2p)}\rangle$$

$$- \langle\langle n_{isl} | n_i\rangle \langle n_{(i+p)sl} | n_{(i+p)}\rangle^2 \langle n_{(i+2p)sl} | n_{(i+2p)}\rangle\rangle]$$

$$= K^{-4} e + K^{-4} N^{-2} \sum_{i=1}^{N} [K^2 \langle n_i n_{(i+p)}\rangle - \langle n_i^2 n_{(i+p)}^2\rangle] \tag{11d}$$

$$+ 2K^{-4} N^{-2} \sum_{i=1}^{N-p} [K\langle n_i n_{(i+p)} n_{(i+2p)}\rangle - \langle n_i n_{(i+p)}^2 n_{(i+2p)}\rangle]$$

$$= K^{-4} e + K^{-4} N^{-2} \sum_{i=1}^{N} [K^2 \langle\langle n_i | \varepsilon_i\rangle \langle n_{(i+p)} | \varepsilon_{i+p}\rangle\rangle \tag{11e}$$

$$- \langle\langle n_i^2 | \varepsilon_i\rangle \langle n_{(i+p)}^2 | \varepsilon_{(i+p)}\rangle\rangle]$$

$$+ 2K^{-4} N^{-2} \sum_{i=1}^{N-p} [K\langle\langle n_i | \varepsilon_i\rangle \langle n_{(i+p)} | \varepsilon_{(i+p)}\rangle \langle n_{(i+2p)} | \varepsilon_{(i+2p)}\rangle\rangle$$

$$-\langle\langle n_i | \varepsilon_i\rangle \langle n_{(i+p)}^2) | \varepsilon_{(i+p)}\rangle \langle n_{(i+2p)} | \varepsilon_{(i+2p)}\rangle\rangle]$$

$$= K^{-4} e + K^{-4} N^{-2} \sum_{i=1}^{N} [K^2 \langle \varepsilon_i \varepsilon_{(i+p)}\rangle - \langle(\varepsilon_i + \varepsilon_i^2)(\varepsilon_{(i+p)} + \varepsilon_{(i+p)}^2\rangle]$$

$$+ 2K^{-4} N^{-2} \sum_{i=1}^{N-p} [K\langle \varepsilon_i \varepsilon_{(i+p)} \varepsilon_{(i+2p)}\rangle - \langle \varepsilon_i (\varepsilon_{i+p} + \varepsilon_{i+p}^2) \varepsilon_{i+2p}\rangle] \tag{11f}$$

In Eq. (11),

$$e = \langle [N^{-1} \sum_{i=1}^{N} n_i n_{(i+p)} - \langle n_i n_{(i+p)}\rangle]^2\rangle \tag{12a}$$

$$= N^{-2} \sum_{i=1}^{N} \sum_{j=1}^{N} [\langle \varepsilon_i \varepsilon_{i+p} \varepsilon_j \varepsilon_{j+p}\rangle - \langle \varepsilon_i \varepsilon_{i+p}\rangle \langle \varepsilon_j \varepsilon_{j+p}\rangle] \tag{12b}$$

$$+ N^{-2} \sum_{i=1}^{N} [\langle(\varepsilon_i + \varepsilon_i^2)(\varepsilon_{i+p} + \varepsilon_{i+p}^2)\rangle - \langle \varepsilon_i^2 \varepsilon_{i+p}^2\rangle]$$

$$+ 2N^{-2} \sum_{i=1}^{N-p} [\langle \varepsilon_i (\varepsilon_{i+p} + \varepsilon_{i+p}^2) \varepsilon_{i+2p}\rangle - \langle \varepsilon_i \varepsilon_{i+p}^2 \varepsilon_{i+2p}\rangle]$$

is the variance of the unclipped correlation estimator formed using full products. Equations (11) and (12) are invalid for $p=0$, but this case is not of interest here.

The rather long series of steps needed to arrive at (11f) will be explained briefly, because similar manipulations are needed for results below. Equation (11b) is found by recasting the square of the sum in (11a) as a double sum, taking the unconditional average in (11a) inside the sums, and writing it as the average of a conditional average in which n_i, n_j, n_{i+p}, and n_{j+p} are given. If $i \neq i+p \neq j \neq j+p$, then the respective photon counts are statistically independent, and the conditional average of the product reduces to the product of the conditional averages, as in the first term of (11c). However, even if $p \neq 0$, this condition is violated if $i=j$ or $i=j+p$ or $j=i+p$. Equation (11c) accounts for these different possibilities by summing the products of the conditional averages in the first term over all i and j as though they were correct for $i=j$, and then introducing a second term in which the correct form for $i=j$ is added and the incorrect form is subtracted for all $i=j$. A similar procedure applied to the cases $i=j+p$ and $j=i+p$ yields the third term in (11c). Equation (11d) is derived from (11c) by application of (9b). The next two equations are found by a sequence of steps that parallel those leading to (11d). Quantities such as $\langle n_i n_{i+p} \rangle$ are recast as averages of conditional averages given the values of the emission parameters, $(\langle\langle n_i n_{i+p} | \varepsilon_i, \varepsilon_{i+p} \rangle\rangle)$, conditional averages of products are reduced to products of conditional averages, $\langle\langle n_i n_{i+p} | \varepsilon_i, \varepsilon_{i+p} \rangle\rangle = \langle\langle n_i | \varepsilon_i \rangle \langle n_{i+p} | \varepsilon_{i+p} \rangle\rangle$, and conditional averages are evaluated in terms of the emission parameters.

3.2 Double-Clipped/Self-Scaled

The estimator for the double-clipped correlation is

$$\frac{1}{N} \sum_{i=1}^{N} n_{i1} n_{(i+p)1} , \tag{13}$$

with mean value

$$\langle n_{i1} n_{(i+p)1} \rangle = \langle\langle n_{i1} | \varepsilon_i \rangle \langle n_{(i+p)1} | \varepsilon_{(i+p)} \rangle\rangle \quad , \quad p \neq 0 \tag{14a}$$

$$= \langle \mu_i \mu_{i+p} \rangle \tag{14b}$$

where

$$\mu = M + \overline{P}\, \varepsilon' \; ; \tag{15}$$

$$M = \langle n_1 | \langle \varepsilon \rangle \rangle , \tag{16}$$

$$\overline{P} = P(n_{th} | \langle \varepsilon \rangle) , \tag{17}$$

$$\varepsilon' = \varepsilon - E , \tag{18}$$

$$E = \langle \varepsilon \rangle \quad . \tag{19}$$

Equation (14b) is obtained from (14a) by using the approximate, linearized result

$$\langle n_1 | \varepsilon \rangle \doteq \mu \tag{20}$$

derived in Ref. [5]. The mean of the double-clipped estimator clearly yields an unbiased measure of the correlation of the fluctuating emission rate $\langle \varepsilon_i' \varepsilon_{i+p}' \rangle$ when this linearization is valid, and this is the self-scaling effect. Fig. 2 shows the locus of 5% linearization error points for the family of curves obtained by varying the threshold level n_{th}. A simple estimate of the range of emission rates for which (20) is accurate for a given value of n_{th} is

$$n_{th} - \sqrt{n_{th}} < \varepsilon < n_{th} + \sqrt{n_{th}} \quad . \tag{21}$$

This is a slight relaxation of an estimate derived earlier on the basis of the behavior of the Poisson statistics for large emission rates [5].

The variance of the double-clipped correlation is

$$e_{11} = \langle [N^{-1} \sum_{i=1}^{N} n_{i1} n_{(i+p)1} - \langle n_{i1} n_{(i+p)1} \rangle]^2 \rangle \tag{22a}$$

$$\doteq N^{-2} \sum_{i=1}^{N} \sum_{j=1}^{N} [\langle \mu_i \mu_{i+p} \mu_j \mu_{j+p} \rangle - \langle \mu_i \mu_{i+p} \rangle \langle \mu_j \mu_{j+p} \rangle] \tag{22b}$$

$$+ N^{-2} \sum_{i=1}^{N} [\langle \mu_i \mu_{i+p} \rangle - \langle \mu_i^2 \mu_{i+p}^2 \rangle]$$

$$+ 2N^{-2} \sum_{i=1}^{N-p} [\langle \mu_i \mu_{i+p} \mu_{i+2p} \rangle - \langle \mu_i \mu_{i+p}^2 \mu_{i+2p} \rangle] \quad .$$

Note that the error in (22a) is defined as the difference with respect to the biased result so that bias errors are excluded from (22a), leaving only sampling errors. Equation (22b) is derived by steps similar to those leading to (11e) using (7) and (20) and the observing that

$$n_{i1}^2 = n_{i1} \ . \tag{23}$$

3.3 Single-Clipped/Single-Scaled

The estimator for the single-clipped and scaled correlation is

$$\frac{1}{N} \sum_{i=1}^{N} n_i n_{(i+p)s1},$$

and the mean value of this estimator is [3]

$$\langle n_i n_{(i+p)s1} \rangle = \langle n_i n_{(i+p)} \rangle / K \ . \tag{24}$$

The variance of the single-clipped and scaled correlation is given by

$$e_{s1} = \langle [N^{-1} \sum_{i=1}^{N} n_i n_{(i+p)s1} - \langle n_i n_{(i+p)} \rangle / K]^2 \rangle \tag{25a}$$

$$= K^{-2} e + K^{-2} N^{-2} \sum_{i=1}^{N} \langle [\varepsilon_i + \varepsilon_i^2][(K-1)\varepsilon_{i+p} - \varepsilon_{i+p}^2] \rangle \ . \tag{25b}$$

In this equation e is the variance of the unclipped estimator, given by (12).

3.4 Single-Clipped/Self-Scaled

The estimator for the single-clipped correlation is

$$\frac{1}{N} \sum_{i=1}^{N} n_i n_{(i+p)1} \ ,$$

with mean value

$$\langle n_i n_{(i+p)1}\rangle = \langle\langle n_i|\varepsilon_i\rangle\langle n_{(i+p)1}|\varepsilon_{(i+p)}\rangle\rangle \,, \qquad p\neq 0 \tag{26a}$$

$$\doteq \langle\varepsilon_i\,\mu_{i+p}\rangle \,. \tag{26b}$$

As in the case of double-clipping, an unbiased value of $\langle\varepsilon_i\varepsilon_{i+p}\rangle$ can be extracted from the above mean when μ_{i+p} is a linear function of ε'_{i+p}, as in Eq. (15).

The variance of the single-clipped correlation is given by

$$e_1 = \langle[N^{-1}\sum_{i=1}^{N} n_i n_{(i+p)1} - \langle n_i n_{(i+p)1}\rangle]^2\rangle \tag{27a}$$

$$\doteq N^{-2}\sum_{i=1}^{N}\sum_{j=1}^{N}[\langle\varepsilon_i\,\mu_{(i+p)}\,\varepsilon_j\,\mu_{j+p}\rangle - \langle\varepsilon_i\,\mu_{i+p}\rangle\langle\varepsilon_j\,\mu_{j+p}\rangle] \tag{27b}$$

$$+ N^{-2}\sum_{i=1}^{N}[\langle(\varepsilon+\varepsilon^2)_i\;\mu_{i+p}\rangle - \langle\varepsilon_i^2\;\mu_{i+p}^2\rangle]$$

$$+ 2N^{-2}\sum_{i=1}^{N-p}[\langle\varepsilon_i\varepsilon_{i+p}(M+\overline{P}+\overline{P}\,n_{th}E^{-1}\varepsilon')_{i+p}\mu_{i+2p}\rangle$$

$$- \langle\varepsilon_i\,\mu_{i+p}\varepsilon_{i+p}\mu_{i+2p}\rangle] \,.$$

The following relationship, derived directly from the Poisson distribution in (5) has been used to obtain the last term in (27b)

$$\langle nn_1|\varepsilon\rangle = \sum_{n_{th}+1}^{\infty} nP(n|\varepsilon) \tag{28a}$$

$$= \varepsilon\;[\langle n_1|\varepsilon\rangle + P(n_{th}|\varepsilon)] \tag{28b}$$

$$\doteq \varepsilon\;[M + \overline{P} + \overline{P}\;n_{th}\;E^{-1}\;\varepsilon'] \tag{28c}$$

4. Comparison of the Correlation Methods

4.1 Single-Clipped

We consider first the single-clipped and scaled correlation variance, and compare it subsequently to the variance of the unclipped correlation and the variance of single-clipped correlation with self-scaling. One of the effects of scaling is to add noise to the signal being correlated. This is seen readily by noting that clipping the signal n_i with respect to a random threshold k_i is equivalent to clipping n_i-k_i with respect to zero. Thus, the scaled signal appears to have superimposed on it noise whose root mean square is approximately $K/\sqrt{12}$ for large K. One expects that this added noise will increase the variance of the correlation estimate. The first term in (25b) is the variance of the full multiplication product estimator, and the second term is the added variance due to scaling. It will be positive if $K-1 > \varepsilon_{i+p}$. But K-1 must be greater than the maximum value of ε in order

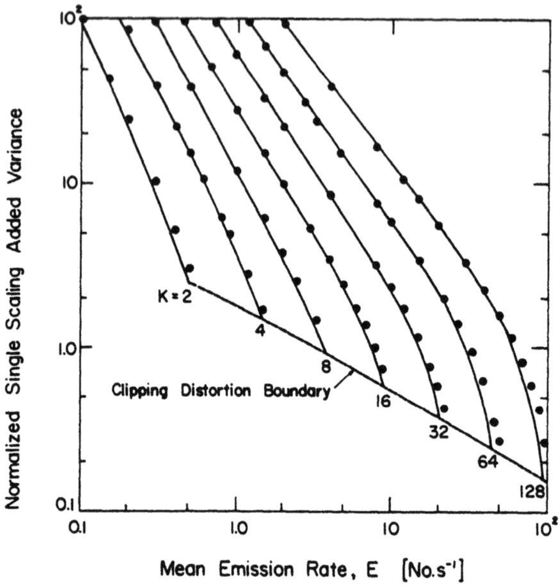

Fig. 3. Added variance due to scaling the single-clipped correlation. Normalized by $NK^2/\langle\varepsilon^2\rangle^2$ and evaluated at $\tau=0^+$.
• $\sigma_\varepsilon/E=0$; $\sigma_\varepsilon/E=0.25$; --- clipping distortion boundary given by $K-1$ = Integer $(\varepsilon + 2\sqrt{\varepsilon})$

to avoid distortion due to clipping; hence scaling always increases the variance.

The normalized added variance for single scaling $NK^2(e_{s1} - K^{-2}e)/\langle\varepsilon^2\rangle^2$ is plotted against the mean emission rate E in Fig. 3, assuming that ε obeys joint normal statistics. The variance is evaluated at correlation time delay $\tau \neq 0^+$ where the single-clipped and scaled correlation has mean value $\langle\varepsilon^2\rangle/K$ and the zero delay noise spike is avoided. Comparison of the results for two ratios of the root mean square ε fluctuation to the mean value, $\sigma_\varepsilon/E = 0$ and $\sigma_\varepsilon/E = 0.25$ shows little difference between these cases. Consequently the simpler case $\sigma_\varepsilon/E = 0$ will be used hereafter for illustration. This represents the physical situation in which a steady light flux produces a constant emission rate equal to the mean. More generally, the results can be interpreted as the asymptotic limit in which σ_ε/E approaches zero. The added variance is large when the mean emission is small compared to the scaling range K because the effective noise introduced by scaling is large compared to the signal. Thus, it is desirable to choose small values of K for the purpose of minimizing the estimation variance. On the other hand K must be large enough to ensure that it exceeds the maximum likely value of the count rate in order to avoid clipping distortion. In this regard, it is desirable to use large values of K so that the signal is properly scaled without distortion over a wide dynamic range of emission rates.

Ideally, the clipping distortion boundary in Fig. 3 would be defined by the locus of ε values at which the probability of n exceeding K-1 reaches some maximum allowable value, $P[n>K-1|\varepsilon] = P_{max}$. Beyond this boundary clipping distortion occurs, and the value of K should be increased. Primarily for purposes of simplicity, a simpler criterion for the clipping distortion boundary has been developed by noting that values of n that are more than a few standard deviations above the mean occur rarely for a Poisson distribution. This suggests selecting the value of K-1 to satisfy

$$K-1 = \mathrm{INT}(\varepsilon + c\sqrt{\varepsilon}) \tag{29}$$

in order to keep the probability of exceeding the clipping range below some acceptable level. As shown in Fig. 4 c=2 allows the probability to increase slowly with emission rate, but maintains it below 2% up to 10 emissions per $\Delta\tau$, whereas the probability has an upper bound of 0.0015 when c=3, certainly adequate to insure minimal distortion. The distortion boundary in Fig. 3 is calculated from (29) using c=2 and $\varepsilon = E$.

The single-scaling added variance

$$\Delta e_{s1} = (e_{s1} - K^{-2} e) \tag{30}$$

is normalized by $N^2 K^2/E^4$ and compared to the unclipped variance in Fig. 5. The heavy line in this figure represents the performance that could be achieved if the clipping range K could be incremented by one each time the distortion boundary was crossed. This could not be accomplished easily in practice unless the range of ε fluctuations was small, lying between two boundary crossings; but it is useful as an optimal mode of operation for purposes of comparison. Along this optimal path Fig. 5 shows that the added variance is small compared to the unclipped variance for mean emission rates below unity, approximately, but the converse is true for rates in excess of five. If K is not adjusted to yield this optimum behavior, the added

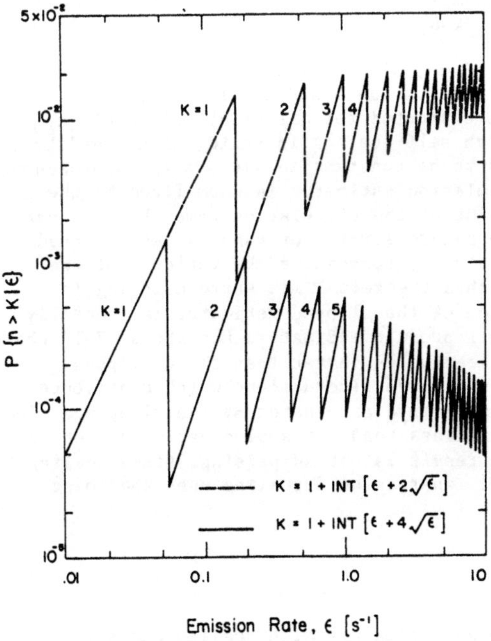

Fig. 4. Probability of clipping distortion given the emission rate

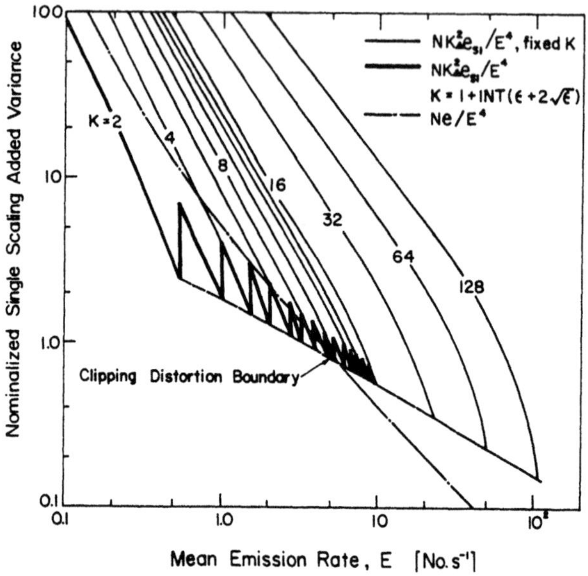

Fig. 5. Added variance Δe_{s1} due to scaling the single-clipped correlations compared to the variance of the unclipped correlation and the variance along $K = 1 + \text{Int}\ (\varepsilon + 2\sqrt{\varepsilon})$. $\sigma_\varepsilon/E=0$, $\tau=0^+$. Clipping distortion boundary is given by $K - 1 = \text{Integer}\ (\varepsilon + 2\sqrt{\varepsilon})$

variance can be much greater; for example, at E=0.3 the normalized added variance is 10 if K=2, and 100 if K=8.

Fig. 6 shows the total variance of the single-clipped/self-scaled correlation in comparison with the unclipped variance and the single-scaled variance. The self-scaled curves are plotted for different values of n_{th} over ranges of emission rate in which self-scaling is valid, as defined by (21). The emission rate is assumed to be constant in i.e. σ_ε/E approaching zero, and the variance of each correlation estimator is normalized by the square of the corresponding mean value of the correlation times N. The curves therefore represent fractional mean square errors per sample. As expected, the single-clipped variance asymptotically approaches the variance of the unclipped correlation as E approaches the zero limit where clipping is irrelevant. Below E=1 the variance of the clipped estimator is generally greater than the variance of the full product estimator, but above E=1 the clipped estimator converges to its mean value faster than the unclipped estimator Apparently, large amplitude noise fluctuations which contribute substantially to the unclipped estimator are attenuated by the clipping. The self-scaled variance is significantly less than the scaled variance over the full range of emission rates. This result is not surprising, since scaling effectively adds noise to the signal, whereas self-scaling uses the noise inherent in the signal.

4.2 Double-Clipped

The variances of the double-clipped estimators with scaling and self-scaling are compared to the full product variance in Fig. 7. The comparison

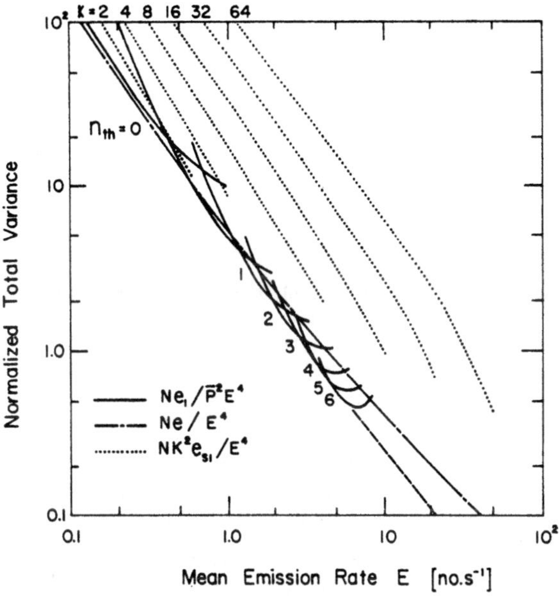

Fig. 6

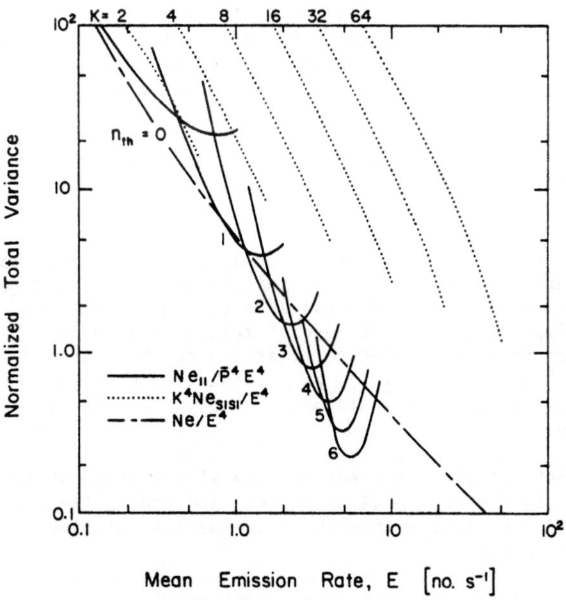

Fig. 7

Fig. 6. Total variance of single-clipped correlations with scaling and self-scaling. The self-scaled curves for each value of clipping threshold are plotted over the range $n_{th} - \sqrt{n_{th}} < \varepsilon < n_{tb} + \sqrt{n_{th}}$, and the scaled curves terminate at the clipping distortion boundary

Fig. 7. Total variance of double-clipped correlations with scaling and self-scaling is for σ_ε/E approaching zero, and variances are normalized by N divided by the square of the mean correlation estimator, as before. Qualitatively, the conclusions to be drawn from Fig. 7 are the same as the preceding conclusions for single-clipping, but the differences between double-scaling, self-scaling and full product multiplication are more pronounced because the scaling works on both channels. Thus, self-scaling has much smaller variance than double-scaling, and the variance is significantly lower than that of full product multiplication at higher mean emission rates.

5. Conclusions

It has been established theoretically that the shot noise fluctuations in the photon pulse count rate act like scaling signals that make it possible to obtain unbiased estimates of the emission rate correlation from correlations of clipped values of the count rate. The statistical stability of clipped correlation is generally worse than that of unclipped correlation at emission rates of order one, but at larger rates or at very small rates the present calculations indicate reduced variances that may even fall below the unclipped variance. Thus, the simplicity of operation that is offered by one-bit clipping does not necessarily imply an inevitable loss of statistical accuracy and/or an increase in sampling time. There is however a penalty that must be paid, and that is a reduction in the dynamic range of emission rate over which self-scaling can successfully unbias the correlation. This range, estimated in (21), becomes narrow relative to the mean emission rate at large rates. For example, with $n_{th}=5$, self-scaling is accomplished only in the range between approximately 2.5 and 7 emissions per $\Delta\tau$. If the mean emission rate were composed of a large, steady rate due to background light plus a small fluctuating signal, this sort of range would be adequate, and clipping with self-scaling would be very useful. If not, then clipping with scaling would be an alternative.

Scaling generally degrades the statistical stability of the clipped correlation significantly, both with respect to the unclipped correlation, and with respect to the clipped and self-scaled correlation. The advantage offered by clipping and scaling is an extension of the allowable range of emission rates down to very small values. That is, scaling only requires that the emission rate must be less than the scaling range without any restriction on the minimum rate. Unfortunately, the added variance due to scaling may become so large that it is impractical to make use of this wide dynamic range. With K=16 double-clipped and scaled correlations can be obtained from 0 up to 10 emisions per $\Delta\tau$ (c.f. Fig. 7), but if this value of K were used at 2 emissions per $\Delta\tau$, the variance would be about 35 times the variance of the unclipped correlation.

In summary, it appears that clipping does not greatly affect the stability of the correlation estimator in the self-scaling regime, and that self-scaling is often preferable to scaled correlation when the range of the emission rate fluctuations is sufficiently small.

References

1. E. R. Pike: Photon Correlation Spectroscopy and Velocimetry, NATO Advanced Study Institute Ser. B, 23, ed. by H. Z. Cummins and E. R. Pike (Plenum, New York 1977)

2. R. J. Adrian: Appl. Optics 21, 1880 (1982)

3. R. J. Adrian: TSI Quart. 8, 3 (1982)

4. C. J. Oliver: Photon Correlation and Light Beating Spectroscopy, NATO Advanced Study Institute Ser. B, 3, ed. by H. Z. Cummins and E. R. Pike (Plenum, New York 1974)

5. R. J. Adrian: Appl. Optics 20, 3901 (1981)

6. A. Papoulis: Probability Random Variables and Stochastic Theory (McGraw-Hill, New York 1965).

24. Hardware and Software Implementation of On-Line Velocity Correlation Measurements in Oscillatory and Turbulent Rotational Couette Flow

G. Pfister, U. Gerdts, A. Lorenzen, and K. Schätzel

Institut für Angewandte Physik der Universität Kiel, Olshausenstraße 40-60
D-2300 Kiel, Fed. Rep. of Germany

Abstract

In hydrodynamics the routes to turbulence and associated phenomena have become an important subject of theoretical and experimental investigation in recent years. Our investigations into this field were carried out in rotational Couette flow where the external parameters can be finely controlled. Theoretical models leading to strange attractors and period doubling are concerned with the non-linear coupling of different flow modes and the development of one single flow mode, respectively. Therefore it is desirable to obtain accurate information about the behaviour of particular flow modes as a function of boundary conditions. This paper reports some measurements concerning specific flow modes in rotational Couette flow using an online velocity correlation method.

1. Introduction

Rotational Couette flow is a hydrodynamic system undergoing a sequence of instabilities on its route to turbulence as the Reynolds number is increased. In our experiments the outer cylinder is at rest and the inner one rotates. One finds for sufficiently small Reynolds numbers that the fluid moves steadily in concentric circles except perhaps for some three-dimensional motion at the fixed ends of the cylinders. As the rotation rate is increased, toroidal counter-rotating steady Taylor vortices are formed, growing smoothly from the ends of the apparatus [1]. When the Reynolds number is increased further the Taylor vortices become unstable and form azimuthal travelling waves. With further increases of the Reynolds number more oscillatory modes appear and noise rises until one may call the flow turbulent [2].

Rotational Couette flow is known to be very sensitive to end effects for stationary flow [3, 4, 1] and also for the onset of oscillatory flow [5, 6, 7]. Therefore extreme care has to be taken in controlling the aspect ratio, i. e. height to width of the gap, and the end conditions.

By varying the aspect ratio three different types of oscillatory modes were obtained separately in our wide gap geometry for a particular cellular mode. Additionally we will show that the aspect ratio has a strong influence on the route to turbulence.

Firstly we shall describe the flow apparatus used and then the LDV system, with which we measured the local velocity. Also we shall describe the data processing system used to compute the velocity auto- and cross-correlation functions and the velocity power spectrum. In the final section we shall present and discuss our experimental results.

2. Couette flow apparatus

The Couette flow apparatus consists of an inner rotating cylinder (r_1 = 11,56 mm) and an outer stationary cylinder (r_2 = 22,82 mm) yielding a gap width of 11,26 mm which was uniform to within 1 % over the entire length. Top and bottom plates were stationary and could be adjusted up to a maximum separation of L = 404 mm, giving a maximum aspect ratio of about 36. By this means the Taylor cell length could be controlled.

The fluid was thermostated to within 0.05 K corresponding to an uncertainty of 0.1 % in the Reynolds number Re, which is defined as

$$Re = \frac{\omega \cdot r_1 \cdot d}{\nu} ,$$

where ω is the rotating rate of the inner cylinder of radius r_1, d is the gap width and ν the kinematic viscosity. The rotation speed of the inner cylinder was controlled by an VCO circuit to within 1 % (short time average) and better than 1 % for long time average. Using silicon oil with various viscosities, we could obtain Reynolds numbers of up to 8000.

3. LDV system

The local velocity was measured by a crossed-beam Laser Doppler Velocimeter. The optical setup is shown in fig. 1. The beam of a 4 mW He-Ne laser was separated by a beam splitter into two parallel beams with equal intensity having a separation of 50 mm. Two Bragg cells shifted the laser frequency by 40 MHz and 40,1 MHz, respectively, shifting the zero-point of velocity to 100 kHz.

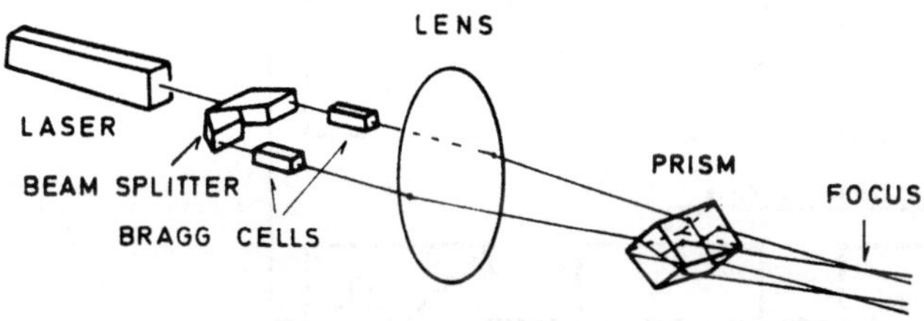

Fig. 1. Cross-beam cross-correlation geometry

A lens (focus 250 mm) causes the beams to intersect giving a fringe system with a spacing of about 3 µm. Thus a velocity of 1 mm/s gives a frequency shift of about 330 Hz.

Additionally a Koester prism could be used to produce two measurement volumes at a variable distance for cross-correlation measurements.

Fig. 2 shows the electronic equipment used in our experiments. The signal from an avalanche diode is fed into a pre-amplifier and then through a band-pass filter to remove the pedestal and noise. The filtered signal is fed into a tracker which gives an analogue voltage proportional to the velocity. This can either be recorded on a x-y plotter or fed into a microprocessor system where estimates of the velocity and acceleration distribution can be obtained. In addition to this analogue signal, the tracker provides a continuous Doppler frequency which can be observed on a frequency meter.

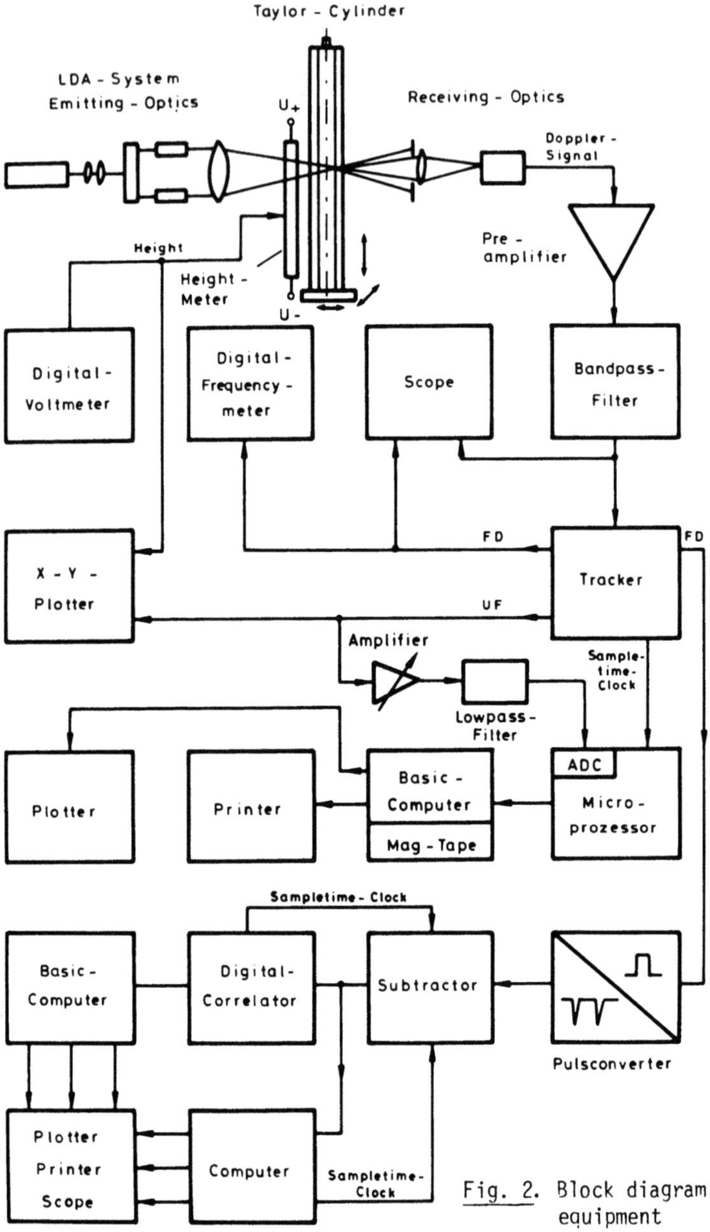

Fig. 2. Block diagram of electronic equipment

To produce on-line velocity correlation functions we used the ERDMANN-GELLERT rate correlation technique [8, 9, 10]. (For details of this technique see K. SCHÄTZEL's contribution in this volume and papers cited therein.)

For measurements of singly periodic states the velocity correlation function was estimated by correlating the rate of zero crossings of the VCO frequency of the tracker using a commercial 128-channel multibit correlator. The advantage of this technique is its sensitivity to time periodic velocity fluctuations.

In the presence of several time periodic states or turbulence, the velocity correlation function was produced by a computer in real time for improved accuracy. The computer was able to estimate 1024 points of the velocity correlation function with a minimum sample time of 10 ms. For higher Reynolds numbers where the tracker could not follow the increased frequencies of the velocity fluctuations, the Doppler bursts were used directly, i. e. without using the tracker. This resulted in longer measuring times but avoided the low pass filter effects of the tracker. The growth of the correlation function could be observed on an oscilloscope in real time. The Fourier transform of the velocity correlation function, performed by the same computer, yields the velocity power spectrum.

4. Experiments and results

A first series of measurements illustrates the importance of the effect of aspect ratio on the qualitative behaviour of the flow. These measurements were obtained with six vortex pairs and various fixed aspect ratios. The aspect ratio is changed to yield different wavelengths of the Taylor cells. Fig. 3 shows that the stationary Taylor vortex flow goes through supercritical bifurcations into three different types of time periodic flow depending upon the size of Taylor cells. At the supercritical bifurcations a critical slowing down occurs [7]. Therefore the time constants diverge at these points so that the measurement times become exceptionally long. For higher Reynolds numbers some of these modes couple, leading to hysteresis effects.

Fig. 4 shows velocity profiles of two time-periodic modes obtained by the analogue tracker [11]. Firstly the wavy mode, where all parts of a vortex are moving up and down and all vortices oscillate in phase. Secondly the jet mode where only the sharp, fast moving outward flow (O) oscillates and the inward flow (I) shows minimal time dependence. Neighbouring vortices oscillate with a phase difference of 180°. The third type

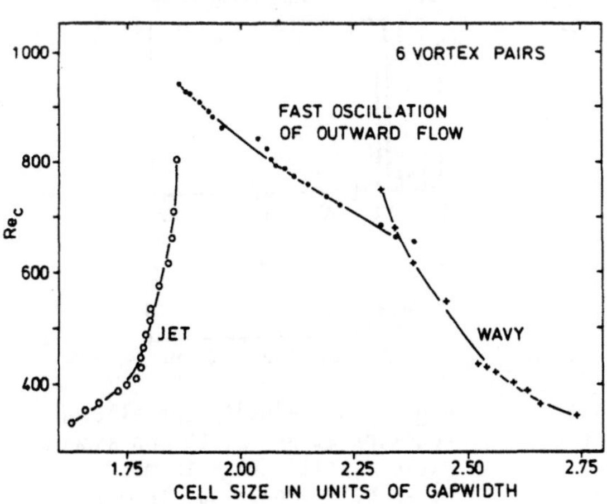

Fig. 3. Critical Reynolds numbers versus cell size for the three types of oscillatory instabilities

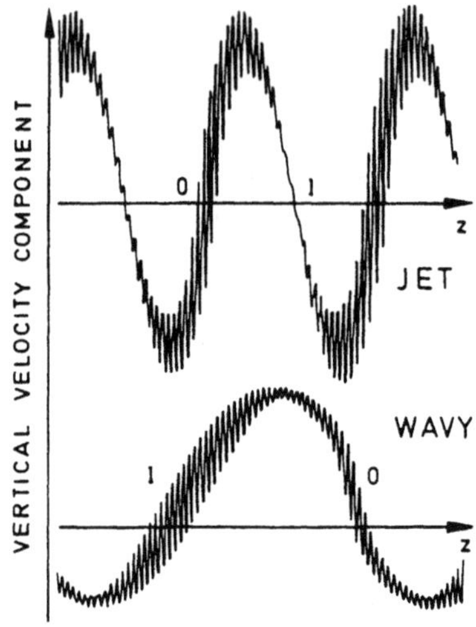

Fig. 4. Velocity profiles of Taylor vortices with jet and wavy mode respectively

of oscillatory mode is characterized by an oscillation of the inward flow only. The normalized frequencies relative to the rotation frequency of the inner cylinder of the wavy-, jet- and fast oscillation outward flow are 0.02, 0.45 and 0.95 respectively.

A second series of measurements was performed at higher Reynolds numbers to investigate the routes to turbulence. This was done to compare the experimental results with models associated with the names of LANDAU, HOPF, LORENZ, RUELLE, TAKENS and FEIGENBAUM.

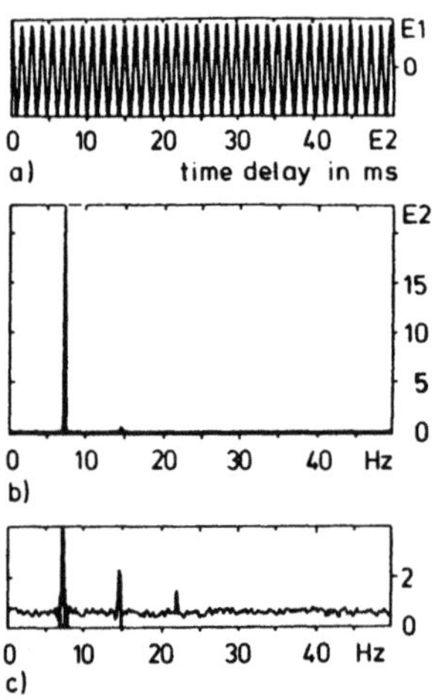

Fig. 5. a) velocity correlation function at Re = 1640 and aspect ratio 2, time delay in ms; b) velocity power spectrum linear scale; c) velocity power spectrum logarithmic scale

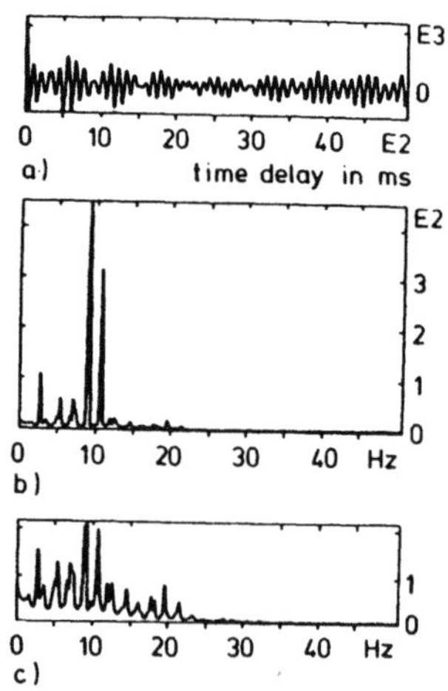

Fig. 6. a) velocity correlation function at Re = 1800 and aspect ratio 2, time delay in ms; b) velocity power spectrum linear scale; c) velocity power spectrum logarithmic scale

For these measurements we used aspect ratios 0.9, 2.0 and 18 giving one single vortex, a vortex pair and 18 vortices respectively. Figures 5 and 6 show examples of computed estimates of the velocity correlation function obtained for an aspect ratio of 2. Both figures show 510 points of the velocity correlation function (a), and velocity power spectrum on both linear (b) and logarithmic (c) scales. Figure 5 shows the behaviour of the flow at a Reynolds number 1640. Only one frequency can be seen in the power spectrum together with two of its harmonics. These are present because we had to measure the velocity modulation in a region with nonlinear spatial velocity distribution. The instrumental noise of the system can be seen in the logarithmic scale, being one ten-thousendth of the maximum signal with a measuring time of about ten minutes.

Figure 6 shows a similar set of plots for the same system with Reynolds number 1800. The power spectrum shows at least three frequencies on a background noise. It is not clear yet if the background noise comes into the power spectrum at the same time as the third oscillatory mode appears and this is a subject for future study. However it does appear to be broad band spreading from zero frequency.

For other aspect ratios the scenario looked qualitatively similar. Even at the highest Reynolds number obtained the velocity power spectrum always contains discrete frequencies together with a monotonically decreasing noise. This may have its origin in the rotational symmetry of the flow geometry.

5. . Conclusions

In this paper we have shown that conventional LDV with analogue velocity estimation combined with rate correlation technique is a powerful tool for investigating time periodic flow and the routes to turbulence. For the onset of isolated oscillatory states the velocity correlation function was produced by a 128-channel 4-bit hardware correlator. The smallness of the oscillation near the critical Reynolds numbers requires only small or unity scaling factors. For the single frequency mode frequency resolution is sufficient. In the presence of many time-periodic states and turbulence, 1024 points of the velocity correlation function were produced by a computer in real time for improved accuracy. With this procedure 1024 channels of an 8 bit x 8 bit signal could be handled in 10 ms. Thus the software version gives better results for broadband velocity fluctuations.

We thank the STIFTUNG VOLKSWAGENWERK for supporting part of the work and T. MULLIN for comments on this manuscript.

References

1 G. Pfister and I. Rehberg; Space dependent order parameter in circular Couette flow transitions; Phys. Letters 83A, 19 (1981)

2 For a recent complete survey of the Taylor flow see the review of: R. C. DiPrima and H. L. Swinney; in "Hydrodynamic instabilities and transitions to turbulence" (edt.: H. L. Swinney and J. P. Gollub), p. 139-180 (Springer-Verlag 1981)

3 T. B. Benjamin and T. Mullin; Anomalus modes in the Taylor experiment; Proc. Roy Soc. London A377, 221 (1981)

4 T. Mullin, G. Pfister and A. Lorenzen; New observations on hysteresis effects in Taylor-Couette flow; Phys. Fluids 25, 1134 (1982)

5 T. Mullin and T. B. Benjamin; Transition to oscillatory motion in the Taylor experiment, Nature 288, 567 (1980)

6 A. Lorenzen, G. Pfister and T. Mullin; End effects on the transition to time-dependent motion in the Taylor experiment, submitted to Phys. Fluids

7 G. Pfister and U. Gerdts; Dynamics of Taylor wavy vortex flow, Phys. Letters 83A, 23 (1981)

8 J. C. Erdmann and R. I. Gellert; Recurrence rate correlation in scattered light intensity; J. Opt. Soc. Am. 68, 787 (1978)

9 R. Vehrenkamp, K. Schätzel, G. Pfister and E. O. Schulz-DuBois; Direct measurement of velocity correlation functions using the Erdmann-Gellert rate correlation technique, J. Phys. E 12, 119 (1979)

10 K. Schätzel; Advances in cross-beam rate correlation, Opt. Acta 27, 45 (1980)

11 G. Pfister; On the structure of non-stationary Taylor vortices; Proc. 4th Conf. on Photon Correlation Techniques in Fluid Mechanics, eds. W. T. Mayo, A. E. Smart, Stanford University, 27-1 (1980)

25. A Study of a Propane Flame as a Deep Phase Screen

R.C. Klewe and P.H. Richards

Marchwood Engineering Laboratories, Central Electricity Generating Board, Marchwood, Southampton, UK

1. Introduction

The mixing of fuel and air in flames relies to a large extent on the creation of a turbulent flow field. High speed photographs show that one consequence of this is the generation of turbulent eddies which appear as balls of hot entrained air moving rapidly through the flame. This so-called thermal "unmixedness" results in a random array of hot-cold interfaces which represents a field in which the refractive index is varying both spatially and temporarily. This feature has been utilised for many years in recording Shadowgraph and Schlieren images but the quantitative information which can be drawn from such data has always been of limited value. The notion that a plane optical wave incident on a flame will suffer different point by point phase changes as it passes through the flame and that the emergent wave front will therefore contain information related to the three-dimensional structure of the flame is only strictly valid if the phase can be measured directly. However, and less restrictively, since the phase variations also cause point by point beam deflection and hence intensity fluctuations, then a measure of the magnitude of this fluctuation should also be dependent on the structure of the flame.

Consider an expanded laser beam passing through a turbulent flame. Simple diffraction theory shows that the intensity fluctuation in the emergent beam at a distance z from the flame will result from eddies with a radius equivalent to the Fresnel radius $\ell_F = (\lambda z)^{\frac{1}{2}}$ where λ is the wavelength of the laser. For example, for $\lambda = 500$ nm, $\ell_F \sim 1$ mm for $z = 2$ m and $\ell_F \sim 4$ mm for $z = 32$ m. Such arguments indicate that measurement of the intensity fluctuation as a function of z could yield the distribution of eddy sizes in the flame and that by repeating these measurements at different heights in the flame could give information leading to the rate of evolution of the eddies.

It should be recognised that a rigorous theoretical approach to this problem would start with the Huygens-Fresnel integral and calculate the scattered field as a function of z as the integral over the perturbed emergent phase front. In this respect the problem is identical to the degradation of laser beams passing through the atmosphere and the twinkling of starlight but with different propagation factors. Considerable progress has been made over the last few years but it seems unlikely that this approach is yet ready to yield much insight into such parameters of immediate interest. A more useful approach has been developed by JAKEMAN and his co-workers [1,2] which involves the identification of discrete scatterers or scattering events. He assumes that the phase screen consists of a finite number N of uniformly illuminated discs or facets and that the phase change introduced by each facet is much larger than any large aperture diffraction effects. Hence the total scattered field at a detector distance r from the phase screen from a light beam passing through the phase screen is given by the sum of the random phases ψ_j from each facet

$$\varepsilon^{+}(\underline{r},t) = \exp(-i\omega t) \sum_{j=1}^{N} a_j(\underline{r},t)\exp(i\psi_j) \quad ,$$

where $\varepsilon^{+}(\underline{r},t)$ is the positive frequency part of the electric field and a_j is the diffraction factor for the jth facet. To evaluate the diffraction factors JAKEMAN [1] assumed that the phase change $\emptyset$ varied linearly over the facets which were characterised by a length scale ξ. The result of interest here is the equation for the relative variance of intensity

$$\frac{\langle I^2(\underline{r},t)\rangle}{\langle I(\underline{r},t)\rangle^2} = 2(1-\frac{1}{N}) + \frac{\overline{\emptyset^2}}{4N}\exp\left(\frac{k^2\xi^2\sin^2\theta}{4\overline{\emptyset^2}}\right) \quad , \tag{1}$$

where k is the wave number $(2\pi/\lambda)$ and θ is the angle of the detector with the line of incidence. It should be noted that this equation is identical to that derived using a direct analytical approach [1] when $N = W_0^2/\xi^2$ where W_0 is the width of the Gaussian intensity profile of the incident beam.

The structure of (1) shows that all the terms containing N quantify deviations from Gaussian statistics. The first term is simply a correction term resulting from there only being a finite number of scatterers but the second term is a single scatterer term and depends on the effects of Fresnel diffraction from each individual facet. A form similar to (1) has been plotted by JAKEMAN and McWHIRTER [2] showing the second moment as a function of distance z from the phase screen and it is clear that very strong intensity fluctuations can result from these non-Gaussian effects. The position of the peak in the second moment occurs at the diffraction maximum for a single facet, i.e. $z/k\xi^2 \sim 1$ and the effective area of the phase screen contributing to this Fresnel 'focus' is determined by slope distribution of the facets. The RMS tilt of a facet is of order $\emptyset_0/k\xi$ so that the length of screen which can contribute to a point at distance z will be $z\emptyset_0/k\xi$. This implies a speckle size

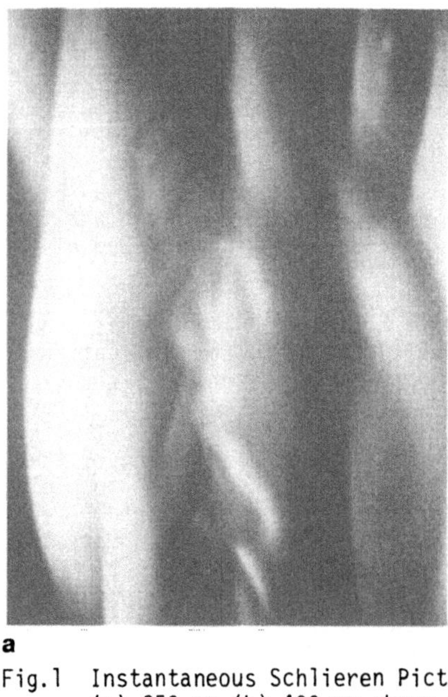

a

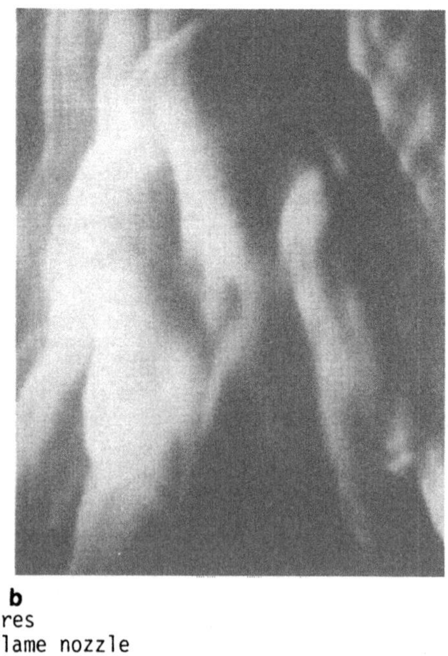

b

Fig.1 Instantaneous Schlieren Pictures
(a) 250 mm (b) 400 mm above flame nozzle

of $\xi/\emptyset_0$ and a number of scatterers $z\emptyset_0/k\xi^2$ and since the peak of the focus results from a single facet then

$$\xi^2 = z\emptyset_0/k$$

which gives a measure of the correlation length in the phase screen.

The purpose in describing the discrete scatterer model is that it would appear to be particularly useful in describing the qualitative behaviour of turbulent flames. High speed Schlieren photographs (Fig.1) show that a flame with a diameter of a few centimetres will consist of turbulent eddies with dimensions of several millimetres giving only 5 or 6 eddies along any line of sight through the flame. It is the purpose of this paper therefore to discuss the measurement of the relative variance of the intensity resulting when a laser beam is passed through a flame. Some point correlation measurements have also been carried out to see if further information relating to the motion of the turbulent eddies can be obtained.

2. Experimental Arrangement

The flame used in these measurements was a typical laboratory propane flame. There was no pre-mixing of the gas and air and the flame nozzle diameter was 2 mm. Visual examination showed that the height of the flame was about 0.5 m and that the mean diameter at half height was about 50 mm. The flame appeared to be laminar for the first 80 mm but was totally mixed out above 100 mm. The intensity fluctuation measurements were carried out at two heights, 250 mm and 400 mm.

High speed Schlieren photographs (Fig.1) taken at 2000 pps show the change of structure in the refractive index field at these positions. The temperature of the flame was measured using a Ni-Cr/Ni-Aℓ thermocouple and a typical radial profile was recorded. The maximum temperature was 1000 K close to the edges of the flame and dropped smoothly to 800 K at the centre. This effect gradually disappeared with height up the flame where the temperature profile became flatter and eventually showed a peak.

The primary problem in setting up the experiment was to arrange the path length of the expanded laser beam after it had passed through the flame to be as long as possible within the constraint of reasonable access for measurement. The experience of PARRY ET AL [3] using a convective flow phase screen was that a path length of 40 m was needed to ensure reaching the focussing region and 60-70 m to be in the Gaussian speckle region. We chose to keep the beam as parallel as possible although this did mean that there was some degradation in the quality of the beam profile at large distances.

The experimental arrangement used is shown in Fig.2. The beam from a 2 mW He-Ne laser was passed through a spatial filter and collimated to give an expanded parallel beam with an e^{-1} diameter of 35 mm. The beam then passed through the propane flame onto mirrors M_1 and M_2 which directed the beam onto two large plane Schlieren mirrors M_3 and M_4. By multiple reflections between M_3 and M_4 a total path length of about 80 m was obtained. This path could be interrupted at any desired distance from the flame and the image deflected onto a plane mount holding a 250 μm fibre optic cable. For the intensity fluctuation measurements a single fibre was used which was fixed directly onto the photocathode of a 56 AVP photomultiplier. The amplified signal was then fed into two Solatron digital voltmeters so that the root mean square and the mean of the fluctuating (analogue) signal could be measured simultaneously.

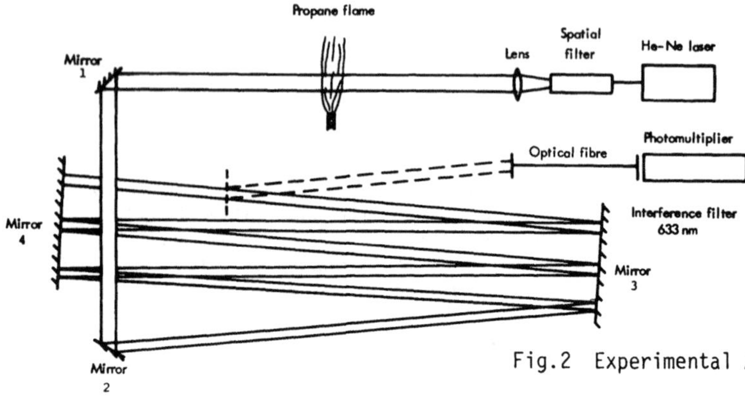

Fig.2 Experimental Arrangement

For the correlation measurements two optical fibres were used and each fed directly onto the photocathode of an EMI 9863 photon-counting photomultiplier. The counting rate was adjusted so that the train of signal pulses were truly digital and reflected the statistics of the optical field. The cross-correlation function was then measured using the Malvern K7023 digital correlator.

3. Results

Photographs of the changing intensity patterns of the scattered light were taken at distances of 1 m, 30 m and 60 m from the flame using a lensless camera with a shutter speed of 0.5 ms. They showed variations very similar to those shown by PARRY ET AL [3]. Close to the flame the intensity fluctuations were negligible indicating that the emergent wavefront was still, partially at least, correlated with the input plane wave. At 30 m the image showed the well-defined caustics characteristic of the focus region where the intensity fluctuations are a maximum. Beyond this focus region at 60 m the image showed the speckle effect expected from the interference of light scattered from all the different eddies in the field of view.

For the measurement of the intensity fluctuations, the root mean square and the mean of the intensity signal were measured simultaneously and the ratio

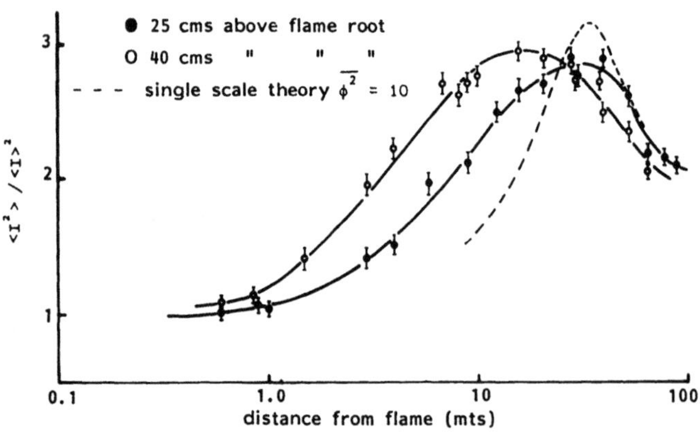

Fig.3 Normalized Second Moment as a Function of Distance from the Flame

$<I^2>/<I>^2$, the normalised second moment of the intensity, was calculated for a wide range of values of distance z. The results are shown in Fig.3 for two different positions in the flame. Both curves clearly show a broad peak indicating the extent of the focussing region of the phase screen. The value of $<I^2>/<I>^2$ at the peak does not change as significantly as might have been expected and the only difference between the two curves is that the results obtained at the higher position in the flame show the focussing region to be closest to the flame by about 20 m. Both curves show a decrease in the far field to the Gaussian limiting value of 2 consistent with the theoretical predictions.

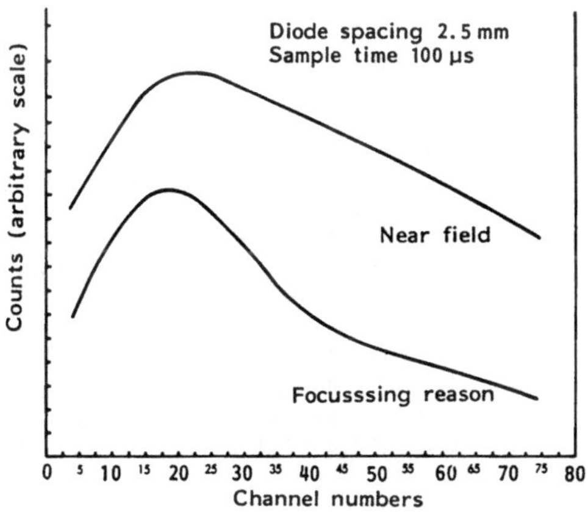

Fig. 4. Examples of the cross-correlation measurement at different distances from the phase screen

The cross-correlation measurements were carried out at the 250 mm height for a range of diode spacings and at a number of different distances from the phase screen. It is possible to give only typical examples of the measured data here and Fig.4 shows two superimposed results obtained in the near field and in the focussing region. It is interesting to note that the peaks do not occur at the same delay time meaning that the velocity of the intensity feature being sampled in each case is not the same. This will be discussed further in the next section. Also, the results obtained in the far field showed correlation functions which were very broad but which still had a recognisable maximum. This indicates that the turbulence is to some extent 'frozen-in' so that the refractive index field is simply moving, relatively undisturbed on the time scales being considered, through the flame with a velocity equal to the convection velocity. The high speed Schlieren photographs, where each frame is separated by 500 μs, certainly indicate that very little change occurs in the refractive index field over three or four frames. For the velocities calculated from the cross-correlation peaks measured closer to the flame the results in the focussing region were always lower than those obtained in the far field and the velocities obtained in the near field were always lower than the results in the focussing region. The fact that these 'correlation' velocities differed from the convection velocity contradicts the idea of frozen-in turbulence but before rejecting this concept we must discuss, in the next section, exactly what these correlation results refer to.

4. Discussion

The high speed Schlieren photographs show that the instantaneous refractive index field results from regions of hot gases which have dimensions of order

1 cm separated by regions of hot air which has been entrained into flow and which have dimensions of a few millimetres. Just above the mixing region the hot air appears to be confined to a central column but this gradually diffuses until further upstream it fills the whole flame area. This picture is confirmed by the temperature profile measurements which show the temperature to be about 200 K lower on the axis than at the edge of the flame for the 250 mm measuring position but which is almost flat at the 400 mm height. The magnitude of the phase fluctuations at 250 mm can be estimated by considering the phase change across an interface between burning propane at 1000 K and hot air at 800 K. The refractive indices can be calculated from the composition of the burning mixture and are approximately $0.82.10^{-4}$ for the burning gas and $0.92.10^{-4}$ for hot air. For facets 5 mm thick, the optical path length change is $\sim 0.1\ \lambda$. If there are five facets along the line of sight through the flame then the total path difference would give a phase fluctuation $\bar{\phi} = 3$.

This value of $\bar{\phi}$ can be used to compare Jakeman's theoretical results [2] with the experimental results. These show that for $\bar{\phi}^2 = 10$ the peak of the second moment of intensity for a single-scale Gaussian model is approximately 3.2 compared to 3.1 for the experimental results. It is therefore reasonable to superimpose the two peaks as shown in Fig.3. The disagreement between the results is most apparent in the Fresnel region where the theory considerably underestimates the measured results. This has become a common feature of such comparisons [3] and has led to recent theoretical consideration of multi-scale models. For example JAKEMAN and McWHIRTER [2] show that the results of PARRY ET AL [3] can be more closely represented by a two-scale model than by the single-scale model used [3]. The discussion above describing the Schlieren photographs indicates clearly that such a model should be applied to the flame results in the future.

It was shown in Section 1 how the position of the peak of the second moment curve could be used to give an estimate of the facet sizes. The peak occurs at the diffraction maximum for a single facet and hence the correlation length $\xi = (z\bar{\phi}/k)^{\frac{1}{2}}$. This gives a facet size of 7 mm for the 250 mm results and 5 mm for the 400 mm results. This demonstrates the evolution of mean eddy sizes as they move upstream.

Velocity measurements coupled with the eddy size evolution measurements enable such parameters as the burning rate to be calculated. In this case it is the velocity of the individual facets which would be required, i.e. the correlation velocity in the focussing region. In the far field, as discussed above, a correlation velocity would indicate frozen-in turbulence whereas in the near field intensity fluctuations result from the turbulence micro-scale and the correlation velocity reflects the motion of this structure. An example of the velocity results obtained for a diode spacing of 1 mm is: far-field 1.8 m/s; focus region 1.6 m/s; near field 1.2 m/s. These and other results show that although the spread in velocities is not large the focus region is consistently dominated by the slower moving eddies. Also the component of velocity in the direction of the two diodes in the near field is always less than in the other regions and presumably this is due to the small-scale randomness associated with micro-turbulence. None of the results obtained completely deny the concept of frozen-in turbulence but it is clear that further cross-correlation measurements with horizontal diodes as well as vertical diodes would have provided further useful information.

5. Conclusions

The advantages of intensity fluctuation measurements over existing techniques are 1) they can provide qualitative data for the velocity, mean eddy sizes and eddy evolution rate which are important in determining a number of combustion parameters, 2) they are relatively simple to carry out and can be adopted whenever there is reasonable access to a flame - which will often include large-scale industrial furnaces.

Further theoretical effort on multi-scale models would be particularly useful for flame measurements and greatly enhance the possibility of acceptance of the technique by combustion engineers.

6. References

1. E. Jakeman, P. N. Pusey: J.Phys.A.(Math.Gen.) 8, 369, (1975)
2. E. Jakeman, J. G. McWhirter: Appl.Phys. B26, 125, (1981)
3. G. Parry, P. N. Pusey, E. Jakeman, J. G. McWhirter: Coherence and Quantum Optics Vol. IV, ed. by L. Mandel and E. Wolf (Plenum Press, New York 1978) p.351.

26. Dynamic Correlation on a Schlieren Image in a Transonic Airflow

P.J. Bryanston-Cross and J-J. Camus

Whittle Laboratory, Department of Engineering, University of Cambridge, UK

P.H. Richards

Central Electricity Generating Board, Marchwood Laboratories, Southampton, UK

1. Introduction

Photographic data produced by schlieren and other opticalsystems often appear to have a regular structure. This is particularly true in the transonic flow region which is sensitive to periodic and quasi-periodic fluctuations. The objective of this method is to provide a simple signal averaging technique which can quantify this structure in optical images by temporal and spatial correlation. As an example of the technique measurements have been made in a two-dimensional linear transonic turbine cascade. The auto-and cross-correlations obtained over a Mach number range 0.5 to 1.25 show a strong coherent vortex structure in the trailing edge region of the blades. Using the method of dynamic correlation it has also been possible to show the persistence of this vortex structure despite the presence of a trailing edge shock for the supersonic Mach numbers. In further more recent work the technique has been combined with holographic interferometry to visualize the vortex street in the trailing edge region on an isolated aerofoil with a blunt trailing edge. The combination of these methods gives both the density distribution of the vortex and its periodicity in the flow.

2. Experimental Set-Up

The experimental system used for both experiments Fig. 1 was a standard single pass schlieren set-up as described by Holder and North [1]. The photographic plate had been replaced by a beamsplitter. This allowed the flow field to be imaged onto two separate independently traversable screens. At the centre of each screen a 0.25 mm hole fed light through a fibre optic link to a photomultiplier which in turn was connected to a Malvern Instruments type K 7023 digital correlator. This correlator has a very high bandwidth in the region of 10 MHz, and is thus suitable for detecting the high frequency fluctuations found in transonic flows. By using point detectors placed in the schlieren image plane auto-and cross-correlation functions are measured by the photon correlator from which the size, velocity and periodicity of the vortical flow can be measured.

3. Auto-Correlation Measurements

Auto-correlation measurements have been made on two linear cascades both tested in the high speed facility at the Whittle Laboratory [2]. These measurements have previously been described in greater detail in [3]. A single point detector was placed in the schlieren image of the trailing edge flow. The actual position of the wake could easily be seen and the

Fig. 1

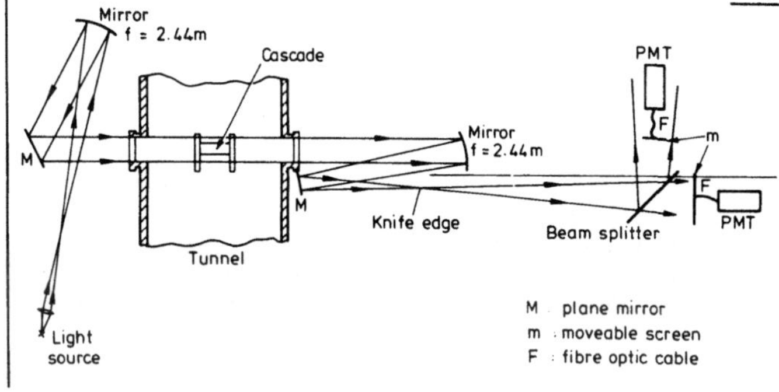

SCHLIEREN SYSTEM

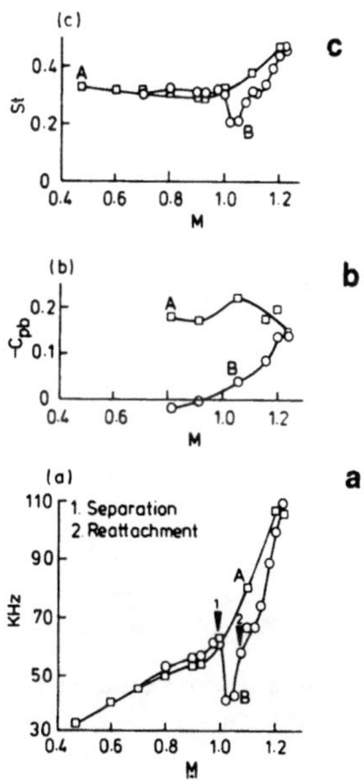

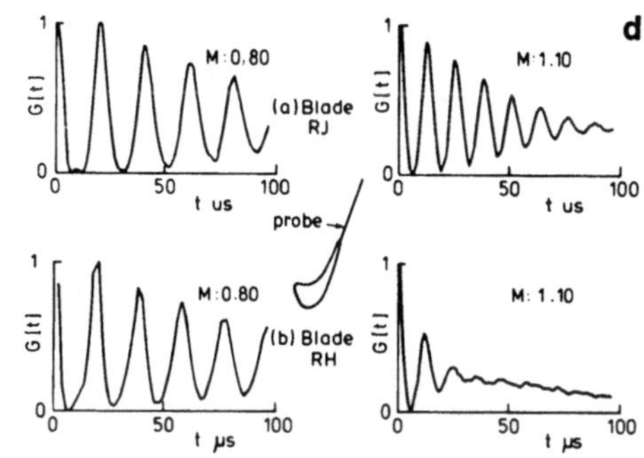

Fig. 2a. Vortex shedding frequency

Fig. 2b. Base Pressure

Fig. 2c. Strouhal number variation with Mach number

Fig. 2d. Density autocorrelations for A & B turbine profiles

detector moved across it to optimize the signal. The Mach number in the cascade was varied from M = 0.5 to 1.2 at a constant Reynolds number of 7×10^5. It can be seen in Fig. 2 that there was a strong periodic vortex flow, which at subsonic Mach number was related to the Strouhal number. Of the two blade profiles tested it has also been found in one case a loss of correlation was related to a subsonic boundary layer separation near the trailing edge of the blade. The profiles were identical apart from a small difference on the suction surface at this point.

For both blade profiles the Mach number was linearly related to the vortex shedding frequency. As the flow changed from the subsonic to the supersonic case the rate of change of the frequency increased. It is not possible to determine at present if this is a transition between stable states as often seen in low speed flow, [4] or if the change in frequency is being controlled by a new parameter i.e. the width of the confluence region.

Vortex frequencies from 30 kHz to 120 kHz were measured; and from the measurement made by Brown and Richards, [5] it appears possible that much higher frequencies occur with higher Mach numbers. It is in this quasi-periodic flow region that signal averaging techniques can provide quantitative data and where the randomness in the signal precludes the use of simple analogue techniques.

A second quite separate series of measurements have been made on an isolated aerofoil with a blunt trailing edge. Those results [6] were obtained in the supersonic Cambridge blowdown tunnel. The measurements have been made in the transonic flow range M = 0.6 to 1.4 at a Reynolds number of 2 x 10^6. A similar type of correlation to that of the turbine cascade Fig. 3a has been obtained; but in this case the flow is separated on one side of the blade, Fig. 3b. As the shock is formed between the aerofoil and the tunnel side wall the correlation falls and has not been detected at the higher supersonic Mach numbers. Holographic measurements have also been made on this blade giving useful complementary results. In the first case it is possible to obtain a density profile of the vortex. A more interesting observation is that it is possible to observe pictures of vorticity in regions where the correlator records little or no periodicity. This is a point which will be taken up again later in the paper.

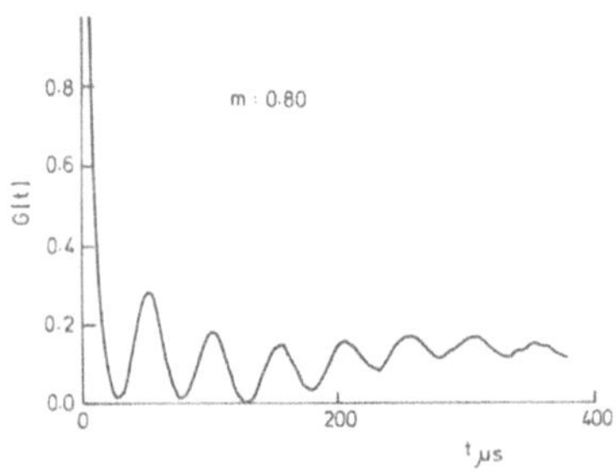

Fig. 3a Density Autocorrelation for a NACA aerofoil section

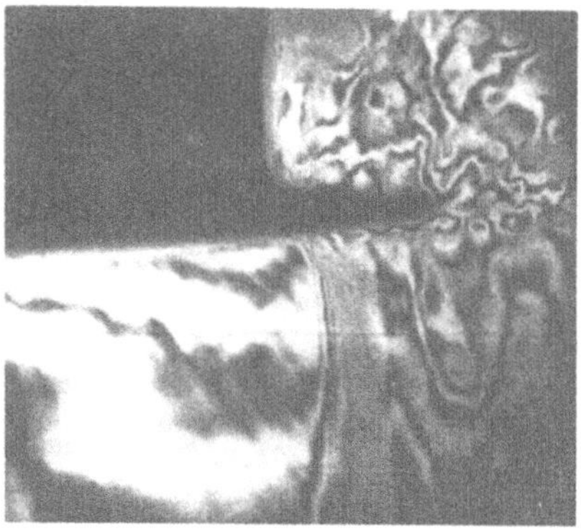

Fig. 3b. Holographic reconstruction showing vortex structure

4. Cross-Correlation Measurements

Two types of cross-correlation method have been used. In the first one detector was placed directly downstream of the other. The time delay of the peak in the cross-correlation corresponded to the time taken for the vortex to travel between the detectors. At a downstream averaged Mach number of 0.91 the velocity was measured as 225 m/sec compared with 210 m/sec ± 10 m/sec calculated from a series of schlieren photographs. The second result was obtained with two detectors placed behind adjacent blades. It was found that a cross-correlation only occurred when a shock was present in the blade passage; and in particular the correlation was strongest just as the shock was formed. The time delay between the cross-correlation peaks decreased with increasing Mach number, 135 μsec at M = 0.98 to 68 μsec at M = 1.2.

5. Discussion of Autocorrelation Results

As an illustration of the method of dynamic schlieren correlation several measurements have been made in the trailing edge region of two linear turbine cascades and a two-dimensional isolated aerofoil section. The primary measurements were made of the vortex shedding frequency over a transonic range of Mach numbers. Similar high frequency results have been reported by Lawaczeck, Heinemann and Bütefisch [7] and [8]. The present method extends these results into the separated flow. Also of importance are measurements which have been made of the contrast [3] of the signal and its rate of decay. These indicate the spatial (spanwise) and temporal coherence of the vortex emission. In the case of the turbine cascades the contrast is very high and the decay rate small implying a spanwise coherence to the vorticity and a temporal randomness of the order of 5% in the subsonic region rising to 20% in the supersonic region. There is, however, a fall in contrast and an increase in the decay rate when flow separation occurs. This indicates a randomization of the vortex structure and not a loss of vorticity. This argument is supported by the fact that there is no corresponding change in the base pressure coefficient Cpb which could have otherwise been expected, as found by [9]. Also in the case of the isolated aerofoil vorticity has been observed holographically Fig. 3 with only a weak accompanying auto-correlogram. It would be of interest to apply the type of investigation made by [10] to these high frequency disturbances to determine if a randomized vorticity maintains a discrete spanwise cell structure. Finally the density distribution of the shed vortices have been measured interferometrically [6] and can be approximated as

$$I(t) = I_o \{1 - \cos (wt + \phi n)\}$$

for the intensity level detected on the face of the fibre optic, Where w is the shedding angular frequency of the vortex and ϕ_n is a measure of the randomness of the shedding. The autocorrelation of this function is then given by

$$R_{12} \;\alpha \int_0^\infty I(t)\; I(t + \tau)\; dt \quad .$$

In order to estimate the magnitude of the randomness, a Monte Carlo method [11] was used to obtain ϕ_n for a numerical simulation of the autocorrelation function. A temporal simulated randomness of between 5% in the subsonic flow region increasing to 20% in the supersonic region produced a similar decay of the autocorrelation as that experimentally observed. Measurement of the autocorrelation function was found to be of the form

$$R_{12} \;\alpha\; e^{-A(\tau/tc)^2} \cos w\tau$$

where tc was the time taken for the vortex to cross the detector face and A is a variable associated with the randomness of the vortex shedding.

6. Discussion of Cross-Correlation Results

The cross-correlation results have previously been discussed in [3], it is perhaps relevant to point out that the measurement of velocity is a time-of-flight method and similar to that used in two-spot or transit anemometry [12]. What is of more interest and not possible to achieve using a conventional anemometer is that dynamic schlieren correlation can be used to form a cross-correlation between any two points within the field of view. This is because the method is based upon changes in refractive index and does not rely on particle following. Hence it has been possible to cross-correlate between two adjacent trailing edges in the turbine cascade. Further it is possible to cross-correlate a digitised analogue signal produced for example by a heat transfer gauge or pressure transducer with an optical signal.

7. Conclusions

Autocorrelograms have been obtained of periodic structures in transonic airflows. Measurements have been made on the trailing edge vortex structure despite the presence of randomness and flow separation. From the autocorrelograms obtained it is possible to deduce the vortex frequency and to attempt a simple description of its behaviour.

Cross- correlograms have been used to calculate the vortex velocity. They have also been used to examine the flow relationship between two adjacent trailing edges.

The technique of dynamic correlation could easily be combined either with a focused schlieren system or used to form cross-correlations between non-optical and optical signals. It could also be used with shadowgraph and interferometric systems.

1. Holder, D.W. and North, R.J.
"Schlieren Methods" NPL notes on Applied Science No. 30 1963.

2. Gostelow, J.P. and Watson, P.J.
"A closed Circuit Variable Density Air Supply for Turbomachinery Research"
ASME Paper No. 76-GT-62 (1976).

3. Bryanston-Cross and Camus, J-J.
"Auto and Cross Correlation Measurements in a Turbine Cascade using a Digital Correlator"
ASME Paper No. 82-GT-132 (1982).

4. Herbert, M.V.
"Base drag of single aerofoils and turbine blades at low speed.
National Gas Turbine Establishment, April 1973.

5. Brown, R.G., and Richards, P.H.
"Oscillatory Correlograms Measured with a Laser Transit Anemometer"
Letter to Optica Acta (to be published).

6. Bryanston-Cross, P.J.
"Optical measurements in High Speed Flow"
ECOSA 1982 7-10 September, University of Edinburgh.

7. Lawaczeck, O., and Heinemann, H.J.,
von Karman Vortex Streets in the Wakes of Subsonic and Transonic Cascades.
AGARD - PEP meeting, Unsteady Phenomena in Turbomachinery
AGARD Conference proceedings No. 177, 1975.

8. Lawaczeck, O., Bütefish, K.A., and Heinemann, H-J.
"Vortex Streets in the Wakes of Subsonic and Transonic Turbine Cascades"
Revue Francaise de Méchanique supp. to proceedings of IUTAM Symposium of Aeroelasticity in Turbomachines ENSTA, Paris, 1976.

9. Nash, J.F., Quincey, V.G., and Callinan, J.
"Experiments on Two Dimensional Base Flow at Subsonic and Transonic Speeds"
NPL Aero report 1070, ARC 25070 F.M. 3356 National Physics Laboratory 1963.

10. Brow, F.K., and Troutt, T.R.
A note on Spanwise Structure in the Two-Dimensional Mixing Layer
Journal of Fluid Mechanics, Vol. 917, 1980, pp. 771.

11. Hamming, R.W.
"Numerical Methods for Scientists and Engineers"
McGraw-Hill, 1973.

12. Moore, C.J., Jones, D.G., Haxell, C.F., Bryanston-Cross, P.J. and Parker, R.J.
"Optical Methods of Flow Diagonistics in Turbomachinery"
ICIASF 81
IEE 81 CH1 712/9/8/40000 0244.

27. Phase and Phase-Difference Statistics of Turbulent Thermal Convection in Air

Jürgen Merz[1] **and Klaus Schätzel**

Institut für Angewandte Physik der Universität Kiel, Olshausenstraße 40-60
D-2300 Kiel, Fed. Rep. of Germany

Abstract

An extensive series of phase measurements using different optical and signal processing techniques was performed on free thermal plumes. These included measurements of probability densities, temporal and spatial correlations and structure functions of absolute phases, phase differences, and phase gradients. One-beam measurements were made with an interferometer using a reference beam, which was passed through a water-cooled tube parallel to the measurement beam. This greatly reduces influences by thermal expansion of the optical setup. Two-beam measurements were carried out with the same interferometer without the cooling tube.

The very small variance for shorter beam distances down to 1 mm required a new $\pi/64$-phase counter. For beam distances smaller than 1 mm we replaced the interferometer by a novel Moiré setup which yields spatial phase gradients.

All experiments show a very good agreement, of probability density and moments with the corresponding gamma and K distributions for single-and double-beam measurements, respectively. Hydrodynamic interpretations of these distributions and their parameters can be given.

1. Introduction

For many years light propagation through a turbulent atmosphere has been a very active area of research. Some interesting examples are the stellar scintillation or the propagation of laser beams over long distances through the atmosphere. Gaussian phase statistics have been a widely used assumption. It is valid only if the scattering medium consists of a very large number of independent scattering layers. Recently, systems have been investigated that do not comply with this assumption, either as a result of statistical dependences between scattering layers in turbulent air flows or due to a small number of contributing turbulent elements.

In the following we will describe laser light scattering experiments, carried out on a laboratory scale phase screen. Beyond that the validity of model distributions for phase, phase difference and phase gradient will be examined.

---

[1] After January 1982 Pharmazeutisches Institut der Universität Kiel, West Germany

2. Generation of the Phase Screen

Operating an electric heater between two aluminum plates produces an ascending stream of heated air. We investigated the turbulent mixing of this plume with cold laboratory air above the heating arrangement (shown in Fig. 1). The length of the plume in the direction of the laser beam is small, that is, comparable to the size of the gap above the heater.

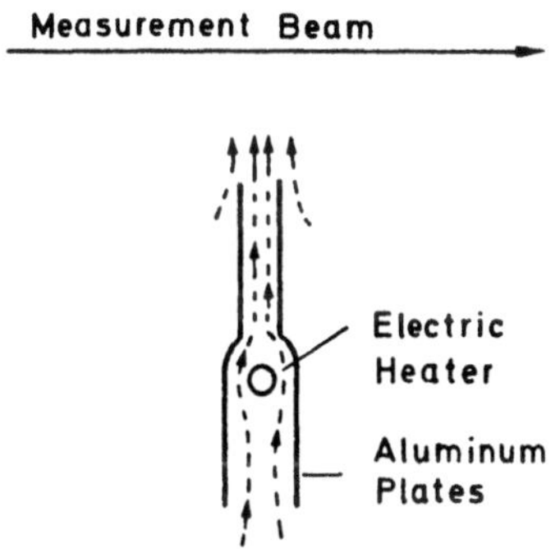

Fig. 1. Generation of thermal plumes

3. Optical Setup

3.1 Four-Detector Interferometer

For measurements of absolute phases we used a four-detector interferometer with polarization coding [1]. After emerging from the laser vertically polarized light is passed through a $\lambda/2$ plate in order to rotate the plane of polarization by 45°. A Kösters prism splits the beam. In one interferometer arm the reference beam propagates through a water-cooled tube, in the other it passes the turbulent air, and for the sake of polarization coding, twice a $\lambda/8$-plate. This produces a circularly polarized return beam. Behind the polarization beam splitter in Fig. 2 the four detector signals are

$$E_{1,3} = A\, e^{i\phi} \mp B\, e^{i\phi'} , \qquad (1)$$

$$E_{2,4} = A\, e^{i\phi} \pm B\, e^{i(\phi'-\pi/2)} , \qquad (2)$$

where ϕ is the phase of the probe beam, ϕ' the phase of the reference beam. With $\gamma = 2\,AB/I_o$, $I_o = A^2 + B^2$, $\Delta\phi = \phi - \phi'$ the detected intensities read

$$I_1 = I_o\,(1 - \gamma \cos \Delta\phi) , \qquad (3)$$

$$I_2 = I_o\,(1 - \gamma \sin \Delta\phi) , \qquad (4)$$

$$I_3 = I_o\,(1 + \gamma \cos \Delta\phi) , \qquad (5)$$

$$I_4 = I_o\,(1 + \gamma \sin \Delta\phi) . \qquad (6)$$

Using two difference-amplifiers, (see Fig. 3), we obtain

$$I_3 - I_1 = 2\, I_o\gamma \cos \Delta\phi , \qquad (7)$$

$$I_4 - I_2 = 2\, I_o\gamma \sin \Delta\phi . \qquad (8)$$

Fig. 2. 4-detector interferometer with polarization coding

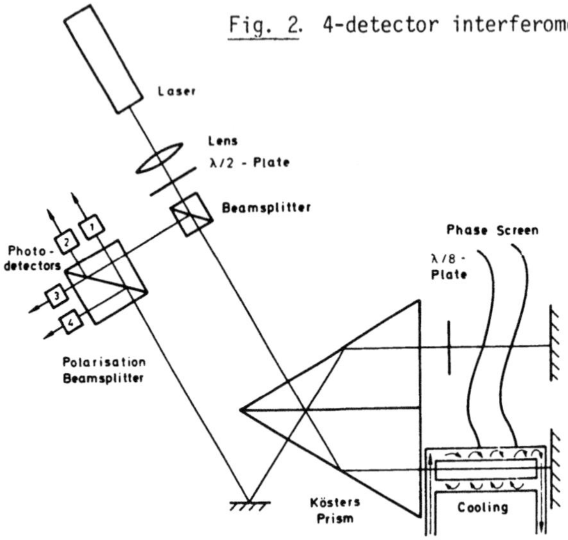

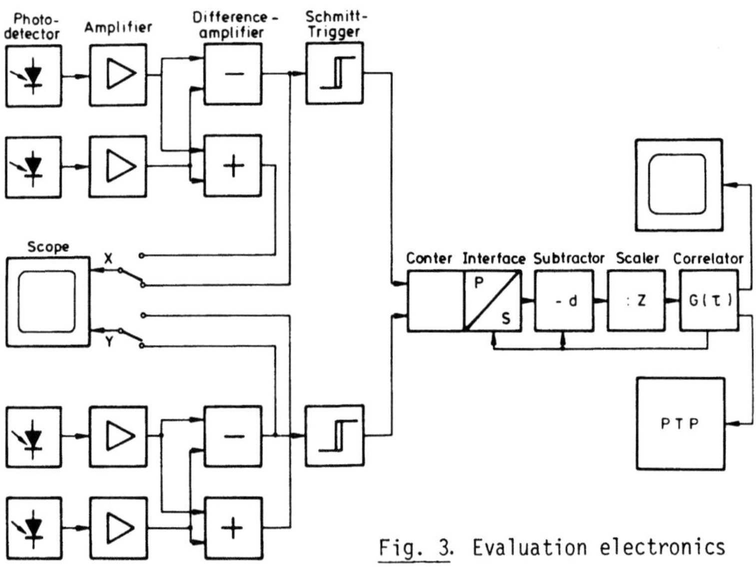

Fig. 3. Evaluation electronics

This greatly reduces the influence of a fluctuating laser intensity. Hard-clipping of (7) and (8) completely eliminates all intensity fluctuations. The electronic evaluation of these signals in a bidirectional phase counter yields the phase with $\pi/4$ resolution.

For phase difference measurements with beam distances between 6 mm and 150 mm we used the same interferometer without cooling tube and the same electronics.

3.1.1 $\pi/64$ Phase Counter

The very small variance of the phase differences for beam distances below 1 mm required a higher resolution of the phase counting. For this purpose we developed a new $\pi/64$ phase counter (shown in Fig. 4).

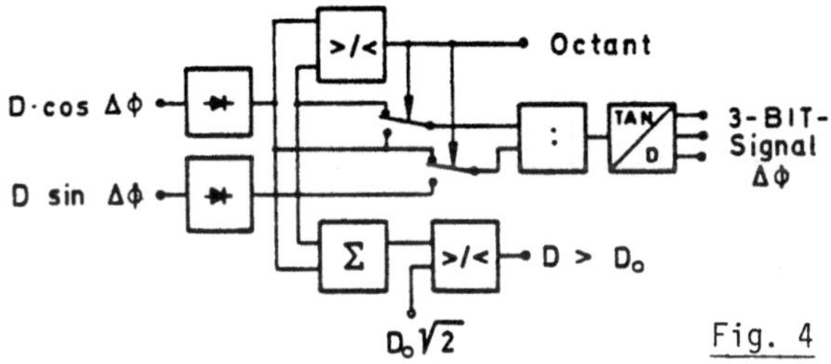

Fig. 4 $\pi/64$-phase counter

After rectification the difference amplifier signals (7) and (8) are adjusted to equal amplitudes. Then it is checked whether the sum of these rectified signals is above a level that is high compared to noise and the hysteresis of the comparator. Everytime the intensity falls below this level, i. e. the amplitude drops severely, the measurement is stopped. A circuitry recognizes whether the sin signal exeeds the cos signal or conversely. In the first case the cos signal is divided through the sin signal, in the other just the other way around. So the output is tan $\Delta\phi$ for $\Delta\phi \in [0, \Pi/4]$, cot $\Delta\phi$ for $\Delta\phi \in [\pi/4, \pi/2]$ and so on. Switching from one case to the other is indicated by a flag bit. Evaluation of the output signal with a 3-bit-$\tan^{-1}$-A/D-converter allows a $\pi/64$ phase resolution. It essentially consists of a voltage divider with suitably spaced taps together with comparators and priority encoders.

3.2 Moiré interferometer

Instead of phase difference measurements with smaller beam distances than 1 mm, that require even more extensive screening against mechanical vibration, phase gradient measurements appear more advisable [4]. For these we used a new Moiré interferometer (Fig. 5), which was developed in cooperation with Dr. Gareth Parry, RSRE, Malvern.

For $L = (2d^2n)/\lambda$ (n = integer, d = grating spacing) the diffraction orders of the first grating are shifted perpendicularly to the optical axis by a whole multiple of d in the plane of the second grating and so reproduces the intensity distribution of the first grating (Talbot effect). A phase gradient at the place of the first grating displaces these orders by

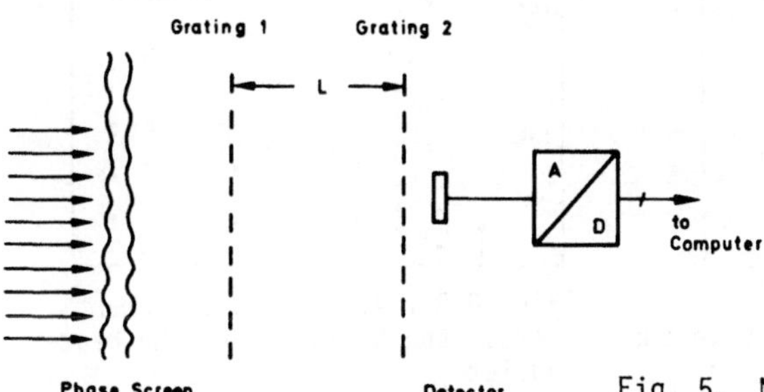

Fig. 5. Moiré interferometer

$$s = \frac{\lambda L}{2\pi} \cdot \frac{\partial\phi}{\partial x} . \qquad (9)$$

The spatialy averaged detected intensity behind the second grating corresponds to the spatial autocorrelation function of the grating transmission, that itself is proportional to the phase gradient within certain ranges of $\partial\phi/\partial x$, as (9) shows.

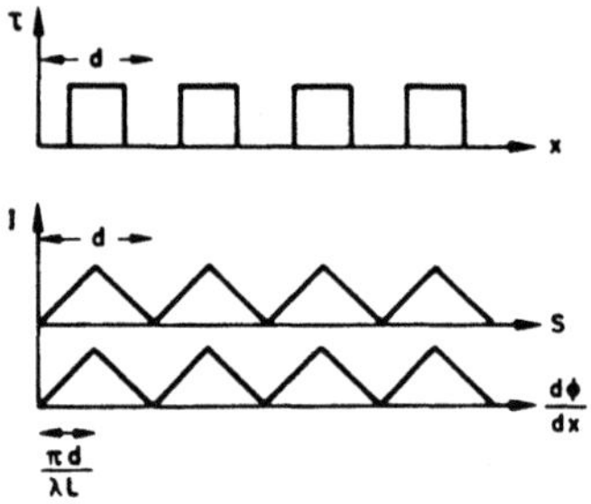

Fig. 6 Grating transmission and autocorrelation function

Within these ranges of size $(\pi d)/(\lambda L)$ the intensity behind the second grating depends linearly upon the phase gradient.

A calibration is easily possible because a shift of the second grating perpendicular to optical axis and grating lines has the same effect as a displacement of the diffraction orders by a phase gradient. So by shifting the second grating with a stepping motor translation stage we can calibrate the detected intensities in units of the phase gradient. Now, using an A/D-converter, the measured intensities can be directly analyzed in a computer.

4. Results

The probability density and the moments of the absolute phase show a very good agreement with the gamma distribution (see Fig. 7 and 8) [1]

$$f_{\alpha,\mu}(\phi) = 1/(\mu\Gamma|\alpha|)\ (\phi/\mu)^{\alpha-1}\ e^{-\phi/\mu} =: \Gamma_{\alpha,\mu}(\phi) . \qquad (10)$$

α can be interpreted as a number of scattering turbulence elements and μ as the influence of one element [1].

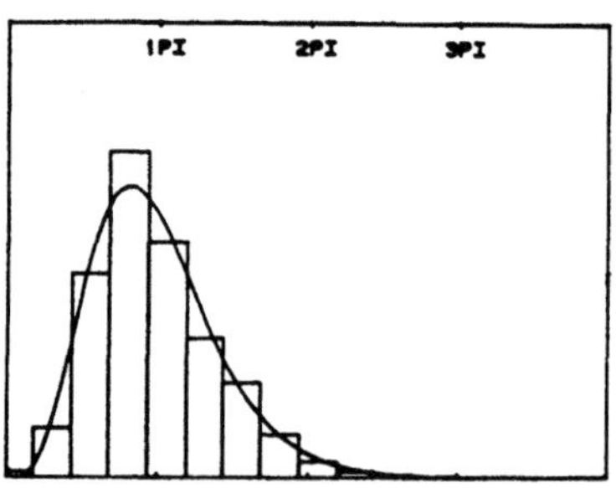

Fig. 7. Probability density of absolute phase at a height above the heater outlet H = 10 mm and the corresponding gamma distribution

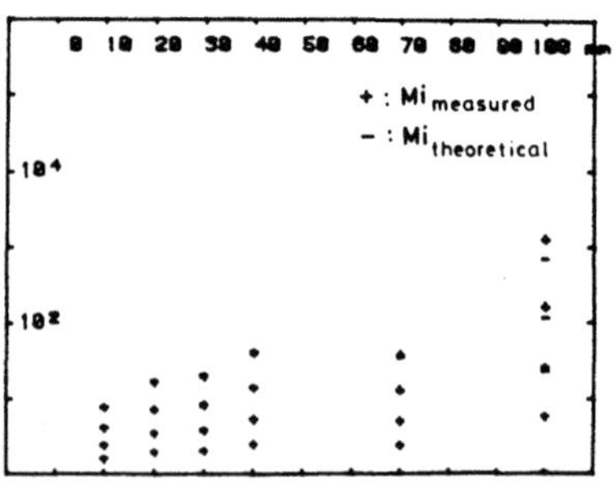

Fig. 8. 3., 4., 5. and 6. moments versus the height above the heater outlet

In the same way the measured probability density of phase differences agrees very well with the K distribution [1]

$$f_{\alpha,\mu}(\Delta\phi) = 1/\left(\pi\Gamma(\alpha)\mu\right)\,(\Delta\phi/2\mu)^{\alpha-1/2}\,K_{\alpha-1/2}(\Delta\phi/\mu) =: K_{\alpha,\mu}(\Delta\phi)\,. \qquad (11)$$

Fig. 9 and 10 show the phase difference density measured with the $\pi/64$ phase counter. The continuous lines represent K distributions.

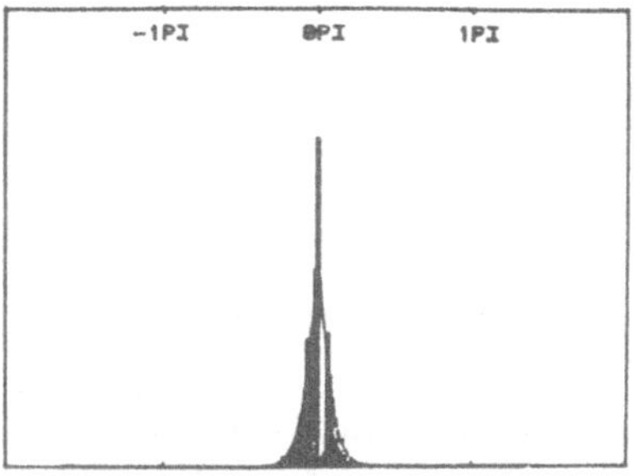

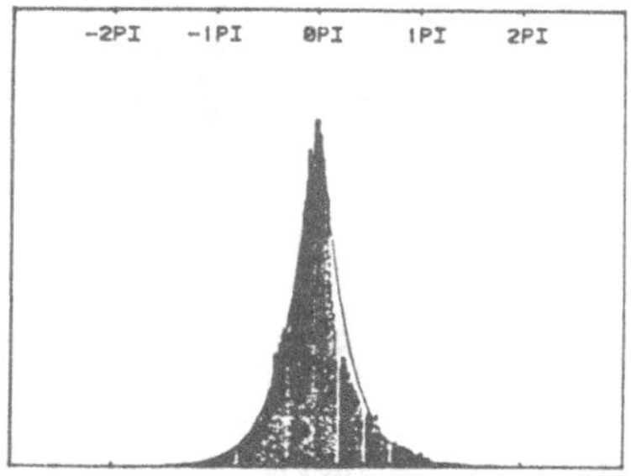

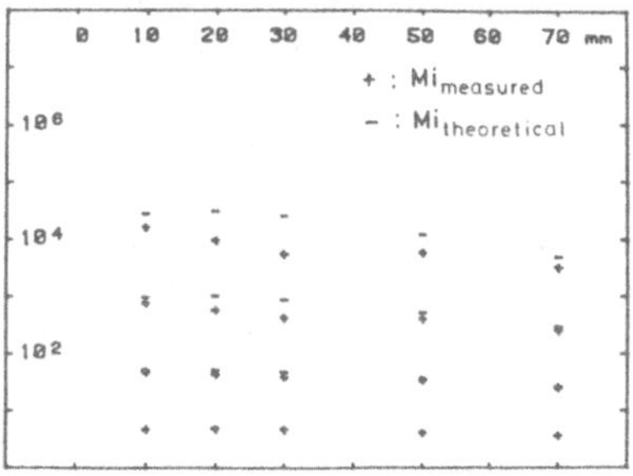

Fig. 9 probability density of phase difference for 1 mm beam distance and 80 mm height above the heater outlet

Fig. 10 probability density of phase difference for 5 mm beam distance and 20 mm height above the heater outlet

Fig. 11. 4., 6., 8. and 10. moments versus the height above the heater outlet

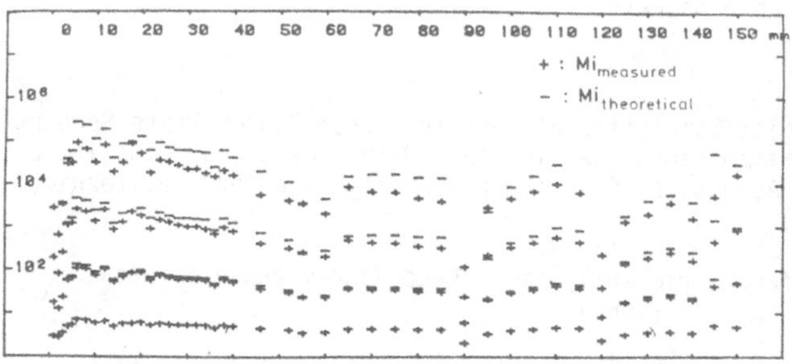

Fig. 12. 4., 6., 8. and 10. moments versus the beam distance

The moments show a very good agreement with the corresponding K moments (Fig. 11 and 12), as well.

The probability density of the phase gradient (see Fig. 13) is approximated very well by the exponential distribution, a special case of the K distribution with $\alpha = 1$. This agrees with the interpretation of α as a parameter describing the number of turbulence elements that influence the scattering process.

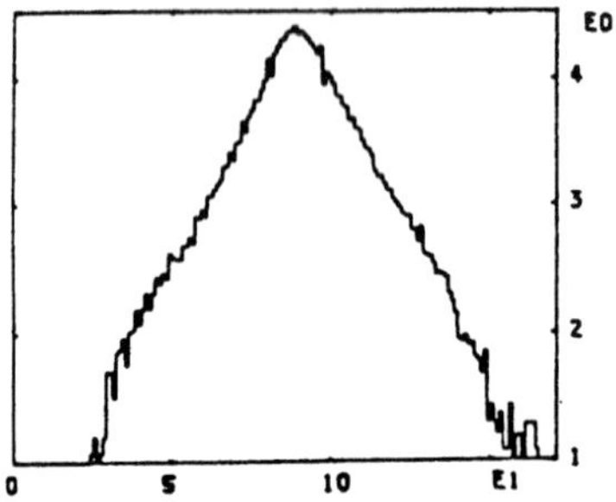

Fig. 13 Probability density of phase gradient at 65 mm height above the heater outlet

5. Conclusions

The measurements carried out with the four-detector interferometer show a very good agreement of the phase distribution with the gamma distribution for a large range of heights above the heater outlet. Similarly the phase difference distribution agrees very well with the K distribution for a large range of beam distances and heights above the heater outlet. Again the probability density and the moments of phase gradient agree well with those of K-distributions, here with a special case, the exponential distribution.

The new $\pi/64$ phase counter allowed high resolution phase difference measurements down to 1 mm beam distance.

The novel Moiré interferometer for phase gradient measurements proved to be a good and simply designed alternative to very complex interferometers for phase difference measurements with beam distances smaller than 1 mm.

Acknowledgements

The authors thank Prof. Dr. E.O. Schulz-DuBois and Dr. G. Parry for stimulating discussions and practical help. We acknowledge financial support by the Deutsche Forschungsgemeinschaft.

References

1. K. Schätzel: Two Point Statistics of Turbulent Deep Random Phase Screens, in: "Photon Correlation Techniques in Fluid Mechanics", Joint Institute of Aeronautics and Acoustics, Stanford University, Stanford California, 23-1 - 23-12 (1980)

2. K. Schätzel: Interferometric Analysis of Deep Random Phase Screens, Opt. Lett., Vol. 5. Nov. 9 (1980)

3. K. Schätzel, E.O. Schulz-DuBois, R. Vehrenkamp: Optical Correlation Techniques in Fluid Dynamics, Optics and Laser Technology (1981)

4. K. Schätzel and G. Parry: Real-time Moiré-Measurement of Phase Gradient, subm. to Optica Acta

5. E. Jakeman: On the Statictics of K-Distributed Noise, J. Phys. A: Gen., 13 (1980)

6. E. Jakeman, J. C. Oliver, E. R. Pike and P. N. Pusey: Correlation of Scaled Photon Counting Fluctuactions, J. Phys. A: Gen., 5, L 93 - L 96 (1972)

7. E. Jakeman and P. N. Pusey: Non-Gaussian Fluctuations in Electromagnetic Radiation Scattered by a Random Phase Screen, I. Theory, J. Phys. A: Math. Gen., Vol. 8, No. 3 (1975)

8. G. Parry and G. Gray: Optical Propagation through Random Phase Screens, in "Optics 78", Bath (1978)

9. G. Parry, P. N. Pusey, E. Jakeman, J. G. McWhirter: Focussing by a Random Phase Screen, Opt. Comm. 22, 195-201 (1977)

10. G. Parry, P.N. Pusey, E. Jakeman, J. G. McWhirter: The Statistical and Correlation Properties of Light Scattered by a Random Phase Screen, Coherence and Quantum Optics IV, ed. L. Mandel, E. Wolf, Plenum Press, New York, 351-361 (1978)

11. G. Parry, P. N. Pusey: K-Distribution in Atmospheric Propagation of Laser Light, J. Opt. Soc. Am. 69, 796-798 (1979)

12. P. N. Pusey and E. Jakeman: Non-Gaussian Fluctuations in Electromagnetic Radiation Scattered by a Random Phase-Screen, II. Applications to Dynamic Scattering in a Liquid Crystal, J. Phys. A: Math. Gen., Vol. 8, No. 3 (1975)

13. P. N. Pusey: Statistical Properties of Scattered Radiation, in "Photocorrelation Spectroscopy and Velocimetry", ed. H. Z. Cummins and E. R. Pike, Plenum Press, New York (1973)

Part III

Photon Correlation Spectroscopy of Brownian Motion: Polydispersity Analysis and Studies of Particle Dynamics

28. Data Reduction in Polydisperse Diffusion Studies

Nicole Ostrowsky and Didier Sornette

Laboratoire de Physique de la Matière Condensée (associé au CNRS LA 190)
Université de Nice, Parc Valrose, F-06034 Nice Cedex, France

Light-scattering studies of suspensions of particles bring in irrevocably the difficult and long discussed [1-4] problem of poly-dispersity. This short review will try to summarize the state of the art in 1982 and serve as an introduction to the more specialized papers which will follow up.

Under the name of polydispersity is actually gathered a great variety of phenomena. In order to deal with the problem of data reduction in polydisperse studies, it seems useful to try first and review the different origins of polydispersity (§I) which can be due either to the complex motion of the individual scattering particles, or to some heterogeneity in their geometrical or optical properties.

We shall then classify the different approaches to polydisperse analysis into two main sections :

- The empirical methods (§II) which include the well-known and most used moment expansion, the various parametrized distributions, and the more recent histogram decomposition [5]

- The methods facing the ill-conditioned nature of the problem (§ III) amongst which we shall mainly describe the exponential sampling method [6] as most of the others will be discussed by their original authors thereafter.

The discussion will end with a few practical and technical remarks.

I. Introduction : Different Origins of Polydispersity

When studying, by Quasi-Elastic Light Scattering, particles suspensions, one rarely encounters what we call a monodisperse spectrum. A first-order correlation function of the scattered electric field, such as the one measured directly in a heterodyne (or extracted from a homodyne) Q.E.L.S. experiment is said to be monodisperse if it decays exponentially with one single time constant $1/\Gamma$,i.e. if it can be written, in its normalized form as :

$$g_{monodisperse}(t) = \exp(-\Gamma t) \quad . \qquad (1)$$

Such a monodisperse correlation function is occasionally observed when studying the Brownian translational motion of rigid, spherical, all identical particles suspended at low concentration in a dust free solvant. In that ideal case, $\Gamma = D_T q^2$ is simply related to the translational diffusion constant of the particles, $D_T = kT/6\pi\eta R$, of radius R suspended in the solution of viscosity η , and to the wave vector q of the scattering fluctuation : $q = (4\pi/\lambda)\ n \sin\theta/2$

(n: solution's index of refraction, λ: incident wave length, θ: scattering angle).

In most experiments, polydispersity will occur for one or several of the following reasons.

I.1 Complex Motion of the Scattering Particles [7]

a) In addition to translational diffusion, the particles undergo a rotational diffusion motion. If the particles have a shape or an optical anisotropy, this additional motion will alter the spectrum in a predictable fashion. For example, for rigid rod-shaped particles of lengh L :

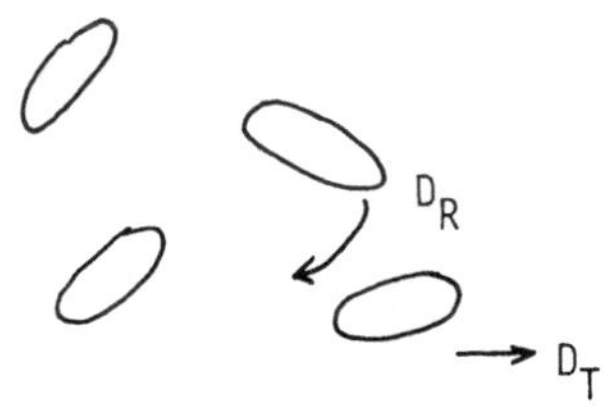

$$g(t) = \exp(-D_T q^2 t)\left[B_o(qL) + \sum_{l\ \mathrm{even}} B_l(ql)\exp(-l(l+1)D_R t)\right] , \quad (2)$$

where the B_l can be analytically expressed in terms of the spherical Bessel functions of order l . By measuring spectra at small and at large q , this expression has been used for extracting both D_T and D_R of macromolecules.

b) If the scattering particles are not rigid, their internal motion may be decomposed into normal modes, each having a characteristic decay time τ_j , decreasing as $1/j^2$. This complex motion leads to a polydisperse spectrum whose first terms have been numerically calculated :

$$g(t) = \exp(-D_T q^2 t)\left[S_o(q^2 R_G^2) + S_2(q^2 R_G^2)\exp(-2t/\tau_1)\right] , \quad (3)$$

where R_G is the radius of gyration of the flexible molecule.

Dealing with the above-mentioned internal motion of the scattering particle is in principle feasible as analytical or numerical expressions for Q.E.S.L. spectra have or can be obtained.

However, when the scattering particles do not all have the same geometrical sizes, the matter becomes more complicated

I.2 Size Polydispersity

Let us first consider the case of a dilute suspension of rigid, spherical particles, having however a radii distribution function N(R)dR which can be due to the sample preparation method, or be the result of some coagulation kinetics of initially monodisperse spheres, etc . The Q.E.L.S. spectrum is now the sum of the spectra due to each particle (see equ.1), weighted by the intensity I(R,q) it scatters at the given scattering angle, i.e. at the given wave vector q. The normalized measured spectrum is thus :

$$g(t) = \frac{\int N(R)\ I(q,R)\ \exp(-D(R)q^2 t)\ dR}{\int N(R)\ I(q,R)\ dR} . \quad (4)$$

Further complications will occur if the particles'concentration is such that interactions between particles can no longer be neglected [8].

If the suspension is now formed of non-spherical, eventually flexible particles of different sizes, it is now (2) or (3) which must be weighted by the intensity scattered by a given size of particles, and integrated over all present sizes.

A number of methods, which will be discussed below, have attempted to deal with such an integral spectrum, written in a general form as :

$$g(t) = \int G(\Gamma)\exp(-\Gamma t)\,d\Gamma \quad . \tag{5}$$

In the most simple case of a size distribution of spherical, rigid particles, we have the correspondance :

$$\Gamma = \frac{kT}{6\pi\eta R} q^2 \quad \text{and } G(\Gamma)\, d\Gamma = N(R)I(q,R)(d\Gamma/dR)\, dR \tag{6}$$

which clearly shows that the shape of the distribution functions $G(\Gamma)$ and $N(R)$ may be quite different (see fig.1)

All these methods can be divided into two main types which will be discussed in turn :

- The empirical methods
- The methods facing the ill-conditioned nature of the problem.

II. The Empirical Methods

II.1 The Expansion Method

One of the very first methods used to analyse the general spectrum (5) was first described by Koppel in 1972.

Using the Taylor series expansion of $\exp(-\Gamma t)$ around $\bar{\Gamma} = \int \Gamma G(\Gamma) d\Gamma$ we can rewrite (5) as :

$$g(t) = \exp(-\bar{\Gamma}t)\left[1 + \sum_{n=1}^{\infty} \mu_n t^n/n!\right] \quad , \tag{7}$$

where the μ_n are the nth moment of the distribution $G(\Gamma)$ around $\bar{\Gamma}$:

$$\mu_n = \int (\Gamma - \bar{\Gamma})^n G(\Gamma)\, d\Gamma \quad .$$

The fitting of the experimental data to the theorêtical form (7) is generally done in a semi-logarithmic scale, i.e. fitting $\ln(g_{exp}(t))$ to :

$$\ln g(t) = -\bar{\Gamma}t + \frac{1}{2!}\mu_2 t^2 - \frac{1}{3!}\mu_3 t^3 + \frac{1}{4!}(\mu_4 - 3\mu_2^{\,2})\, t^4 + \ldots \tag{8}$$

In most experiments, this expansion is limited to the quadratic term, occasionnally to the cubic one. It is therefore only valid for small polydispersity, typically for $V = \mu_2/\bar{\Gamma}^2 \lesssim 0.1$.

When (8) is actually used for larger variances, the parameters $\bar{\Gamma}$ and V extracted from the data should not be used to characterize the peak and width of the distribution $G(\Gamma)$. They can only be considered as numerical parameters (V is often called the "polydispersity index"), useful in relative measurements, i.e. comparison of different samples.

The validity range of expansion (8) may however be increased, provided one assumes for $G(\Gamma)$ a particular shape as is explained below.

II.2 Parametrized Distributions.

- Gaussian Distribution :

If one assumes for $G(\Gamma)$ a Gaussian shape :

$$G(\Gamma) = \frac{1}{(2\pi V \bar{\Gamma}^2)^{1/2}} \exp - \frac{(\Gamma - \bar{\Gamma})^2}{2 V \bar{\Gamma}^2} \qquad (9)$$

all odd moments of expansion (7) vanish and all even moments can be expressed as a function of $V = \mu_2/\bar{\Gamma}^2$.

Furthermore, for values of $V < 0.3$ and $\bar{\Gamma}\, t_{max} < 3$ one can limit with less than a few percent relative error, expansion (6) to the quadratic term [9]. The data can then be fitted either in a semi-logarithmic scale to this truncated expansion, or directly to the simple analytical form :

$$g(t) = \exp(-\bar{\Gamma} t) \exp(+ V/2 (\bar{\Gamma} t)^2) \qquad (10)$$

which can be readily done on a microcomputer on-line with the correlator.

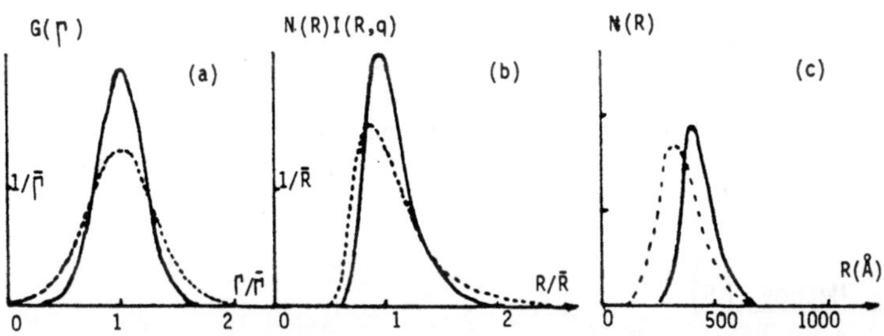

Fig. 1. Example of Gaussian distributions $G(\Gamma)$ (a) and their corresponding intensity distribution N(R)I(R,q) (b) and size distribution N(R) (c). The solid and dashed lines correspond respectively to a variance of 0.04 and 0.09 . Curve (c) was computed assuming $\bar{R}$=500 Å and the wave vector q equal to 2.4 10^{-3} $Å^{-1}$

To account for a wider variety of distribution shapes, more elaborated parametrized functions have been used, such as Pearson's distributions [4], or the Schultz distribution [10], described below.

- The Schultz distribution :

This widely used two-parameter (z and $\bar{r}$) distribution has the following analytical form :

$$f_z(r/\bar{r}) = 1/z! \left[(z+1)/\bar{r}\right]^{z+1} r^z \exp(-(z+1)r/\bar{r}) \quad . \qquad (11)$$

It has the peculiarity of changing its shape as its width increases, going from a narrow gaussian shape (large z) to a wide distribution with a skew towards larger sizes (small z) . This last feature makes it quite useful in biologically related studies, where small quantities of aggregates are often present in the suspension, and are accounted for by this non-symmetrical distribution.

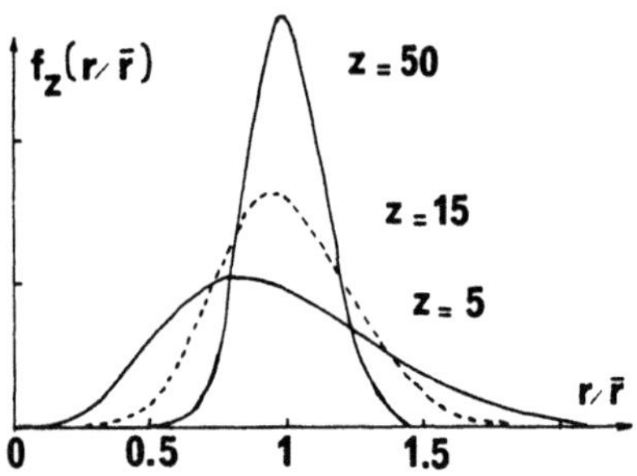

Fig. 2. The Schultz distribution (see ref. [10] $f_z(r/\bar{r})$ as a function of the reduced variable $r/\bar{r}$

Nevertheless the need for a more general data processing method became more pressing as more and more applications of the Q.E.L.S towards size distribution measurements were found. Improvements of the method could be searched mainly in two directions :

a) Try and put into the resolution as little a priori knowledge as possible, which rules out any parametrized distribution, as well as the cumulant expansion valid only for small polydispersity.

b) Establish what is the maximum information that can be recovered from the data and find the best (and fastest) way to extract it.

A number of research teams have been working on these problems [5,6,11-19]. An improvement in direction a) came with the use for $G(\Gamma)$ of a histogram type of function, which we shall now discuss.

II.3 The histogram Method [5]

In this method, the distribution function $G(\Gamma)$ is approximated by a "staircase" type of function (see fig.3), the intervals $\Gamma_1, \Gamma_2, \ldots, \Gamma_i, \ldots, \Gamma_N$ being fixed a priori, the only adjustable parameters being the heights $a_1, a_2, \ldots a_i, \ldots a_{N-1}$ of each step. These N-1 parameters are computed by a least square fit of the experimental (heterodyne) correlation function to the following "theoretical" function :

$$g_{hist.}(t) = \sum_{i=1}^{N-1} a_i \int_{\Gamma_i}^{\Gamma_{i+1}} \exp(-\Gamma t)\, d\Gamma$$

$$= \sum_{i=1}^{N-1} a_i/t \left[(\exp(-\Gamma_i t) - \exp(-\Gamma_{i+1} t)) \right] . \quad (12)$$

This approach has been successfully used for the study of quite wide size distributions. The determination of the intervals $\Gamma_1, \ldots, \Gamma_N$ is however empirical, both in spacing and in number. In case a logarithmic scale is chosen

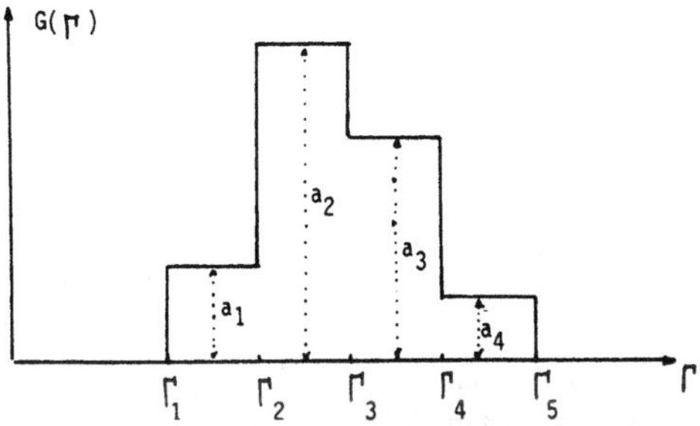

Fig. 3 The histogram type of function. The width of the intervals (4 in this example) are defined by $\Gamma_1, \ldots, \Gamma_5$ which are chosen and fixed in advance. The heights $a_1, \ldots, a_4$ of the steps are adjustable parameters

($\ln \Gamma_i = \ln \Gamma_{i-1} + \Delta$) , the histogram method bears some similarity with the exponential sampling method (see below). Note however that to enforce the positivity of the a_i coefficients, the least square fit of (12) to the data points is often carried setting $a_i = p_i^2$. Care must be taken with such a procedure, which may furnish, in case of over-resolution, a strongly biased solution. Furthermore and as we shall explain below, in order to avoid unreliable results in the data processing, the true $G(\Gamma)$ function must be approached by a truncated function $\overline{G(\Gamma)}$ which has no high frequency fourier compo nent. A staircase-like function is thus not a good choice and may bring in artificial additionnal instability in the solution.

These considerations lead us to the second type of approach dealing explicitly with the ill-conditioned nature of the problem, and which brings in improvements both in direction a)(little a priori knowledge of $G(\Gamma)$) and in direction b) (what is the maximum information recoverable from the data ?).

III. Methods Facing the Ill-Conditioned Nature of the Problem

The problem of inverting (5) to recover $G(\Gamma)$ from the measured correlation function g(t) is well known to be ill-conditioned : a small deviation of g(t), due to noise for example, leads to large distortion in the recovered $G(\Gamma)$ function. This has been the underlying problem of all previous methods, limiting in some inexplicit fashion the number of parameters which could safely be extracted from the data.

We shall mainly describe the simplest of these methods, the exponential sampling technique [6] , pointing out the origin of the ill-conditioning and a possible way to simply circumvent the problem and extract as much information as possible from data having a given noise level.

A more quantitative way of limiting the recoverable information given the allowed Γ range for $G(\Gamma)$ and the covered t range for the measured correlation function requires the use of the singular value approach [3] , as will be explained by E.R.PIKE in his paper[17] .

The problem may also be solved using regularization constraints techniques as will be described by G.R DANOVICH and I.N SERDYUK [18] , and by S.PROVENCHER [13,14,19] whose group has provided a very complete computer package for solving a very wide class of inversion problems.

III.1 Where Does the Ill-Conditioning Come From ?

The question was easily answered once E.R PIKE and J.Mc WHIRTER [15] gave the analytical expressions of the eigenfunctions and eigenvalues of the Laplace transformation (see fig.4). It was then recognized that since λ_ω vanishes as $\omega \to \infty$, it meant that all high components of $G(\Gamma)$ were so weakly transmitted through the Laplace transformation that they appeared in g(t) with an intensity below the noise level. Their information content could thus not be extracted.

In a more mathematical form, this reasoning can be written as follows. The decomposition of the true $G(\Gamma)$ function on the eigen functions of the Laplace transform $\psi_\omega(\Gamma)$ is simply :

$$G_{true}(\Gamma) = \int_{-\infty}^{+\infty} a_\omega \, \psi_\omega(\Gamma) \, d\omega \quad . \tag{13}$$

Upon going through the Laplace transformation (see equ.5), this function becomes :

$$g_{true}(t) = \int_{-\infty}^{+\infty} a_\omega \, \psi_\omega(t) \, d\omega \quad . \tag{14}$$

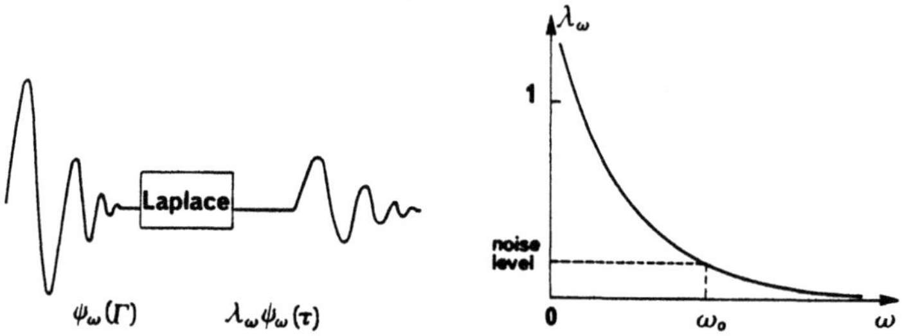

Fig. 4. Schematic representation of the eigenfunctions ψ_ω and eigenvalues λ_ω of the Laplace tranform . Beyond a cut-off frequency ω_0, the eigenfunctions $\psi_\omega(\Gamma)$ are so weakly transmitted through the Laplace transform that they are lost in the noise of the experimentally measured correlation function

We actually measure experimentally a correlation function :

$$g_{exp}(t) = g_{true}(t) + \text{noise} \quad .$$

Information on $G(\Gamma)$ can only be recovered from that part of the data which is above the noise level :

$$g_{exp}(t) = \int_{|\omega|<\omega_0} \lambda_\omega a_\omega \psi_\omega(t) \, d\omega + \int_{|\omega|>\omega_0} \lambda_\omega a_\omega \psi_\omega(t) \, d\omega + \text{noise}, \tag{15}$$

i.e. from the first term (" the relevant part of the data ") on the right-hand side of (15) . This term, inverted back into the Γ space , yields the"truncated" $G(\Gamma)$ function :

$$G(\Gamma) = \int_{|\omega|<\omega_0} a_\omega \psi_\omega(\Gamma) d\omega \qquad (16)$$

which is thus a filtered distribution function which has no high ω components.

The ill-conditioning thus comes from the fact that the high frequency part of $G(\Gamma)$ is lost in the noise of $g_{true}(t)$.

III.2 How Should One Deal With the Problem

From the above analysis the answer is clear. One should only extract from the data the filtered distribution $\overline{G(\Gamma)}$ (see equ.16), which contains the most information one can safely recover from the data.

How should one do this in practice ?

A first idea could be to try and extract from the data the series of coefficients a_ω up to ω_0 so as to use (16) to reconstruct $\overline{G(\Gamma)}$. This has been done [15] but is impractical as it is mathematically quite involved.

A simpler and more practical solution to avoid over-resolution, i.e. only extract $\overline{G(\Gamma)}$ from the data, is explained below.

Explicating the form of the eigenfunctions, the "filtered" distribution function becomes :

$$\overline{G(\Gamma)} = \exp(-\ln\Gamma/2) \int_{|\omega|<\omega_0} F(\omega) \exp(i\omega \ln\Gamma) \quad .$$

We thus see that the Fourier transform of $\overline{G(\ln\Gamma)}$ is limited to frequencies smaller than ω_0 . This gives us the clue as to how we should limit the information we extract from the data :

To Use Only The Relevant Part Of The Data
means
Do Not Ask For Details in $G(\ln\Gamma)$ Finer Than π/ω_0

Using the sampling theorem, we can then reconstruct the filtered function $\overline{G(\ln\Gamma)}$:

$$\overline{G(\ln\Gamma)} = \overline{G(\ln\Gamma_n)} \frac{\sin\omega_0 \ln(\Gamma/\Gamma_n)}{\omega_0 \ln(\Gamma/\Gamma_n)} \left(\frac{\Gamma}{\Gamma_n}\right)^{\frac{1}{2}} \qquad (17)$$

from the knowledge of $\overline{G(\ln\Gamma)}$ at points Γ_n such that :

$$\Gamma_n = \Gamma_{n-1} \times \boxed{\exp(\pi/\omega_0)} \qquad (18)$$

where $\exp(\pi/\omega_0)$ is called the dilation factor.

We have shown [6] that replacing in (5) the true function $G(\Gamma)$ by its truncated version (17) amounts to fitting the data to :

$$g_{th}(t) = \sum_{n=1}^{N} a_n \exp(-\Gamma_n t) \qquad , \qquad (19)$$

the parameters $a_n = \frac{\pi}{\omega_0} G(\ln\Gamma_n)$ being computed by minimizing over the M experimental points:

$$\sum_{j=1}^{M} \left[g_{ex}(t_j) - g_{th}(t_j)\right]^2 \qquad . \qquad (20)$$

The "filtered" distribution function $\overline{G(\ln\Gamma)}$ can then be reconstructed by :

- interpolation (see equ.17) using the a_n' values, i.e. the $\overline{G(\ln\Gamma_n)}$ values, found by fitting the data to (19).

- shifting the set of Γ_n'in (19) and fitting again the data to a new expression (19). This yields a different set of coefficients a_n' and after several shiftings and fittings, the whole range of $\overline{G(\ln\Gamma)}$ will be covered.

Some examples of applications of this method are shown on fig.5.

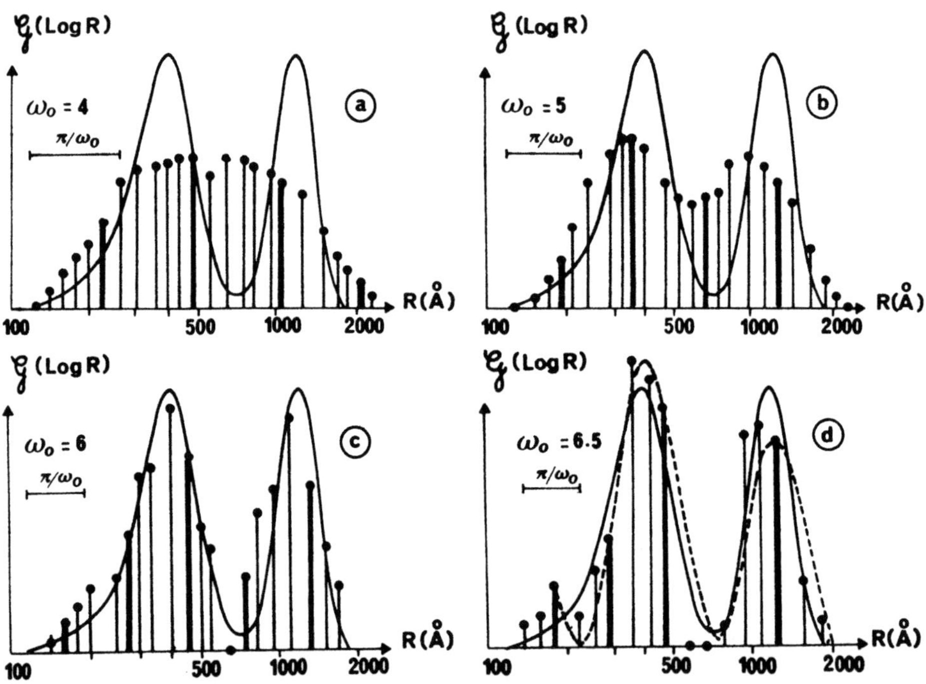

Fig. 5. Examples of reconstruction, by shifting the sets of Γ_n's , of an intensity distribution $\overline{G}$, expressed as a function of lnR (which in case of spherical particles is proportionnal to $\ln\Gamma$), taken from [6] .
The data were simulated from an a priori chosen two-peaked distribution function (solid lines). The results are presented for increasing values of ω_0 up to the over-resolution limit. The dashed line on (d) was obtained by interpolation

It was experimentally seen that the measured correlation function has to be recorded over a very wide time domain (often using several correlograms recorded with different sampling times and matched together) in order to improve the resolution on $G(\Gamma)$, i.e. increase the cut-off value ω_0. In a similar way, the resolution was also strongly dependent on the allowed Γ range for $G(\Gamma)$. Any a priori knowledge which enabled us to reduce this range greatly improved the resolution.

A quantitative study of these dependences of ω_0 on the range t of g(t) and Γ range of $G(\Gamma)$ cannot be done in the formalism of the eigenfunctions of the Laplace transform for which it is assumed that both supports of the $G(\Gamma)$ and g(t) functions are identical, continuous and range from 0 to ∞. As the correlation function is actually measured at a finite number of discrete time values, and as the Γ range covered by $G(\Gamma)$ is finite in a true experiment, the singular value formalism [17] must be used and provides analytical or numerical expressions for the dependence of the dilation factor $\exp(\pi / \omega_0)$ with the Γ range, the number of data points and the sampling time.

The other class of methods, using regularization constraints, basically works with the following leading ideas.

Amongst all possible solutions for $G(\Gamma)$, choose the one which has the least irregularities ("parcimony principle" of S. PROVENCHER [20]). This choice is ensured by minimizing the sum of two terms. The first one is the usual least square term which fits the experimental data to a multi-exponential function, $\sum x_j \exp(-\Gamma_j t)$. The second one, the regularization term, consists of some combination of the x_j's, which penalizes solutions going against the parcimony principle (or against some a priori known shape for $G(\Gamma)$). The operational part of this method is quite elaborated and contains many attractive features, such as sophisticated positivity constraints, a confidence margin determination, a statistical evaluation of the data, pointing out systematical errors, etc . It can be applied to a wide variety of inversion methods and has in particular been used for polydisperse data analysis of macromolecular suspensions. It is however mathematically quite heavy and for the time being cannot be implemented on microcomputers or even small minicomputers.

There are some basic similarities between the regularization methods and the singular value approach, inasmuch as both methods use as an apriori knowledge the finite time interval scanned by the measured correlation function, and possibly the finite range allowed for the Γ's. Both methods favour solutions for $G(\Gamma)$ with no high frequency fourier components, the singular value approach explicitly giving the cut-off value ω_0 as a function of noise, and $\Gamma_{max}/\Gamma_{min}$.

One basic difference is however the positivity constraint added in the regularization method, and the fact that contrary to the singular value approach, the resolution of $G(\ln\Gamma)$ along the $\ln\Gamma$ axis must not be constant but depends on the distribution of the data points and their noise.

Nevertheless, both methods seem to yield similar results except in the case of single very narrow peaked distribution which can be better recognized by the regularization method.

A possible improvement and simplification of these methods may come from a slightly different approach, suggested by BERTERO and PIKE [21].

Instead of trying to recover the distribution $G(\Gamma)$ from the data, look for the cumulative distribution :

$$F(\Gamma) = \int_0^{\Gamma} G(\Gamma')\, d\Gamma' \quad .$$

The recovery of this monotonic (actually increasing) function of Γ from the data is in principle no longer ill-conditioned [22] . It has been successfully applied to the study of vesicle's coagulation kinetics [23] as is shown, for example, on Fig.6.

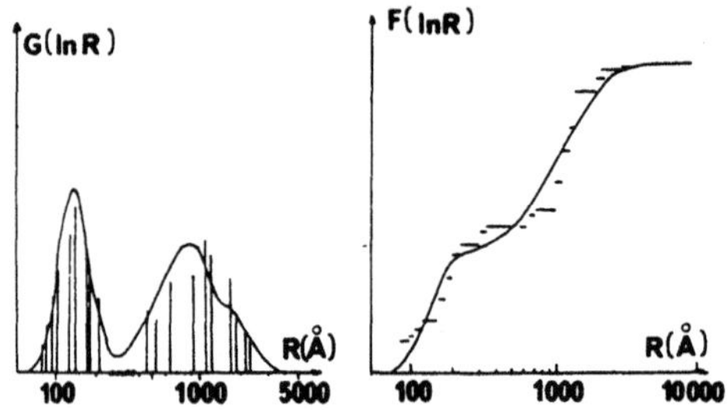

Fig. 6. Example of the reconstruction of an intensity distribution function G(ln R) and its cumulative form F(ln R). The data were taken on a suspension of vesicles undergoing coagulation, and the solid lines were computed from a fusion kinetics model (see [23])

Before concluding, let us point out a few practical facts of importance in polydisperse diffusion studies.

Our experience has shown that it is useless to try and determine more than two parameters from data which have a relative noise of a few per cent. An order of magnitude must be gained to try and use a more sophisticated analysis.

The problem of heterodyne versus homodyne becomes crucial in this accuracy domain and the slightest dust contamination ($\simeq$ 1%) produces distortion well above the acceptable threshold. We have thus systematically used controlled heterodyning (i.e. with a separate reference beam) and are actually constructing an optical fiber angular spectrometer [24] device for that purpose.

In conclusion, we hope to have shown that polydispersity may have a great variety of origins. Although in many cases it appears as an unwanted experimental complication - to be dealt with rapidly if not ignored - it can also be the main source of information. In experiments such as particles' size distribution measurements, particularly if the distributions are time dependent (kinetic studies of aggregation, coagulation, fusion etc .) polydisperse analysis of Q.E.L.S. spectra is certainly a useful, non-perturbative method of investigation.

A whole panoply of methods can be used in the data processing, ranging from the simple cumulant expansion, to the sophisticated regularization methods. The nature of the problem, the amount of a priori knowledge, and the quality of the data will finally determine the best choice of method.

1 "Photon Correlation and Light Beating Spectroscopy"
Ed.by H.Z Cummins and E.R Pike, Plenum Press, New York (1974)
p.285-330 "Applications of Light Beating Spectroscopy to Biology"
by H.Z Cummins.

2 "Photon Correlation Spectroscopy and Velocimetry" ed. by H.Z Cummins and E.R Pike, Plenum Press, N.Y (1977) p .164-199 "Dynamics of Macromolecular Motions" by H.Z.Cummins and P.N Pusey.

3 "Scattering Techniques Applied to Supramolecular and Non Equilibrium Systems," ed. by S.H Chen, B.Chu and R.Nossal (1981) p.179-200 "The Analysis of Polydisperse Scattering Data" by E.R Pike.

4 "The Application of laser Light Scattering to the Study of Biological Motion", NATO ASI Maratea, June 1982,see "Correlation Function profile Analysis in Laser Light Scattering" by B.Chu.

5 Es.Gulari, Er.Gulari, Y.Tsunashima and B.Chu, J.Chem.Phys. 70, 3965 (1979).

6 N.Ostrowsky, D. Sornette, P.Parker and E.R.Pike, Optica Acta, 28, 1059 (1981).

7 See for example :"Dynamic Light Scattering" by B.Berne and R.Pecora, J.Wiley ed. (1976).

8 See "Brownian Motion of Interacting Particles" by P.N.Pusey in this book.

9 P.Bezot, N.Ostrowsky, C.Hesse-Bezot, Optics Com. 25, 14 (1978).

10 S.R.Aragon and R.Pecora, J.Chem.Phys. 64, 2395 (1976).

11 D.G.Gardner, J.C.Gardner, G.Laush and W.W.Meinke, J.Chem.Phys. 31 978, (1959).

12 D.G Gardner, Ann. N.Y. Acad.Sci. 108,195 (1963).

13 S.W.Provencher, Makromol.Chem. 180, 201-309 (1979).

14 S.W.Provencher and V.G.Dovi, Jour.of Biochem.and Biophys.Methods, 1 (1979) 313-318.

15 J.G.Mc Whirter and E.R.Pike, J.Phys.A, 11,1729 (1978).

16 J.G.Mc Whirter, Optica Acta 27, 83 (1980).

17 See "On the Extraction of Polydispersity Information in Photon Correlation Spectroscopy" by E.R.Pike and M.Bertero, this book.

18 See :"A Solution of an Ill-Posed Inverse Problem : Approximation of Polydisperse Scattering Data as a Sum of Many Exponents" by G.R.Danovich and I.N.Serdyuk, this book.

19 See :"A General Purpose Constrained Regularization Method for Inverting Photon Correlation Data" by S.Proventcher, this book.

20 S.W.Provencher, Contin Users Manual, Technical Report EMBL-DA02 (Nov.1980).

21 M.Bertero and E.R.Pike, Private Communication.

22 A.V.Goncarskii and A.G.Jagola, Soviet Math.Dokl. 10 (1969) p.155.

23 D.Sornette and N.Ostrowsky, Lund Symposium on"Surfactants in Solution",June 1982,to be published by Plenum Press.

24 This experiment is carried out with the help of an AMTEC Angular Photon Correlation Spectrometer.

29. On the Extraction of Polydispersity Information in Photon Correlation Spectroscopy

M. Bertero

Istituto Matematico dell 'Università and
Istituto Nazionale di Fisica Nucleare, Genoa, Italy

E.R. Pike

Royal Signals and Radar Establishment, St. Andrews Road,
Malvern, Worcestershire, UK

1. Introduction - Eigenfunction Analysis

It is well known that the problem of recovery of polydispersity information in photon correlation spectroscopy is preceded by the necessity to invert a Laplace transform in which the data is necessarily sampled, of finite extent and corrupted by noise [1]. There is a large literature on the numerical inversion of the Laplace transform but a new approach was initiated in 1978 by McWHIRTER and PIKE [2] who discovered the eigenvalues and eigenfunctions of the transformation and were thus able to construct an "information theoretic" inversion procedure based on the well-known Nyquist sampling ideas of standard information theory. This approach has been pursued and applied to the polydispersity problem by OSTROWSKY et al. [3]. In this application the polydisperse distribution of diffusion constants $p(\Gamma)$ is reconstructed at a set of sample points

$$\Gamma_n = e^{\frac{n\pi}{\omega_0}} = \delta_0^n \quad , \tag{1}$$

where ω_0 is a "frequency cut-off" which limits the sampling rate, as does the Nyquist criterion in the analogous sampling theory case, and δ_0 is a "resolution ratio". The value of ω_0, unlike the Nyquist theory, was shown in [2] to be a function of the noise on the data. The method, based on a simple linear least-squares procedure, and discussed with examples in [3,4] is known as the "exponential sampling" method.

2. Finite "object" - Singular function analysis

The authors of [3] noted that the resolution criterion (1) could be improved by utilising an a priori knowledge of the support of the function $p(\Gamma)$, rather than allowing a large unknown range of Γ values to be possible. In a sequence of more recent papers [5-7] a theoretical basis for this observation has been established and the improvement possible in resolution quantified. This work has entailed an extension of the concepts of information theory using the theory of singular values and singular functions in place of eigenvalues and eigenfunctions, which fail to be applicable over variable domains of support of the recovered function. It was found that the new resolution limit δ_s was a function only of the ratio γ of the upper and lower bounds of the (simply connected) support and, of course, the signal to noise ratio, E/ε. Analytical results from [6] are shown in Fig.1.

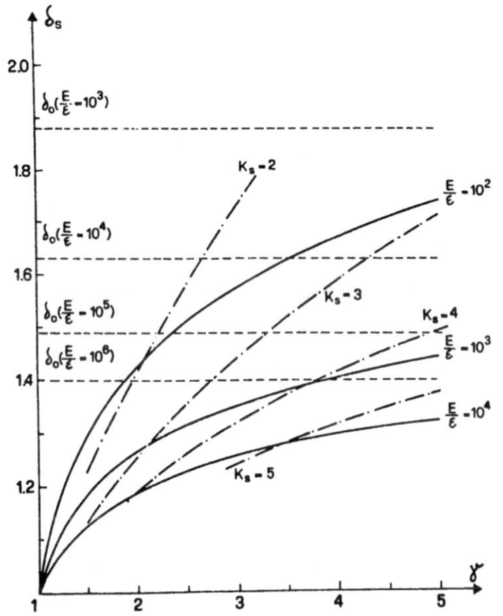

Fig.1. The resolution ratio δ_S as a function of the ratio γ of upper to lower bound of support of p (Γ) and of the "signal to noise" ratio E/ε. The horizontal dotted lines are the values for infinite support. The dash-dotted lines give the number of exponentials which may be recovered

3. Finite object and finite sampled data - Condition number analysis

The results displayed in Fig.1 are based on an analysis which assumes that the data, although noisy, is known over its full (infinite) domain of support. In [7] the effect of finite and sampled data is investigated both for fixed numbers of linearly and "exponentially sampled" data points. Optimum placing of these points is derived by minimisation of the condition number (ratio of highest to lowest singular value) using numerical techniques. It is found that the resolution given in Fig.1 is only slightly degraded in the practical case of 100 linearly spaced experimental points optimally placed. A somewhat lower number of geometrically spaced points has been found to achieve the same objective in the photon-noise limited case but further calculations are required to discuss fully the signal-noise-limited (high photon number) case, where correlated noise over the coherence time of any particular component will reduce drastically the number of channels of data required to approach the theoretical resolution values of Fig.1. This fact is used in the recently designed "log-lin" correlator(1) which can be switched from a normal linear delay spacing to a geometrical series of ratio $\sqrt{2}$ which covers, at low cost and without essential loss of information, a linear range equivalent to 8196 channels of delay.

4. Positivity and constraints

The results of the analytical and numerical work described above confirm and quantify what has been known empirically in photon correlation spectroscopy and other fields for many years, namely, that in practice one is not able to resolve more than at best about three fairly well spaced components. This unfortunate outcome can be roughly derived, in fact, by very much more simple calculations, of the type first considered by PUSEY et al.[8]. In

(1) Malvern Instruments Ltd Model 7027

these the difference of data values obtained with a single exponential and with some fraction of that exponential removed and redistributed 50-50 at lower and higher geometrically spaced points is compared with expected experimental noise. One can show, for example [5], that for an experiment lasting 10^5 coherence times, ten per cent of a single exponential can be removed and redistributed at equal weights on adjacent geometrically spaced points, the higher one at twice the ordinate of the lower one, without disturbing the correlation function within the limits of the noise. An experiment of this duration would give an RMS error in the data of $\sim 2.6\ 10^{-3}$ which is on the high end of normal experimental accuracy.

If, using the exponential sampling method, for example, one continues to request more components of the inversion than are allowed by Fig.1 one will find that the problem becomes ill-conditioned and that the permissible lower resolution inversions have added to them oscillatory components, which, according to the simple calculations above, do not affect the data within the noise level but which may result in spurious fine structure or even negative values in the result.

An alternative inversion technique to the "information theoretic" approach described here takes no notice of the theoretical resolution limit which we have shown is in the nature of the problem and requests larger numbers of independent "object" points than are intrinsically available from the data. However, a priori assumptions of smoothness and positivity are built in which restrict the then unavoidable unphysical effects. The approach, if given constraints equivalent to restricting the number of components requested to the correct theoretical value of Fig.1, should give satisfactory results but the method requires non-linear fitting and is very computer-intensive compared with the exponential sampling technique. Both methods can be illustrated using the geometrical diagrams of [4]. The object vector f_0 in space F is mapped into the noiseless data vector g_0 in space G (Fig.2). Noise induces an uncertainty in the data which is represented by a spherical domain about g_0.

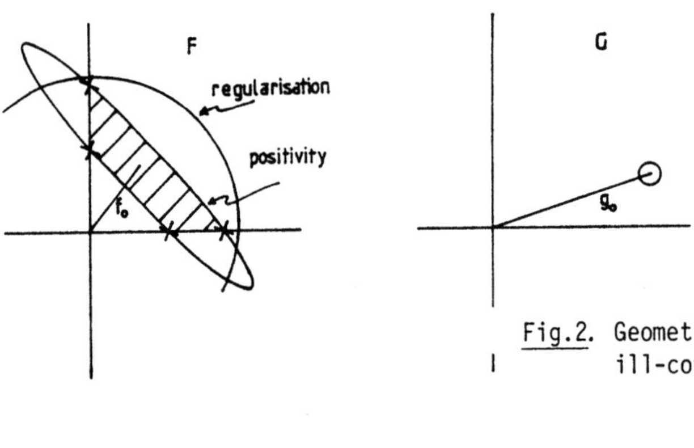

Fig.2. Geometrical representation of ill-conditioning

When the dimension (number of components requested) of the inversion space is too high this spherical domain in G maps back into an elliptical domain in F as illustrated, allowing solutions far from the object vector f_0 which cannot be distinguished without using additional constraints. The simplest and most common "regularisation" technique ensures smoothness by constraining the total "power" in the object. This corresponds to only accepting solutions which lie in a sphere about the origin as shown.

Positivity is normally ensured using the "Simplex" method which will arrive at one of the vertices marked with a cross on the figure, but in any case only constraining the solution to the hatched "quadrant". Of course, if the correct dimension is used for inversion the domain in F will be small and close to the required vector f_0, thus removing the ill-conditioning in a satisfactory manner.

A striking example of the impossibility of positivity per se to combat ill-conditioning is given in [5]. We reproduce in Fig.3 two distributions which have the same least-square RMS error per channel of $2.08\,10^{-3}$, similarly distributed, from a simulated experiment of duration 10^5 mean coherence times with two equally weighted components of 150 and 527 nm diameter, respectively.

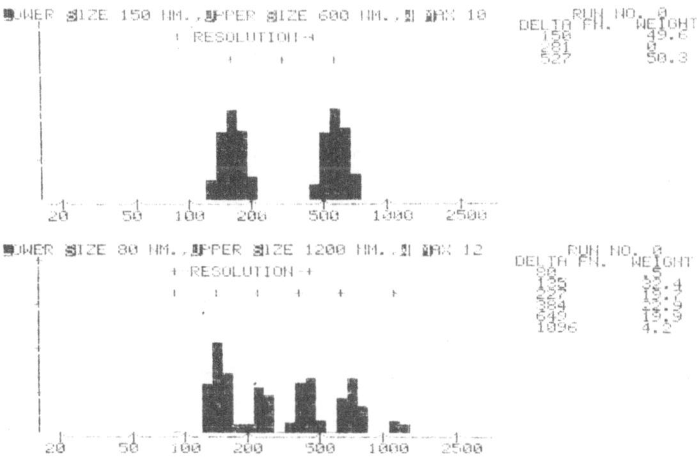

Fig.3. Two distributions indistinguishable from each other by use of the data

Without further a priori knowledge both of these possible positive solutions cannot be distinguished.

Further examples of simulated and real experiments with two and three component mixtures are given in the literature cited and there seems to be no doubt about the basic conclusions which are displayed in Fig.1. This may therefore serve as a guide for experiment and can be used to remove ill-conditioning from inversion in the way described above. A great benefit of this method as compared with constrained fitting is the economical use of computing, the examples of Fig.3 for instance were carried out in a few minutes on a PET personal desk computer.

Acknowledgements

We are pleased to acknowledge valuable contributions to this work over several years by Miss P. BOCCACCI of the Univ. of Genoa, Dr. J.McWHIRTER and Mrs.P. PARKER of RSRE Malvern, Dr. F. McNEIL WATSON, Mr. D. WATSON and Mr. M. ALDRITT of Malvern Instruments and Mme. N. OSTROWSKY and Dr. D. SORNETTE of the University of Nice. Dr. P. PUSEY has provided and characterised by cumulants mixtures of polystyrene spheres in aqueous suspension for testing the new "Log-lin" correlator system and has participated in valuable discussions.

References

1. Photon correlation and light beating spectroscopy, 1974, Eds. H.Z. Cummins and E.R. Pike (Plenum Press: New York)

2. McWhirter, J.G. and Pike,.E.R., 1978, J. Phys. A: Math Gen., 11, 1729

3. Ostrowsky, N., Sornette, D., Parker, P. and Pike, E.R., 1981, Optica Acta, 28, 1059

4. Pike, E.R., 1981, in Scattering techniques applied to supra-molecular and non-equilibrium systems NATO ASI Wellesley 1980 Eds. S.H. Chen, R. Nossal, and B. Chu (Plenum Press: New York)

5. Pike, E.R., Watson, D. and McNeil Watson, F., 1982 Proc. 1st Conference American Aerosol Association Santa Monica 1982 Ed. B. Dahneke (Wiley: New York)

6. Bertero, M., Boccacci, P. and Pike, E.R., 1982, On the recovery and resolution of exponential relaxation rates from experimental data - A singular value analysis of the Laplace transform inversion in the presence of noise. Proc. Roy. Soc. A, London, to be published

7. Bertero, M., Boccacci, P. and Pike, E.R, 1983, On the recovery and resolution of exponential relaxation rates from experimental data II - The optimum choice of experimental sampling points for Laplace transform inversion, to be published

8. Pusey, P.N., Koppel, D.E., Schaefer, D.W. and Camerini-Otero, R.D., 1972. IBM Res Rep RC 3924 (No 17791)

30. Correlation Function Profile Analysis in Laser Light Scattering. III. An Iterative Procedure

J.R. Ford and B. Chu

Department of Chemistry, State University of New York,
Stony Brook, NY 11794, USA

1. Introduction

In photoelectron correlation function profile analysis, the inversion of the Laplace transform

$$|g^{(1)}(\tau)| = \int_0^\infty G(\Gamma)e^{-\Gamma\tau}d\Gamma \qquad (1)$$

to obtain the normalized linewidth distribution function $G(\Gamma)$ from the net electric field correlation function $g^{(1)}(\tau)$ is essentially an unresolved ill-conditioned problem, where Γ and τ are, respectively, the characteristic linewidth and the delay time. In practice, $g^{(1)}(\tau)$ contains noise and the integral has upper (b) and lower (a) bounds. Consequently, in order to remove the ill-conditioning, we need to have estimates of both the signal-to-noise ratio and the width, in terms of the support ratio $\gamma(\equiv b/a)$, of the linewidth distribution function. However, as $g^{(1)}(\tau)$ depends upon the delay time range of our experiment, we now encounter a problem whereby our experimental conditions and the results we hope to obtain are interactive. Then, the success of a laser light scattering experiment depends upon (1) a proper choice of experimental conditions, as well as (2) appropriate inversion of the measured $g^{(1)}(\tau)$ to obtain $G(\Gamma)$. Thirdly, further analysis of $G(\Gamma)$ is often required to obtain the desired information, such as molecular weight distribution, internal motions, etc. These three requirements are highly interdependent and the experimenter must be aware of the uncertainties introduced at each step. In this article, we propose an iterative procedure that tries to meet the above requirements.

A proper choice of experimental conditions refers to all aspects of the actual measurement. The choice of equipment will influence the quality of the measured correlation function - pulse saturation and after-pulsing should be considered. Fluctuations in the laser intensity will degrade the baseline and should be minimized. The detection optics will determine the coherence area and thus affect the signal to noise ratio.

The sample must be homogeneous in composition and temperature. Fluctuations in either on the time scale of the experiment will both degrade the baseline and be convoluted with the desired signal, rendering interpretation of the resulting correlogram difficult. In addition, the ever-present dust problem will seriously affect the baseline.

The choice of dwell times is critical. The motions of interest in the sample have a finite bandwidth, or support. Obviously measurements of the correlation function must be made to cover this range, even though many have previously ignored this implicit but necessary requirement. The number of channels available and their spacing will determine the dynamic range of the correlator accessible to the experimenter. Furthermore, the intended method of analysis should be considered when choosing a spacing scheme in the delay time - for an expansion in terms of the eigenfunctions of the Laplace transform, exponentially spaced dwell times allow extrapolation of the necessary intermediate values of $g^{(1)}(\tau)$ without introducing additional uncertainties in the data. For linearly spaced sampling of the correlation function and well-behaved unimodal linewidth distribution functions we have found the singular value decomposition technique [1] to be particularly applicable, and it is described below.

Due to the ill-posed nature of the inversion process, the signal-to-noise ratio S/N for a given support determines the maximum resolution obtainable from inversion of $g^{(1)}(\tau)$ [2], and therefore all possible means of maximizing this quantity should be sought. Whether or not one has achieved a sufficiently large S/N can often only be seen after the inversion has been attempted; again, some iterative process is indicated.

The success of the inversion depends upon integration of the conditions of the experiment with the method of inversion. In addition, the more sophisticated approaches usually require inputs of initial estimates on unknown parameters as starting values; these estimates can often be computed from simpler methods of data analysis.

It can be quite misleading to apply the more sophisticated multi-exponential or histogram fitting procedure to data whose S/N does not warrant such treatment. The unwary experimenter often forgets that the inversion techniques will almost always yield parameters which fit the experimental data. However, overfitting of experimental data will render those parameters without physical meaning. This is the very nature of the ill-conditioned problem. In this article we try to provide guidelines as to how we can avoid some of the pitfalls.

Finally, it must be kept in mind that $G(\Gamma)$ is not usually an end in itself, and that subsequent analysis of the linewidth distribution function is likely to be performed. In molecular polydispersity studies, for example, we are concerned with obtaining the molecular weight distribution $f(M)$. Thus, we must be aware of the possible existence of rotational and internal motions (entanglement, bending, twisting, etc.) and particle interactions, as the presence of any of these components in $G(\Gamma)$ will affect the resultant $f(M)$ unless they can be corrected for. Therefore, an understanding of the uncertainties of the linewidth distribution function and how these uncertainties map through the subsequent transform to $f(M)$ is a prerequisite to evaluating the significance of the final result.

2. Theoretical Background

The measured self-beating photocount autocorrelation function has the form

$$G^{(2)}(\tau) = A(1+\beta|g^{(1)}(\tau)|^2) , \tag{2}$$

where β is a spatial coherence factor usually considered to be an unknown parameter in the fitting process and the background A can be computed. Equation (1) represents $g^{(1)}(\tau)$ for a solution of polydisperse particles and is a Fredholm integral equation of the first kind with a Laplace kernel which contributes to the notoriously ill-conditioned problem [3-6]; the amount of information that can be recovered about $G(\Gamma)$ is limited.

Analytical expressions for the eigenfunctions and eigenvalues of the Laplace transform have been worked out by McWhirter and Pike [5]. These eigenfunctions signify the fundamental elements of information transmitted through the integral operator of equation (1), scaled by the corresponding eigenvalues.

Unfortunately, the eigenvalues decay rapidly to zero for the Laplace transform. Only those information elements whose values are above the noise level of the data can be recovered. In principle it is possible to expand $g^{(1)}(\tau)$ in terms of the eigenfunctions and solve for $G(\Gamma)$ to the resolution dictated by the noise. In practice it is difficult to determine the exact location of the noise cutoff; furthermore the zero crossings of the eigenfunctions are not evenly spaced in time, requiring extrapolation of $g^{(1)}(\tau)$ over short and long delay times [6].

This discouraging analysis of the limit of resolution is slightly mollified by accounting for the finite extent, or support, of $G(\Gamma)$. Bertero et al. [2] have used singular value analysis to show that as the support of $G(\Gamma)$ is decreased the resolution obtainable increases. In particular, the resolution ratio δ_s (spacing in log space) is given by

$$\delta_s = \gamma^{1/K_s} , \tag{3}$$

where the support ratio $\gamma (\equiv b/a)$ is $\Gamma_{max}/\Gamma_{min}$, and K_s is the number of singular values satisfying the inequality

$$\alpha_k \geq \varepsilon/E. \tag{4}$$

Here the singular values α_k can be approximated by

$$\alpha_k^2 \sim \pi/\cosh(\pi\omega_k) \tag{5}$$

with

$$\omega_k = k\pi/\ln\gamma \quad \text{for} \quad k = 0,1,2\ldots, \tag{6}$$

and the noise parameter ε/E is given by

$$\varepsilon/E = \eta\sqrt{d(\gamma-1)}, \tag{7}$$

with η being the mean square error of the sampled data points, and d, the (constant) spacing between data points. For the generated data described below, we have $\eta = 1.32$ and $\gamma = 20$, giving a value of 5.0×10^{-3} for ε/E. Comparison of this quantity with the singular values calculated from equations (5) and (6) and listed in Table 1 shows that we can expect to recover around 4

Table 1. Singular values calculated from eqs. (5) and (6) for $\gamma = 20$

k	α_k
0	1.772
1	4.824×10^{-1}
2	9.296×10^{-2}
3	1.790×10^{-2}
4	3.447×10^{-3}
5	6.639×10^{-4}
6	1.278×10^{-4}
7	2.462×10^{-5}
8	4.741×10^{-6}
9	9.130×10^{-7}

resolution elements, since $\alpha_4 = 3.45 \times 10^{-3} \leq 5 \times 10^{-3}$. Unfortunately, the quantities η and γ are not available a priori, and the fitting procedure must necessarily consist of a successive approximation technique beginning with a simple cumulants expansion to obtain the initial parameters, and continuing through more elaborate approximations until one feels that the available information has been recovered.

3. Methods of Data Analysis

Inversion of $g^{(1)}(\tau)$ involves two distinct steps. An approximation for $G(\Gamma)$ is chosen - here we will discuss cumulants expansion, exponential, and histogram approximations - and the parameters of the approximation are found according to minimization of some norm of the residual vector. The Euclidean norm of the residual vector, χ^2, is usually chosen as the minimization criterion, giving rise to the least squares techniques. Other approaches, notably the Simplex algorithm based on minimization of the L^1 norm, deserve mention but are outside the scope of this article. Many algorithms are available for arriving at the parameters of the least squares fit. These generally fall into two classes, linear and nonlinear. We have found a modified Marquardt-Levenberg approach useful in the solution of nonlinear least squares problems, and will elaborate below on the benefits of using singular value decomposition for the solution of the linear problems arising from multiexponential and histogram approximations.

The particular approximations for $G(\Gamma)$ that we will consider here are the cumulants expansion, the multiexponential, and the histogram techniques. They are discussed in detail in the following sections.

3.a Cumulants Method

Here

$$\ln|g^{(1)}(\tau)| = -\bar{\Gamma}\tau + \frac{1}{2!}(\mu_2/\bar{\Gamma}^2)(\bar{\Gamma}\tau)^2 - \frac{1}{3!}(\mu_3/\bar{\Gamma}^3)(\bar{\Gamma}\tau)^3 + \ldots \tag{8}$$

with

$$\bar{\Gamma} = \int_0^\infty G(\Gamma)d\Gamma \tag{9}$$

and

$$\mu_i = \int_0^\infty (\Gamma-\bar{\Gamma})^i G(\Gamma)d\Gamma \quad . \tag{10}$$

The series is truncated after the 2nd, 3rd, etc., moments, giving rise to the 2nd, 3rd, etc., order cumulants methods. A nonlinear least squares minimization of the residual norm χ^2 with respect to the μ_i is performed.

3.b Double Exponential Method

The double exponential method is singled out as a special case of the multiexponential approximations. It is our only nonlinear multiexponential method. The linewidth distribution function is represented by

$$G(\Gamma) = B_1\delta(\Gamma-\Gamma_1) + B_2\delta(\Gamma-\Gamma_2) \tag{11}$$

and thus

$$|g^{(1)}(\tau)| = \sum_{i=1}^{2} B_i e^{-\Gamma_i\tau} \quad . \tag{12}$$

Both Γ_i and B_i are floated to minimize χ^2.

3.c Multiexponential Method

The multiexponential approach models $G(\Gamma)$ as

$$G(\Gamma) = \sum_{i=1}^{N} B_i \delta(\Gamma-\Gamma_i) \tag{13}$$

giving

$$|g^{(1)}(\tau)| = \sum_{i=1}^{N} B_i e^{-\Gamma_i \tau} \tag{14}$$

where we fix the Γ_i values and fit for values of B_i. As was shown by Ostrowsky et al.[10], the natural spacing of the Laplace transform indicates that the delta functions should be spaced exponentially, i.e.,

$$\ln\Gamma_{j+1} = \ln\Gamma_j + \delta \quad . \tag{15}$$

Using the singular value decomposition procedure, the choice of the spacing parameter δ is not critical. However, for a straightforward solution of the least squares problem, δ must be determined carefully at this point, as it should be identified with δ_s. Unfortunately, according to equations (3) - (7), it is difficult to estimate δ_s because we usually do not have sufficient a priori information to determine the support of $G(\Gamma)$ and the signal-to-noise parameter ε/E.

3.d Histogram Method

The histogram approximation assumes that $G(\Gamma)$ is constant over a range from Γ_j to Γ_{j+1} for each of N boxes. This gives

$$|g^{(1)}(\tau)| = \sum_{i=1}^{N} \left(\frac{B_i}{\tau}\right)\left(e^{-\Gamma_{i+1}\tau} - e^{-\Gamma_i \tau}\right) \tag{16}$$

where the values of Γ_i are held constant with the spacing factor δ selected as described for the multiexponential method.

4. Simulated Data

To clarify the procedures discussed below, we have generated a linewidth distribution modeled on a normalized Pearsons type I distribution, or

$$P(\Gamma) = C\left(\frac{\Gamma - \Gamma_{min}}{\Gamma_{max} - \Gamma_{min}}\right)^{M_1}\left(\frac{\Gamma_{max} - \Gamma}{\Gamma_{max} - \Gamma_{min}}\right)^{M_2} \quad , \tag{17}$$

where the parameters of the generated distribution are given in Table 2. This distribution was generated at 1000 equally spaced points along the Γ axis. The Laplace transform of the distribution was calculated and $g^{(1)}(\tau)$ generated at 92 equally spaced delay times ($\Delta\tau$ = 750 ns.). The resulting correlation function was corrupted by noise according to the formula [7]

$$\mathrm{var}\, G^{(2)}(\tau) = \frac{1}{2N_c}\left[1 + 8e^{-x} - e^{-2x}(5 + 2x)\right] \tag{18}$$

with $x = 2\tau/\tau_o$. $N_c = 10^3$ was chosen to give somewhat noisier data than would usually be encountered in a careful experiment. We have taken τ_o to be the equivalent (single-exponential) characteristic decay time as represented by $1/\bar{\Gamma}$.

Table 2. Parameters of the simulated data

Γ_{min}	5000
Γ_{max}	100000
M_1	1.00
M_2	2.50
92 channels, $\Delta\tau = 750$ ns.	
$\bar{\Gamma}$	39456
μ_2	$3.21\text{x}10^8$
$\bar{\Gamma}\tau_{max}$	2.7

5. An Iterative Approach

We describe an iterative approach to the selection of experimental conditions and method of inversion that we have found to be helpful in the Laplace inversion and subsequent analysis of molecular polydispersity.

Preliminary measurements are performed using some initial guess for the dwell time. This is adjusted until the resulting correlogram appears to cover approximately three decay periods ($\Gamma\tau_{max} \sim 3$ with τ_{max} being the maximum delay time). A short measurement (on the order of a few minutes duration) is taken, transferred to the on-line computer, and a second-order cumulants fit is performed yielding a value for $\bar{\Gamma}\tau_{max}$. The dwell time is then adjusted so that the $\bar{\Gamma}\tau_{max}$ value is in the range of ~ 3, and the short experiment repeated. The resulting $\mu_2/\bar{\Gamma}^2$ value is used to suggest a choice on the eventual inversion method.

Once the experimental parameters have been selected (first approximation), the experiment is performed until the desired S/N is obtained. This criterion is often reconfirmed by the variance obtained from the cumulants fits - if the distribution appears to be wide or composed of more than one motion, the S/N must be sufficiently large to be able to resolve the desired components (some indication of N/S ($\sim \varepsilon/E$) can be estimated from (7)).

The proper analysis begins with second or third-order cumulants fits. The $\bar{\Gamma}$ and variance from these calculations are used as initial values for the parameters of a double-exponential fit. The double-exponential results often give better values for $\bar{\Gamma}$ and variance than the cumulants fit, for reasons mentioned in Paper I of this series [8]. The results of the cumulants and double-exponential fits on the simulated data are given in Table 3.

If estimates of the signal-to-noise ratio and the support of $G(\Gamma)$ reveal that we may be able to retrieve more information than $\bar{\Gamma}$ and $\mu_2/\bar{\Gamma}^2$, then we proceed with either multiple-exponential or histogram analysis. The choice of histogram over multi-exponential seems to be largely a matter of personal taste; we find it easier to interpret and normalize the results of the histogram fit.

Table 3. Results of cumulants and double-exponential (B_1 = 0.452, Γ_1 = 20730 sec^{-1}, B_2 = 0.547, Γ_2 = 54530 sec^{-1}) fits using simulated data with parameters listed in Table 1

	Actual Value	Cumulants	Double Exponential
$\bar{\Gamma}$	39456	39360	39198
μ_2	3.21×10^8	3.1×10^8	2.8×10^8
$\mu_2/\bar{\Gamma}^2$	0.206	0.200	0.184

The histogram fit (and the multi-exponential, although we will restrict our attention to the former) involves the solution of a set of over-determined equations, which is done by least-squares techniques. In the simplest case, the locations of the boxes in Γ space are held constant, and the heights of the boxes, i.e., relative contribution from each interval in Γ space, are solved for. There are usually no restrictions on the parameters. Specifically, they are not constrained non-negative although there is no physical meaning to a negative box. The boxes are generally spaced logarithmically in Γ space to maximize the transmission of information across the inversion procedure.

The resulting linear least-squares problem can be represented as

$$Ax \overset{\sim}{=} b \, , \tag{19}$$

where A is the m by n curvature matrix with m and n being, respectively, the number of data points and the number of adjustable parameters; x is the parameter vector of length n, and b is the data vector of length m. The problem is solved using a singular-value decomposition technique which involves the transformation of (19) to the form

$$USV^{-1}x \overset{\sim}{=} b \tag{20}$$

with U and V^{-1} being orthogonal matrices and S, the diagonal matrix. The diagonal elements of S (the singular values) are monotonically decreasing. By retaining the first k+1(k = 0, 1, 2 ...) of these singular values, with α_k being the highest term, and setting the k+1, k+2, ... singular values to zero, we obtain a set of "candidate solutions", each solution corresponding to a particular choice of cutoff. The number of singular values to retain is determined by the S/N ratio of the experiment; this quantity plays the same role as the ω_{max} parameter in the eigenfunction solution [5]. Although there are some general guidelines (e.g.,(3) through (7); see also [1]) for the choice of cutoff, one usually finds it necessary to try several candidate solutions and choose the "best" one. The number of singular values retained can be examined on the basis of the appearance of the solution vectors - those solutions whose residual norms are "sufficiently small" without invoking "appreciably large" norms of the solution vectors themselves can often be used as a reasonable criterion for the correct cutoff. The terms in quotes are, of course, subjective. Regularization techniques such as those used by Provencher [9] could be used to automate the choice of cutoff.

The histogram technique requires as input the number and locations of the boxes in G(Γ). In the singular value decomposition technique, the diagonal matrix of singular values can be examined midway through the problem, and the important elements of information retained. This rank reduction technique allows the problem to be phrased in terms of a large number of parameters, thus assuring a better chance of covering the correct range, but gives the user an opportunity to reduce the number of resolution elements to include only the significant terms. The resulting solution vector is a linear combination of the resolution elements selected at the rank reduction step. The initial number of parameters can be made as high as feasible, restricted mainly by the amount of computer time and memory available for the inversion. This point may be misleading - choosing a large number of boxes does not imply that high resolution is going to be obtained. The final resolution depends on the number of singular values retained in the expansion of the solution vector. This technique gives a relatively smooth curve as the output from the analysis, bypassing the need for interpolation schemes or sliding the boxes along the Γ axis on subsequent fits [10].

Continuing with the analysis procedure, we have implicitly assumed a knowledge of the support of $G(\Gamma)$ in choosing the boxes of the histogram model. Generally, the initial approximations should overestimate the support range. The resultant fit will indicate this condition by forcing the boxes outside the support range to become negative. Experience has shown that the crossover points on the Γ axis are reasonably good estimates of the actual support, and the accuracy of these estimates increases as the model distribution approaches the actual distribution. Therefore, the histogram fit is recalculated using the modified estimate of the support range. The procedure is repeated until there is only negligible contribution to $G(\Gamma)$ from negative boxes.

The progress of the histogram fitting is indicated by Figures 1-5. The initial range of $G(\Gamma)$ was chosen large enough to cover the relatively wide distribution indicated by the variance of the cumulants expansion fit. For example, with $\bar{\Gamma}$ = 39000 and $\mu_2 / \bar{\Gamma}^2$ = 0.2, 95% of an equivalent normal distribution would fall in the range Γ = 3000 to Γ = 75000. To illustrate the concept of crossover points serving as boundary estimates of $G(\Gamma)$, we overestimate the support for the first fit, and take Γ_{min} = 2000 and Γ_{max} = 200,000. In Fig. 1 the candidate solutions for k = 2, 3, 4 and 5 are shown. Relative deviations plots shown in Fig. 2 rule out the k = 2 solution, as the deviations are highly correlated. The norm of the residual vector is

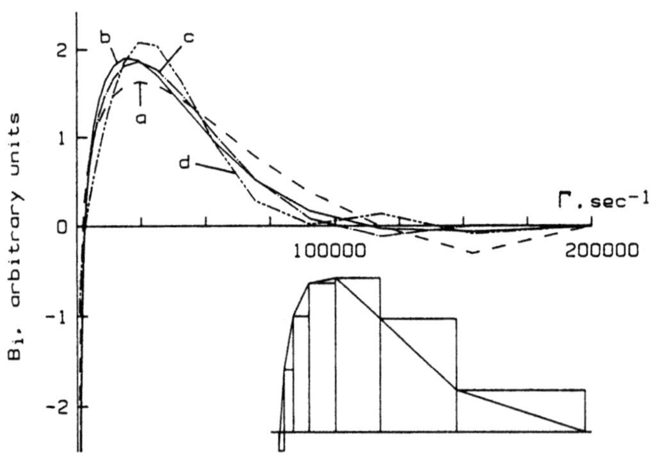

Fig. 1. Histogram analysis of sample data for a = 1000 sec^{-1}, b = 200000 sec^{-1} Curve a) represents the k = 2 candidate solution. Curves b) through d) show the candidate solutions for k = 3, 4, and 5. The inset illustrates the method of plotting the histogram fits (B_i versus Γ_i) that is used to clarify the superposition of several different solutions.

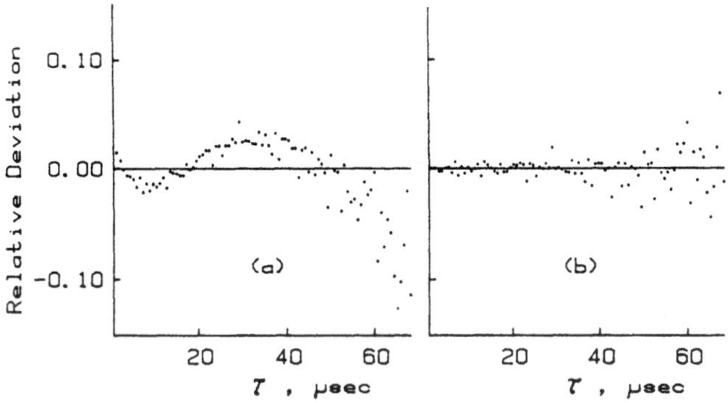

Fig. 2

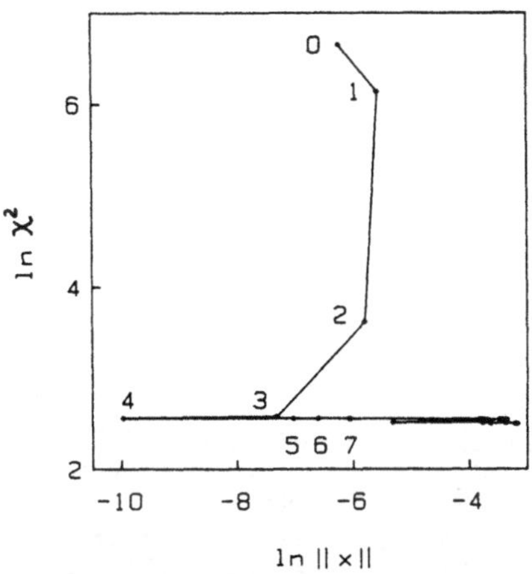

Fig. 3

Fig. 2. Relative deviations plots, $(|g^{(1)}(\tau)|^2_{data} - |g^{(1)}(\tau)|^2_{calc})/|g^{(1)}(\tau)|^2_{data}$ vs delay time τ) for the k = 2 and k = 3 (curves a and b, respectively) candidate solutions of Fig. 1

Fig. 3. Log-log plot of χ^2 vs. the norm of the solution vectors ($||x||$) for k = 0 through k = 20, for the fits of Fig. 1. The first 8 points are labeled with the corresponding k values. χ^2 is the norm of the residual vector $||r|| = (\sum_1^M r_i^2)^{1/2}$ with M = the number of channels in the correlogram (92) and $r_i = |g^{(1)}(\tau)|^2_{data} - |g^{(1)}(\tau)|^2_{calc}$

approximately constant for the other solutions (Fig. 3); the k = 4 solution has the lowest solution norm and is therefore postulated as the optimal solution for this range of G(Γ). From the zero crossings of the resultant distribution, a diminished support for G(Γ) was chosen for a second fit. This process was repeated (Fig. 4 a-c) until the negative boxes were negligible. Figure 4d indicates the effect of underestimating the support of G(Γ) - the solutions have a truncated appearance, and the solution vectors do not exhibit as much stability with respect to the inclusion of more singular values. Figure 5 shows the optimal solution of Fig. 4c overlaid with the parent distribution function.

We observe that 4 resolution elements were recovered, in agreement with the estimate obtained from (3) through (7). Of course, direct calculation of K_S is

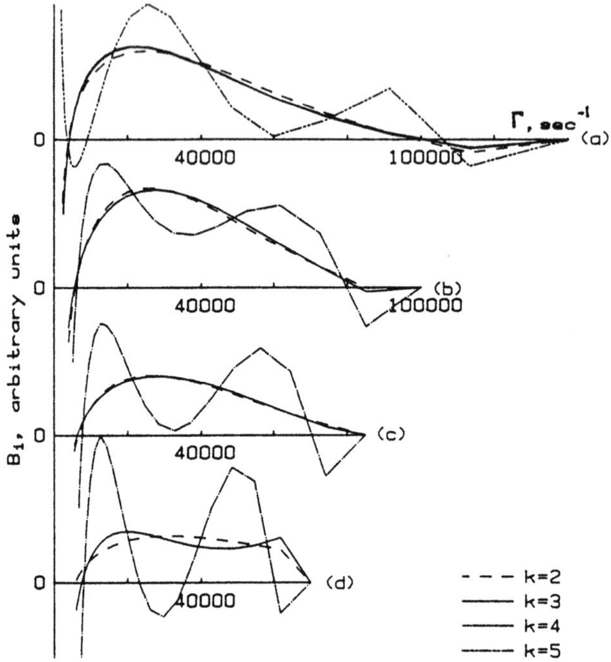

Fig. 4. Progress of the histogram analysis for simulated data of Table 1. a) Γ_{min} = 2000 sec^{-1}, Γ_{max} = 140000 sec^{-1}. b) Γ_{min} = 4000 sec^{-1}, Γ_{max} = 100000 sec^{-1}. c) Γ_{min} = 5500 sec^{-1}, Γ_{max} = 85000 sec^{-1}. d) Γ_{min} = 6000 sec^{-1}, Γ_{max} = 70000 sec^{-1}. Note that the vertical scale of each plot has been adjusted according to the magnitude of the largest peak, and is not the same for all plots

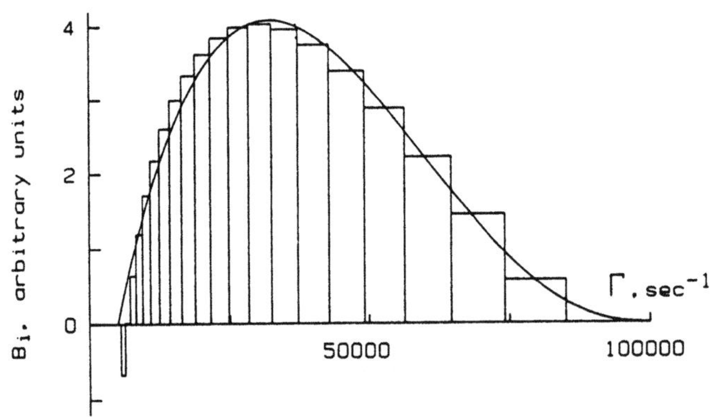

Fig. 5. Histogram analysis of sample data for a = 5500 sec^{-1}, b = 85000 sec^{-1}. The k = 3 candidate solution is represented by the histogram distribution. The smooth curve is the parent Pearsons distribution (Table 2) used to generate the data

not usually possible. Overestimation of the information present may occasionally lead one to suspect a bimodal distribution where none exists. Unfortunately, other criteria which could be used to select the appropriate solution vector often require an a priori knowledge of the data uncertainties. As can be seen from the example above (Fig. 2), the relative deviations plots can sometimes aid in making the decision.

It is possible to make our approach more sophisticated by floating the locations of the boxes, or by constraining them to be non-negative [11]. These approaches, however, make the least squares problem nonlinear and may introduce complexities that offset the advantages of the refined model.

When the cumulants, double exponential, or histogram procedures indicate the presence of unresolved components of $G(\Gamma)$, or a distribution whose support cannot be covered by the range of the data collected, it is necessary to repeat the measurements. The dynamic range can be extended by the use of a dual-delay [12] or logarithmically-spaced correlator. In the absence of one of these, the dynamic range of the available correlator can be extended by performing several experiments and fitting the composite correlogram. The multiplexing approach, however, can be subject to serious problems [8].

6. Conclusion

We have shown that the eventual success of a laser light scattering experiment depends upon a proper choice of experimental conditions as well as data inversion methods. Indeed, an iterative approach to optimizing all of these is necessary. Such an approach benefits greatly by the existence of an on-line computer capable of sophisticated methods of data analysis. Furthermore, it is imperative that the investigator understands the limitations of experimental data and data inversion methods.

Acknowledgements

BC wishes to thank the National Science Foundation, Polymers Program (DMR 8016521) and the U.S. Army Research Office, Durham, for financial support; and E. R. Pike and E. F. Grabowski and I. D. Morrison for sending him their preprints.

References

1. C. L. Lawson and R. J. Hanson, "Solving Least Squares Problems," Prentice-Hall, New Jersey (1974).

2. M. Bertero, P. Boccacci and E. R. Pike, Proc. Roy. Soc. A, to be published.

3. D. L. Phillips, J. Assoc. Comput. Mach., 9:84 (1962).

4. R. J. Hanson, SIAM J. Numer. Anal., 8:616 (1971).

5. J. G. McWhirter and E. R. Pike, J. Phys. A11:1729 (1978).

6. J. G. McWhirter, Optica Acta 27:83 (1980).

7. E. Jakeman, E. R. Pike and S. Swain, J. Phys. A4:517 (1971).

8. B. Chu, "Correlation Function Profile Analysis in Laser Light Scattering. I. General Review on Methods of Data Analysis," Proceedings of the NATO ASI on the Application of Biological Motions, Plenum Press, 1982, to be published.

9. S. W. Provencher, Makromol. Chem. 180:201 (1979).

10. N. Ostrowsky, D. Sornette, P. Parker, and E. R. Pike, Optica Acta 28:1059 (1981).

11. E. F. Grabowski and I. D. Morrison, "Particle Size Distributions from the Analysis of Quasielastic Light Scattering Data," presented at the First National Aerosol Symposium, Santa Monica, California, 1982.

12. K. M. Abbey, J. Shook and B. Chu, "Correlation Function Profile Analysis in Laser Light Scattering. II. A Hybrid Photon Correlation Spectrometer," Proceedings of the NATO ASI on the Application of Biological Motions, Plenum Press, 1982, to be published.

31. A Solution of an Ill-Posed Inverse Problem: Approximation of Polydisperse Scattering Data as a Sum of Many Exponential (Lorentzian) Curves

Grigory R. Danovich and Igor N. Serdyuk

Institute of Protein Research, Academy of Sciences of the USSR,
142292 Poustchino, Moscow Region, USSR

1. Introduction

Experiments in which the method of photon correlation was used serve as an example of physical experiments where the effect function is measured with the aim of determining the cause function.

A mathematical model of such experiments enables us by using a Fredholm integral equation of the first kind, to relay the effect function (the known correlation function f(t)) with the cause function (the unknown distribution of exponential curves $c(\tau)$ in it)

$$\int_{\tau_1}^{\tau_2} A(t,\tau)\ c(\tau)\ d\tau = f(t), \qquad (1)$$

where $A(t,\tau) = \exp(-t/\tau)$.

It is common knowledge that the solution (1) is unstable with respect to minor changes in f(t) and, strictly speaking, does not exist while conducting measurements of f(t) in a real noise-imposed experiment. This is attributable to the fact that under all conditions (non-existence, non-uniqueness, instability) the solution will be an ill-posed (in the Hadamard sense) problem.

In experiments f(t) is measured in a finite number of points $\{t_j;\ j=1,N\}$. We then substitute a quadrature grid with k points $\{\tau_i;\ i=1,K;\ K \leq N\}$ for the integral in (1) to obtain a set of linear algebraic equations

$$\sum_{i=1}^{k} A_{ij}\ c_i = f_j;\ j = 1,N; \qquad (2)$$

where $A_{ij} = A(t_j,\tau_i)\cdot w_i$, w_i is a weight of the ith point in the quadrature formula;
$f_j = f(t_j)$ is an experimental value measured at point t_j;
$c_i = c(\tau_i)$ is treated as an unknown value of the distribution component at point τ_i.

The solving of (2) is an ill-posed problem, since (2) is a discrete analogue of (1). In a number of investigations [1-5] a method has been developed for solving the ill-posed problem by means of substituting a similar well-posed problem for it. This technique was called the regularization method [2,3]. In essence it provides a vector function $\{c_i\}^\alpha$ that minimizes the functional M^α (consisting of two terms) to solve (2)

$$M^\alpha = \sum_{j=1}^{N} R_j\ (f_j - \sum_{i=1}^{k} A_{ij}c_i)^2 + \alpha\Omega(c), \qquad (3)$$

where $R_j>0$ is the weight coefficient of the jth experimental point and $\alpha>0$ is a regularization parameter.

The first term of (3) proceeds from the least-squares method. The second term which was called a stabilizing functional in [2,3] can be generally represented as

$$\Omega(c) = \sum_{[L]} \left(\sum_{i=1}^{k} \rho_{li}\{(c_i - c_{oi})^{(l)}\}^2 \right),$$

where $\{c_{oi}\}$ is the frame vector function;
$(c_i-c_{oi})^{(l)}$ is the numerical value of the lth derivative of the difference function $(c(\tau)-c_o(\tau))$ at the ith grid point at τ (here $(c_i-c_{oi})^{(o)} = c_i-c_{oi}$);
$\rho_{li}>0$ is the weight coefficient and [L] is a set of non-negative l indices.

The stabilizing functional characterizes the deviation of the unknown vector function $\{c_i\}$ from the frame vector function $\{c_{oi}\}$ either directly and/or with respect to smoothness. Therefore, the variation of α from 0 to ∞ tends to reduce the problem to the least-squares method (independently of the frame vector function) or to obtaining a function which is similar to the frame vector function in the sense of $\Omega(c)$ (regardless of the experimentally obtained $\{f_j\}$). In the first case (with $\alpha=0$) the solution is unstable, in the second case (with $\alpha=\infty$) it is meaningless. The solution, only for an optimal α, satisfies the least-squares test and at the same time it is similar to $\{c_{oi}\}$ (in the sense of $\Omega(c)$).

Several authors have shown [2,3,5] that the solution $\{c_i\}^\alpha$ exists and is unique for all $\{f_j\}$ and tends to be precise when the noise level decreases. Thus, the solution of (2) obtained by minimizing (3) is well-posed (in the Hadamard sense).

If we impose additional limitations on the class of solution, the stability of solution can be improved. The search for $\{c_i\}^\alpha$ in a non-negative area may be considered as an intrinsic extra limitation when interpreting the photon correlation experiments [6,7].

2. The "SIPP" Programme Package to Solve Ill-Posed Problems

The realization of the developed programme package is subdivided into two stages.

Stage 1. The grids at t and τ, the kernel $\{A_{ij}\}$ and the stabilizing functional $\Omega(c)$ are chosen. The experimental data are introduced from outside or simulated with or without noise. Then a regularization parameter α is determined or assigned.

Stage 2. The solution is done by using the chosen algorithm and the result is then visualized. Almost all the operations above are covered by various subroutines allowing for an easy adaptation of the programme package to solve problems by various modifications of the regularization method.

The most important part of the procedure, a quadratic programming algorithm for search of $\{c_i\}$ in a non-negative area, is outlined in the Appendix.

Determination of the optimal regularization parameter α is realized for the chosen class of experimental curves by solving simulated examples with a previously known distribution of the exponential curves in the experimental ones. Simulated examples are chosen proceeding from a priori information. In the case when such information is absent the regularization parameter α may be deduced from the variants of the solution with different α that are to be checked for stability. A minimum value of α which still can provide stability is regarded as an optimal value.

It should be stressed that we deliberately declined to automatize the selection of α, since errors in determining α in this manner are often indicated in computer experiments.

3. Computer Experiments

In computer experiments (2) was tested by the "SIPP" package programme. To obtain the vector functions $\{f_j\}$ by means of (2) we selected different vector functions $\{c_i\}$. They were used as experimental data (noise-imposed at a 0.3% level of noise or noise free). The weights $R_j \approx 1/f_j$ for (3) were chosen.

The solution was analyzed for four constructed distribution functions $\{c_i\}$. The first distribution function represented a "straight line" (Fig.1), the second distribution represented a "wide peak" (Fig. 2), the third - "two wide peaks" (Fig.3), and the fourth - "two narrow peaks" corresponding to two exponential curves (Fig.4).

The following problems were analyzed in conducting experiments with the first three distribution functions.

Choice of grid at t and τ. We used the grid at t with a uniform step having an initial value $t_1=0$, which corresponds to conditions of a real experiment. It was established that the grid at τ must be denser with the decrease of τ to provide for an effective reconstruction of the components of $\{c_i\}$ corresponding to small τ values. Although the use of the logarithmic grid at τ is more theoretically justified [8], we have chosen a quadratic grid:

$$\tau_i = \tau_1 + [\Delta_o + D\cdot(i-2)]\cdot(i-1), \qquad (4)$$

where i is the point number,

$\Delta_o>0$ and $D>0$ are parameters for controlling the grid step.

In contrast to a logarithmic grid, the quadratic grid being denser at small values of τ provides for a sufficient number of points at large values of τ.

Choice of the stabilizing functional and regularization parameter α. We tested the stabilizing functionals $\Omega(c)$ for the following [L] sets: [0]; [0,1]; [1]; [2]; [3]; [4]. We employed the following terms to denote the use of these functionals: the zero-order regularization, the mixed both the zero and first-order regularization, the first-order regularization, etc. In all the cases we choose $\{c_{oi}=0\}$, as we have not analyzed the procedure for choosing a frame vector function differing from zero. The findings of computer experiments are presented in the Table. Analysis of the Table leads to the conclusion that the first and higher regularization orders are better as a whole than the zero- and mixed regularization orders. The absence of an appropriate criterion for the quality of solution hampers a detailed comparison between the regularization of higher orders. We used a uniform metric ($|\Delta c_i|_{max} = \max_i|c_i - c_i^{\alpha}|$), traditionally employed as such a criterion, although this metric is not quite adequate to the problem. The fact that the optimal value of regularization parameter α has the tendency to grow with the increase in the regularization order also follows from the Table. As the increase in the regularization parameter causes disturbance of the original problem, it seems preferable to choose the first-order regularization, provided all other conditions are equal.

Table The Quality of Reconstruction $|\Delta c|_{max}$ of Distribution Function and the Optimal Regularization Parameter α_{opt} for Two Types of Distribution Function Constructed Using Various Regularization Orders

	"Straight line"						"Wide peak"						
regular. order	0	0,1	1	1*	2	3	0	0,1	1	1*	2	3	4
$\|\Delta c\|_{max}$	1.6	1.9	0.21	0.054	0.77	0.79	4.1	4.0	3.4	1.6	3.2	3.0	2.8
lg α_{opt}	-1.6	-0.6	1.7	3.6	1.7	1.3	-1.7	-1.8	-0.3	-0.5	0.3	0.5	0.3

* with unequal weights

The results of computer experiments have allowed us to establish an increasing instability of solution at small τ values, causing the appearance of a high-frequency component in the solution (peaks or shoulder, see Figs.1,2).

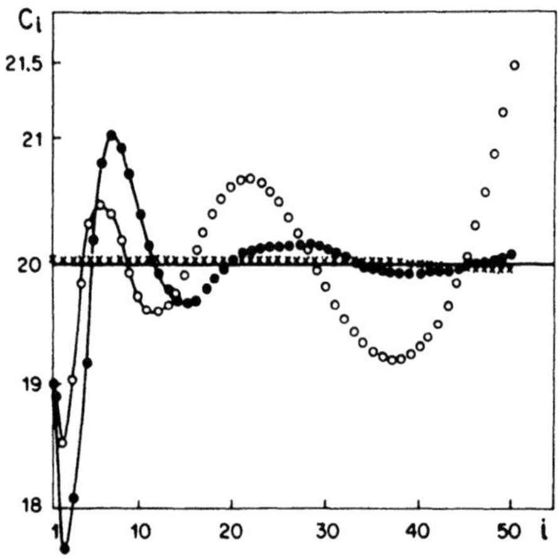

Fig.1. A constructed distribution function in the form of a "straight line" (solid curve). Reconstructed distribution functions: the zero-order regularization with equal(●●●) and unequal(ooo) weights, the first-order regularization with unequal weights (xxx)

To increase the stability of solution at small τ, we used unequal weights in the stabilizing functional for the first-order regularization, i.e.

$$\Omega(c) = \sum_{i=2}^{k} \rho_i (c_i - c_{i-1})^2, \qquad (5)$$

where $\rho_i = \rho_2 q^{i-2}$, $0<q<1$. This sufficiently improved the reconstruction of the "straight line" as well as of the "wide peak", which was intentionally shifted to the right (i.e. to the domain of large τ values, see Fig.2).

However, in the case of the "straight line" distribution, application of unequal weights caused an excessive increase of the regularization parameter α, which is already large enough for cases of equal weights (see Table). It would be reasonable to suppose that this is connected with the characteristics of the above distribution. When such large values of α are used

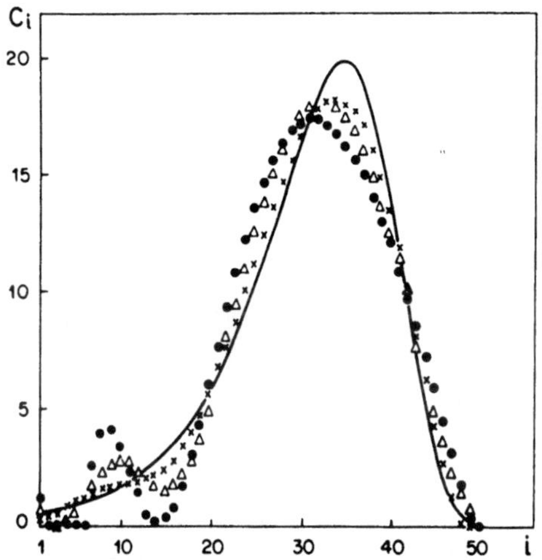

Fig. 2. A constructed distribution function in the form of a "wide peak intentionally shifted to the right" (solid curve). Reconstructed distribution functions: the zero-order regularization (●●●), the third-order regularization (ΔΔΔ) and the first-order regularization with unequal weights (xxx)

at reconstruction, we get a nearly "straight line" distribution which is independent of the original data. This can be accounted for by the fact that regularizations of higher orders are closely connected with suppression of high frequencies in the reconstructed distribution functions. This assumption can be supported by the fact that for distribution functions which differ from the "straight line" distribution function, the optimal value of α is by far smaller than for the "straight line" distribution function (see Table).

It should be noted that the first-order regularization provides good results for the third constructed distribution function in the form of "two wide peaks" (Fig. 3).

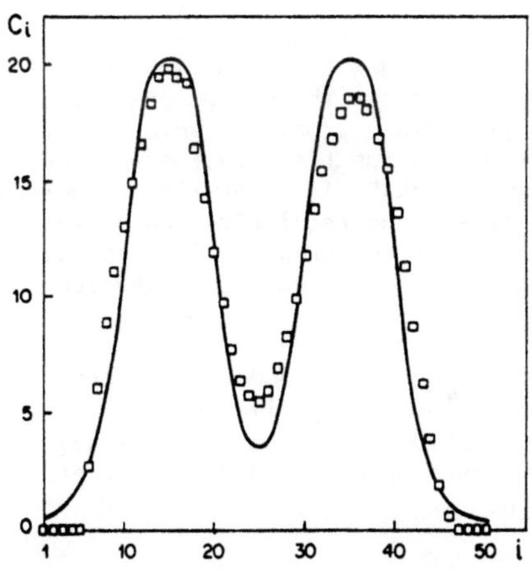

Fig. 3. A constructed distribution function in the form of "two wide peaks". A reconstructed distribution function: the first-order regularization (□□□)

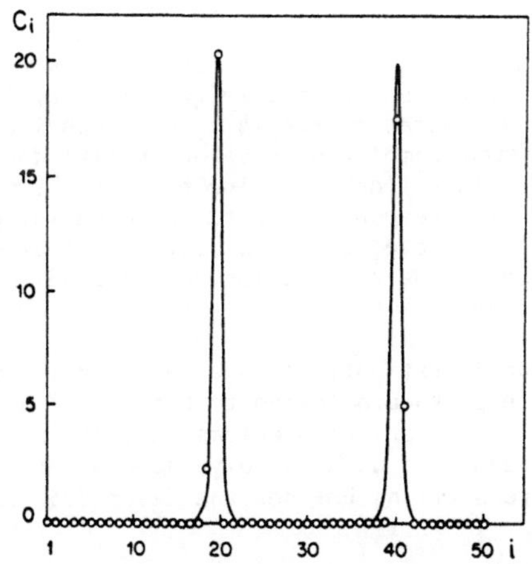

Fig. 4. A constructed distribution function in the form of "two narrow peaks" corresponding to the exponential curves. A reconstructed distribution function: the zero-order regularization (ooo)

In experiments with the fourth constructed distribution function (two narrow peaks corresponding to the two exponential curves, Fig.4) a uniform grid at τ was chosen for their identical analysis conditions. An optimal value of α for this distribution function proved to be negligible. In this case the regularization is achieved due to a considerable reduction in the dimension of the problem owing to the restriction on the sign of the unknown distribution function. Computer experiments have shown that at a 0.3% level of noise, the two exponential curves separate down to the ratio of 2:1 between the indices of the exponential curves.

The problem of separating the exponential curves into individual ones is very sensitive as regards the level of noise. Indeed, let us compare two close functions: $f_1=\exp(-t/\tau)$ and $f_2=0.5\{\exp[-t/(\tau+\varepsilon)]+\exp[-t/(\tau-\varepsilon)]\}$. The peak of their difference

$$f_2(t) - f_1(t) = \exp(-t/\tau)[\frac{1}{2}\left(\frac{t\varepsilon}{\tau^2}\right)^2 + \frac{1}{24}\left(\frac{t\varepsilon}{\tau^2}\right)^4 + \dots]$$

is in the domain $t=2\tau$ and approaches zero with $t\to 0$ and $t\to\infty$. Thus, the interval of τ values where the separation of close exponential curves takes place is rather small. There is a possibility to objectively distinguish between $f_1(t)$ and $f_2(t)$ only if the level of noise in this interval is considerably below the difference $(f_2(t)-f_1(t))$.

Computer experiments analogous to those described above were performed with an integral equation with a Lorentzian kernel corresponding to the method of frequency spectrum in dynamic light scattering and yielded similar results.

4. Conclusions

The computer experiments reported here have shown that the extraction of polydispersity data obtained from experiments with dynamic light scattering may be solved using the regularization method and the restriction on the sign of the unknown distribution function. In comparing various modifications of this method the best results were obtained with the first-order regularization and the weights in $\Omega(c)$ (5) increase with the decrease of τ on the quadratic grid (4). This technique provides a promising solution of simulated examples possessing initial distribution functions represented by a "straight line", a "wide peak" or "two wide peaks". If we obtain two narrow peaks (corresponding to two exponential curves), the regularization of solving is achieved due to a decrease of dimension of the problem. This decrease is due to the restriction on the sign imposed on the unknown distribution function.

Our investigations have been carried out using a developed "SIPP" programme package allowing to both realize the computer experiments and the solving of various problems on data interpretation. This programme package has been also used to solve some other ill-posed problems [DANOVICH et al., in preparation] besides the described above.

Appendix

To search the solution with a restriction on the sign of the unknown distribution function we used an algorithm of suboptimization on varieties with conjugated gradients for minimization. This algorithm provides for a successive solving of subproblems to minimize (3) in varieties when some components from $\{c_i\}$ are equal to 0.

The method of reseting of the subproblems allows for a monotonical decrease of the functional and results in a rapid obtainment of the unknown variety. The use of conjugated gradients for minimization permits to solve every subproblem using such number of steps that does not exceed the dimension of the subproblem. The average time needed for the work of the algorithm is within the range of 20 to 30 seconds (with 50 variables, computer ES-1040).

For a general idea of the algorithm and evidence for its convergence see [9,10].

References

1. Phillips D. L., A technique for the numerical solution of certain integral equations of the first kind. - J. Assoc. Comput. Mach., 1962, 9, No.1, pp.84-97.

2. Tikhonov A. N., On the solution of ill-posed problems and the method of regularization - Dokl. Akad. Nauk SSSR, 1963, 151, No.3, pp.501-504.

3. Tikhonov A. N., On the regularization of ill-posed problems. - Dokl. Akad. Nauk SSSR, 1963, 153, No.1, pp.45-52.

4. Twomey S., On the numerical solution of Fredholm integral equations of the first kind by the inversion of the linear system produced by quadratiere. - J. Assoc. Comput. Mach., 1963, 10, No.1.

5. Ivanov V. K., On the approximate operator equations of the first kind. - Zhurnal vychislitelnoy matematiki i matem. fiziki, 1966, 6, No.6, pp.1089-1094.

6. Gurari L. M., Rukman G. I., Tolpina S. P., On the solution of the Fredholm equation of the first kind with the Lorentz kernel in Dopler spectroscopy problems. - Metrologiya, 1977, No.7, pp.26-30.

7. Provencher S. W., Inverse problems in polymer characterization: direct analysis of polydispersity with photon correlation spectroscopy. - Makromol. Chem., 1979, 180, pp.201-209.

8. Ostrovsky N., Sornette D., Parker P., Pike E. R., Exponential sampling method for light scattering polydispersity analysis. - Optica Acta, 1981, 28, pp.1059-1070.

9. Lashenko I. N., Linear and non-linear programming. - Kiev, 1975.

10. Zangwill W. I., Nonlinear programming: a unified approach. - Prentice Hall, 1969.

32. A General-Purpose Constrained Regularization Method for Inverting Photon Correlation Data

Stephen W. Provencher

European Molecular Biology Laboratory (EMBL),
D-6900 Heidelberg, Fed. Rep. of Germany

Introduction

In photon correlation spectroscopy (PCS), there are relatively few cases where there is prior knowledge that the scattered field autocorrelation function contains only one component. Even in these favorable cases, every data analysis procedure should have the capability of accounting for extra unexpected components, e.g., due to instrumentation problems, aggregation, strongly scattering impurities, or even an important undiscovered effect. Often the main purpose of the experiment is to study polydispersity anyway. The first-order correlation function can often be approximated as a linear superposition of apparently independent contributions, yielding a Fredholm integral equation of the first kind,

$$y_j = \int_a^b s(\lambda)K(\lambda,t_j)d\lambda + \varepsilon_j, \qquad j=1,\ldots N_y, \tag{1}$$

where $K(\lambda,t_j)$ is known [e.g., $\exp(-\lambda t_j)$], and $s(\lambda)$ (e.g., a velocity or particle radius distribution) is to be estimated from the data y_j, containing unknown noise components ε_j.

The estimation of $s(\lambda)$ is an <u>ill-posed problem</u>. This is best illustrated by an example [1] using the version of the Riemann-Lebesque lemma that says that in the usual case $K(\lambda,t)$ is absolutely integrable over λ,

$$\lim_{\omega\to\infty} \int_a^b K(\lambda,t)\sin(\omega\lambda)d\lambda = 0. \tag{2}$$

This means that even for arbitrarily small $\varepsilon_j \neq 0$ and an arbitrarily large amplitude A, there still exists an ω such that $s(\lambda)+A\sin(\omega\lambda)$ still satisfies (1) to within the experimental errors, ε_j. Thus there exists a large (generally infinite) set, Ω, of possible solutions, all satisfying (1) to within experimental error. Even worse, the members of Ω can have arbitrarily large deviations from the true solution (as illustrated with the arbitrary A above); i.e., the errors are unbounded.

Three types of strategy for handling this apparently hopeless nonuniqueness are discussed in detail elsewhere [2]. <u>Absolute prior knowledge</u> eliminates members of Ω that violate basic truths, such as the nonnegativity of mass or number distribution functions. The overwhelming majority of members of Ω have negative lobes because of (2), and we shall see that imposing nonnegativity can enormously improve the resolution and reliability of the solution. Statistical prior knowledge of the mean and covariance of $s(\lambda)$ allows an optimal solution (i.e., minimum expected mean squared error for any unbiased estimator). However, such precise knowledge is seldom available in PCS, unless one is testing the hypothesis that an $s(\lambda)$ significantly deviates from an expected distribution, as in testing the reproducibility of a polymerization or the homogeneity of a batch product. Usually it is necessary to use the principle of parsimony. This says that, of all members of Ω that have not been eliminated by absolute prior knowledge, take the simplest solution, i.e., the one that reveals the least amount of new information or detail. In PCS, peaks and shoulders usually reveal significant information. Therefore the appropriate definition of "simplest" is generally "smoothest". In this way a bias against peaks and shoulders is imposed and any that do occur, despite this bias, are demanded by the data and therefore much less likely to be artifacts. Thus, while the most parsimonious solution may very well not contain all the detail of the true solution, the detail that it does have is more likely to be true.

Most methods attempt to somehow bias toward smoothness. For example, when $K(x,t)=k(xt)$, $a=0$, and $b=\infty$, (1) has a formal solution directly obtainable using the Fourier convolution theorem or, equivalently, the well known fact that the eigenfunctions of a convolution kernel over the real axis are just the complex Fourier components [3]. However, applications [4-6] of this to real data have serious problems because the formal solution assumes that the continuous function $y(t)$ is available from $t=0$ to $t=\infty$. This in turn implies that the resolution of the solution is completely uniform along the $\ln\lambda$ axis, that the Whittaker-Shannon sampling theorem is relevant, and that a simple low-pass Fourier filter can be applied to the solution. However, the data are discrete, not continuous, covering only a restricted range of t, usually with signal-dependent noise. Therefore the maximum attainable resolution will depend strongly on λ and on the data. Simple Fourier filters must inevitably sacrifice resolution in one part of the $\ln\lambda$ axis to avoid instabilities and wild oscillations in other parts. The problem of the finite range of t was eliminated by expanding the solution in eigenfunctions whose domain matches that of the data [7], but the nonstationarity of the noise and discreteness of the data were still unaccounted for. Most important, it is difficult to impose nonnegativity with these methods.

Constrained Regularization

Over the last six years we have applied a constrained regularization method to a wide range of ill-posed problems, including polydispersity in PCS [8,9]. A portable Fortran IV program implementing the method has recently become available, either from EMBL or the CPC library [2] . The method is described in detail elsewhere [2] and will only be outlined here.

For a given regularization parameter α, the constrained regularized solution is the $s(\lambda)$ that satisfies

$$\mathrm{SUMSQ}(\alpha) + \mathrm{REG}(\alpha) = \text{minimum}, \tag{3}$$

subject to

$$s(\lambda) \geq 0, \tag{4}$$

where

$$\mathrm{SUMSQ}(\alpha) \equiv \sum_{j=1}^{N_Y} [y_j - \int K(\lambda,t_j)s(\lambda)d\lambda]^2/\sigma_j^2, \tag{5}$$

$$\mathrm{REG}(\alpha) \equiv \alpha^2 \int [d^2s/dz^2]^2 dz, \tag{6}$$

σ_j is the estimated standard deviation in the noise in y_j. Because of the wide range of λ in PCS, we typically use as an abscissa $z=\ln\lambda$. The integrals in (5) and derivatives in (6) are usually well approximated by numerical integration and finite differences, but they can also be evaluated exactly by representing $s(\lambda)$ as an expansion in basis functions, e.g., cubic B splines. In either case the unique global solution to (3) and (4) is obtained by a numerically stable quadratic programming procedure [2].

With $\alpha=0$, only the SUMSQ term is present in (3), which would then be the classical weighted least squares condition. According to the Gauss-Markov theorem [10], of all linear unbiased estimates, this yields the solution with the minimum expected mean squared error. However, with such ill-conditioned systems, this error can still be extremely large, and HOERL and KENNARD [11] have shown that there always exists a biased estimator of the form of (3) with a smaller expected mean squared error. With (6), this bias is toward parsimony, since peaks and shoulders have second derivatives of large magnitude. The exact form of the smoothing regularizor in (6) is not critical, but in certain special cases (e.g., interpolation) the second-derivative regularizor has optimal properties [12].

It is very important that the resolution along the $\ln\lambda$ axis is not fixed beforehand; it is automatically determined by the discrete statistically weighted data in (3). In regions of the $\ln\lambda$ axis where there is a lot of information in the data, the SUMSQ term will dominate and there will be high resolution. In regions of low information, the REG term will dominate and spurious peaks tend to be suppressed because any reduction they would bring to SUMSQ is outweighted by the increase in REG in (3).

CONTIN delivers a series of solutions with α increasing from a very small (effectively zero) value to a very large value. Classical confidence regions can be constructed about the classical least squares solution (i.e., with $\alpha \approx 0$). Thus a probability level, PROB1(α), is assigned to each solution; [1-PROB1(α)] is approximately the fraction of times that the increase in SUMSQ(α) relative to SUMSQ(0) would be due to chance (random noise) alone and not to oversmoothing. Solutions in the range $0.01 \lesssim$ PROB1(α) $\lesssim 0.99$ can all be considered significant possibilities. The one with PROB1 nearest 0.5 is recommended as a good approximation to the most parsimonious solution consistent with the data and is always the one shown in this paper. Since only ratios of SUMSQ are used, the σ_j must only be specified to within a proportionality constant. For example, if it is known that the σ_j are constant, $\sigma_j = 1.0$ can be used. "Error bars", based on the covariance matrix of the solution, are also plotted. These analogs to confidence regions and standard error estimates are very useful in objectively assessing the uncertainties and stability of the entire analysis. Their derivation and interpretation is discussed in detail elsewhere [2].

The general quadratic programming problem in (3) and (4) is reduced to a ridge regression problem, whose statistical properties have been widely studied. This permits a fairly precise definition of the number of degrees of freedom in a solution [9,2], which is useful in assessing the information content of the data and in preventing overinterpretation of the solution.

CONTIN is very flexible. Nearly all aspects of the problem are specified in small simple subprograms that can be easily modified for other applications, e.g., velocimetry. For comparative studies, it is also easy to implement other methods, e.g., by suppressing the regularization and nonnegativity and imposing a coarse grid or histogram.

Results and Discussion

In this section we illustrate the strong dependence of the resolution on λ and the importance of nonnegativity. This is most clearly done with simulations where the simulated $s(\lambda)$ is a Dirac δ distribution at $\lambda=\lambda_o$, $\delta(\lambda-\lambda_o)$. In analogy with optics we call the resultant solution the point-spread function, PSF($\lambda;\lambda_o$), which represents the spreading of an infinitely sharp distribution due to the regularized analysis of the noisy data. It represents the resolution limit of the data, since we could never hope to detect polydispersity if the $s(\lambda)$ were as sharp as the PSF. For linear algorithms (i.e., without nonnegativity), the PSF has fundamental significance (not just for δ distributions) since any solution $\tilde{s}(\lambda)$ will be simply a linear combination of PSFs,

$$\tilde{s}(\lambda) = \int s(\lambda_o) \mathrm{PSF}(\lambda;\lambda_o) d\lambda_o . \tag{7}$$

The default version of CONTIN was used to simulate a noisy second-order correlation function,

$$g(t_j) = B[1+\gamma^2 y^2(t_j)] \tag{8}$$

with $K(\lambda,t_j)=\exp(-\lambda t_j)$, $t_j=1,\ldots,88$ channels, the variance in the noise given by

$$\mathrm{Var}[g(t_j)] = g(t_j),$$

$B=25\times10^6$, and $\gamma=0.7$. Many other types of simulations were performed and the qualitative conclusions were unchanged. Eq. (5) was solved for the $y(t_j)$ and the statistics of the noise in the $y(t_j)$ were accounted for as described elsewhere [9].

Figure 1 shows the variation of the resolution along the $\ln\lambda$ axis. Strictly speaking, the limits in (1) should be from $a=0$ to $b=\infty$. In Fig.1 this can be seen to be nearly the case, since the upper limit of $\lambda=10$ channels^{-1} corresponds to a lifetime of only 0.1 channel, which is hardly distinguishable from $\lambda\sim\infty$. Similarly, the lower limit corresponds to $\lambda t_{max}<0.01$, where t_{max} is the maximum t_j, and this is practically indistinguishable from $\lambda\sim0$. Thus practically the entire observable range of λ can be covered and the regularization and nonnegativity are very effective at keeping the solution zero in regions where it should be, even in the very unstable end regions where wild oscillations tend to occur with Fourier methods. This possibility of analyzing the data without subjectively eliminating any regions of λ a priori is very useful because of the importance of unexpected results mentioned in the introduction. Fourier methods assume a uniform resolution along the entire $\ln\lambda$ axis, and neither such a wide range of λ nor the resolution attained by the $\lambda_o=0.02$ PSF would be possible.

Although the $\lambda_o=0.001$ and 0.7 are broad, these solutions are not misleading. The number of degrees of freedom for both solutions is only two. This implies that only about two parameters can be determined by the data, roughly speaking the mean λ and the area under $s(\lambda)$, but not the width. In fact the normalized first moments (e.g., z-average diffusion coefficent) for these two broad peaks are only in error by 1% and 6%, respectively.

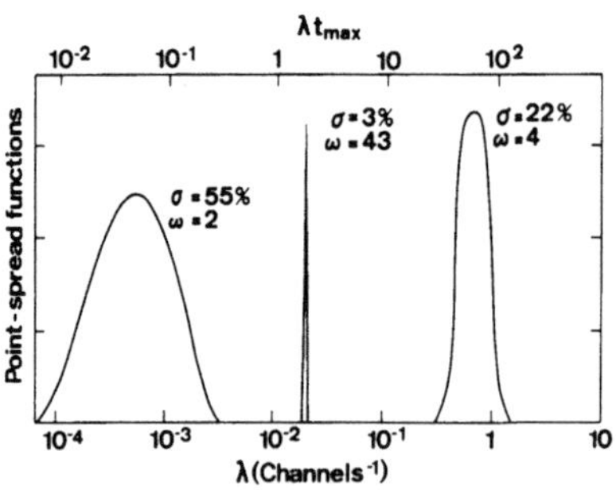

Fig.1. Results of three separate analyses with simulated δ distributions centered at $\lambda_o=0.001$, 0.02, and 0.7 channels^{-1}. The entire range of the $\ln\lambda$ axis shown was used for each analysis. The standard deviaton of each peak, as percent of the mean is given by σ. The approximate cutoff frequency needed to resolve each peak by Fourier methods is given by ω. With a Fourier method an $\omega=43$ would require a normalized first-order correlation function with noise levels less than about 10^{-29} [13]. Ordinate units are arbitrary; the $\lambda_o=0.7$ solution was scaled by a factor of 500 relative to the others for display

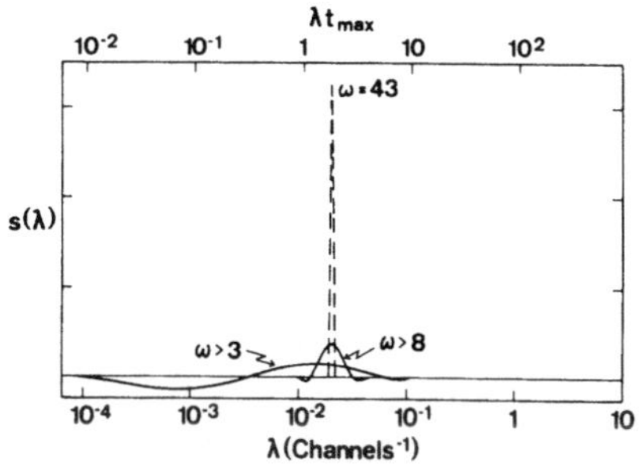

Fig.2. Analysis of a simulated δ distribution centered at 0.02. By only removing the nonnegativity constraint, the solution marked $\omega>3$ was obtained instead of the solution with broken lines (which is the same as in Fig.1). By drastically restricting the range to $0.01\leq\lambda\leq0.04$ the solution marked $\omega>8$ was obtained

Figure 2 illustrates the necessity of the nonnegativity constraint. When only the nonnegativity constraint is removed, the resolution can be seen to deteriorate drastically. The regularization nevertheless keeps the oscillations down to a reasonable magnitude, especially in view of the fact that, with a $\ln\lambda$ abscissa, $\lambda s(\lambda)$ is often plotted instead of $s(\lambda)$. Cutting one decade off each end of the range did not lead to significant improvement. Only by drastically reducing the λ range was the $\omega>8$ solution obtained.

Numerous other simulated and real data sets have been analyzed with CONTIN. In particular, simulation of the experiments of STELZER et al. [14] led to results similar to theirs. The optional Rayleigh-Debye form factors for spheres in CONTIN were used in both cases. The reduced resolution of the bimodal distributions compared with the unimodal case in this paper is to be expected. Excellent resolution of bimodal distributions can be achieved with DISCRETE [15], a program that assumes a priori that $s(\lambda)$ is a sum of δ distributions. However, obtaining a bimodal distribution that fits the data well with a method that assumes this from the start is not very convincing evidence for bimodality. Only when a solution with CONTIN is bimodal despite a parsimonious bias against extra peaks is there good evidence for bimodality. Of course the peaks are the broadest possible that will still fit the data. However, a good two-component analysis with DISCRETE and the sharp bimodal solutions with PROB1(α)<0.5 clearly indicate that such solutions are significant alternatives. It is important that CONTIN is automatic and objective in that one laboratory will get the same results with the same data as any other. The interpretation of the range of possible solutions as something analogous to a confidence region and of the most parsimonious solution as a safe extreme member of this region is clear. Typical subjective inputs concerning the range of λ, cutoff frequency ω, or the number and placement of bins in a histogram representation are unnecessary.

Acknowledgments

I thank Ernst Stelzer for helpful discussions, Roland Kaplan for assistance with the computing, and Ines Benner for typing the manuscript.

References

1. D.L. Phillips, J. Assoc. Comput. Mach. 9, 84 (1962)
2. S.W. Provencher, Comput. Phys. Commun. 27 (in press) (1982)
3. See, e.g., P.M. Morse, H. Feshbach, Methods of Theoretical Physics (McGraw-Hill, New York, 1953) pp. 960-967
4. S.W. Provencher, W. Gobush, Characterization of Macromolecular Structure, D.A. McIntyre, ed. (National Academy of Sciences, 1968) p. 143
5. S.W. Provencher, Biophys. J. 16, 27 (1976)
6. J.G. McWhirter, E.R. Pike, J. Phys. A 11, 1729 (1978)
7. S.W. Provencher, J. Chem. Phys. 64, 2772 (1976)
8. S.W. Provencher, J. Hendrix, L. De Maeyer, N. Paulussen, J. Chem. Phys. 69, 4273 (1978)
9. S.W. Provencher, Makromol. Chem. 180, 201 (1979)
10. See, e.g., S.D. Silvey, Statistical Inference (Halsted Press, New York, 1975), p. 51
11. A.E. Hoerl, R.W. Kennard, Technometrics 12, 55 (1970)
12. T.N.E. Greville, Mathematical Methods for Digital Computers, A. Ralston, H.S. Wilf, eds. (Wiley, New York, 1967) Chap. 8
13. N. Ostrowsky, D. Sornette, P. Parker, E.R. Pike, Optica Acta 28, 1059 (1981)
14. E. Stelzer, H. Ruf, E. Grell, in this volume
15. S.W. Provencher, R.H. Vogel, Math. Biosci. 50, 251 (1980)

33. Analysis and Resolution of Polydisperse Systems

E. Stelzer, H. Ruf, and E. Grell

Max-Planck-Institut für Biophysik, Heinrich-Hoffmann-Straße 7
D-6000 Frankfurt a.M. 71, Fed. Rep. of Germany

Abstract

A new method of evaluating photon correlation spectra allowing quantitative determination of the size distribution of spherical particles has been tested successfully with calibrated DOW latex particles. This contribution demonstrates the applicability of this general method for the characterization of phospholipid bilayer vesicles.

Introduction

Phospholipid vesicles are widely used for the study of membrane properties [1]. Vesicles may be regarded as hollow spheres separating the inner aqueous compartment from the bulk solution. Their radius can be as low as 10 nm [2] and as high as 1000 nm [3]. The phospholipid bilayer membrane has a thickness of approximately 3 nm. Many systems of biological interest, such as natural and artificial vesicles, exhibit polydisperse properties. Although photon correlation spectroscopy (PCS) [4] is an extremely powerful method of determining the average particle size in solution [5], the data reduction schemes available up until now [6] allow quantitative determination of polydispersity only with severe restrictions. In 1978 PROVENCHER [7] introduced a new strategy of data reduction into the field of PCS and realized it with CONTIN. This program applies a quasi-continuous Laplace inversion on the normalized autocorrelation function (ACF) using nonnegativity to preclude nonphysical results. An additional regularizer prevents uncorrelated oscillations of the resultant spectra. These features reduce the number of effective degrees of freedom in distribution analyses. CONTIN can therefore give an estimate of the distribution independently of the inversion range and the grid spacing. In the experiments described in the following, the program was tested with calibrated DOW latex particles (characterized by electron microscopy). Both monomodal (one-component) and bimodal (two-component) solutions were analyzed. In addition, the program was applied to the characterization of phospholipid vesicle preparations.

Materials and Methods

PCS measurements were performed on an ALV spectrometer (ALV Laser Vertriebsgesellschaft mbH, D-6070 Langen) equipped with a K 7023 correlator (Malvern Ltd., Malvern Worcestershire, UK) which has 88+4 real-time channels. A single measurement lasted between 70 and 400 seconds. The spectra were normalized using the monitor channels and analyzed using all 92 channels. The scattering angle was 90°.

The program CONTIN was obtained from S. W. PROVENCHER (EMBL, D-6900 Heidelberg). It was modified to fit on a Hewlett Packard System 1000 minicomputer and is a part of a complete on-line evaluation system.

The "Standard Polystyrene Latex Particles" were obtained from Serva (D-6900 Heidelberg) and are produced by DOW Chemical Co. The latex solutions were prepared by mixing the original material with doubly distilled water and filtering it through Millipore filters (220 nm and 450 nm) into round cuvettes.

Pure egg lecithin was prepared according to the method of SINGLETON et al. [8]. Vesicle preparation was carried out according to BRUNNER et al. [9]. The phosphorus content to the vesicle solution was determined by atomic emission spectroscopy [10] employing a Spectraspan IV (SMI Spectrametrics Inc., Andover, Ma., USA).

Results

A large number of measurements were performed with various latex particles differing both in size and size distribution. This made it possible to vary the mean radius (x) and the standard deviation (s) in monomodal solutions, and the ratio of the mean radii in bimodal solutions. Four analyses outline the general tendency and the limits of CONTIN, thus giving an idea of its capabilities. Two examples pertain to monomodal solutions and two to bimodal solutions.

In the first example (Fig. 1) CONTIN estimates a mean of 21.2 nm and a standard deviation of 3.9 nm. The specifications are 22.8 nm for the mean and 3.5 nm for the standard deviation. A second example (Fig. 2), which was measured with a much narrower peak, gives 89 nm and 3.9 nm, while 88 nm and 1.2 nm are specified.

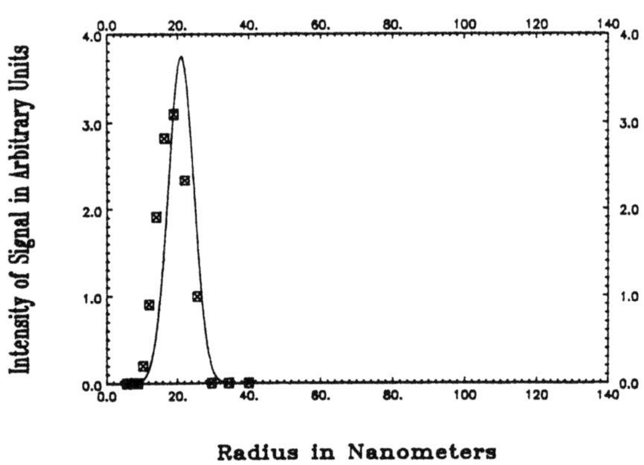

Fig. 1 Comparison of the distribution of 22.8 nm (0) DOW latex particles (solid line) with the distribution as analyzed by CONTIN (crossed boxes). The DOW specifications were recomputed massweighted for purposes of comparison with CONTIN's output. The concentration (c) was $2 \cdot 10^{-5}$ g/l; the temperature (T) 22.8 ^{o}C and the total duration of the measurement (t) 400 s. The grid spacing is logarithmic.

Figures 3 and 4 show bimodal distributions. Figure 3 shows the analysis of a mixture of 38.5 nm and 88 nm latex particles. The specifications for the 85 nm (0) latex particles could not be confirmed. Instead of 43.5 nm for the weighted average radius, 38.5 nm was found. The latter value was therefore used throughout the measurements. See also TROTTER and PINDER

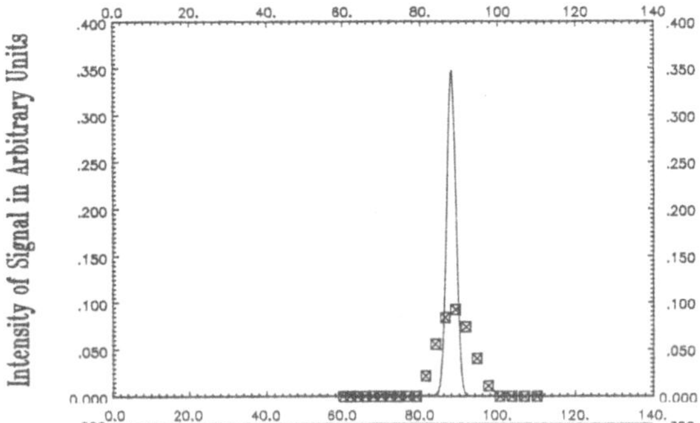

Fig. 2

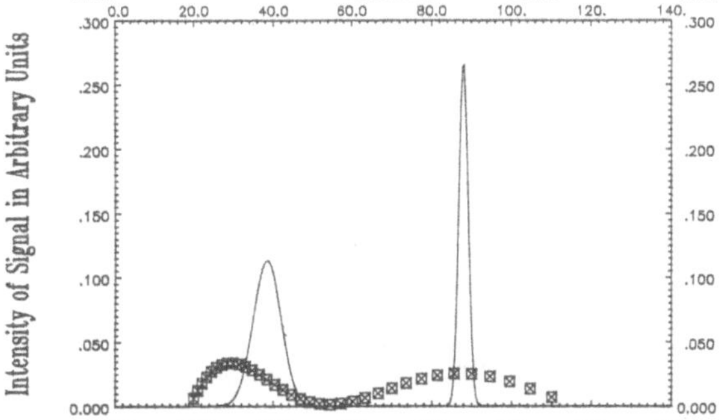

Fig. 3

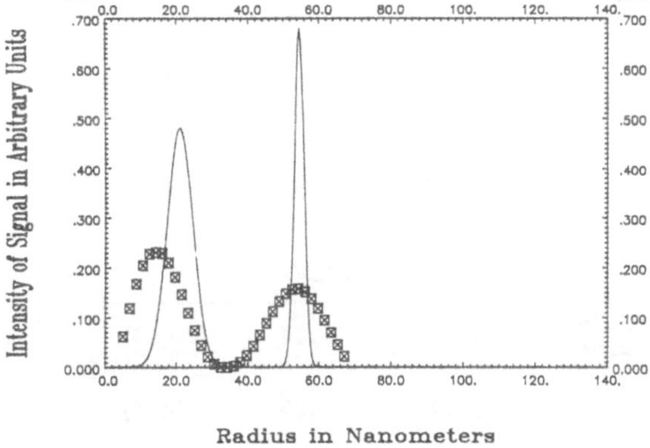

Fig. 4

Fig. 2 Comparison of the distribution of 89 nm (0) DOW latex particles with the distribution as analyzed by CONTIN ($c = 2 \cdot 10^{-5}$ g/l, $T = 23.32^{0}C$, $t = 400$ s).

Fig. 3 Comparison of the distribution of a mixture of 38.5 nm and 88 nm (0) DOW latex particles with the distribution obtained by analysis with CONTIN. The concentration is 10^{-5} g/l for both components. The inversion grid is logarithmic ($T = 23.3^{0}C$, $t = 350$ s).

Fig. 4 Comparison of the distribution of a mixture of 22.8 nm and 54.7 nm (0) DOW latex particles with the distribution obtained by analysis with CONTIN. The concentrations are 10^{-5} and $2 \cdot 10^{-5}$ g/l respectively. The inversion grid is linear ($T = 22.3^{0}C$, $t = 250$ s).

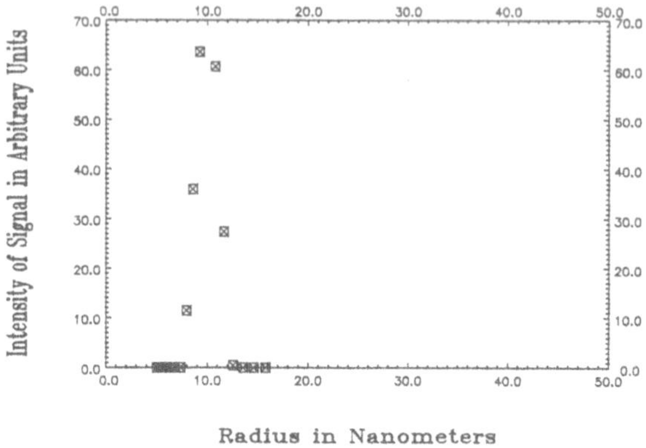

Fig. 5 Analysis of egg lecithin vesicles prepared according to BRUNNER et al. [9]. The data was analyzed with a logarithmic grid spacing ($c_{Phosphorus} = 5\cdot10^{-3}$ M, T = 23 °C, t = 200 s).

Table 1

A comparison of DOW specifications with results obtained by CONTIN. Columns 1 and 2 give the data, column 3 shows the appropriate qualifier, and columns 4 and 5 are error estimates. "DOW" refers to specifications as reported by DOW Chemical Co.

	x[nm]	s[nm]	s/x[%]	Δx/x[%]	Δs/s[%]	
DOW	22.8	3.55	15.57	7.50	8.97	(Fig.1)
CONTIN	21.22	3.9	18.4			
DOW	88	1.2	1.36	1.14	225	(Fig.2)
CONTIN	89	3.9	4.38			
	x_1[nm]	x_2[nm]	x_2/x_1[1]	$\Delta x_1/x_1$[%]	$\Delta x_2/x_2$[%]	
DOW	38.5	88	2.29	22.08	1.14	(Fig.3)
CONTIN	30	89	2.97			
DOW	22.8	54.7	2.4	19.43	0.55	(Fig.4)
CONTIN	17	55	3.24			

[11]. Analysis by CONTIN yields 30 nm and 89 nm. Another example (Fig. 4) shows an analysis of mixed components of 22.8 nm and 54.7 nm. The program estimates 55 nm and 17 nm.

Figure 5 shows the results of measurements on vesicles. The mean radius is 10.4 nm, the standard deviation 1.2 nm. Table 1 summarizes the results shown previously in figures 1 to 4 and discussed in the text.

Discussion

These results clearly indicate that CONTIN was very successful in the analysis of ACF. The mean radius was always in good agreement with the specifications. The ratio (standard deviation/mean radius) is an essential qualifier for the program's ability to fit the data. Whereas the first example (Fig. 1) fits easily (3.9/21.2 = 18 %), in the second the ratio is 1.4 % and CONTIN gives 4.4 %. According to extensive simulations [12] the lower limit for the ratio is 3 %. The latter may therefore not be compatible.

The characterization of bimodal solutions shows the real benefits of CONTIN. The qualifier for this type of analysis is the spacing of the two peaks and can be expressed by the ratio of the mean radii. As can be seen from the example, CONTIN is able to detect both components. The information on the widths of the peaks is lost. However, the mean radii given by the program are still correct within a reasonable margin of error. The general tendency is that the larger component receives the best fit. This is because the program applies second-order regularizations during analysis. The basic concept is to have as few minima (The least amount of information) as possible. This smooths the distribution and prevents wild oscillations, but also broadens the distributions. From all possible solutions CONTIN selects the "most parsimonious".

Such a vesicle suspension has been analyzed by various methods [9]. The resultant mean radius is reported to lie between 10.5 and 15 nm. The peak width was always found to be very small (5 %). BRUNNER et al. [9] chose 15 nm to be the best estimate for their vesicles. Their preference could not be confirmed by our experiments. Averaging over a large number of measurements, we obtained a mean radius of 10-11 nm and a peak width between 1 and 2 nm.

The method of evaluation illustrated above allows a detailed analysis of vesicular suspensions. This includes stability of size and size distribution, as well as their dependence on temperature and time.

Acknowledgement

We are indebted to Dr. S. W. PROVENCHER for providing us with CONTIN and for stimulating discussions. We wish to thank Mr. G. SCHIMNACK for his expert technical assistance.

References

1 F. Szoka, D. Papahadjopoulos:
Annu. Rev. Biophys. Bioeng. *9*, 467 (1980)
2 Chien-Hsien Huang:
Biochemistry *8*, 344 (1969)
3 A. Darszon, C. A. Vanderberg, M. H. Ellisman, M. J. Montal:
J. Cell Biol. *81*, 446 (1979)
4 H. Z. Cummins, E. R. Pike (eds.):
Photon Correlation and Light Beating spectroscopy
(Plenum, New York 1974)
5 P. N. Pusey, D. E. Koppel, D. E. Schaefer, R. D. Camerini-Otero, S. H. Koenig:
Biochemistry *13*, 952 (1974)
6 a: N. Ostrowsky, D. Sornette, P. Parker, E. R. Pike:
Opt. Acta *8*, 1059 (1981)
b: B. Chu, E. S. Gulari, E. R. Gulari:
Phys. Scr. *19*, 476 (1979)
7 a: S. W. Provencher, J. Hendrix, L. De Maeyer, N. Paulussen:
J. Chem. Phys. *69*, 4273 (1978)
b: S. W. Provencher:
Macromol. Chem. *180*, 201 (1979)
8 W. S. Singleton et al.:
J. Am. Oil Chem. Soc. *42*, 53 (1965)

9 J. Brunner, P. Skrabal, H. Hauser:
Biochim. Biophys. Acta *455*, 322 (1976)
10 E. Lewitzki, E. Grell:
In Preparation
11 C. M. Trotter, D. N. Pinder:
J. Chem. Phys. *75*, 118 (1981)
12 S. W. Provencher: this volume

34. The Dynamics of Interacting Brownian Particles [1]

P.N. Pusey

Royal Signals and Radar Establishment, St. Andrews Road,
Malvern, Worcestershire WR14 3 PS, England

1. Introduction

Suspensions of microscopic particles in a liquid occur widely in nature and are important in many industrial processes. In very dilute suspensions, the particles can be regarded as effectively non-interacting; thus a particle can take up any position in, and diffuse through, the suspension without interference from other particles. At higher concentrations, however, interactions (for example, short-ranged "hard-sphere" or longer-ranged Coulombic) between the particles become important and the positions and motions of different particles become correlated. Detailed theoretical understanding of the structure and dynamics of concentrated suspensions constitutes an interesting and challenging many-body problem to which modern statistical mechanical and hydrodynamical theories can be applied.

Over the past 10 years, photon correlation dynamic light scattering (DLS) has proved a powerful experimental technique for studying the natural Brownian motion of interacting particles. The aim of this article is to give a brief description of the principles and techniques of DLS and its application to the study of interacting Brownian particles. The bibliography is not exhaustive but is intended to provide a newcomer to the subject with a reasonable point of entry to a rapidly growing literature. The article draws heavily on recent reviews by PUSEY and TOUGH [1], HESS [2], both of which contain more complete bibliographies, and PUSEY [3,4]. Refs. [5-9] can also be consulted for a general introduction to dynamic light scattering.

In the next section we outline briefly the principles and techniques of dynamic light scattering. Section 3 describes the simple case of identical, non-interacting particles whereas in section 4 we consider, from a theoretical viewpoint, the properties of a suspension of interacting particles which can be obtained from DLS data. In section 5 the effects of particle polydispersity are discussed. Finally these theoretical considerations are illustrated briefly with experimental data in section 6.

2. Dynamic Light Scattering - Principles and Techniques (taken largely from [4])

Consider a large number N of spherically symmetrical Brownian particles dispersed in a liquid and illuminated by laser light of wavelength λ (in the

---

[1] This article is a modified version of a paper by the author and R J A TOUGH delivered at the IUTAM/IUPAC symposium on Interaction of Particles in Colloidal Dispersions held in Canberra, March 1981, which will be published in Advances in Colloid and Interface Science.

liquid). We assume that the intensity of light scattered by the liquid is negligible compared to that scattered quasielastically by the particles. The instantaneous value of the complex (field) amplitude of the light single-scattered into the far field at angle θ to the laser beam can then be written

$$E(K,t) \propto \sum_{i=1}^{N} a_i \exp\left[i\underline{K}\cdot\underline{r}_i(t)\right] \quad , \tag{1}$$

where a_i is the field amplitude of the light scattered by particle i, $\underline{r}_i(t)$ is the position of its centre at time t and K is the usual scattering vector ($K \equiv |\underline{K}| = (4\pi/\lambda) \sin \theta/2$). The factor $\underline{K}\cdot\underline{r}_i(t)$ represents the phase shift in the light scattered by particle i relative to that scattered by a particle at the (arbitrary) origin $\underline{r} = 0$. Because laser light is coherent, phase relationships between the light scattered by different particles are maintained in the scattering process so that E(K,t), considered as a function of K (or detection position), describes the instantaneous random diffraction or "speckle" pattern of the random arrangement of particles at time t. As the particle positions $\{\underline{r}_i(t)\}$ change due to Brownian motion, E(K,t) (and hence the detected light intensity) fluctuates randomly in time. The simplest function by which a randomly fluctuating quantity may be characterised is its first-order correlation function which, in normalised form, can be written

$$g^{(1)}(K,\tau) \equiv \langle E(K,0)E^*(K,\tau)\rangle / \langle |E(K,0)|^2\rangle \tag{2}$$

where the $\langle\ \rangle$ indicate ensemble or time averages. Substitution of (1) into (2) gives

$$g^{(1)}(K,\tau) = F(K,\tau)/S(K) \tag{3}$$

where the "dynamic structure factor"

$$F(K,\tau) = \left(N\overline{a^2}\right)^{-1} \sum_i \sum_j \left\langle a_i a_j \exp\left\{i\underline{K}\cdot\left[\underline{r}_i(0) - \underline{r}_j(\tau)\right]\right\}\right\rangle \tag{4}$$

and the "static structure factor"

$$S(K) = F(K,0) \quad ; \tag{5}$$

the $\overline{}$ indicates an average over the distribution of particle scattering amplitudes $\{a_i\}$. The functional dependence of $F(K,\tau)$ on K and τ contains information on the Brownian motions of the particles.

The most common experimental arrangement used nowadays to process the fluctuating scattered light intensity comprises a photomultiplier operating in the photon-counting mode followed by fast digital electronics (resolving time about 10^{-8} to 10^{-7} sec) to construct the temporal correlation function of the detected photons. Detectors based on the photoelectric effect are sensitive to the intensity $|E(K,t)|^2$ of the incident light field and DLS measures the intensity, rather than the field, correlation function. However, when, as in all the situations considered below, the number N of scatterers is large enough, E(K,t) is a complex Gaussian random variable, in which case there is a simple relation between the field and intensity correlation functions. DLS then provides an experiment estimate of $c^{\frac{1}{2}} g^{(1)}(K,\tau)$ for a number (usually 50-100) of discrete values of τ (see, for example, Fig.2); c is a constant of order one determined by the experimental conditions.

3. Identical, Non-interacting Particles

The simplest case commonly encountered is that of identical particles in dilute enough suspension that interparticle interactions can be neglected. Then, in (4), all the $\{a_i\}$ are equal and cross-terms $i \neq j$ vanish on average so that

$$F(K,\tau) = \langle \exp[-i\underline{K}\cdot\Delta\underline{r}(\tau)]\rangle \tag{6}$$

where the particle displacement

$$\Delta\underline{r}(\tau) \equiv \underline{r}(\tau) - \underline{r}(0) \quad . \tag{7}$$

$F(K,\tau)$ thus depends on the average motion of single particles. With certain assumptions [4] (6) becomes

$$F(K,\tau) = \exp\left[-(K^2/6)\langle\Delta r^2(\tau)\rangle\right] \quad , \tag{8}$$

where, for particles in free Brownian motion, the mean-square displacement $\langle\Delta r^2(\tau)\rangle$ is given by

$$\langle\Delta r^2(\tau)\rangle = 6D_o\tau \quad , \tag{9}$$

D_o being the "free-particle" translational diffusion coefficient. Thus

$$F(K,\tau) = \exp(-D_oK^2\tau) \quad . \tag{10}$$

For spherical particles

$$D_o = kT/f_o \quad , \quad f_o = 6\pi\eta R \quad , \tag{11}$$

where k is Boltzmann's constant, T the absolute temperature, f_o the "free-particle" friction coefficient, η the shear viscosity of the liquid and R the particle radius. By use of (10) and (11), DLS is frequently exploited as a method of sizing submicron spherical particles. The decay time τ_c of $F(K,\tau)$, roughly speaking the fluctuation time of the light intensity, is given by (see (10))

$$\tau_c = (D_oK^2)^{-1} \tag{12}$$

and is in the μsec to msec range for submicron particles and typical values of K.

4. Identical Interacting Particles - Theoretical Interpretation of DLS Data

4.1 Introduction

Interactions between the particles in a suspension affect both the structure, the average spatial arrangement of the particles, and their dynamics or motions. When these interactions are strong enough, the structure factor

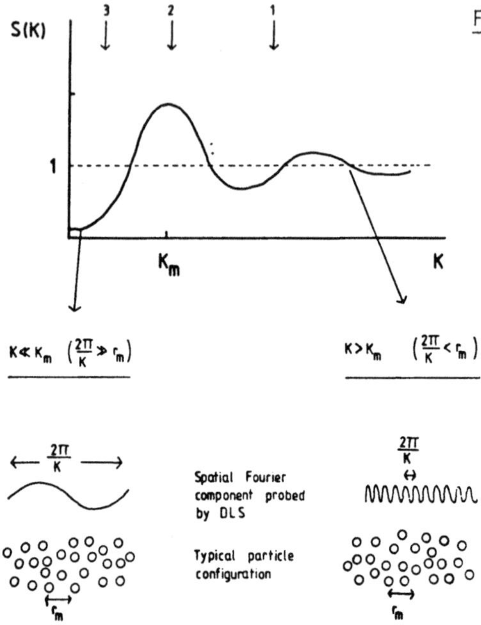

Fig.1. Top: Typical structure factor S(K) for suspension of strongly interacting particles. Bottom: Schematic representation of spatial Fourier components probed by DLS for $K \ll K_m$ and $K > K_m$

S(K) (see (5)), which essentially describes the angle- or K-dependence of the time-averaged scattered intensity (after correction for the scattering profile of a single particle) and characterises average structure, can have the form shown schematically in Fig.1. This occurs, for example, for suspensions of "hard spheres" at volume fractions $\phi \gtrsim 0.15$ and for suspensions of charged particles in a low electrolyte solution at $\phi \approx 10^{-3}$ or less. Very roughly the main peak in S(K) at $K = K_m$ can be interpreted as a "Bragg reflection" from planes of particles separated by the mean interparticle spacing r_m which can be taken as

$$r_m \approx \rho^{-1/3} \tag{13}$$

where ρ is the number of particles per unit volume (the particle number density). The "Bragg" condition gives

$$K_m r_m \approx 2\pi \tag{14}$$

and r_m provides a natural spatial scale in the suspension.

Several temporal scales characterise the particle dynamics. The two most important are τ_B, the fluctuation time of the particle velocities caused by the rapidly fluctuating solvent forces, and τ_I, the fluctuation time of the particle velocities caused by the direct interparticle interactions; the latter can be taken roughly as the lifetime of the nearest-neighbour shell of particles surrounding a given "test" particle, that is, as the "structural relaxation" time. An important property of colloidal suspensions in general, which stems from the large size difference between the particles and the liquid molecules, is the wide separation of the times τ_B and τ_I: for submicron particles τ_B is typically less than 10^{-7} sec whereas τ_I can be in the range 10^{-6} sec $\lesssim \tau_I \lesssim 10^{-2}$ sec. An important feature of DLS as a technique for studying colloidal dynamics is that it operates on a timescale τ almost always much greater than τ_B but, in many cases, spanning τ_I. The solvent forces can thus be regarded as inducing a "random walk" or diffusive motion in the particle system (Brownian motion) whereas the motions caused by the interparticle forces must often be treated more explicitly. There is, however, a second effect of the solvent: the motions of different particles can be coupled through the fluctuating hydrodynamic velocity fields in the solvent. These indirect "hydrodynamic interactions" are important at volume fractions $\phi \gtrsim 10^{-2}$; they are treated in more detail elsewhere in this volume and will only be mentioned briefly when relevant here.

As outlined in section 2 DLS measures the temporal autocorrelation function of the scattered field amplitude $E(K,t)$. Inspection of (1) shows that for identical particles, $E(K,t)$ is simply the spatial Fourier transform of the instantaneous particle number density. Speaking roughly, then, DLS observes the fluctuating growth and decay of sinusoidal spatial variations in number density of wavelength $2\pi/K$. Associated with this wavelength is the characteristic decay time τ_c of the dynamic structure factor $F(K,\tau)$; for non-interacting particles τ_c is given by (12). The properties of a particle suspension which are measured in a particular DLS experiment depend on the ratios of the characteristic lengths, $r_m/(2\pi/K) = K/K_m$ (see (14)), and times, τ_I/τ_c, of the suspension and technique respectively. In sections 4.2-4.5 we consider various important regimes of K/K_m and τ_I/τ_c, some of which are illustrated in Fig.1.

4.2 Collective Diffusion, $K \ll K_m$, $\tau_c \gg \tau_I$

At very small scattering vector $K \ll K_m$ (usually attained by use of small scattering angle θ), where $S(K) \approx S(0)$, the spatial wavelength $2\pi/K$ is much greater than the typical interparticle spacing r_m and $\tau_c \gg \tau_I$ ((12) at small K). In this limit, therefore, DLS observes slow fluctuations of macroscopic spatial extent and, not surprisingly, is expected to measure the same collective diffusion coefficient as conventional gradient diffusion measurements. The dynamic structure factor is then

$$F(K,\tau) = S(0) \exp(-D_c K^2 \tau) \tag{15}$$

where the collective diffusion coefficient D_c is given by

$$D_c = kT/S(0)f_c = (\partial\Pi/\partial\rho)/f_c \quad . \tag{16}$$

Here Π is the osmotic pressure of the suspension and $\rho^{-1}(\partial\rho/\partial\Pi)$ can be regarded as an "osmotic compressibility"; f_c is the collective friction coefficient which is expected to be the same as that describing sedimentation. In its second form (16) can be recognised as the usual expression for the gradient diffusion coefficient; in its first form it is sometimes called the "generalised Stokes-Einstein equation" [10]. It is possible to give a microscopic expression for D_c [2,11]

$$D_c = \lim_{K\to 0} \left[NK^2 S(0)\right]^{-1} \sum_i \sum_j \langle \underline{K}\cdot\underline{\underline{D}}_{ij}\cdot\underline{K}\rangle \tag{17}$$

where the diffusion tensor $\underline{\underline{D}}_{ij}$ is

$$\underline{\underline{D}}_{ij} = \int_0^\infty \langle \underline{v}_i(0)\underline{v}_j(t)\rangle \, dt \tag{18}$$

and $v_i(t)$ is the velocity of particle i at time t. In the general case the velocity cross-correlation functions in (18) are influenced by both direct and hydrodynamic interactions [12,13,14].

4.3 Self Diffusion, $K > K_m$, $\tau \approx \tau_I$

The opposite extreme from that considered in section 4.2 is $K \gg K_m$, $\tau_c \ll \tau_I$. Here, because $2\pi/K$ is much smaller than the typical interparticle spacing $\bar{r}_m$, DLS observes essentially the motions of individual particles over distances much smaller than r_m. However, in some cases, it is possible to choose a value of K somewhat greater than K_m where S(K) is roughly one (see Fig.1). Here cross terms, $i \neq j$, in (4) are close to zero on average so that, for identical particles, (4) becomes

$$F(K,\tau) \approx \langle \exp[-i\underline{K}\cdot\Delta\underline{r}(\tau)]\rangle \equiv F_s(K,\tau) \quad , \tag{19}$$

the self-dynamic structure factor. It is important to realise that in this intermediate regime, $K > K_m$, $S(K) \approx 1$, it is possible to make measurements at correlation delay times τ comparable to the structural fluctuation time τ_I. With certain assumptions (not always accurately fulfilled) (19) can be written

$$F_s(K,\tau) = \exp\left[-(K^2/6)\langle\Delta r^2(\tau)\rangle\right] \tag{20}$$

where, as before, $\langle\Delta r^2(\tau)\rangle$ is the mean-square displacement. Thus DLS provides some estimate of the average motion of a test particle in an interacting system; on the timescale $\tau \gtrsim \tau_I$, $\langle\Delta r^2(\tau)\rangle$ can be strongly affected by interparticle interactions (see section 6.3 for an example).

4.4 Cumulants, all K, $\tau \ll \tau_I$

An attractive feature of the DLS technique is that it is often possible to make quite accurate measurements (at all K excluding the true $K \to 0$ limit) of the initial (short-term) decay of $F(K,\tau)$ and obtain thereby the cumulants, the coefficients in an expansion of $\ln F(K,\tau)$ as a power series in τ. The timescale of this "initial" decay is still much greater than the Brownian velocity fluctuation time τ_B though much less than the structural fluctuation time τ_I; thus $\tau_B \ll \tau \ll \tau_I$ and the cumulants characterise the initial motion of the particle system (over distances much less than the interparticle spacing) away from a given spatial configuration, averaged over all such starting configurations. The cumulants are related to the direct derivatives $d^nF(K,\tau)/d\tau^n|_{\tau_B \ll \tau \ll \tau_I}$ which can, in principle, be calculated exactly (provided $\tau_I \gg \tau_B$). Here we quote the first derivative including the effect of hydrodynamic interactions [15]

$$\frac{dF}{d\tau}(K,\tau)\bigg|_{\tau_B \ll \tau \ll \tau_I} = -\frac{1}{N}\sum_{i,j} \langle \underline{K}\cdot\underline{\underline{D}}'_{ij}\cdot\underline{K} \exp\left\{i\underline{K}\cdot\left[\underline{r}_i(0) - \underline{r}_j(0)\right]\right\}\rangle \tag{21}$$

where the diffusion tensors $\underline{\underline{D}}'_{ij}$ are given by

$$\underline{\underline{D}}'_{ij} = \int_0^{\tau_B \ll \tau \ll \tau_I} \langle \underline{v}_i(0)\underline{v}_j(t)\rangle \, dt \tag{22}$$

and differ from the $\underline{\underline{D}}_{ij}$ of (18) in the upper limit of integration. There have been several measurements of the first cumulant in interacting systems (see section 6.2). The second derivative, neglecting hydrodynamic interactions, is [15]

$$\left.\frac{d^2F}{d\tau^2}(K,\tau)\right|_{\tau_B \ll \tau \ll \tau_I} = D_0^2K^4 + \frac{D_0}{Nf_0}\sum_{i,j} \langle \{\exp i\underline{K}\cdot[\underline{r}_i(0) - \underline{r}_j(0)\}$$

$$(\underline{K}\cdot\nabla_i)(\underline{K}\cdot\nabla_j)U[\{\underline{r}_i\}]\rangle \ . \tag{23}$$

An important property of the second cumulant is that it contains the interparticle potential $U[\{\underline{r}_i\}]$ explicitly so that accurate measurements can (in principle, though not yet in practice) be inverted to provide information on U [16].

4.5 $K \approx K_m$, $\tau \approx \tau_I$

A complicated regime, interesting from a theoretical viewpoint, is $K \approx K_m$, $\tau \approx \tau_I$, where DLS probes spatial scales comparable to the interparticle spacing and temporal scales comparable to the structural relaxation time. This regime has been studied in detail by the Konstanz group, both theoretically [2,17,18] and experimentally [19,20], but will not be considered further here.

5. Non-Identical Interacting Spheres - Polydispersity [3,21]

In sections 3 and 4 we assumed the particles to be identical so that the scattering amplitudes $\{a_i\}$ (section 2) were the same for all particles. If this is not the case, for example with spheres of different sizes, the measured dynamic structure factor is given by (4). In a suspension of interacting hard spheres, for example, the $\{a_i\}$ depend on the particle radii $\{R_i\}$, (for homogeneous spheres, $a_i \propto R_i^3$ for $R_i \ll 2\pi/K$), as do the instantaneous interparticle separations $\underline{r}_i - \underline{r}_j$; thus the latter are correlated with the $\{a_i\}$. Formally $F(K,\tau)$ can be written as a sum of many partial dynamic structure factors. However theoretical understanding of these is severely limited at present.

Nevertheless valuable insight into the behaviour of $F(K,\tau)$ in a polydisperse system can be obtained if it is assumed that the particles are identical in all respects except scattering power. Then it is possible to perform separately in (4) the average over particle positions and the average over the distribution of the $\{a_i\}$. By substituting the obvious identity

$$\overline{a_i a_j} = \bar{a}^2 + \delta_{ij}(\overline{a^2} - \bar{a}^2) \tag{24}$$

into (4) we get

$$F(K,\tau) = \frac{\bar{a}^2}{\overline{a^2}} F^I(K,\tau) + \left(1 - \frac{\bar{a}^2}{\overline{a^2}}\right) F_s{}^I(K,\tau) \quad , \tag{25}$$

where $F^I(K,\tau)$ and $F_s^I(K,\tau)$ are respectively the full and self-dynamic structure factors which would be found for truly identical particles. Thus in the presence of "scattering power polydispersity" the measured dynamic structure factor contains terms which depend both on collective motions (F^I) and single-particle motions (F_s^I). This is formally similar to the case of coherent and incoherent inelastic neutron scattering [3,22].

As mentioned above, the more common case where there is a distribution of size as well as of scattering power is more complicated. Nevertheless, there is growing evidence that provided the size distribution is fairly narrow, the measured dynamic structure factor can be written in a form similar to (25)

$$F(K,\tau) = (1 - x)\overline{F^I}(K,\tau) + \overline{xF_s^I}(K,\tau) \quad , \tag{26}$$

where x will now be a complicated function of the particle size and scattering power distributions and $\overline{F^I}$ and $\overline{F_s^I}$ are "average" dynamic structure factors; however the nature of this average is not well defined at present. By identifying the collective term as a stochastic "compression-expansion" mode and the self term as an "exchange" or concentration-fluctuation mode VRIJ [23] has calculated the amplitude x in terms of the particle size distribution for hard spheres in the Percus-Yevick approximation.

6. Experiment

6.1 Introduction

In this section the theoretical considerations of sections 4 and 5 are illustrated with experimental data obtained by the author and colleagues. Fig.2 shows log correlation functions, plotted against $K^2\tau$, for a dilute sample, $\phi \approx 10^{-3}$, of charged spheres of radius about 250 Å suspended in electrolyte-free water [24]. The particles interacted through a long-ranged Coulombic repulsion and showed a structure factor similar to that of Fig.1.

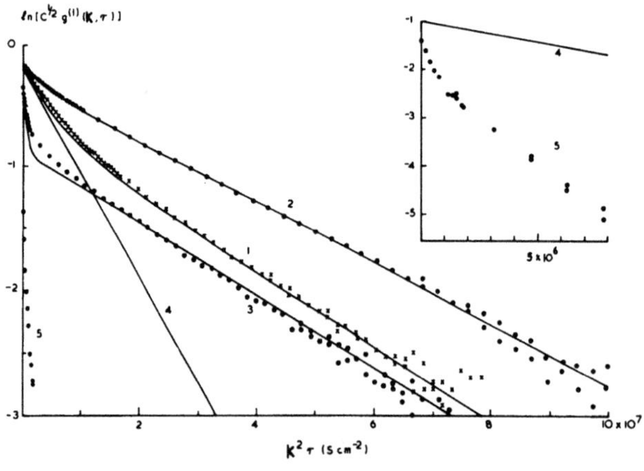

Fig.2. DLS correlation functions for charged particles (radius about 250 Å) suspended in electrolyte-free water at volume fraction about 10^{-3}. Curve 1, $K > K_m$; curve 2, $K \approx K_m$; curve 3, $K < K_m$ (see Fig.1); curve 4, "free particle" behaviour; curve 5, result of subtracting slow, self-diffusion mode from curve 3 (see section 6.4)

Correlation functions are shown for three values of K/K_m corresponding roughly to the arrows in Fig.1. If the particles were non-interacting (which could be approximated experimentally by adding electrolyte to shield the particle charges), (10) predicts that the data should all lie on the straight line (4) which has slope D_o. It is immediately apparent that interparticle interactions have a strong effect on the form of the correlation functions. Before detailed discussion of the data two further points should be made. Firstly, because of the low volume fraction $\phi \approx 10^{-3}$ in this system, hydrodynamic interactions have only a small effect which will be neglected. Secondly, the particles were significantly polydisperse with a standard deviation of radius about 20% of the mean. Nevertheless, with one exception, the data shown in Fig.2 can be analysed approximately as if the system were monodisperse. The exception concerns measurements at $K \ll K_m$ (curve 3 and section 6.4) where $F^I(K,0) \equiv S(K) \ll 1$ so that the amplitude of the first term in (26) is suppressed and the relative effect of the second term becomes appreciable. By contrast, for $K \gtrsim K_m$, the first term is much larger and, in any case, F^I and $F_s{}^I$ have more similar forms since $F^I \to F_s{}^I$ as K/K_m becomes large (section 4.3).

6.2 Charged Particles - the First Cumulant

If hydrodynamic interactions are neglected, $\underline{K}\cdot\underline{\underline{D}}'_{ij}\cdot\underline{K} = D_o K^2 \delta_{ij}$ [25], and (21) becomes simply

$$\left.\frac{dF}{d\tau}(K,\tau)\right|_{\tau_B \ll \tau \ll \tau_I} = -D_o K^2 \tag{27}$$

so that

$$\left.F(K,\tau)\right|_{\tau_B \ll \tau \ll \tau_I} = S(K)\left[1 - \frac{D_o K^2 \tau}{S(K)} + \ldots\right] \tag{28}$$

and one can define an effective diffusion coefficient

$$D_{eff} = D_o/S(K) \tag{29}$$

which describes the initial ($\tau_B \ll \tau \ll \tau_I$) decay of $F(K,\tau)$. Thus direct interactions only enter D_{eff} through their effect on $S(K)$. The data of Fig.2 are clearly in qualitative agreement with (29): for curve 1, where $K > K_m$, $S(K) \sim 1$, the initial slope D_{eff} of $\ln g^{(1)}(K,\tau)$ is close to D_o, the free particle result (curve 4); for curve 2, where $K \approx K_m$, $S(K) \approx 2$, D_{eff} is significantly smaller than D_o; curve 3, for $K \ll K_m$, $S(K) \ll 1$, decays initially much more rapidly than the free particle curve 4. More quantitative tests of (29) have been given by BROWN et al. [26], DALBERG et al. [27] and GRÜNER and LEHMANN [20]. Although the form of (29) is deceptively simple it should be mentioned that there has been some dispute concerning its physical content. In particular ACKERSON [28] has challenged the assertion [3] that $S(K)$ enters (29) simply through a "trivial normalisation".

6.3 Charged Particles - Self Diffusion

Curve 1 of Fig.2 shows a measurement made above the peak in $S(K)$ where $S(K) \approx 1$. The discussion of section 4.3 shows that these data represent approximately the average motion of a test particle and that $\ln g^{(1)}(K,\tau) \approx \ln F_s(K,\tau) = (-K^2/6)\,\langle\Delta r^2(\tau)\rangle$ (see (2), (3), (19) and (20)). At small τ, $\tau \ll \tau_I$, we see that the mean-square displacement $\langle\Delta r^2(\tau)\rangle$ increases at a rate similar to that of a freely diffusing particle (curve 4). However, at longer times, $\tau \approx \tau_I$, the rate of increase of $\langle \Delta r^2(\tau) \rangle$ becomes smaller and ultimately, for $\tau \gg \tau_I$, $\langle \Delta r^2(\tau) \rangle$ increases roughly linearly with τ but at a rate much less than for free diffusion. These observations can be given the following qualitative explanation [24]. Initially a particle feels a strong, rapidly-fluctuating force due to collisions with the thermally-agitated solvent molecules and moves off on an essentially free diffusive random walk. The direct interparticle forces are much weaker but act in a given direction for a much longer time (they fluctuate on the timescale τ_I of the structural relaxation, much more slowly than the solvent forces). Thus, after a while, whatever the direction of the initial displacement of the particle, it will tend to push up against a neighbour which will retard its progress. On a long enough timescale, however, the particle will "see" many different sets of neighbours and it will again perform a diffusive random walk which is nevertheless retarded on average by the direct interactions. Thus the final, long-time slope of curve 1 in Fig.1 can be identified with a macroscopic self-diffusion coefficient D_s which can be written

$$D_s = \frac{1}{3} \int_0^\infty \langle \underline{v}_i(0) \cdot \underline{v}_i(t) \rangle \, dt \quad ; \tag{30}$$

this coefficient is expected to be the same as would be measured by a macroscopic tracer diffusion technique.

These simple ideas have been put on a more quantitative basis (see, for example, [1,2,4,17,18,24,29]).

6.4 Charged Particles - Collective Diffusion

Curve 3 in Fig.2 shows measurements made at $K \approx 0.4\ K_m$. While this is not strictly in the limit $K \ll K_m$, the structure factor $S(K)$ was of order 0.15, not far from its limiting value $S(0)$, so that, following section 4.2, one would expect these data to represent, reasonably well, collective diffusion. An expression for the collective diffusion coefficient is given by (17). If hydrodynamic interactions are neglected and direct interactions taken to be pairwise additive, (17) simplifies considerably to give [1,2,29,30]

$$D_c = D_o/S(0) \quad . \tag{31}$$

Thus, under these conditions, D_c is predicted to be identical to the $K \to 0$ limit of the effective diffusion coefficient (29) describing the initial decay of the correlation function which should therefore be a rapidly decaying ($D_c \gg D_o$, since $S(K) \ll 1$) single exponential at all times $\tau \gg \tau_B$. Inspection of curve 3, Fig.1, shows that this is clearly not the case, the measured correlation function showing a long-time decay much slower than for free diffusion. This slowly-decaying low-K mode, whose origin was for some time obscure, was attributed by WEISSMAN [21] to polydispersity through the considerations outlined in section 5 (i.e. to the second term in (26)). If the slow mode is subtracted from the measured $g^{(1)}(K,\tau)$ of curve 3 the result is curve 5, assumed now to give an estimate of the full dynamic structure factor $F^I(K,\tau)$ in (26), which clearly decays rapidly at all times. The curvature still apparent in curve 5 is probably due to the fact that the measurement was not made in the true $K \ll K_m$ limit. This identification of the slow mode in curve 3 with the self-dynamic structure factor is supported by its similarity which slow decay of curve 1 which has already been attributed to self diffusion (section 6.3).

6.5 Microemulsions

We conclude this cursory survey of experiment with a brief discussion of data obtained recently from a water-in-oil microemulsion [31]. This consisted of small (< 100 Å) surfactant-stabilised water droplets dispersed in xylene. Dynamic light-scattering measurements were made at effective volume fractions of water ranging up to $\phi \sim 0.5$. Because of the small particle size the mean interparticle spacing r_m was also small and measurements at all scattering angles were in the limit $K \ll K_m$. It was found that the light-scattering correlation functions $g^{(1)}(K,\tau)$ were reasonably approximated by single exponentials at small ϕ but became distinctly non-exponential for $\phi > 0.1$ and could be fitted by a sum of two exponentials for $\phi > 0.2$. Again we attribute this to polydispersity so that (26) applies. In the small K limit we expect

$$F^I(K,\tau) = S^I(0) \exp(-D_c K^2 \tau) \tag{32}$$

and

$$F_s^I(K,\tau) = \exp(-D_s K^2 \tau) \tag{33}$$

so that analysis of the two-exponential fits should give the collective and self-diffusion coefficients D_c and D_s. These are plotted (relative to D_0 to account for particle size changes on varying ϕ) in Fig.3. It is seen that D_c/D_0 is fairly independent of concentration up to $\phi \approx 0.5$ whereas D_s/D_0 decreases rapidly. Although no theories exist for D_c and D_s at such high volume fractions these results can be discussed qualitatively.

Inspection of (16) shows that collective diffusion can be regarded as being driven by osmotic pressure and retarded by interparticle friction. It is known [12,13] that, for hard spheres, when hydrodynamic interactions are taken into account these two factors have opposite and roughly equal effects to first order in ϕ. The data of Fig.3 imply that a considerable degree of cancellation continues to high volume fraction so that D_c shows little dependence on concentration even though $\partial\Pi/\partial\rho$ and f_c, taken separately, are strong functions of concentration. By contrast, for self diffusion, the osmotic term is absent (see (30)) and interparticle interactions can severely reduce the particle mobility at high concentration.

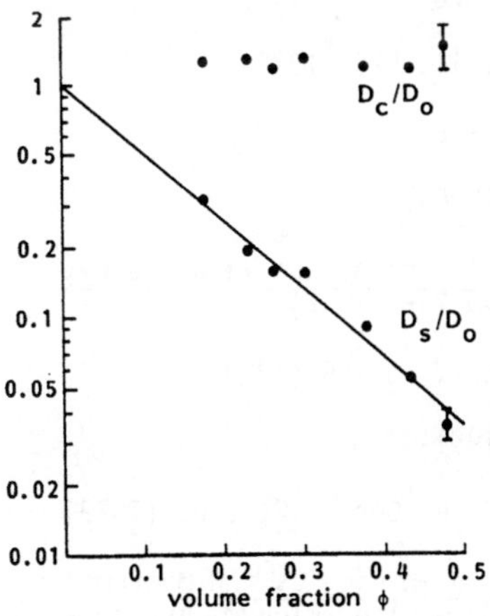

Fig.3. Concentration dependence of collective D_c and self D_s diffusion coefficients for water-in-oil microemulsions (droplet radius $\lesssim$ 100 Å)

Acknowledgement

I am grateful to my colleague Dr. R.J.A.TOUGH for many valuable discussions on the subject matter of this article.

References

1. P.N. Pusey, R.J.A. Tough: in Dynamic Light Scattering and Velocimetry: Applications of Photon Correlation Spectroscopy, ed. R. Pecora (Plenum, New York, in press).

2. W. Hess: in Light Scattering in Liquids and Macromolecular Solutions, eds. V. Degiorgio, M. Corti, M. Giglio (Plenum, New York 1980).

3. P.N. Pusey: in Light Scattering in Liquids and Macromolecular Solutions, eds. V. Degiorgio, M. Corti, M. Giglio (Plenum, New York 1980).

4. P.N. Pusey: Phil. Trans. Roy. Soc. Lond. A, 293, 429 (1979).

5. P.N. Pusey, J.M. Vaughan: in Dielectric and Related Molecular Processes, Volume 2, ed. M. Davies (The Chemical Society, London 1977).

6. H.Z. Cummins, E.R. Pike (eds.): Photon Correlation and Light Beating Spectroscopy (Plenum, New York 1974); Photon Correlation Spectroscopy and Velocimetry (Plenum, New York 1977).

7. B.J. Berne, R. Pecora: Dynamic Light Scattering (Wiley, New York 1976).

8. B. Chu: Laser Light Scattering (Academic, New York 1974).

9. J.M. Shurr: C.R.C Critical Reviews in Biochemistry, 371 (November 1977).

10. G.D.J. Phillies: J. Chem. Phys. 60, 976 (1974).

11. A.R. Altenberger: Acta Physica Polonica, A46, 661 (1974).

12. G.K. Batchelor: J. Fluid Mech., 74, 1 (1976).

13. B.U. Felderhof: J. Phys. A: Math. Gen., 11, 929 (1978).

14. A.R. Altenberger: Chem. Phys., 15, 269 (1976).

15. B.J. Ackerson: J. Chem. Phys., 64, 242 (1976).

16. B.J. Ackerson: in Photon Correlation Spectroscopy and Velocimetry, eds. H.Z. Cummins, E.R. Pike (Plenum, New York 1977).

17. W. Hess, R. Klein: J. Phys. A: Math. Gen., 13, L5 (1980).

18. W. Hess, R. Klein: Physica, 105A, 552 (1981).

19. F. Gruner and W. Lehmann: J. Phys. A: Math. Gen., 12, L303 (1979).

20. F. Gruner, W. Lehmann: in Light Scattering in Liquids and Macromolecular Solutions, eds. V. Degiorgio, M. Corti, M. Giglio (Plenum, New York 1980).

21. M.B. Weissman, J. Chem. Phys., 72, 231 (1980).

22. S.W. Lovesey, T. Springer (eds): Dynamics of Solids and Liquids by Neutron Scattering, (Springer, Berlin 1977)

23. P.N. Pusey, H.M. Fijnaut, A. Vrij: J. Chem. Phys., (in press).

24. P.N. Pusey: J. Phys. A: Math. Gen., 11, 119 (1978).

25. P.N. Pusey: J. Phys. A: Math. Gen., 8, 1433 (1975).

26. J.C. Brown, P.N. Pusey, J.W. Goodwin, R.H. Ottewill: J. Phys. A: Math. Gen., 8, 664 (1975).

27. P.S. Dalberg, A. Bøe, K.A. Strand, T. Sikkeland: J. Chem. Phys., 69, 5473 (1978).

28. B.J. Ackerson, P.N. Pusey, R.J.A. Tough: J. Chem. Phys. 76, 1279 (1982).

29. B.J. Ackerson: J. Chem. Phys., 69, 684 (1978).

30. P.N. Pusey, R.J.A. Tough: J. Phys. A: Math. Gen., 15, 1291 (1982).

31. D.J. Cebula, R.H. Ottewill, J. Ralston, P.N. Pusey: J. Chem. Soc. Faraday Trans. I, 77, 2585 (1981).

35. Dynamics of Na^+-Polystyrene Sulfonate in Solution at Low Ionic Strength

F. Grüner, W.P. Lehmann, H. Fahlbusch, and R. Weber

Fakultät für Physik, Universität Konstanz, Bücklestraße 13
D-7750 Konstanz, Fed. Rep. of Germany

Abstract

The dynamics of Na^+-Polystyrene-Sulfonate (Na^+-PSS) of molecular weight 10^5 in water of low ionic strength was investigated by photon correlation spectroscopy. The cooperative diffusion coefficient obtained from the first cumulant was measured in a concentration range of five decades. Concentrations range from the dilute ($c < c^*$) to the semidilute ($c^* < c < c^{**}$) to the concentrated regime. The overlap concentration is $c^* \approx 10^{-4}$ g/cm^3 which corresponds to 3×10^{-7} monomers/Å^3. It is in good agreement with the overlap concentration calculated for fully extended rodlike molecules. In the dilute regime, the diffusion coefficient is smaller than the one of a gaussian coil. This may be explained by the assumption of a rodlike shape of the molecule. In the semidilute regime the cooperative diffusion coefficient follows a power law $D_{eff} \approx c^{\,0.52 \pm 0.02}$ in good agreement with theoretical predictions for polyelectrolytes.

When the screening length gets equal to the electrostatic persistence length

$$L_e = \frac{1}{16\pi\, Q\cdot a\cdot c}$$

(Monomerlength $a \approx 2.5$ Å, Bjerrumlength $Q = 7$ Å) at the concentration $c^{**} \approx 4.8\cdot 10^{-3}/(Q^2\cdot a)$ ($c^{**} \approx 10^{-2}$ g/cm^3), D_{eff} stays approximately constant at concentrations $c > c^{**}$. This may be understood by a scaling behavior according to

$$D_{eff} \sim \xi^{-1} \sim L_e^{-1} \left(\frac{c^{**}}{c}\right)^{\mu} .$$

Assuming that ξ does not depend on the momener length a for $L_e < a$ one ends up with the result

$$D_{eff} \sim \xi^{-1} \sim Q^{-1}$$

which is independent of the concentration of the polyelectrolyte and furthermore, independent of the structure of the polyelectrolyte, provided that $a < Q$.

The larger part of this work has been published in

[1] F. Grüner, W. P. Lehmann, H. Fahlbusch and R. Weber, Dynamics of Na^+-polystyrene sulfonate in solution at low ionic strength, J. Phys. A: Math. Gen. 14 (1981) L 307-L 313 (Letter).

36. A Photon Correlation Study of the Dynamics of Strongly Interacting Colloidal Systems: Colloidal Liquids and Colloidal Crystals

F. Grüner and W.P. Lehmann

Fakultät für Physik, Universität Konstanz, Bücklestraße 13
D-7750 Konstanz, Fed. Rep. of Germany

Abstract

Due to the strong electrostatic repulsion of Polystyrene latexes in water, these systems show short-range liquid-like order at low concentrations and long range, crystalline order at higher concentrations. Both phases were investigated by photon correlation spectroscopy. The first dynamic light scattering experiments on single crystals are reported. The dynamics of the colloidal liquids may be described by two diffusion coefficients for short and long times.

This behaviour may be explained by a memory function formalism in a viscoelastic approximation. It is shown that the same formalism describes, in a purely-elastic approximation, the colloidal crystals as well. The theory allows to extract information from the data on the elastic moduli and the sound velocities and, in the crystalline systems, a "phonon" dispersion, which yield, in principle, information on the particle interaction potential. The results on the crystals, however, may not be explained with a simple screened coulomb potential.

For a more complete account of this work, see

[1] F. Grüner and W. P. Lehmann,
Dynamics of colloidal crystals and colloidal liquids,
J. Phys. A: Math. Gen. 15 (1982) 2847-2863.

37. Dynamic Properties of Silica Dispersions in Cyclohexane

M.M. Kops-Werkhoven, A. Vrij*, and H.M. Fijnaut

Van't Hoff Laboratory, State University Utrecht, Padualaan 8, 3584 CH Utrecht, The Netherlands

The dynamic behaviour of hard spherical silica particles of 70 nm radius in cyclohexane has been investigated with Photon Correlation Spectroscopy. In the diluted region, the obtained field autocorrelation functions could be described by single exponentials, but at particle concentrations above 0.30 g/cm^3 deviations of the single exponential behaviour of the functions became pronounced. However, in this concentration range the experimental functions seem to be reasonably well described by a sum of two exponential functions. Recently WEISSMAN and PUSEY suggested that the occurrence of two different relaxation modes in concentrated systems might be due to optical polydispersity of the dispersed particles. One relaxation mode is attributed to the collective diffusion of the particles and is called the collective part, while the other relaxation mode is identified with the self-diffusion of the particles and is called the incoherent part. We were able to test the influence of the optical polydispersity experimentally with our system by performing optical contrast variation experiments in which we altered the difference between the refractive index of the particles and that of the solvent by varying the temperature. The results of these measurements appeared to be consistent with the theories of WEISSMAN and PUSEY. The concentration dependence of the two relaxation modes was studied. In addition the self-diffusion coefficient was determined independently by a new method involving the study of the light scattering of specially prepared tracer particles.

1. Introduction

The study of dynamic properties of concentrated colloidal solutions is of great interest in that it helps us to unravel the influence exerted by direct and indirect interparticle forces on the dynamic behaviour of these solutions.

Information about the dynamic behaviour of colloidal solutions can be obtained with Photon Correlation Spectroscopy (PCS) by measuring intensity autocorrelation functions. Concentrated colloidal solutions, however, will usually show the undesirable effects of multiple scattering. It is therefore of interest that VAN HELDEN et al. [1] have developed well-defined colloidal dispersions in which the optical contrast - that is the difference between the refractive index of (silica) particles and solvent (cyclohexane) - is so small that the scattering intensity even at volume fractions up to 50% remains low enough for multiple scattering effects to be neglected. In a previous paper [2] we reported on experiments done on these silica disper-

---

*to whom correspondence should be addressed.

sions at lower concentrations. The direct interaction could be described by a hard-sphere repulsion and the autocorrelation function could be described by a single exponential.

In this paper we shall summarize some experiments done on concentrated systems. Furthermore we shall mention some recent results obtained with a new experimental technique involving the use of tracer particles.

Concentrated Systems (3,4)

In the concentrated colloidal solutions the correlation functions sometimes showed (large) deviations from single-exponent behaviour. We agree with the suggestion of WEISSMAN [5] and PUSEY [6] that optical polydispersity is the origin of these deviations. Our colloidal system is a particularly good one on which to base a discussion of this effect. In cases of small optical contrast, the statistical deviation (in the optical contrast for individual particles) from the average value may become relatively large. Following this route a collective and self-diffusion coefficient can be obtained.

Tracer Particles (3)

Self-diffusion can be measured when a few tracer particles of large scattering power are mixed with an excess of particles that do not scatter at all. In this way we can compare the self-diffusion results obtained with the two methods.

2. Light-Scattering Equations

In photon correlation spectroscopy (PCS) experiments the autocorrelation function $\phi(K,\tau)$ of the scattered intensity I is measured. The scattering process follows Gaussian statistics and the normalized autocorrelation function of the scattered electrical field $g^{(1)}(K,\tau)$ is obtained using the relation [7]

$$\phi(K,\tau) = \frac{\langle I(K,0)I(K,\tau)\rangle}{\langle I(K,0)\rangle^2} = 1 + G[g^{(1)}(K,\tau)]^2 \quad , \qquad (2.1)$$

where τ is the time and K the absolute value of the scattering wave vector. $K = |\underset{\sim}{K}| = (4\pi/\lambda)\sin(\theta/2)$; here θ is the scattering angle and λ the wavelength of the light in the scattering medium. G in (2.1) is an experimental constant of order one.

The autocorrelation function of the field can be formulated as [7,8],

$$g^{(1)}(K,\tau) = \frac{S^m(K,\tau)}{S^m(K,0)} \quad ; \qquad (2.2)$$

here the symbol $S^m(K,\tau)$ stands for "measured" dynamic structure function. For a dispersion of N spherical particles (radius a_i),

$$S^m(K,\tau) = \frac{1}{Nb^2} \langle \sum_{i=1}^{N} \sum_{j=1}^{N} b_i b_j \exp[i\underset{\sim}{K}\cdot(\underset{\sim}{r}_i(0)-\underset{\sim}{r}_j(\tau))]\rangle \quad . \qquad (2.3)$$

Here b_i is the field amplitude of the light scattered by particle i, which is proportional to the particle volume $V_i \sim a_i^3$ and to the difference in the refractive index of particle n_{p_i} and solvent, n^0.

For $Ka \ll 1$, b_i is independent of K. Furthermore $r_j(\tau)$ is the position of the centre of particle j at time τ, and $\overline{b^2} = \Sigma N_i b_i^2/N$.

The pointed brackets in (2.3) indicate ensemble averaging. If all particles are identical, (2.3) becomes

$$S^I(K,\tau) = \frac{1}{N} < \sum_{i=1}^{N} \sum_{j=1}^{N} \exp[i\underline{K}.(\underline{r}_i(0)-\underline{r}_j(\tau))] > \tag{2.4}$$

For $Ka \ll 1$ this describes large-scale (collective) number density fluctuations. Then one can write [7]

$$S^I(K,\tau) = S^I(K)\exp(-D_c K^2 \tau), \tag{2.5}$$

where

$$S^I(K) \equiv S^I(K,\tau=0) \tag{2.6}$$

is the well-known static structure factor known from liquid state theory. For "hard-spheres", e.g. $S^I(K=0)$ can be approximated [9] by the CARNAHAN-STARLING [10] equation,

$$S^I(0) = \frac{(1-\phi)^4}{(1+2\phi)^2 + (\phi-4)\phi^3} \tag{2.7}$$

where ϕ is the volume fraction $\phi = (4/3)\pi a^3 N/V$. Further D_c is called the collective diffusion coefficient, which in principle is a function of particle concentration and of K.

For $K\to0$, D_c is related to $S^I(0)$ with a generalized Einstein relation [2]

$$D_c(K\to0) = \frac{k_B T}{S^I(0) f_c} \tag{2.8}$$

The collective friction constant f_c is the friction that the particles experience when they move as a "collection"in the same direction. This is the same friction as that which occurs in an ultracentrifuge and which is obtained from the (laboratory fixed) sedimentation velocity.

At infinite dilution,

$$D_o = \frac{k_B T}{6\pi\eta a} \tag{2.9}$$

where η is the solvent viscosity.

For the diluted region and for hard spheres [11]

$$\frac{D_c(K\to0)}{D_o} = 1 + 1.45\phi \tag{2.10}$$

where $\phi = (4/3)\pi a^3 N/V$, i.e. the volume fraction of hard spheres. At $K\neq0$ one may write [12-14],

$$\frac{D_c(K)}{D_o} = \frac{H(K)}{S(K)} \quad , \tag{2.11}$$

where H(K) is a function describing the hydrodynamic interactions.

Polydispersity in Scattering Power

Suppose an (artificial) case where a system contains particles of identical size 50% of which have an amplitude equal to b and 50% an amplitude equal to -b. This could be achieved when particles of the first sort have a refractive index difference between particle and solvent that is exactly equal in magnitude but is opposite in sign to that of particles of the second sort.

Because the b_i and r_i are uncoupled, the averaging over b_i can be done separately. In our example all terms containing $b_i b_j$ with $j \neq i$ average to zero, whereas those containing squares $b_i^2 = b^2$ remain. Then (2.3) becomes

$$S_s^m(K,\tau) = \langle \exp i\underline{K}\cdot(\underline{r}(0)-\underline{r}(\tau))\rangle \quad . \tag{2.12}$$

This result is entirely different from that of (2.4) and describes the motion of a single particle. For small K this process is known as self-diffusion and

$$S_s^m(K,\tau) = \exp(-D_s K^2 \tau) \tag{2.13}$$

where D_s is the self-diffusion coefficient.

In the more general case where particles are identical in size but have a distribution of the scattering amplitudes b_i, it can be shown that

$$S^m(K,\tau) = \frac{\bar{b}^2}{\overline{b^2}} S^I(K) \exp(-D_c K^2 \tau) + \frac{\overline{b^2}-\bar{b}^2}{\overline{b^2}} \exp(-D_s K^2 \tau) \ , \tag{2.14}$$

With $\bar{b} = \Sigma N_i b_i / N$. Equation (2.14) reduces to (2.5) for $\overline{b^2} = \bar{b}^2$ and to (2.13) for $\bar{b} = 0$, as expected). Eq. (2.14) expresses the interesting notion that "optical polydispersity" ($\overline{b^2} > \bar{b}^2$) allows the separate detection of two different rate processes in the system: collective- and self-diffusion [5,6], particularly in concentrated systems where $S^I(K\approx 0)$ is small (see (2.7) and D_c and D_s differ greatly in magnitude.

Tracer Particles

For a system with two particle components a and b one has

$$\bar{b} = (N_a b_a + N_b b_b)/(N_a + N_b) \ ,$$
$$\overline{b^2} = (N_a b_a^2 + N_b b_b^2)/(N_a + N_b) \ . \tag{2.15}$$

The particular case where the particles a are nearly "invisible" ($b_a \simeq 0$) and a small number of particles b makes a predominating scattering contribution, $N_b b_b^2 \gg N_a b_a^2$, will be a clear example of self-diffusion measurement. For $K \to 0$ the self-diffusion coefficient can be equated to [8]

$$D_s(K=0) = \frac{k_B T}{f_s} \quad . \tag{2.16}$$

Here f_s is called the self-friction coefficient of the particles. According to BATCHELOR [11], for small concentrations,

$$\frac{D_s}{D_0} = 1 - 1.83\phi \quad , \tag{2.17}$$

but in this calculation the relaxation of surrounding particles is neglected and thus (2.17) only applies at $Ka \gg 1$. For $Ka \ll 1$ the relaxation of the particles surrounding the "central" particle that is observed must be taken into account. BATCHELOR [15] suggested for K=0,

$$\frac{D_s(K=0)}{D_0} = 1 - 2.68\phi \quad . \tag{2.18}$$

As far as we know no further theories that take account of this effect [16] are available in the literature.

3. Results and Discussion

Materials

Silica particles were prepared according to the method of VAN HELDEN et al.[1]. The particles consist of a spherical core of silicon dioxide (SiO_2) onto which stearylalcohol molecules are terminally bonded so that the particles become soluble in non-aqueous solvents.

Concentrated Systems

Particles with a hydrodynamic radius of 71 nm (standard deviation 9%) and a refractive index $n_p = 1.434$ ($\lambda = 546$ nm, 25 °C) were studied in cyclohexane. $n^0 = 1.425$ ($\lambda = 546$ nm, 25 °C).

The contrast, $n_p - n^0$, and thus the scattering power, could be varied by altering the temperature ($|dn^0/dT| > |dn_p/dT|$). The contrast was found to increase with increasing T.

The dynamic light scattering experiments were performed at concentrations varying from c = 0.01 to 0.87 g cm^{-3}, corresponding to hard-sphere volume fractions of $\phi = 0.006$ to 0.49.
For $c < 0.32$ g cm^{-3} the measured autocorrelation functions could be fitted to a single exponential. Above this concentration significant deviations from this behaviour were found and the data could be fitted to a sum of two exponentials,

$$G^{\frac{1}{2}}\, g^{(1)}(K,\tau) = A_1 \exp(-D_1K^2\tau) + A_2\exp(-D_2K^2\tau) \tag{3.1}$$

for $c > 0.32$ g cm^{-3}, and

$$G^{\frac{1}{2}} g^{(1)}(K,\tau) = A_3 \exp(-D_3K^2\tau) \tag{3.2}$$

for $c < 0.32$ g cm^{-3}.

For a given concentration and observation angle θ about ten correlation functions with different sample times were measured and combined. In this way $S(K,\tau)$ could be obtained over a wide time interval.
In so doing, we obtained values of A_1, A_2, D_1 and D_2 within an accuracy of about 10%. D values for several concentrations and scattering angles (K) are given in Fig. 1.

The upper three curves are for D_1, the three lower ones for D_2. One observes that D_1 is K dependent and for $K \gg 0$ hardly depends on the concentration. On the other hand, D_2 decreases strongly with increasing concentration and is K independent. This implies that according to (2.14) and (2.16) D_s,

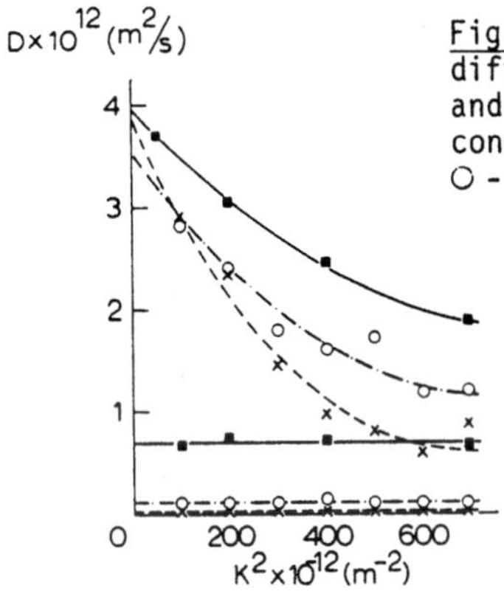

Fig.1. The $\underset{\sim}{K}$-dependence of the experimental diffusion coefficients D_1 (upper points) and D_2 (lower points) for three different concentrations: X - 0.86 g cm^{-3}; ○ - 0.77 g cm^{-3} and ■ - 0.60 g cm^{-3}

the self-diffusion coefficient, decreases strongly and the self-friction coefficient f_s increases strongly when the solution becomes more concentrated, this is understandable. On the other hand $D_1(K=0)$ is equal to the collective diffusion coefficient D_c (K=0), (see (2.14), and (2.8) predicts that because S^I (0) decreases and f_c increases with c, D_c is expected to /ary vary much less; this is indeed what we find.

The K dependence of D_c can be related to that of $S^I(K)$, (see (2.11)) and indeed follows the increase of this function with K. However, the function H(K) would have to be considered as well to complete the argument.

Temperature Variation

In Fig. 2 the autocorrelation function $g^{(1)}(K,\tau)$ is plotted for different temperatures. It shows the gradual disappearance of the fast component (collective diffusion) when T is lowered. According to (2.14) this implies that $(\overline{b^2} - \bar{b}^2)/\overline{b^2}$ increases with decreasing T. This is just what one would expect. The scattering amplitudes b_i of the individual particles are distributed around a mean value $\bar{b}$. When $\bar{b}$ drops to small values the relative fluctuation in b will increase.

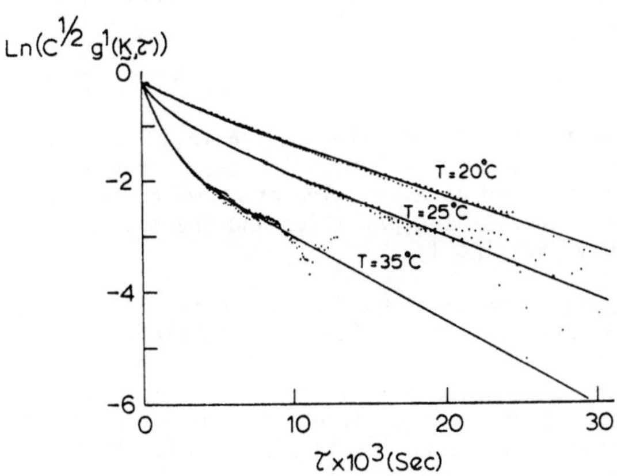

Fig.2. The logarithm of the field autocorrelation functions $\ln[G^{\frac{1}{2}} g^{(1)}(K,\tau)]$ as a function of the time τ from a silica dispersion of c = 0.63 g cm^{-3} at different temperatures, as indicated in the figure

Lower Concentrations

We mentioned above that for non-concentrated systems only a single exponential (3.2) is observed. The value of the "average" diffusion coefficient D_3, hereafter called D_m, reflects the observation of a mixture of collective- and self-diffusion.

In Fig. 3, D_m is plotted as a function of c for several temperatures and at $K \to 0$.

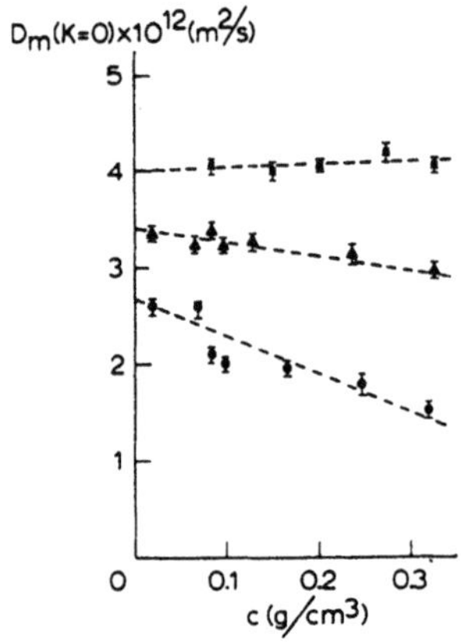

Fig. 3

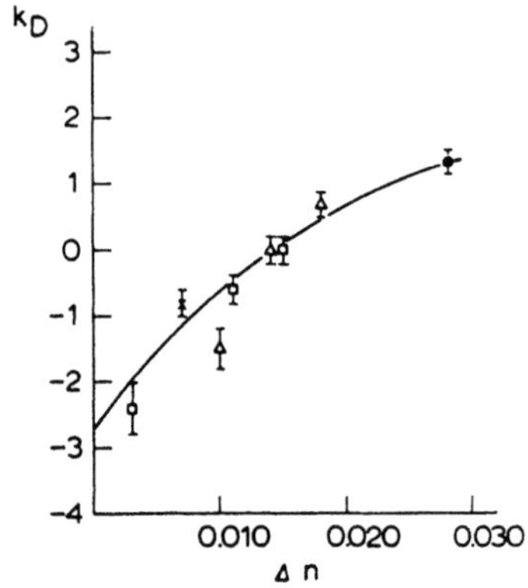

Fig. 4

Fig.3. Concentration dependence of $D_m(K=0)$ at different temperatures: * - T = 34.4 °C; ▲ - T = 25 °C and ● - T = 12 °C

Fig. 4. Values of the constant k_D from (3.3) as a function of the optical contrast represented by the difference in the (mean) refractive index of particles and solvent Δn from different silica dispersions.
□ - silica in cyclohexane a_h = 71 nm; x - silica in cycloheptane a_h = 71 nm;
Δ - silica in cyclohexane a_h = 46 nm; ● - silica in cycloheptane a_h = 23 nm

From theoretical considerations we expect that far away from the matching point $D_m(K=0) \to D_c(K=0)$ and the concentration dependence will be given by (2.10). Near the matching point ($\bar{b} = 0$) on the other hand $D_m(K=0) \to D_s(K=0)$ and the concentration dependence will be different from that described by (2.18). The slopes of the plots,

$$D_m(K=0)/D_o = 1 + k_D\phi \tag{3.3}$$

are expected to be situated between the extreme values given in (2.10) and (2.18). It is apparent from Fig. 3 that this is indeed the case. This is also seen from Fig. 4 where k_D is plotted as a function of Δn which is the difference between the refractive index of the particles and the solvent. (The variation in Δn is achieved by varying T).

One observes that the extrapolated value of k_D for $\Delta n \to 0$ is -2.7 ± 0.5, which is indeed a range that includes the value -2.68 of Batchelor.

For a discussion of e.g. the K-dependence of D_m, the c- and K-dependence of A_1 and A_2, and also for a discussion of independent friction measurements obtained from sedimentation and viscosity experiments [3] we refer to a future publication [17].

Tracer Experiments

Two preparations of particles were used with radii a_a = 33 nm and a_b = 38 nm and refractive indices n_{p_a} = 1.455 and n_{p_b} = 1.430. Solutions were made in cycloheptane which contained the nearly isorefractive particles of preparation a. Mixtures of a and b were made with $N_a > N_b$ and $N_a b_a^2 \ll N_b b_b^2$, so self-diffusion was measured. The results are given in Fig. 5.

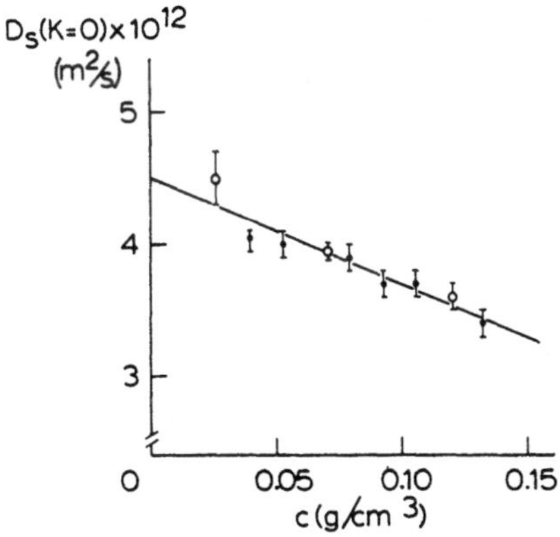

Fig. 5. Dependence of the self-diffusion coefficient at K = 0 on the total weight concentration of the silica particles in solutions.
● corresponds to the series where c_b was kept constant (c_b = 0.0183 g/cm)
○ corresponds to the experiments with varying values of c_b. T = 25 °C

From the slope of this plot a value k_D = -2.7 ± 0.3 is obtained in the same range as found earlier with the contrast variation experiments. There is a small mismatch of radii, i.e. 33 vs 38 nm but we do not expect this to cause more than a 0.3 change in k_D.

More details will be given in a future paper [18] in which we shall also report on temperature variation experiments, average light scattering intensity and on sedimentation experiments performed on mixtures of particles a and b.

4. Conclusions

We have shown that in light scattering studies contrast variation yields results for the correlation functions of PCS which are consistent with the fact that collective- and self-diffusion make separate contributions, as first noted by WEISSMAN and PUSEY. It proved possible to measure collective and self-diffusion coefficients of (lyophilized) silica spheres in some non-polar solvents. We devised an independent method for measuring the self-diffusion coefficient by making use of specially prepared tracer particles. The results were in accordance with those we obtained with the contrast variation method.

Acknowledgement

The authors are indebted to Drs. C.G. de Kruif and J. Dhont for assisting in the preparation of the final version of the manuscript.

This work is part of the research programme of the Foundation for Fundamental Research of Matter (FOM) and was carried out with financial support from the Netherlands Organization for Pure Research (ZWO).

References

1. A.K. van Helden, J.W. Jansen and A. Vrij, *J. Colloid Interface Sci.* 81, 354 (1981).
2. M.M. Kops-Werkhoven and H.M. Fijnaut, *J. Chem. Phys.* 74, 1618 (1981).
3. M.M. Kops-Werkhoven, dissertation, University of Utrecht, 1982.
4. M.M. Kops-Werkhoven, H.J. Mos, P.N. Pusey and H.M. Fijnaut, *Chem. Phys. Letters* 81, 365 (1981).
5. M.B. Weissman, *J. Chem. Phys.*72, 231 (1980).
6. D.L. Cebula, R.H. Ottewill, J. Ralston and P.N. Pusey, *J. Chem. Soc. Faraday Trans II*, to be published.
7. P.N. Pusey and R.J.A. Tough, "Particle Interactions" (to appear in *"Dynamic Light Scattering and Velocimetry: Applications of Photon Correlation Spectroscopy"*, Ed. R. Pecora, Plenum Press, New York, 1981).
8. D.L. Cebula, R.H. Ottewill, J. Ralston and P.N. Pusey, *J. Chem. Soc. Trans I* 77, 2585 (1981).
9. A.K. van Helden and A. Vrij, *J. Colloid Interface Sci.* 78, 312 (1980).
10. N.F. Carnahan and K.E. Starling, *J. Chem. Phys.* 51, 635 (1969).
11. G.K. Batchelor, *J. Fluid Mech.* 52, 245 (1972); 74, 1 (1976).
12. H.M. Fijnaut, C. Pathmamanoharan, E.A. Nieuwenhuis and A. Vrij, *Chem. Phys. Letters* 59, 351 (1978).
13. H.M. Fijnaut, *J. Chem. Phys.* 74, 6857 (1981).
14. W.B. Russel and A.B. Glendining, *J. Chem. Phys.* 74, 948 (1981).
15. G.K. Batchelor, private communication.
16. The situation given here can be compared to the long known "relaxation effect" of the electrical double layer in electrophoresis and self-diffusion. When the central particle is moving, the surrounding (ionic) atmosphere becomes asymmetrical because it takes some time to break down the old atmosphere and to build a new one. The asymmetrical atmosphere exerts a retarding force on the central particle.
17. M.M. Kops-Werkhoven and H.M. Fijnaut, *J. Chem. Phys.*, accepted for publication.
18. M.M. Kops-Werkhoven, C. Pathmamanoharan, A. Vrij and H.M. Fijnaut, *J. Chem. Phys.*, accepted for publication.

38. The Application of Number Fluctuation Spectroscopy to Aggregating Systems and the Measurement of Rotational (Anti-) Correlation Functions of Low-Order Aggregates

J.G. Rarity

Royal Signals and Radar Establishment, St. Andrews Road,
Malvern, Worcestershire, UK

K.J. Randle

Royal Military College of Science, Shrivenham, Wiltshire, UK

1. Introduction

The use of number fluctuation techniques to investigate polydispersity and/or chemical kinetics has been proposed by several authors [1,2,3]. To date the theories developed by these authors have been put to little practical use excepting the technique of Fluorescence correlation spectroscopy [4]. This work presents a further practical application of these theories to the study of Brownian or Smoluchowski [5] aggregation. Assuming quasistationarity the moments of the measured intensity distribution can be related to the moments of the aggregate intensity distribution. Using an accurate approximation for the intensity scattered by an aggregate, information about the aggregate distribution itself can be obtained. In practice the poor statistical accuracy of higher moments measurements limits the amount of information obtainable. This application uses the rate of increase of the second moment to measure the aggregation rate constant (section 4) assuming Smoluchowski [5] kinetics (section 2) and a simple RGD[1] approximation for the aggregate scattering (section 3).

Recently a two-detector cross-correlation technique [6] was used to study the rotation of asymmetric particles in the number fluctuation regime. The use of two detectors ensures no residual component of the field correlation function is seen and an anticorrelation effect occurs when the detectors are symmetrically opposed at scattering angles close to 90^0. The form of this cross-correlation from RGD doublets is evaluated in section 5. Estimates of the rotational diffusion constant of doublet particles have been obtained from cross-correlation measurements made in the early stages of an aggregation. These measurements are presented in section 8 along with moments measurements taken from a system of aggregating polystyrene spheres.

2. Stability and Aggregation

The stability of charged particles in aqueous suspension is very dependent on the ionic strength of the bulk solution. Increasing the ionic strength compresses the ionic double layer surrounding the particle. The range of the repulsive potential between particles is measured by the thickness of this layer. Above a critical ionic strength the normally shorter range Van der Vaal's attractive force between particles can dominate causing irreversible aggregation.

---

[1] RGD = Rayleigh Gans Debye approximation.

The coagulation process follows second-order kinetics, the rate of decrease of particle concentration n with time being given by

$$-\frac{dn}{dt} = K_o n^2 \quad . \tag{1}$$

The rate constant K_o depends on the rate at which particles diffuse into each others sphere of influence [5,7]

$$K_o = \frac{4K_B T}{3\eta W} \quad , \tag{2}$$

where $K_B T$ is the Boltzmann energy, η is the viscosity of the supporting medium and W is a stability factor included to account for residual repulsive potential and hydrodynamic effects at close approach.

Assuming the rate constants for collisions between arbitrary sized aggregates to be equal SMOLUCHOWSKI [5] developed equations for the number of aggregates of order K after time t

$$n_K = \frac{n_o\left(\frac{t}{t_{\frac{1}{2}}}\right)^{k-1}}{\left(1 + \frac{t}{t_{\frac{1}{2}}}\right)^{k+1}} \quad , \tag{3}$$

where $t_{\frac{1}{2}} = 1/K_o n_o$ is the half life of the aggregation, the time taken for the total number of particles to drop to half its original value n_o. Many experimental studies have confirmed the validity of these results [8] and have provided valuable information on the interaction forces between colloidal particles. Early approaches to the measurement of aggregation kinetics tended to be invasive (eg. coulter counters and flow ultramicroscopes) or non-absolute (eg. turbidity measurements). Recently LIPS et al. [9] used an RGD approximation for the scattering from aggregates to accurately predict the change in mean scattered intensity during an aggregation. This approximation is described in the following section. LIPS et al. [10] were able to obtain absolute rate constants from high angle and low angle measurements of the mean scattered intensity.

3. Aggregate Scattering

In the RGD approximation the scatter from a K-fold aggregate is given by [9]

$$I_K(\theta) = I_1\, K\left(1 + \frac{2}{K} A_k(\theta)\right) \quad , \tag{4}$$

where I_1 is the singlet scattered intensity, θ is the scattering angle and

$$A_k(\theta) = \sum_{i}^{k} \sum_{>\,j}^{k} \frac{\sin qr_{ij}}{qr_{ij}} \quad ; \tag{5}$$

q is the modulus of the scattering vector ($4\pi n/\lambda \sin\theta/2$) and r_{ij} the separation of particles i and j within the aggregate. LIPS et al. [9] have shown that this approximation is valid even when individual particles scatter in the Mie regime. If particles are in contact within the aggregate a major contribution to $A_k(\theta)$ will arise from terms of the form sin(nqd)/nqd where n is non-zero integer and d is the particle diameter. Arranging the scattering angle such that $qd = m\pi$, $I_k(\theta)$ can be approximated by

$$I_k(\theta) = KI_1 \quad . \tag{6}$$

On viewing an aggregation at this angle no change in the mean intensity is detected. It is often dubbed "the angle of no change". Equation (6) and equation (3) form the basis of the aggregate intensity distribution to be used in evaluating the number fluctuation intensity moments.

4. Normalised Factorial Moments (n(r)) from an Aggregating System

The general form of the normalised intensity moments, equivalent to the normalised factorial moments of the photocounts distribution, assuming full incoherence, are given by [1,11]

$$\bar{n} = \eta\langle I\rangle = \beta\langle m\rangle\langle a^2\rangle \tag{7a}$$

$$n^{(2)} = 1 + \frac{c_2 a_2}{\langle m\rangle} \tag{7b}$$

$$n^{(3)} = 1 + \frac{3c_2 a_2}{\langle m\rangle} + \frac{c_3 a_3}{\langle m\rangle^2} \tag{7c}$$

$$n^{(4)} = 1 + \frac{6c_2 a_2}{\langle m\rangle} + \frac{3c_2^2 a_2^2}{\langle m\rangle^2} + \frac{4c_3 a_3}{\langle m\rangle^2} + \frac{c_4 a_4}{\langle m\rangle^2} \quad . \tag{7d}$$

Here η is the detector efficiency and β is a constant of proportionality. $\langle I\rangle$ is the mean intensity incident on the detector, $\langle m\rangle$ is the mean number of scatterers in the volume and $\langle a^2\rangle$ is the mean scattered intensity per particle. C_n are constants dependent on the scattering volume intensity profile. When the intensity profile is approximated to a gaussian-ended gaussian cylinder with the effective scattering volume V lying within the $1/e^2$ intensity contours $c_2 = 1$ and $c_3 = (4/3)^{3/2}$. The a_n are the normalised moments of the individual particle scattering intensities $p(a^2)$, ie.

$$a_n = \frac{\langle a^{2n}\rangle}{\langle a^2\rangle^n} = \frac{\int_0^\infty a^{2n} p(a^2)\,da^2}{\left[\int_0^\infty a^2 p(a^2)\,da^2\right]^n} \quad . \tag{8}$$

In this application $p(a^2)$ is directly related to the aggregate distribution, ie

$$P(I_k(\theta) = a^2) = \frac{\dfrac{n_o E^{k+1}}{(1+E)^{k-1}}}{\sum\limits_{k=1}^{k_{max}} \dfrac{n_o E^{k+1}}{(1+E)^{k-1}}} = \frac{E^{k+1}}{(1+E)^{k-2}} \quad . \tag{9}$$

Here we have introduced a dimensionless extent of aggregation $E = t/t_{\frac{1}{2}}$. Combining equations (6-9) yields the expressions for the first three incoherent moments at the angle of no change [11]

$$\bar{n} = \beta\langle m_o\rangle \tag{10a}$$

$$n^{(2)} = 1 + \frac{1}{\langle m_o\rangle}(1+2E) \tag{10b}$$

$$n^{(3)} = 1 + \frac{3}{\langle m_o\rangle}(1+2E) + c_3\frac{1}{\langle m_o\rangle^2}(1+6E+6E^2) \quad , \tag{10c}$$

where $\langle m_o\rangle = (1+E)\langle m\rangle$ is the original mean number of particles in the scattering volume. With a knowledge of the scattering volume (V) a direct estimate of the rate constant can be obtained from the gradient of a plot of $n^{(2)}$ versus time

$$n^{(2)} = 1 + \frac{1}{\langle m_o\rangle} + \frac{2K_o}{V}\cdot t \quad . \tag{11}$$

The gradient of equation (11) is independent of the initial level of aggregation in the sample and only slightly affected by the levels of polydispersity normally encountered in aggregation measurements [11]. The two main limits of the accuracy with which K_o can be determined are the statistical accuracy of the second-moment measurement and the accuracy of the measurement of V.

5. Rotational Correlation

The aggregate scattering formulae (equation (4)) are orientation averaged; thus the sample time for moments measurements had to be made longer than the decay time of any rotational correlation. The scattered intensity at specific orientation is given in the RGD approximation by

$$I_k(\underline{q},t) = kI_1\left(1 + \frac{2}{K}\sum_{i}^{k} {}_{>} \sum_{j}^{k} \cos \underline{q}\cdot\underline{r}_{ij}(t)\right) \tag{12}$$

which simplifies to

$$I_2(\underline{q},t) = 2I_1(1 + \cos \underline{q}\cdot\underline{r}_{12}(t)) \tag{13}$$

in the case of doublet scattering.

The measured cross-correlation function of intensity $g^{(2)}(\underline{q}_1,\underline{q}_2,\tau)$ can be written [1]

$$g^{(2)}(\underline{q}_1,\underline{q}_2,\tau) = 1 + \beta|g^{(1)}(\tau)|^2 + g_N(\tau)g_R(\tau) \quad . \tag{14}$$

When $\underline{q}_1 \neq \underline{q}_2$, $\beta = 0$ and the correlation function consists of a number fluctuation term $g_N(\tau)$ and a term due to polydispersity and asymmetry $g_R(\tau)$. In this experiment adequate sampling of the aggregate distribution was achieved by translating the sample across the scattering volume; thus the number fluctuation term $g_N(\tau)$ is given by [1,11]

$$g_N(\tau) = \frac{1}{<m>} e^{-V^2\tau^2/\sigma^2} \quad , \tag{15}$$

where V is the translation speed and σ is the $1/e^2$ beam radius. In an experiment viewing only doublets the rotational component is

$$g_R(\tau) = \frac{<I_2(\underline{q}_1,0)\ I_2(\underline{q}_2,\tau)>}{<I_2(\underline{q}_1)>\ <I_2(\underline{q}_2)>} \tag{16}$$

with $I_2(\underline{q},t)$ given by equation (13). It involves a time-dependent component

$$c(\tau) = \cos \underline{q}_1 \cdot \underline{r}_{12}(0) \cos \underline{q}_2 \cdot \underline{r}_{12}(\tau) \quad . \tag{17}$$

This is evaluated following a procedure similar to that used by Berne and Pecora [12] (rigid rod, coherent case). Both the cosine terms and the solution of the rotational diffusion equation are expanded in terms of the spherical harmonics. Averaging over initial and final orientations and use of the orthogonality relations of the spherical harmonics leads to

$$g_R(\underline{q}_1,\underline{q}_2,\tau) = 1 + \frac{1}{A} \sum_{\substack{\ell>0 \\ \text{even}}} 2\ell+1\ j_\ell(q_1 r_{12}) j_\ell(q_2 r_{12}) \times P_\ell(\cos\theta) \exp^{-\ell(\ell+1)D_R\tau} \tag{18}$$

where

$$A = (1+j_o(q_1 r_{12}))(1+j_o(q_2 r_{12})) \quad . \tag{19}$$

j_ℓ are spherical bessel functions of order ℓ, $P_\ell(\cos\theta)$ are legendre polynomials, θ is the angle between scattering vectors $\underline{q}_1$ and $\underline{q}_2$ and D_R is the rotational diffusion constant.

In the early stages of an aggregation ($E < 0.05$) the system can be assumed to consist only of doublets and singlets. Furthermore with detectors symmetrically placed at angles where $q\ r_{12} = \pi$, the $\ell = 2$ component in equation (18) dominates leading to

$$g^{(2)}(\underline{q}_1,\underline{q}_2,\tau) = 1 + \frac{(1+2E)}{<m_o>} e^{-V^2\tau^2/\sigma^2} (1+C\cdot P_2(\cos\theta) e^{-6D_R\tau}) \tag{20}$$

where

$$C = \frac{180}{\pi^4} \frac{E}{(1+E)^4(1+2E)} \quad . \tag{21}$$

When the angle of no change is close to 90° and $E \leqslant .05$ the modulation of the rotational component ($cP_2(\cos\theta)$) is less than 4% and the contribution due to triplets and higher-order aggregates amounts to less than 10% of this. At these angles $P_2(\cos\theta)$ is negative thus an "anticorrelation" [6] is seen.

6. Apparatus

The apparatus is shown in figure 1. Light from a 25 mW He/Ne laser passes through a microscope objective into the scattering cell. The scattering cell

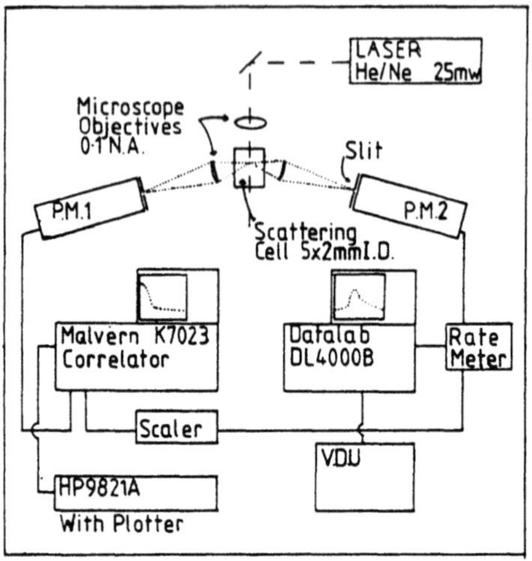

Fig.1. General layout of the apparatus

(2 mm x 5 mm) is mounted in a thermostattable brass holder on a motorised translation stage capable of moving at speeds up to 25 μ/sec (~ 2 mm/minute). The scattered light can be viewed through ports in the sides of the holder. Microscope objectives (NA ~·1) are used to transfer the beam image onto slits placed over the photocathodes of two symmetrically placed photon counting photomultipliers (PMs). The slit image along the beam defined a scattering volume of length roughly 80 μm. This combined with a beamwaist 18 μm in diameter yields a scattering volume of ca. 2×10^{-8} cm^3. The PMs were mounted on X Y Z stages. Motion along the beam was provided by micrometer driven microposition stages of accuracy better than 2 μm (Ealing Beck). The sample volume was located on the beamwaist by maximising the second moment from a test sample. Maximisation of the cross-correlation function intercept led to at least 80% volume overlap in the two detector experiments. This alignment appeared stable for periods up to 12 hours.

The photopulses from one PM passed to a junction box/ratemeter. One output from this junction led to a Data lab DL4000B system used in 4096 point histogram analysis configuration. Calculation of the n(r) could be carried out internally and results displayed on an attached VDU. The other output led to the Malvern K7023 correlator where it could be cross-correlated with the signal from the other PM. Analysis of the correlation functions was carried out on an HP9821A desktop computer with attached plotter.

7. Methods

A reasonably large sample volume (in terms of number fluctuation experiments) allowed the use of low initial number densities while retaining $\langle m_o \rangle \sim .5$. This ensured $t_{\frac{1}{2}}$ lay in the region of $\frac{1}{2}$-1 hr. Translation of the volume at ca 1.8 mm/sec meant that reasonable accuracy could be obtained in a measurement time of 7 minutes. Background counts were non-negligible (~ 2% of the signal) but could be corrected for [1] after their estimation using a blank sample (pure filtered water).

The particle used in this experiment was a 192 nm charge stabilised polystyrene latex of narrow size distribution and negligible preaggregation. With 632.8 nm (HeNe) illumination the angle of no change for this system is 77.2^o which implies a PM viewing angle of 73°.

Rigorously clean sample preparation techniques were needed as the presence of dust can seriously distort the intensity distribution and cause havoc with the higher moments. Clean cells were rinsed in filtered (.22 μm) distilled water, dried in Filtered nitrogen, and sealed. Latex samples were filtered directly into the sample cell (.45 μm millipore) and aggregation was initiated by filtering (.22 μm) a solution of NaCl into the sample and mixing (with a precleaned paddle).

8. Results and Discussion

The factorial moments measured during an aggregation experiment are shown in figures 2-4. The sample time used was 35 ms, much longer than the timescale of rotational components seen in correlation measurements. The mean countrate (figure 2) remains stable over the experimental duration thus confirming the detector is aligned on the angle of no change. A straight line fit to the second moment data yields $t_{\frac{1}{2}}$ = 31.3 minutes. Using the measured gradient and $V = 2.08 \times 10^{-8}\ cm^3$ (calibrated using a sample of known number density) in equation (11)

$$K_0 = 3.34 \pm .16 \times 10^{-18}\ m^2\ s^{-1} \qquad (T = 22 \pm .5^{0}C)$$

is obtained in reasonable agreement with that measured using other techniques [13,10]. The $n^{(3)}$ data (figure 4) is compared with that predicted from

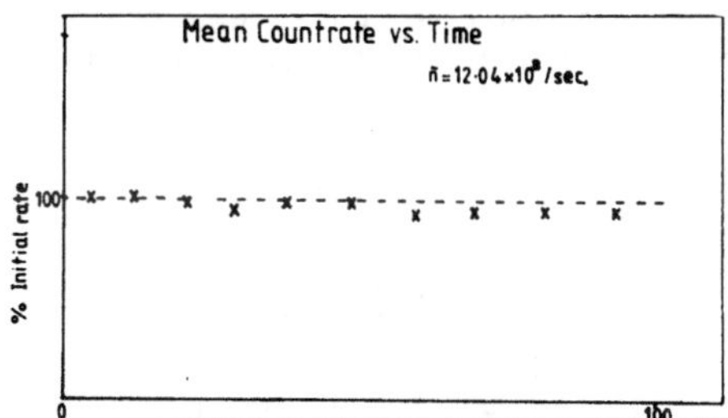

Fig.2. Mean countrate vs time during an aggregation. Plotted as a percentage of the initial countrate. ---- is the predicted (constant) countrate, x-x experimental points

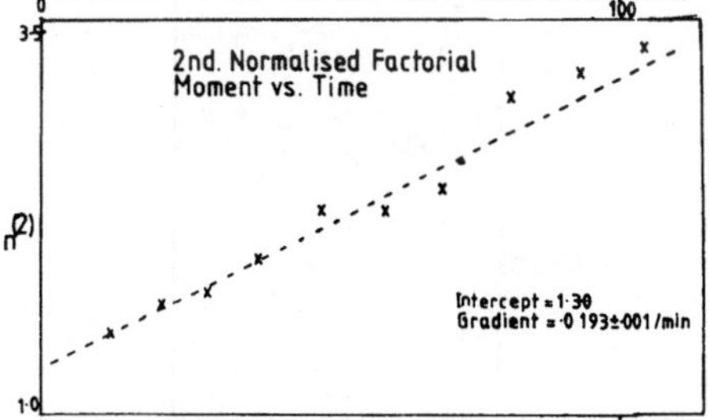

Fig.3. Second normalised factorial moment ($n^{(2)}$) vs time. x-x are data points and ---- is fitted straight line with intercept and gradient as shown

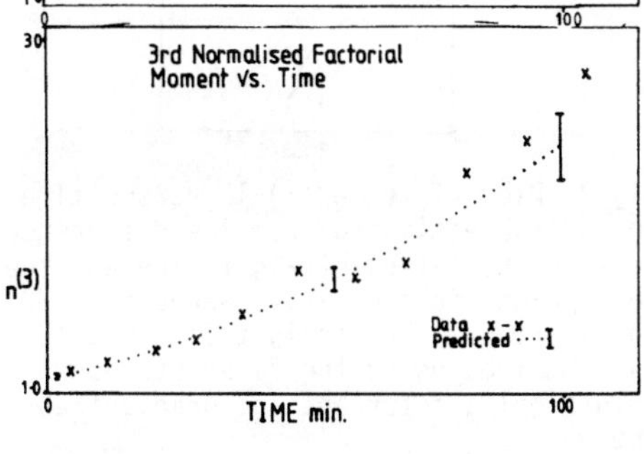

Fig.4. Third normalised factorial moment ($n^{(3)}$) vs time. x-x are data points and ---- is the predicted $n^{(3)}$ using values of E taken from the second moment fitted line and a $c_3 = 1.7$ (see equation (11)). Errors bars (I) predicted from [11]

equation (10c) using the measured value of t_1 and a value of $c_3 = 1.7$ estimated from measurements on the unaggregated latex. Methods of estimating the predicted error indicated by the error bars are discussed elsewhere [11]. The points all lie within these error estimates. The statistical error is in fact too large to allow reliable measurement of the a_3 factor (equation (7c)). Use of a longer measurement time and lower number density would reduce this uncertainty but would be more susceptible to dust and background light levels. The high value of the measured c_3 in this experiment is probably due to the non-ideal shape of the scattering volume at this scattering angle. This could also be enhanced by slight polydispersity of the latex. Preaggregation was ruled out as cross-correlation measurements on the original sample showed no short timescale correlations.

The normalised cross-correlation function measured in the early stages of a slower aggregation is shown in figure 5. The sample time was 0.5 ms thus only the flat top of the number fluctuation term $g_N(\tau)$ is seen. The barely detectable dip at short sample times is seen much more clearly in a plot of $\ln(g^{(2)}(\underline{q}_1,\underline{q}_2,\tau) - 1)$ versus τ^2 (figure 6). At longer delay times the linear decay arising from translation is fitted with a straight line. At short delay times the difference between the straight line and the measured points

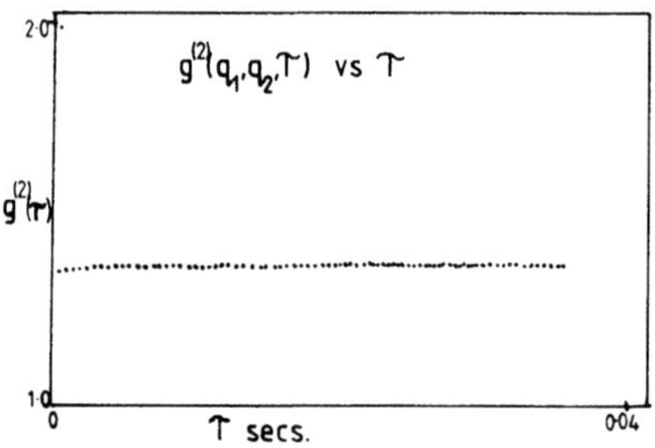

Fig.5. The intensity cross-correlation function $g^{(2)}(\underline{q}_1\underline{q}_2,\tau)$ at short delay times taken during the early stages of an aggregation (E < .05). Note the imperceptible dip in the early channels

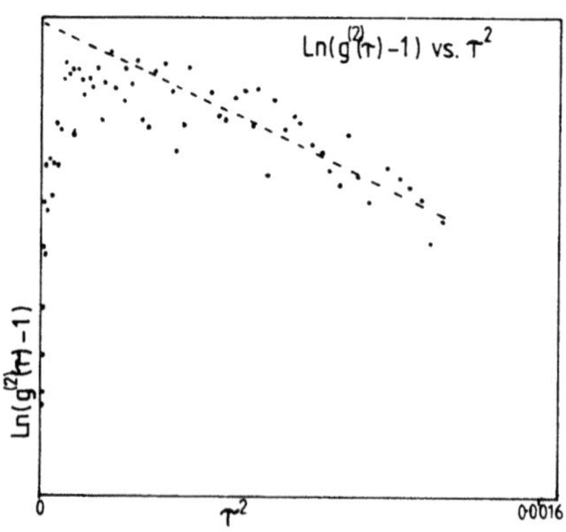

Fig.6. Logarithm of the data in figure 5 showing the full extent of the anticorrelation component. ---- is the fitted line to later channels (40-72), the linear decay being due to the gaussian form of $g_N(\tau)$

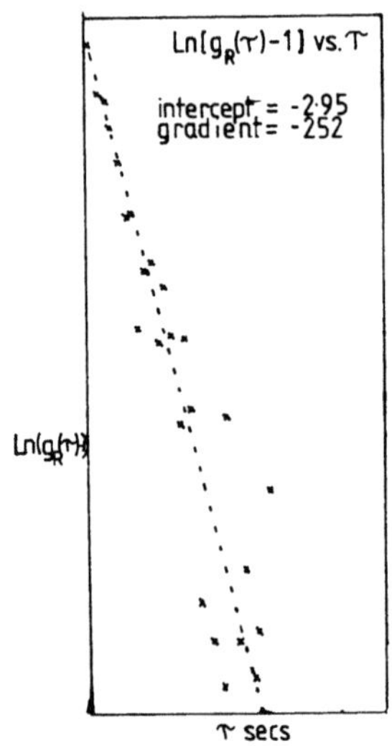

Fig.7. Plot of $\ln[g_R(\tau)-1]$ versus time. $\ln[g_R(\tau)]$ estimated from the difference between the fitted straight line and the data points in the early channels (1-30) in figure 6. The straight line fit (---) was obtained using the first sixteen points only; intercept and gradient as shown

provides an estimate of $\ell n\ g_R(\tau)$. $\ell n(g_R(\tau) - 1)$ is plotted against τ in figure 7 with fitted straight line. The gradient yields an estimate of $-6D_R$. D_R is related to the radius of the particles in a contacting doublet by the equation [14]

$$D_R = \frac{K_B T}{29.9\pi\eta r^3} \tag{22}$$

and estimates of r from 8 results taken at times when $E < 0.05$ gave

$2r = 196 \pm 6$ nm

in good agreement with the PCS determined diameter [11] of

$2r = 192 \pm 3$ nm.

9. Discussion

These results prove the feasibility of using number fluctuation techniques to study irreversible Brownian aggregation and, in general, colloidal reaction kinetics. However number fluctuation measurements require long durations to ensure good statistical accuracy. This can be improved by translation of the sample but often as in this case where it is necessary integrate out rotational effects (moments measurement) there is an upper limit to this increase in accuracy.

The measurement of particle diameter to an accuracy of 3% from the rotational correlation functions may appear surprising when one regards the noise in the data. However, due to the r^{-3} dependence of the rotational diffusion coefficient an error in D_R of 10% can be tolerated before the error in r reaches 3.3%. In general the rotational diffusion constant is dependent on volume rather than linear dimension and as such is much more sensitive to polydispersity than the translational diffusion coefficient. The difficulties inherent in its measurement, however, tend to counteract this advantage.

References

1. P.N. Pusey, p.45 in "Photon Correlation Spectroscopy and Velocimetry", eds. H.Z. Cummins and E.R. Pike, Plenum (1977).

2. R. Barakat and J. Blake, Phys. Rev. A13, 1122-1130 (1976).

3. B.J. Berne and R. Pecora, "Dynamic Light Scattering", ch. 6, Wiley (1976).

4. D. Madge, E.L. Elson and W. Webb, Phys. Rev. Letts. 29, 705 (1972).

5. M. von Smoluchowski, Z. Phys. Chem., 92, 129 (1917).

6. W.G. Griffin and P.N. Pusey, Phys. Rev. Letts. 43, 1100 (1979).

7. A. Spielman, J. Coll. Interf. Sci. 33, 562 (1970).

8. H.R. Kruyt, "Colloid Science", Vol. I Chapters 7 and 8, Elsevier (1952).

9. A. Lips, C. Smart and E. Willis, Trans. Faraday Soc. 67, 2979 (1971).

10. A. Lips and E. Willis, Faraday Trans. 1, 69, 1226 (1973).

11. J.G. Rarity, Ph.D. thesis in preparation.

12. Ref. 3 - Chapters 5 and 7.

13. W. Hatton, P. McFadyen and A.L. Smith, Faraday Trans. I, 70, 655 (1974).

14. E.B. Vadas, H.L. Goldsmith and S.G. Mason, J. Colloid. Interface Sci. 43, 630 (1973).

39. A Study of Fibrin Polymerization Combining Static and Dynamic Light Scattering with the Measurement of Fibrinopeptide Release

P. Wiltzius, G. Dietler, and W. Känzig

Laboratorium für Festkörperphysik, CH-8093 Zürich, Switzerland

A. Häberli, P.W. Straub

Thrombose-Laboratorium, Medizinische Universitätklinik, Inselspital, CH-3010 Bern, Switzerland

The enzyme thrombin catalyzes the release of small peptide groups (fibrinopeptides FPA and FPB) from the plasma protein fibrinogen (M_w 340'000). This release transforms the soluble fibrinogen molecule into the fibrin monomer which polymerizes under physiological conditions. The monomers aggregate spontaneously to form the so-called intermediate polymers and finally a gel. This process plays a major role in blood coagulation. In the present work we undertook an investigation of the pre-gelation phase by combining dynamic and static light scattering with measurements of the release of the fibrinopeptides A (which plays the dominant role in the aggregation process since the release of FPB is delayed). Low-angle experiments ($5^o \leq \Theta \leq 13^o$) yield the weight-averaged molecular weight M_w and the z average of the diffusion coefficient $\bar{D}_z$, whereas measurements at $\Theta = 90^o$ yield an apparent $\bar{D}_{z\ app}$. The number of released FPA is a measure of the extent of the polymerization. Without making any assumptions about the polymer distribution one is led to the following conclusions:

- The polymers are rod-like .

- At the gel point (where M_w diverges) there are still a large number of low oligomers in the solution coexisting with the infinitely large cluster.

- The Flory-Stockmayer model distribution is to be preferred over distributions provided by percolation theories, i.e. cyclic bonds are negligible, critical exponent $\gamma = 1.05 \pm 0.1$.

- In the early stages the apparent functionality is $f = 2$, and it increases with increasing release of fibrinopeptides A, approaching eventually the value $f = 4$.

A full account of this work is in press: P. Wiltzius et al., Biopolymers.

40. Quasielastic Light-Scattering Study of Mixed Micelles

Mario Corti
CISE S.p.A., P.O. Box 12081, Milano 20100, Italy
Vittorio Degiorgio
Istituto di Fisica Applicata, Università di Pavia, Pavia 27100, Italy

Introduction

The study of aggregation phenomena of amphiphilic molecules in aqueous solution is an increasingly interesting field of application of the static and dynamic light-scattering technique. Amphiphilic molecules are made with a non-polar hydrophobic "tail" (usually one or two hydrocarbon chains containing various numbers of methylene groups), which is joined to a polar hydrophilic head group. Such molecules are also called surfactants or detergents. Amphiphilic molecules present great technological importance because they are used in many industrial processes (detergency, tertiary oil recovery, polymerization reactions, lubrification, cosmetics, pharmaceutics, etc.). They are also quite interesting in biology as basic constituents of living matter, like phospholipids in membranes. Finally amphiphilic molecules present important physical problems, the understanding of which is quite stimulating.

Before entering into a specific discussion of a light-scattering experiment on the mixing effects of two amphiphilic molecules in water, we prefer to give an introductory overview of the physical problems involved in micellar solutions.

At low concentration in water, amphiphilic molecules exist as monomers. At higher concentrations globular aggregates are formed with a hydrocarbon core and a polar or ionic surface which shields the core from the surrounding water [1]. An optimum size is reached when the unfavourable exposure of the hydrocarbon chains to water is minimized. Typical aggregation numbers are of the order of a hundred with a micelle radius of 20-30 Å.

A representative phase diagram of an amphiphile-water solution is shown in Fig.1 in the range of low concentrations of amphiphile. In phase I the concentration is so low that the amphiphilic molecules exist in solution as monomers. In a narrow concentration region between phase I and II micelles are formed as a dynamical association-dissociation equilibrium process. The concentration at this phase boundary is called the critical micelle concentration (c.m.c.). The sharpness of the transition depends on the particular amphiphilic molecule. Micelle formation can be monitored by a variety of experimental techniques: surface tension measurements, electrical conductivity, osmometry, static light scattering, etc. Dynamic laser light-scattering experiments have also been performed to measure the micelle diffusion coefficient close

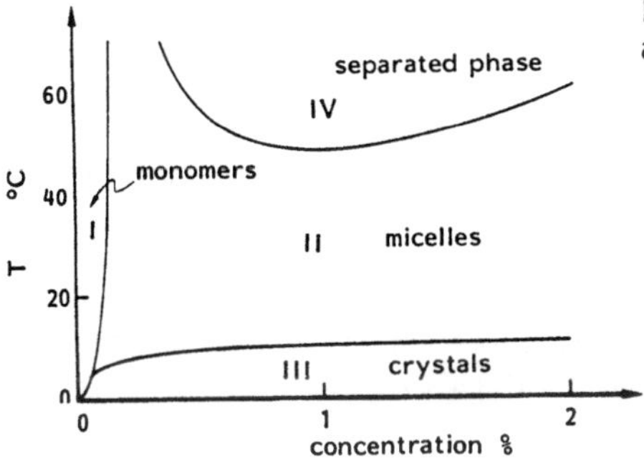

Fig.1 Representative phase diagram of an amphiphile-water solution

to the transition [2]. Experiments are difficult because the available scattered intensity is low and even very small quantities of insoluble impurities may largely affect the light-scattering measurements at the transition. Droplets of insoluble material are probably formed, which are then solubilized above the c.m.c. [3]. Inside phase II micelles coexist with monomers. Adding detergent to the solution, the number of micelles increases while the free monomer concentration remains roughly constant at the c.m.c. value. In this region it is interesting to study the properties of individual micelles and micelle-micelle interactions, with both static and dynamic light-scattering measurements as a function of the solution concentration. From the measured concentration dependence of the apparent molecular weight M and of the mutual diffusion coefficient D, it is possible to derive a pair intermicellar interaction potential V. This potential consists of a hard-sphere repulsive part, of an electrostatic long-range repulsion and of a London-Van der Waals short-range attraction. It contains only two unknown parameters, the electric potential at the shear surface of the micelle and the Hamaker constant. The idea of adding an attractive contribution to the interaction potential, although usual in colloid theory, is rather recent for micellar solutions [4]. For charged micelles (ionic detergents), the electrostatic part is dominant at low ionic strength of the solution.

At increasing ionic strength, the Coulomb potential is screened more and more effectively, the Debye-Hückel length is smaller and the short-range London-Van der Waals attraction becomes important. Analysing the light-scattering results with the above procedure, reliable data on the micellar charge in solution can be obtained [4,5]. Conversely micelles may offer a simple and flexible model to study interaction effects in solution, since different micellar radius, surface charge and shape are easily obtainable with a judicious choice of the amphiphilic molecule. Indeed interaction effects are important also for macromolecules in solution, like for instance proteins and polymers. High-ionic-strength solutions of ionic micelles have been investigated by several groups using quasielastic light scattering [6-10]. The observed formation of large aggregates with a size dependent on the temperature, and on the salt and amphiphile concentration has been discussed by most authors in terms of a sphere-to-rod transition of micelles.

Coming back to Fig.1, at the boundary of phase II and III a first-order transition is observed. The detergent precipitates to form very hydrated crystals. The kinetics of phase III has not been investigated extensively, except for some phenomenological papers appeared in the chemical literature. Nevertheless there is room for light-scattering investigations in order to understand for instance the role of the phase boundary line on the observed increase of the apparent molecular weight of the ionic micelle in phase II when approaching phase III in high-ionic-strength solutions. Another important point could be to understand whether the hydrated crystals are formed starting from aggregation of micelles or from monomeric detergent molecules.

The last portion of the diagram, phase IV, is typical of non-ionic detergents. The homogeneous micellar phase II breaks into two separate phases. There is a lower critical consolution point, called cloud point, and the transition is of the second-order type like in liquid binary mixtures. Critical effects are observable even at tens of degrees away from the critical temperature. Reaching phase IV from phase II at a given concentration of detergent, the initial homogeneous micellar solution will separate in two homogeneous micellar solutions at different concentrations, one below and the other above the initial concentration. It is quite interesting to understand the physics of the second-order transition in micellar solutions. Recent work has been done with static and dynamic light scattering on some non-ionic detergents [11] . One finds clear evidence of the critical behaviour, but the measured critical exponents are not directly understandable on the basis of current theories of second-order phase transitions [12].

Mixed Micelles Experiment

We present a light-scattering experiment on a solution containing two different amphiphiles, the ganglioside GM1 and the non-ionic detergent, n-dodecyl octaoxyethylene glycolmonoether ($C_{12}E_8$). The ganglioside GM1 is a glycolipid occurring in neuronal plasma membranes [13] . It has a hydrophilic part made of an oligosaccharide chain containing one molecule of sialic acid and a hydrophobic part made of two long aliphatic chains belonging to a sphingosine and a fatty acid residue. Both GM1 and $C_{12}E_8$ form micelles in solution with a hydrodynamic radius R_H=57 Å and 35 Å, respectively. The micellar molecular weights differ by about 8 times. Therefore these two systems are quite suitable for a physicochemical study of the mixed micelles formed by mixing the two amphiphiles in different proportions. Little information exists in the literature about the size and the aggregation number of mixed micelles. We have measured the hydrodynamic radius R_H and the molecular weight M of the mixed micelle by varying the molar concentration C_1 of $C_{12}E_8$ from 0 to 50 mM with the molar concentration C_2 of GM1 constant at 0.8 mM. The $C_{12}E_8$ was gas-liquid cromatographic-grade material from Nikko (Tokyo, Japan). The ganglioside GM1 isolated from calf brain according to the method of TETTAMANTI et al. [15] and purified following the indications of SONNINO et al. [16], was kindly supplied to us by the Biochemistry Group of the Medical School of the University of Milan. The mixture of two amphiphiles was dissolved in 25 mM sodium phosphate - 5 mM Na_2EDTA buffer (pH 7.0).

The light-scattering apparatus was equipped with an Argon-ion laser on the green 5145 Å line and a scattering cell temperature controlled within 0.002°C. The scattering angle was 90°. Extrapolation of the data to zero angle are not necessary because micelles are small. The intensity correlation function was measured by a 108-channels digital correlator. More details of the apparatus may be found in Ref.14. The Z-average mass diffusion coefficient is derived by applying the Koppel fit to the logarithm of the time-dependent part of the intensity correlation function. Polydispersity is small for all the measurements. The hydrodynamic radius R_H is derived from D by using the Stokes-Einstein relation and the viscosity of the solvent. In the upper part of Fig.2 the experimentally determined values of R_H are drawn as a function of the $C_{12}E_8$ concentration, the concentration of GM1 being fixed at 0.8 mM. It is seen that R_H decreases gradually from 57 Å, the value of pure GM1 micelles, to 35 Å which is just slightly greater than that found for pure $C_{12}E_8$ micelles. The data are taken at a temperature of 37°C.

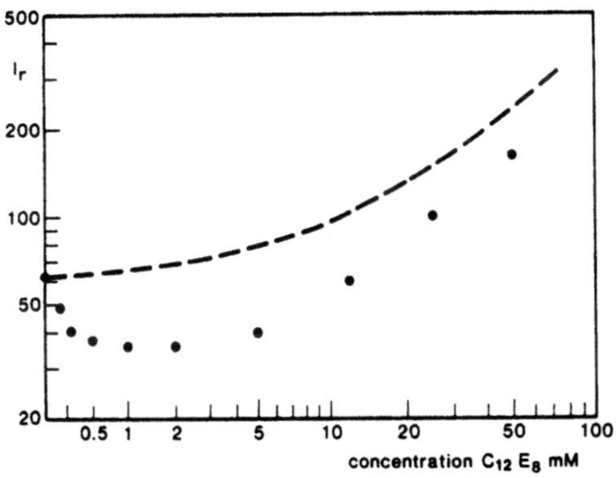

Fig.2. Scattered-light intensity from mixed GM1-$C_{12}E_8$ micelles at 37°C as a function of $C_{12}E_8$ molar concentration with fixed 0.8 mM GM1 concentration. The dashed line represents the scattered-light intensity expected for unmixed micelles.

The average scattered intensity measurements are not so simple to interpret. First of all one has to know the amount of amphiphilic material which does not go into micelles. This is usually the c.m.c. value for micellar solutions. In our case we have two amphiphiles with different c.m.c.'s, 10^{-9} M for GM1 and $7 \cdot 10^{-5}$ M for $C_{12}E_8$, and the c.m.c. of the mixed micelle can be calculated to be intermediate [17].

The free monomer concentration can also be calculated for various mole ratios of the two amphiphiles and qualitatively checked by surface tension measurements [18]. It is found that it can be neglected in our case.
A second problem in interpreting intensity data for mixed micelles comes from the fact that the two amphiphiles may have different optical contrast, represented by their refractive index increments. Therefore we have to extend the usual formula

$$I_r = A\left(\frac{dn}{dc}\right)^2 M(c-c_0) \qquad (1)$$

which relates the scattered intensity to the molecular weight of homogeneous non-interacting particles, with $I_r=(I_S-I_W)/I_W$, I_S and I_W being the scattered intensities by the solution and the solvent, A the calibration constant of the apparatus (equal to 5.86 cm^{-3} in our case), dn/dc the refractive index increment, M the micellar molecular weight, c the concentration of the amphiphile in solution in gr/cm^3 and c_0 the amphiphile c.m.c. in gr/cm^3. In order to extend formula (1) to the mixed micelle case, we consider that the light intensity scattered by N homogeneous non-interacting micelles in the scattering volume is proportional to

$$N\left[\left(\frac{dn}{dc}\right) M\right]^2 \quad . \tag{2}$$

The quantity in square brackets is proportional to electric field scattered by the single micelle. Formula (2) can be immediately extended to the mixed micelle as

$$N\left[\left(\frac{dn}{dc}\right)_1 n_1 m_1 + \left(\frac{dn}{dc}\right)_2 n_2 m_2\right]^2 \quad , \tag{3}$$

where m_1= 539 and m_2=1510 are the molecular weights of $C_{12}E_8$ and GM1 and n_1 and n_2 are the number of $C_{12}E_8$ and GM1 monomers in the mixed micelle. The refractive index increments are (dn/dc) =0.134 cm^3/gr for $C_{12}E_8$ and $(dn/dc)_2$= 0.143 cm^3/gr for GM1. Expressing the number N of mixed micelles in terms of the amphiphile concentration and the mixed micelle molecular weight, we obtain [18]

$$I_r = gA\left(\frac{dn}{dc}\right)_2^2 M(c-c_0) \tag{4}$$

where g is a dimensionless factor given by

$$g = \left(\frac{1+\beta\alpha X}{1+\alpha X}\right)^2 \tag{5}$$

with $\alpha=m_1/m_2$, $X=n_1/n_2$ and $\beta=(dn/dc)_1/(dn/dc)_2$. It should be noticed that g=1 independently of X if $(dn/dc)_1=(dn/dc)_2$. This means that in this case the mixed micelle is optically homogeneous and Eqs.(4) and (1) become identical. The measured values of the relative scattered intensity I_r for 0.8 mM GM1 solutions at 37°C are reported in Fig.2 as a function of the added concentration of $C_{12}E_8$. The intensity I_r at first decreases upon addition of $C_{12}E_8$ and then starts to increase again. This is a clear indication that mixed micelles are formed. If GM1 and $C_{12}E_8$ micellize separately, the scattered intensity would not go through a minimum, when $C_{12}E_8$ is added, but would increase monotonically like the dashed line in Fig. 2 which is calculated for the unmixed case.

From the intensity data of Fig.2, the molecular weight of the mixed micelles is calculated with Eq.(4). Neglecting the free monomer concentration, X is simply C_1/C_2, the molar ratio between the two amphiphiles in solution. The dots in the lower part of Fig.3 represent the molecular weight of the mixed micelles as a function of the $C_{12}E_8$ concentration. As expected, we find that M decreases smoothly from the value 420,000 appropriate for the GM1 micelle, to that typical of the $C_{12}E_8$ micelle about 66,000.

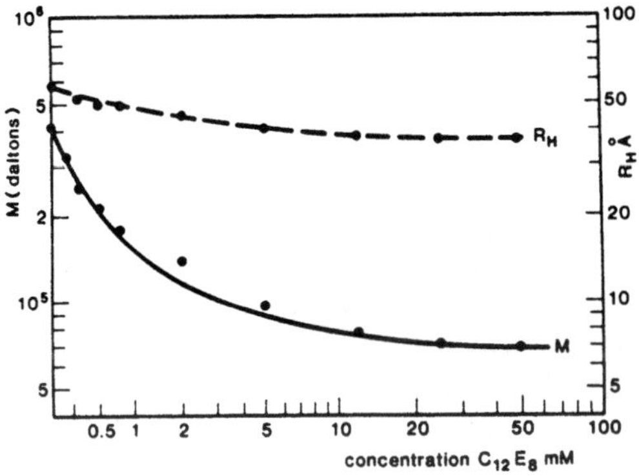

Fig.3. Molecular weight M and hydrodynamic radius R_H of mixed GM1-$C_{12}E_8$ micelles at 37°C as a function of $C_{12}E_8$ molecular concentration with fixed 0.8 mM GM1 concentration

The number of monomers of each component in the micelle can be determined from M by using the equation

$$M = n_1 m_1 + n_2 m_2 \tag{6}$$

and by considering that the ratio $n_1/n_2 = X$ is known for each value of M.

It is interesting to note that the molecular weight data of Fig.3 can be reproduced by a simple model of mixed micelle. We assume that $C_{12}E_8$ and GM1 monomers contribute on the average with the same amount of surface curvature regardless of whether they are in the mixed micelle or not. This is equivalent to say that in the mixed micelle each monomer takes the same solid angle taken in the homogeneous micelle, regardless of the molar ratio of the two amphiphiles, that is

$$4\pi = (4\pi/n_{10})n_1 + (4\pi/n_{20})n_2 \quad , \tag{7}$$

where n_{10} and n_{20} are the number of monomers in the separate micelles of $C_{12}E_8$ and GM1. Combining Eqs.(6) and (7) we can express the molecular weight M of the mixed micelle in terms of the ratio $X = n_1/n_2$ as follows

$$M = m_2 n_{20} \frac{1 + \alpha X}{1 + \gamma X} \tag{8}$$

where $\alpha = m_1/m_2$ and $\gamma = n_{20}/n_{10}$. The full line in Fig.3, drawn according to (8), confirm a quite good agreement with the proposed model.

As a conclusion we like to stress that whenever possible, it is important to perform static and dynamic light-scattering experiments on the same system at the same time. Indeed the example of the mixed micelles shows that from average scattered intensity measurements one can have some additional information on the system which is not obtainable from the hydrodynamic radius only.

References

1. C.Tanford: *The Hydrophobic Effect* (Wiley, New York 1980)
2. M.Corti, V.Degiorgio: Chem. Phys. Lett. 53, 232 (1978)
3. M.Corti, V.Degiorgio: Chem. Phys. Lett. 49, 141 (1977)
4. M.Corti, V.Degiorgio: J. Phys. Chem. 85, 711 (1981)
5. R.Dorshow, J.Briggs, C.A.Bunton, D.F.Nicoli: J. Phys. Chem. 86, 2388 (1982)
6. N.A.Mazer, J.B.Benedek, M.C.Carey: J. Phys. Chem. 80, 1075 (1976)
7. M.Corti, V.Degiorgio: In *Solution Chemistry of Surfactants*, ed. by K.L.McHal (Plenum Press, New York, 1979) p.377
8. P.J.Missel, N.A.Mazer, G.B.Benedek, C.Y.Young: J. Phys. Chem. 84, 1044 (1980)
9. G.Porte, J.Appell: J. Phys. Chem. 85, 2511 (1981)
10. S.Ikeda, S.Hayashi, T.Imac: J. Phys Chem. 85, 106 (1981)
11. M.Corti, V.Degiorgio: Phys. Rev. Lett. 45, 1045 (1980) and J; Phys. Chem. 85, 1442 (1981)
12. M.Corti, V.Degiorgio, M.Zulauf: Phys. Rev. Lett. 48, 1617 (1982)
13. H.Wiegandt: Adv. Lipid. Res. 9, 249 (1971)
14. M.Corti, V.Degiorgio: Ann. Phys. (Paris) 3, 303 (1978)
15. C.Tettamanti, F.Bonali, S.Marchesini, V.Zambotti: Biochem. Biophys. Acta 296, 160 (1973)
16. S.Sonnino, R.Ghidoni, G.Galli, G.Tettamanti: J. Neurochem. 31, 947 (1978)
17. J.Clint: J. Chem. Soc. 71, 327 (1975)
18. M.Corti, V.Degiorgio, R.Ghidoni, S.Sonnino: J. Phys. Chem. 86, 2533 (1982)

41. Measuring Thermal Diffusivities Using Two Photomultipliers in Cross-Correlation

Gerd Hiller and Gerhard Simonsohn

Institut für Atom- und Festkörperphysik, Freie Universität Berlin, D-1000 Berlin 33, Fed. Rep. of Germany

1. Introduction

In pure liquids far from the critical point, light scattering leads to a Rayleigh signal arising from entropy fluctuations (see,e.g. [1],[2],[3]). The process is diffusion-like, i.e. an exponential autocorrelation function with a time constant τ_c inversely proportional to the modulus of the scattering vector is to be expected

$$\tau_c = \frac{1}{q^2 D_T} \qquad q = \frac{4\pi}{\lambda_L} \sin \theta/2 \tag{1}$$

(λ_L: wavelength in the liquid, θ: scattering angle) .

The diffusion constant D_T, called thermal diffusivity, is connected with the thermal conductivity Λ, density ρ, and specific heat at constant pressure c_p by the relation $D_T = \Lambda/(\rho \cdot c_p)$.

Dynamic light scattering in combination with correlation spectroscopy is an interesting alternative to conventional methods for measuring D_T or Λ, c_p separately, since the measurement is performed without any macroscopic temperature gradient. An accuracy of 1% or more is desirable for a critical comparison of values.

Compared with other diffusion processes studied by photon correlation spectroscopy the cross-section is extremely small and the process is fast, i.e. D_T is large. Special experimental problems arise. In particular, self-correlation of the detector system must be reduced to a greater extent than otherwise. Therefore a set-up with two photomultiplier tubes (PMT's) in cross-correlation is used. This, in addition, offers the possibility of including the first point of the complete autocorrelation function, $\Gamma(0)$, into the measurement by means of a delay in one of the channels.

Experimental examples are shown. There are usually deviations from a single exponential. The role of Brillouin contributions as a possible source of error is discussed.

2. Experimental Conditions and Set-Up

On account of the large value of D_T, scattering angles of only a few degrees have to be used; time constants τ_c in the μs region, suitable for correlation spectroscopy, are then obtained. The low scattering angles in conjunction with the small cross-section lead almost inevitably to heterodyne conditions.

Apart from the second PMT the experimental arrangement is the same as that used by OLIVER et al. [4] for their early measurements on carbon tetrachloride (see Fig. 1). The scattered radiation does not pass any lens. Both the aperture solid angle (usually less than the coherence angle) and the scattering volume are defined by two stops about 1.5m apart, one of them in front of the PMT's and the other one a few cm behind the scattering cell at the end of a long tube. The coherence length of the laser radiation (from an Ar ion laser in single mode operation) is large compared with the dimensions of the scattering cell. This is important for producing a *coherent* background, since most of the background originates from the cell walls.

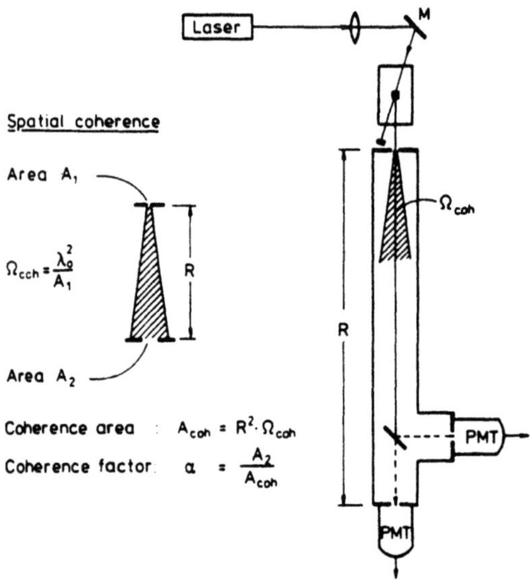

Fig. 1. Optical part of the set-up. Definition and adjustment of spatial coherence by stops in front of the photomultipliers PMT and at the entrance of a long tube ($R \approx 1.5$ m). The signals from the two PMT's are cross-correlated

The signal count rate is typically only 1 kHz and the sample time is necessarily about 1 µs. Under these conditions, i.e. for only one count per 1000 sample times, after-pulses of the PMT or any other self-correlation of the detector system, even if it is small, may become a serious source of error. This critical situation does not essentially change if we add the background of, say, 200 kHz in order to get a heterodyne signal. The effect of self-correlation in comparision with the *total* count rate has been reduced, but we are now looking for the signal of weak contrast (1%) on the background.

After-pulses result in a distortion of the autocorrelation function by a pile-up at the beginning. Dead-time effects may be considered to be another kind of self-correlation. They result in the opposite distortion.

Cross-correlation between two PMT's is used in order to reduce these effects to a negligible amount, even with medium quality tubes and standard equipment (see Fig. 1). The two PMT's receiving the signal from the scattering cell are images of each other with respect to a semi-transparent mirror. The signals from the PMT's are fed (via discriminators and pulseformers) into the two separate input channels of the correlator (Malvern K 7023) which is used in the cross-correlation mode.

The cross-correlation function, being a long-time average, is equivalent to the autocorrelation function of a single tube, except for the after-pulses (or other self-correlations). A simple argument based on the statistical independency of the tubes shows that the influence of after-pulses is considerably reduced. Indeed, as was shown by ARECCHI et al. [5], only a convolution of the whole autocorrelation function with some self-correlation function remains. In our case, this is a negligible effect. The cross-correlation scheme has recently also been applied by BURSTYN et al. [6] in an investigation on critical concentration fluctuations.

We may use the set-up for measuring the complete autocorrelation function including the first point $\Gamma(0)$. To do so, we must insert a delay in one of the two channels. If the correlator has the customary configuration with a shift register, the pulses in the direct branch have to be delayed until the pulses in the other branch have made some steps, or at least one step, through the shift register.

If we would try to do the same with the branched pulses from a single PMT we would fail. $\Gamma(0)$ is then completely covered by the shot-noise singularity expressing the correlation of each pulse with itself. There is not a term like this if we use the two PMT's in cross-correlation. The only correlation between the two sets of pulses is that which we are looking for; the pulses are otherwise independent.

Owing to the statistics of the process the delay should be continuously open. A delay generator which is blocked during the delay period or may, perhaps, store one further pulse in a buffer is not suitable. The large dead time introduced in this way could lead to a significant distortion even within the scope of the residual convolution effect mentioned above.

We have used the simple remedy of a coaxial cable (50 Ω), in conjunction with a broadband amplifier to account for absorption. A small distortion occurs in this case, too. The pulses undergo dispersion on their way through the cable, resulting in an increase in their widths. This amounts to a dead-time effect, but this is typically less than 100 ns for the delay time of some μs employed here.

Figure 2 shows two examples. For the first curve (tetralin) the delay is 3.6 μs, corresponding to 6 sample times. Consequently, the maximum $\Gamma(0)$ appears in the sixth correlator channel. For the second curve (toluene) the delay corresponds to 4 sample times. To the left of the maxima we find the values $\Gamma(-\Delta\tau)$, $\Gamma(-2\cdot\Delta\tau)$, ... The symmetry of the function with respect to $\Gamma(0)$ is a consequence of stationarity. The curves are least-squares fits to all the points beginning with $\Gamma(0)$, the baseline being treated as a free parameter.

3. Deviations from a Strictly Exponential Autocorrelation Function

If we start from fits like those in Fig.2, the D_T values calculated from τ_c and the scattering angle are mainly in good agreement with published data from other light-scattering measurements (see Table 1 for a comparison of toluene values). However, the fit in Fig.2 is not as good as it might appear

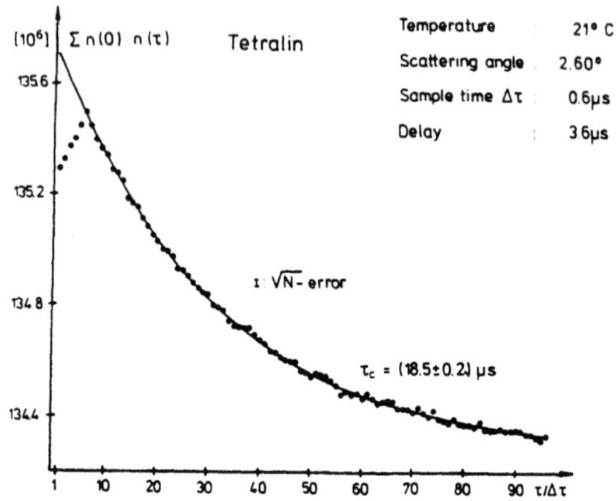

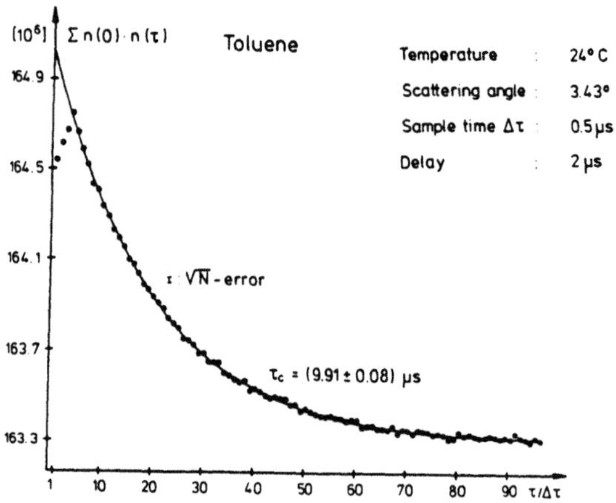

Fig. 2. Results for tetralin and toluene. The maximum represents $\Gamma(0)$ corresponding to a delay of 6 and 4 sample times, respectively. A single exponential is fitted to all the points (correlator channels) beginning with $\Gamma(0)$

Table 1. Thermal diffusivity of toluene corrected for a temperature of 22 $^{\circ}$C; values rounded to two decimals. A temperature coefficient of $-2.8 \cdot 10^{-3}$ (K)$^{-1}$ calculated from data in [7] has been used for the correction

References	[8]	[9]	[10]	This work (derived from Fig.2)
D_T [10^{-3} cm^2 s^{-1}]	0.87±0.03	0.84±0.04	0.87±0.03	0.88±0.01

from a first inspection. In Fig.3 the same measuring points as in Fig.2 (toluene) are evaluated, but the fit is restricted to different parts of the function. The difference between the two time constants is by far outside the limits of error. We find a rapid decay for the beginning of the function (lower curve in Fig.3). If we shift the fit region away from the beginning

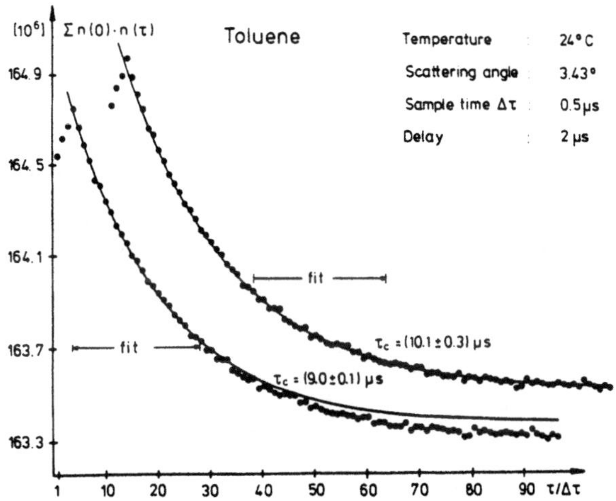

Fig.3. The same set of data as shown in Fig.2 (toluene). A single exponential is fitted to parts of the function. Lower curve: fit domain channels 4 to 28. Upper curve: fit domain channels 28 to 53. The time constants differ appreciably. The curves are displaced the scale relating to the lower one

Table 2. Deviation from an exponential autocorrelation function

Effect	Precaution, control
After-pulsing Dead-time effects	Cross-correlation
Finite second-order linewidth of laser radiation	Measurements without a dynamically scattering medium
Mechanical vibrations	Control of long-time behaviour
Finite angular aperture $\Delta\theta$	Angle θ and scattering volume large; no undue focusing
Homodyne contribution	High variable degree of heterodyning
Brillouin scattering	

the time constant increases (upper curve in Fig.3). With respect to the latter fit the deviation at the beginning shows as a pile-up. This is the situation we most frequently met with our evaluations using measurements with or without a delay.

There are many possible reasons for a distortion like this. They are listed in Table 2 together with some keywords concerning precautions and control measurements. In spite of all our efforts to account for these effects and to perform control measurements we have not yet been able to explain the deviation.

The points listed in Table 2 have been discussed in the literature on various occasions, except for a possible distortion by Brillouin scattering. Therefore some remarks on this point will be added.

4. Brillouin Contribution

The scattered radiation comprises the two Brillouin components, too. Brillouin scattering occurs at spontaneous sound waves. There is a shift of the centre frequency of each component by an amount $\omega_S = c_S \cdot q$ which may be interpreted as a Doppler shift caused by the propagation of the sound waves with a velocity c_S.

A Brillouin component has an autocorrelation function, the time constant of which is between one and two orders of magnitude less than that of the Rayleigh component. This constant reflects the attenuation of the sound waves. The intensity is roughly the same for these two contributions (Landau-Placzek ratio, see [1] [2] [3]).

In a homodyne experiment correlation spectroscopy, being insensitive to the centre frequency of the radiation, could not distinguish between the Rayleigh and Brillouin components. Thus, the superposition of the Brillouin radiation could be a serious drawback in a homodyne experiment aiming at a critical evaluation of Rayleigh scattering.

An important feature of heterodyne detection with a coherent background (derived from the unshifted laser radiation) is that it can distinguish between the Rayleigh and Brillouin components. If only

the Rayleigh signal $a_R(t) \cdot \exp(-i\omega_0 t)$

and the background $a_0 \cdot \exp(-i\omega_0 t)$

are taken into account (and the scattering process is assumed to be Gaussian), then the second-order autocorrelation function has the form (cf., e.g., [11]):

$$\Gamma_2(\tau) = \left[|a_0|^2 + \langle |a_R|^2 \rangle\right]^2 + (\langle |a_R|^2 \rangle)^2 \exp(-2\tau/\tau_c) + 2 \cdot \frac{|a_0|^2}{\langle |a_R|^2 \rangle} (\langle |a_R|^2 \rangle)^2 \exp(-\tau/\tau_c) \quad .$$

There is a constant term (in square brackets), the homodyne term with a decay constant $\tau_c/2$, and the heterodyne term which is "amplified" with respect to the homodyne term by the ratio of background and signal intensities (times 2). The evaluation of a heterodyne experiment is based on this latter term.

The Brillouin radiation does not find a coherent background of its centre frequency among the stray light used as the coherent background. In a generalized formula for $\Gamma_2(\tau)$ the Brillouin contributions appear without the "amplifying factor" $2|a_0|^2/\langle |a_R|^2 \rangle$, but solely in the form of a homodyne term and a constant term. This seems plausible and may be supported by a complete calculation.

Taking this into account and estimating the time constant τ_c to be that expected from ultrasonic data, it seems, at the moment, improbable that a small pile-up in the first channels originates from a Brillouin contribution.

There is an additional effect which tends to reduce the influence of the Brillouin radiation. The frequency shift $\omega_s = c_s \cdot q$ depends on the scattering angle θ. For the small angles considered here it is proportional to θ. Any finite angular width $\Delta\theta$ is connected with a spread $\Delta\omega_s$ of centre frequencies. Putting

$$c_s = 1.3 \cdot 10^3 \text{ ms}^{-1} \text{ (toluene) and } \lambda_L = 2.3 \cdot 10^{-7} \text{ m (cf. (1))},$$

we find a frequency spread $\Delta\omega_s$ as large as 22 MHz even for the small value $\Delta\theta = 10^{-3}$. This results in a beating between frequency components which "washes out" a decay signal of, say, 1 μs.

The same effect impedes any correlation experiment aiming at the very measurement of Brillouin linewidths, the Rayleigh component now being an unwanted contribution. In principle, such an experiment could be performed as a heterodyne experiment with a frequency shifted background, the shifted frequency being adapted to the centre frequency of one of the Brillouin lines. However, very small angular widths $\Delta\theta$ would be necessary. Owing to the fundamental diffraction limit this can only be accomplished by beams of large diameters.

References

1. P. Lallemand: In *Photon Correlation and Light Beating Spectroscopy* (Plenum Press, New York, London 1977) pp. 237-283
2. B. Crosignani, P. Di Porto, M. Bertolotti: *Statistical Properties of Scattered Light* (Academic Press, New York, San Francisco, London 1975) pp. 71-80
3. B.J. Berne, R. Pecora: *Dynamic Light Scattering* (John Wiley & Sons Inc., New York. London, Sydney, Toronto 1976) pp. 233-249
4. C.J. Oliver, E.R. Pike, J.M. Vaughan: In *Coherence and Quantum Optics* (Plenum Press, New York. London 1973) pp. 457-464
5. F.T. Arecchi, M. Corti, V. Degiorgio, S. Donati: Opt. Comm. 3, 284-288 (1971)
6. H.C. Burstyn, R.F. Chang, J.V. Sengers: Phys. Rev. Letters 44, 410-413 (1980)
7. N.B. Vargaftik: *Tables on the Thermophysical Properties of Gases and Liquids*, 2nd ed. (John. Wiley & Sons Inc., New York, London, Sidney, Toronto 1975) p. 348
8. J.B. Lastovka, C.B. Benedek: Phys. Rev. Lett. 17, 1039-1042 (1966)
9. E. Gulari, R.J. Brown, C.J. Pings: AIChE J. (USA) 19, 1196-1204 (1973)
10. M. Corti, V. Degiorgio: J. Phys. C8, 953-960 (1975)
11. C.J. Oliver: In *Photon Correlation and Light Beating Spectroscopy* (Plenum Press, New York, London 1974) pp. 151-223; see p. 204

42. Multiple Rayleigh-Gans-Debye Light Scattering in Colloidal Systems

Jan K.G. Dhont
Van't Hoff Laboratory, University of Utrecht, Padualaan 8, 3584 CH Utrecht, The Netherlands

1. Introduction

To study interaction between Brownian particles,experiments on highly concentrated dispersions must be performed.Using light-scattering techniques the experimental data are difficult to interpret due to multiple scattering effects.Not only photons scattered once ,but also photons scattered more than once are detected.To obtain information about interaction between particles and their internal structure,one needs the first-order scattered intensity due to photons scattered only once .Therefore scattering data should be corrected for these so called higher-order scattered intensities.In this paper general expressions for the nth-order scattered electric field strengths are derived.From these equations,by ensemble averaging,equations for the total average scattered intensity,and the total intensity auto-correlation function are derived,assuming that only single and double scattering events are important.Using these equations,iterative correction procedures are proposed.
Many workers considered the problem of multiple scattering,[1]-[6].However, they do not derive expressions which are useful in practice,and often make assumptions which are not satisfied.It is often assumed that particles do not interact,that the average distance between the particles is large compared to the wavelength and/or the size of the particles is small compared to the wavelength.For highly concentrated dispersions of large particles none of these assumptions is fulfilled.

2. General theory

In this section,equations for n^{th}-order scattered electric field strengths are derived.The assembly of Brownian particles is assumed to be fixed in space,that is,they are supposed not to translate or rotate.Two approximations are made in assuming this fixation. Retardation effects are neglected, and in the Maxwell equations the free current density is omitted. It is not difficult to confirm the validity of these approximations.

Point of departure are the Maxwell equations:

$$\nabla \wedge \vec{E}(\vec{r},t) = -\frac{\partial}{\partial t}\vec{B}(\vec{r},t) \tag{1}$$

$$\nabla \wedge \vec{H}(\vec{r},t) = \frac{\partial}{\partial t}\vec{D}(\vec{r},t) \tag{2}$$

where $\vec{E}(\vec{r},t)$, $\vec{B}(\vec{r},t)$, $\vec{H}(\vec{r},t)$ and $\vec{D}(\vec{r},t)$ are the electric field strength, the magnetic induction, the magnetic field strength and the dielectric displacement, respectively, all depending on the position $\vec{r}$ and the time t. The materials in the scattering system are supposed to behave as linear dielectrics, with a magnetic susceptibility equal to that of vacuum, μ_0:

$$\vec{D}(\vec{r},t) = \varepsilon(\vec{r})\vec{E}(\vec{r},t) \tag{3}$$

$$\vec{B}(\vec{r},t) = \mu_0 \ \vec{H}(\vec{r},t) \quad . \tag{4}$$

Here $\varepsilon(\vec{r})$ is the dielectric constant in the scattering system at $\vec{r}$. Furthermore it is assumed that the incident radiation is a monochromatic plane wave:

$$\vec{E}_0(\vec{r},t) = \vec{E}_0 \exp(i(\vec{k}_0,\vec{r}) - i\omega t) \tag{5}$$

and that the scattering process is elastic. The total electric field strength then is of the form:

$$\vec{E}(\vec{r},t) = \vec{E}(\vec{r})\exp(-i\omega t) \quad . \tag{6}$$

From (1)-(6) one easily derives the following differential equation for the electric field strength:

$$\nabla(\nabla,\vec{E}(\vec{r})) - \nabla^2\vec{E}(\vec{r}) = -\mu_0\,\varepsilon(\vec{r})\,\omega^2\vec{E}(\vec{r}) \quad . \tag{7}$$

Here and in the following (..,..) denotes the $\mathbb{R}^3$ innerproduct. Fourier transforming (7) with respect to $\vec{r}$ yields:

$$[k^2 - \vec{k}\vec{k} - \mu_0\varepsilon\omega^2]\,\vec{E}(\vec{k}) = \mu_0\omega^2 \int (\varepsilon(\vec{r}) - \varepsilon)\vec{E}(\vec{r})\exp(-i(\vec{k},\vec{r}))\,d\vec{r} \tag{8}$$

where

$$\vec{E}(\vec{k}) = \int \vec{E}(\vec{r})\exp(-i(\vec{k},\vec{r}))\,d\vec{r} \tag{9}$$

and ε is the dielectric constant of the solvent. Eq. (8) is transformed into an integral equation in $\vec{E}(\vec{k})$ inverting the tensor $[k^2 - \vec{k}\vec{k} - \mu_0\varepsilon\omega^2]$ and expressing $\vec{E}(\vec{r})$ on the right-hand side in terms of $\vec{E}(\vec{k})$. The inverse of the tensor does not exist in the points where $k = \pm\sqrt{\mu_0\varepsilon}\,\omega = k_0$, the magnitude of the incident wave vector. We apply a technique proposed by BEDEAUX and MAZUR [7], writing for the inverse:

$$\overleftrightarrow{T}_\alpha(\vec{k}) = \frac{\mu_0\varepsilon\omega^2 - \vec{k}\vec{k}}{\mu_0\varepsilon\omega^2\left[k^2 - (\sqrt{\mu_0\varepsilon}\,\omega + i\alpha)^2\right]} \tag{10}$$

where α is a small real number. This number is taken zero later in resulting expressions. The physical interpretation of α is that the solvent is made slightly absorbing the radiation; it is the imaginary component of the dielectric constant of the solvent. Note that ε in (10) is real. It is now possible to obtain the following integral equation in the Fourier transform of the electric field strength $\vec{E}(\vec{r})$:

$$\vec{E}(\vec{k}) = \frac{\mu_0 \varepsilon \omega^2}{(2\pi)^3} \overleftrightarrow{T}_\alpha(\vec{k}) \sum_j \int B_j(\vec{k}'-\vec{k}) \exp(i(\vec{k}'-\vec{k},\vec{r}_j)) \vec{E}(\vec{k}) d\vec{k}' \qquad (11)$$

where the summation runs over all particles, and

$$B_j(\vec{k}) = \int_{V_j} \frac{\Delta\varepsilon_j(\vec{r})}{\varepsilon} \exp(i(\vec{k},\vec{r})) d\vec{r} \quad , \qquad (12)$$

where the integration ranges over the volume V_j of particle j, with respect to a coordinate system with its origin at $\vec{r}_j$, the so-called position coordinate of particle j. In (12) $\Delta\varepsilon_j(\vec{r})$ is the difference in dielectric constant of particle j and the solvent at $\vec{r}$ with respect to the coordinate system with its origin at $\vec{r}_j$. If $\Delta\varepsilon_j(\vec{r})$ is small for all particles, (11) can be solved by iteration, formally applying the Banach fixed point theorem [8]. We take as a starting function for the iteration the Fourier transform of $\vec{E}_0 \exp(i(\vec{k}_0,\vec{r}))(3\varepsilon_0/\varepsilon+2\varepsilon_0)\exp(-\frac{1}{2}\tau d)$, where ε_0 is the dielectric constant of vacuum, τ is the turbidity and d is the distance travelled by the radiation to pass the scattering system. The factor containing the dielectric constants is found solving a boundary value problem for a homogeneous spherical symmetrical particle [9], or by comparing the first term in the iterated series with the Rayleigh-Gans-Debye expression for the scattered electric field strength. The Lambert-Beer attenuation factor is added since one is interested only in the electric field strengths leaving the scattering system. Hence the starting function is:

$$\vec{E}_0(3\varepsilon_0/\varepsilon+2\varepsilon_0)\exp(-\tfrac{1}{2}\tau d)(2\pi)^3 \delta(\vec{k}_0-\vec{k}) \qquad (13)$$

where δ is the Dirac delta distribution. The terms in the iterated series are obtaining by induction. For $n \geqslant 2$ the terms contain integrals ranging over all values of the wave vectors. This does not reflect the elasticity of the scattering process. It must be possible to perform the integration over the magnitude of the wave vectors. Indeed, using the asymptotic behaviour and the symmetry properties of the integrands, noting that $\overleftrightarrow{T}_\alpha(\vec{k})$ has two first-order poles at $k = k_0 \pm i\alpha$, the integration over the magnitudes of the wave vectors can be performed applying the residue theorem [10]. As a final step in the derivation, the equations are Fourier inverted. The resulting equations where α is taken as zero are:

$$\vec{E}(\vec{r}) = \sum_{n=1}^{\infty} \vec{E}_n(\vec{r}) \tag{14}$$

with

$$\vec{E}_1(\vec{r}) = \frac{2\pi^2 K_0^2}{(2\pi)^3} (3\,\varepsilon_0/\,\varepsilon + 2\,\varepsilon_0) \exp(-\tfrac{1}{2}\tau d) \frac{\exp(ik_0 r)}{r} \sum_j^{N_0} B_j(\vec{k}_0 - \vec{k}_s) \exp(i(\vec{k}_0 - \vec{k}_s, \vec{r}_j)) \overleftrightarrow{T}(\vec{k}_s) \vec{E}_0 \tag{15}$$

and

$$\vec{E}_n(\vec{r}) = 2\pi^2 k_0^2 \frac{(k_0 \pi i)^{n-1}}{(2\pi)^{3n}} (3\,\varepsilon_0/\,\varepsilon + 2\,\varepsilon_0) \exp(-\tfrac{1}{2}\tau d) \frac{\exp(ik_o r)}{r}$$

$$\sum_{j_1\, j_2\, j_3 \ldots j_{n-2}\, j_{n-1}\, j_n}^{N_1\, N\, N \ldots N\, N\, N_2} \oint dS_1 \oint dS_2 \ldots \oint dS_{n-2} \oint dS_{n-1}$$

$$B_{j_1}(\vec{k}_0 - \vec{k}_1) B_{j_2}(\vec{k}_1 - \vec{k}_2) \ldots B_{j_{n-1}}(\vec{k}_{n-2} - \vec{k}_{n-1}) B_{j_n}(\vec{k}_{n-1} - \vec{k}_s)$$

$$\exp(i(\vec{k}_0 - \vec{k}_1, \vec{r}_{j_1}))\exp(i(\vec{k}_1 - \vec{k}_2, \vec{r}_{j_2}))\ldots\exp(i(\vec{k}_{n-2} - \vec{k}_{n-1}, \vec{r}_{j_{n-1}}))\exp(i(\vec{k}_{n-1} - \vec{k}_s, \vec{r}_{j_n}))$$

$$\overleftrightarrow{T}(\vec{k}_s)\overleftrightarrow{T}(\vec{k}_{n-1})\overleftrightarrow{T}(\vec{k}_{n-2}) \ldots \overleftrightarrow{T}(\vec{k}_2)\overleftrightarrow{T}(\vec{k}_1)\vec{E}_0 \qquad ;n \geqslant 2 \tag{16}$$

where $\vec{k}_s$ is the scattered wave vector. The integrations range over a half spherical surface with radius k_0, reflecting the elasticity of the scattering process. dS_i is a surface element in $\vec{k}_i$ space. N_1 and N_2 are the number of particles in that part of the scattering system where the incident intensity is unequal to zero, and the part from which scattered radiation can be detected, respectively. N_0 is the number of particles in the intersection of these two regions. N is the total number of particles in the scattering system. The definition of these numbers is depicted in fig. 1. For obvious reasons $\vec{E}_n(\vec{r})$ is interpreted as the nth-order scattered electric field strength. Furthermore:

$$\overleftrightarrow{T}(\vec{k}) = \overleftrightarrow{I} - \frac{\vec{k}\vec{k}}{k^2} \,. \tag{17}$$

3. Static light scattering

Scattered intensities may be obtained ensemble averaging the squared electric field strengths given by (14)-(16). The following four assumptions are made: 1 All particles are spherical symmetrical. The consequence of this assumption is that $B(\vec{k})$ is no longer a stochastic variable, since now rotation of particle j does not change the integration range V_j, if $\vec{r}_j$ is choosen in the centre of symmetry.

2 All particles are identical. Then the amplitudes $B_j(\vec{k})$ are identical for all particles j.

3 The quotient N_0^2/N_1N_2 is much smaller than 1. Due to this assumption the particles N_1 and N_2 may be treated statistically independent.

4 The differences in dielectric constants between the particles and the solvent is small enough to truncate (14) after the second term:

$$\vec{E}(\vec{r})=\vec{E}_1(\vec{r})+\vec{E}_2(\vec{r}) \quad . \tag{18}$$

The last assumption means that only first-and second-order scattering events are considered.

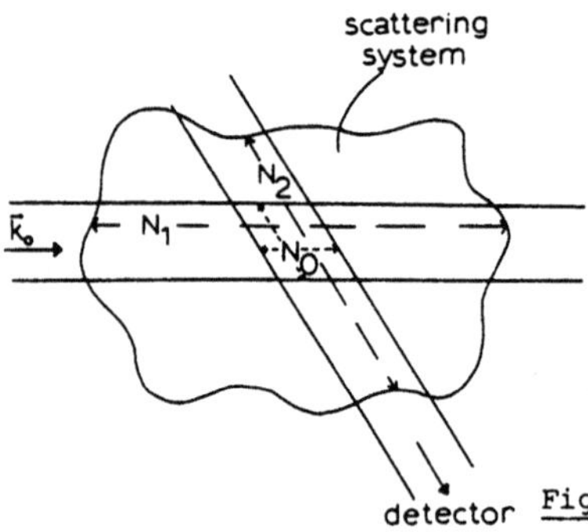

Fig. 1. The definition of the numbers N_1, N_2 and N_0

Ensemble averaging over position coordinates $\vec{r}_j$ then gives the following expression for the total Rayleigh ratio R:

$$R=R_1+R_2 \tag{19}$$

with

$$R_1=\frac{4\pi^4 k_0^4}{(2\pi)^6}\quad(3\,\varepsilon_0/\,\varepsilon\;+2\,\varepsilon_0)^2\exp(-\tau d)$$

$$\rho\, V_p^2 P(\kappa)S(\kappa)\;\;(\overleftrightarrow{T}(\vec{k}_s)\hat{n}_o,\hat{n}_s)^2 \quad , \tag{20}$$

where ρ is the particle number density, V_p is the particle volume, $\hat{n}_o$ is the polarisation of the incident radiation, $\hat{n}_s$ the polarisation of the detected radiation, $P(\kappa)$, with $\kappa = |\vec{k}_0-\vec{k}_s|$, the formfactor, defined as

$$P(\kappa)=|B(\kappa)|^2/V_p^2 \tag{21}$$

and $S(\kappa)$ the structure factor, defined as

$$S(\kappa)=\frac{1}{N}\sum_{i,j=1}^{N}\left\langle\exp(i(\vec{\kappa},\vec{r}_i-\vec{r}_j))\right\rangle \; . \tag{22}$$

R_1 is the well-known Rayleigh-Gans-Debye expression for the scattered intensity. The second-order Rayleigh ratio R_2 is given by:

$$R_2 = \frac{4\pi^6 k_0^{\,6}}{(2\pi)^{12}} \; (3\varepsilon_0/\varepsilon + 2\varepsilon_0)^2 \exp(-\tau d) \; \frac{V_1 V_2}{V_0}$$

$$\oint dS_1 \; \rho V_p^{\,2} P(|\vec{k}_0-\vec{k}_1|) S(|\vec{k}_0-\vec{k}_1|) \; \rho V_p^{\,2} P(|\vec{k}_1-\vec{k}_s|) S(|\vec{k}_1-\vec{k}_s|) \tag{23}$$

$$(\hat{n}_s, \overleftrightarrow{T}(\vec{k}_s)\overleftrightarrow{T}(\vec{k}_1)\hat{n}_0)^2 \quad G(\vec{k}_1, |\vec{k}_0-\vec{k}_s|) \quad ,$$

where V_1, V_2 and V_0 are the volumes containing the N_1, N_2 and N_0 particles, respectively. $G(\vec{k}_1, |\vec{k}_0-\vec{k}_s|)$ is a function completely determined by the geometry of the scattering system and the magnitude of the wave vectors. This function is given by:

$$G(\vec{k}, |\vec{k}_0-\vec{k}_s|) = \oint dS_2 \; \frac{\int_{V_1} \exp(i(\vec{k}_2-\vec{k},\vec{r}))\,d\vec{r}}{V_1} \; \frac{\int_{V_2} \exp(i(\vec{k}-\vec{k}_2,\vec{r}))\,d\vec{r}}{V_2} \quad . \tag{24}$$

Even for the simplest of geometries this function must be numerically evaluated.

These equations suggest the following iterative scheme to obtain R_1 from R, the experimental Rayleigh ratio:

$$R_{1,n} = R - R_{2,n-1} \qquad ; n \geq 1 \quad . \tag{25}$$

Here $R_{2,0}$ is the value of (20) calculated from the function $\rho V_p^{\,2} P(k) S(k)$ determined from the substitution of R for R_1 in (19). $R_{2,n}, n \geq 1$, is the same value obtained from the substitution of $R_{1,n}$ for R_1. The following relations hold:

$$\lim_{n\to\infty} R_{1,n} = R_1 \qquad \lim_{n\to\infty} R_{2,n} = R_2 \quad . \tag{26}$$

In practice we found that for a correction of 40-50% about 6 iterations were necessary.

4. Dynamic light scattering

The same four assumptions as in "static light scattering" are made. The following expression for the total intensity auto-correlation function is found:

$$\varphi(k,t) = \frac{\exp(-2D(k)k^2 t)}{(1+R_2/R_1)^2} \left[1 + \frac{\pi^2 k_0^{\,2}}{(2\pi)^6} \frac{V_1 V_2}{V_0} \right. \tag{27}$$

$$\oint dS_1 \frac{F(|\vec{k}_0-\vec{k}_s|,t)\ F(|\vec{k}_1-\vec{k}_s|,t)}{F(|\vec{k}_0-\vec{k}_s|,t)} \frac{(\overleftrightarrow{T}(\vec{k}_s)\overleftrightarrow{T}(\vec{k}_1)\hat{n}_0,\hat{n}_s)^2}{(\overleftrightarrow{T}(\vec{k}_s)\hat{n}_0,\hat{n}_s)^2} G(\vec{k}_1,|\vec{k}_0-\vec{k}_s|)\Big]^2 .$$

Here $D(\boldsymbol{k})$ is the effective diffusion coefficient,and

$$F(\boldsymbol{k},t) = \int V_p^2 P(\boldsymbol{k})S(\boldsymbol{k}) \exp(-D(\boldsymbol{k})\boldsymbol{k}^2 t) \quad . \tag{28}$$

Eq. (27) is valid only when first order scattered intensities are detected, that is,when $R_1 \neq 0$.Equations for purely second-order correlation functions can be obtained.These correlation functions are directly proportional to a surface integral,ranging again over a half-spherical surface with radius k_0. Notice the similarity of (27) and (19),(20),(23).A similar iterative procedure to obtain the first-order intensity autocorrelation function

$$\varphi_1(\boldsymbol{k},t) = \exp(-2D(\boldsymbol{k})\boldsymbol{k}^2 t) \tag{29}$$

from the experimental correlation function $\varphi(\boldsymbol{k},t)$ may be applied.The iterative scheme is:

$$\varphi_{1,n}(\boldsymbol{k},t) = \varphi(\boldsymbol{k},t)/\Delta_{n-1}(\boldsymbol{k},t) \quad ;n \geqslant 1 \ . \tag{30}$$

Here $\Delta_0(\boldsymbol{k},t)$ is the proportionality factor between $\varphi(\boldsymbol{k},t)$ and $\varphi_1(\boldsymbol{k},t)$ in (27),calculated from (28),after substitution of the experimental diffusion coefficients (determined from a least square fit of $\varphi(\boldsymbol{k},t)$). In (27) and (28) corrected static light-scattering data are needed.These corrected data are obtained from experimental data as described in "static light scattering".From $\varphi_{1,1}(\boldsymbol{k},t)$ a new set of diffusion coefficients is calculated by means oi a least square fit.These new diffusion coefficients are used to calculate $\varphi_1(\boldsymbol{k},t)$,etc. The following relation holds:

$$\lim_{n\to\infty} \varphi_{1,n}(\boldsymbol{k},t) = \varphi_1(\boldsymbol{k},t) \quad . \tag{31}$$

Eq. (31) yields the, for double scattering corrected, effective diffusion coefficients.

5. Discussion

Explicit expressions for the total scattered intensity and intensity auto-correlation function are derived,assuming that only first and double scattering events are important.These expressions can be used to correct data for double scattering iteratively.

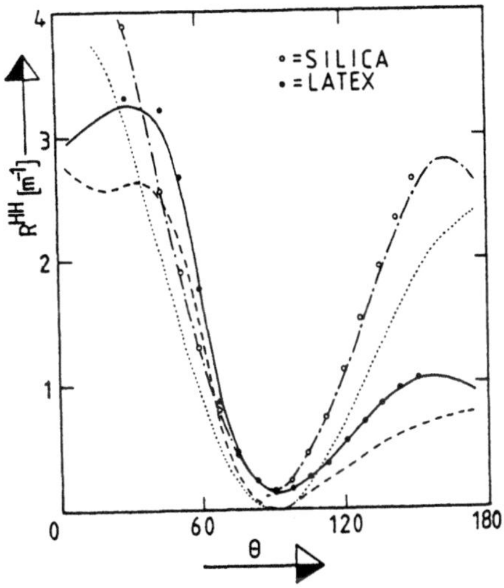

Fig.2. HH Rayleigh ratios as a function of the scattering angle.
Points represent experimental data.
······· and - - - - are calculated first-order Rayleigh ratios,
—·— and ——— are calculated first-plus second-order
Rayleigh ratios for silica and latex respectively

As far as the static light scattering is concerned, the equations are verified experimentally. For example, VV measurements (V stands for vertical; the first symbol represents the polarisation of the incident radiation, the second of the detected radiation) through the iteration described in "static light scattering", the function $\rho V_p^2 P(\kappa) S(\kappa)$ is found. Then VH, HV and HH (H stands for horizontal) Rayleigh ratios can be calculated from (20) and (23), and directly compared to experimental ratios. For a latex and silica sample the HH results are given in fig.2.

At this time we cannot comment on dynamic light-scattering experiments, since these experiments are not yet concluded.

6. References

1 D.W.Oxtoby and W.M.Gelbart, J.Chem.Phys., 60, 3359 (1974)

2 A.Bøe and T.Sikkeland, Phys.Rev.A, 16, 2105 (1977)

3 C.M.Sorensen et. all., Phys.Rev.A, 14, 1520 (1976)

4 V.L.Kuzmin, Opt.Spectrosc., 39, 306 (1975)

5 J.Böheim et. all., Z.Physik B, 32, 237 (1979)

6 G.Koopmans et. all.,J.Phys.A:Math.gen.,12,581(1979)

7 D.Bedeaux and P.Mazur,Physica,67,32(1973)

8 E.Kreyszig,Introductory functional analysis with applications,(J.Wiley &Sons,1978)pp.299

9 J.D.Jackson,Classical electrodynamics,(J.Wiley & Sons,1975)pp.147

10 M.R.Spiegel (Schaum's outline series),Complex variables,(McGraw-Hill, 1964)pp.172

43. Fluorescence Correlation Spectroscopy in Solutions

P. Kask, T. Kändler, P. Piksarv, M. Pooga, and E. Lippmaa
Institute of Chemical Physics and Biophysics, Estonian Academy of Sciences, 10 Lenini puiestee, Tallinn 200001, USSR

1. Introduction

A number of problems in contemporary biophysics is related to the investigation of some special dynamic properties of complicated samples, such as solutions of biologically important macromolecules or their mixtures. To distinguish between the different molecules and active sites, fluorescence reveals more specific properties than light scattering. This determines the experimentators' interest to combine the specificity of fluorescence detection with the dynamic capabilities of concentration correlation experiments.

Unfortunately, fluorescence is very sensitive to changes in the molecular vibrational and rotational states, occurring as a result of molecular collisions, which, as a rule, are too fast for most correlation experiments. The fluorescence actually measured is practically incoherent emission and an ordinary light beating experiment fails to work. However, this difficulty has been circumvented in fluorescence correlation (FC) experiments [1] by exposing a very small sample volume to obtain the number fluctuations of certain molecules inside this sample volume. The same has been done in some scattering correlation experiments, too [2].

There are several different processes that can cause stochastic number fluctuations of the fluorescent molecules inside the sample volume, *e.g.* the flow of the solution through this volume [3], translational and rotational diffusion of dye molecules or molecular complexes [4,5], chemical reactions of the dye molecules with molecules of another species [6], intramolecular reactions of dyes or labeled molecules [6], including conformational changes [7] and physical relaxation processes of the molecular excited states [8]. All these processes can be studied in FC experiments.

In spite of the ten years which have passed from the pioneering experiments [1], FC spectroscopy has not yet found as wide a use as the conventional light beating spectroscopy, mostly because of experimental difficulties. FC spectroscopy is relatively widely used to study translational diffusion in biophysical systems [9-11]. Much interest has been expressed about the chemical reaction and rotational diffusion experiments, but here the research has not yet advanced past the preliminary methodological stage.

2. The Practical Limits of Utilization

The strictest condition for a successful FC experiment appears to be the necessity of photochemical stability of the dye. Exciting light destroys the dye molecules under study, produces unwanted substances which can influence the behavior of the whole sample, and forces the molecules to occupy metastable excited states. All these mechanisms produce distortions of the measured correlation function. If the excitation intensity is too high, the distortions are unacceptably large. If the intensity is too low, low signal to noise ratio results.

In the following, we shall estimate practical limits for the quantum yield of photobleaching ϕ_b of the dye used in a FC experiment. Let α be the bleaching probability per incident dye molecule per its average time of diffusion out of the beam, τ_d:

$$\alpha = I_{ex}\sigma_a\phi_b\tau_d , \tag{1}$$

where I_{ex} is the exciting beam intensity [number of photons per second per cm^2] and σ_a is the absorption cross section of a dye molecule. At low intensities the signal-to-noise ratio μ is [12]

$$\mu = I_{ex}\sigma_a\phi_f\chi\eta(T\theta)^{1/2} , \tag{2}$$

where ϕ_f is the quantum yield of fluorescence, χ - the emission collection factor of experiment, η - the photon detection quantum yield, θ - the correlation channel width, T - the full run time of the experiment. Inserting the expression for I_{ex} from (1) into (2) we obtain

$$\mu = \alpha\phi_f\chi\eta(T\theta)^{1/2} / \phi_b\tau_d . \tag{3}$$

For the practical values of the quantities: $\phi_f = 0.5$, $\chi\eta = 0.01$, $T = 1000$ s, $\theta = 10$ µs, $\tau_d = 10$ ms, if we wish $\mu \geq 100$ and $\alpha \leq 0.02$, we obtain $\phi_b \leq 10^{-5}$.

To estimate the photobleaching quantum yield of the real dyes, we have observed the rate of disappearance of the absorption bands of a number of dyes produced by the exposition of a 10^{-5} to 10^{-4} M dye solution to a broad argon laser beam with power from 0.1 to 2 W/cm². In air-saturated water solutions the quantum yield of photobleaching is about 10^{-7} to 10^{-6} for Oxazine dyes (Cresyl Violet, Nile Blue, Oxazine-1), Rhodamines and Pyronines, increases to 10^{-5}-10^{-4} for Ethidium and Propidium salts, Fluoresceins, Safranins and Thionine, to 10^{-4}-10^{-3} for Acridine dyes and to 2×10^{-3} for Riboflavine. Besides, a tendency exists that the longer wavelength dyes have lower ϕ_b values.

Although 10^{-5} for ϕ_b is not a magical number, most of the fluorescent substances are unsuitable for FC spectroscopy. The art of FC spectroscopy lies in the full use of a limited number of suitably stable dyes.

3. The Methodology of Experiments in Solutions

The experimental difficulties, concerned mainly with the design of a suitable microfluorometer, have been overcome to a large extent at present. Contrary to suggestions in a number of papers on FC spectroscopy, the main demands - high emission-collecting efficiency, low blank (background) emission level, small sample volume, possibility of convenient adjustment and checking, insensibility to mechanical vibrations - can be fulfilled at once without any special construction elements used except a mediocre optical table, a stable laser and photon correlator electronics. A sketch of the optical setup used in our experiment is given in Fig.1. The two microscope objectives, one to focus the expanded laser beam, the other to collect the fluorescent emission, define by their common focus a cylindrical sampling volume with its radius changeable from about 2 to 5 μm (defined by the focused laser beam) and with the length chosen to equal about ten radii as defined by an extra aperture at the output focal plane of the emission collection microscope. To collect the emission, the use of large-aperture optics is needed. We have used 100 × 1.30 oil immersion standard objective without ocular. The collection efficiency of the oil immersion objectives exceed 30%. The angle between the two objective axes is about 95°. The output cell window is a standard microscope slide, glued at the top of a bypass 2 × 2 × 8 mm³ cuvette. If the adjustment elements of the cuvette and of the two objectives are tightly bound to each other, moderate mechanical vibrations do not disturb the experiment.

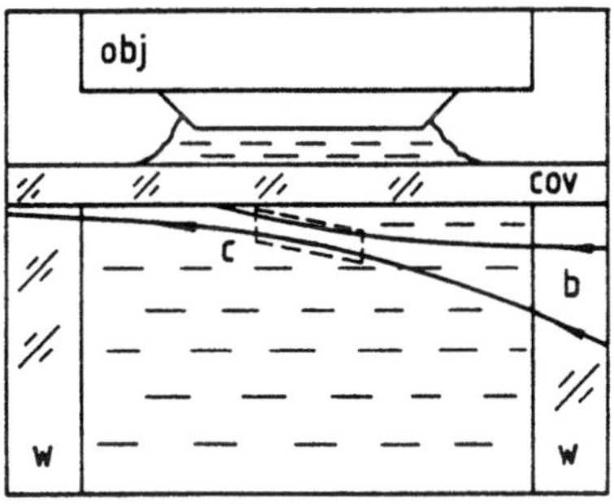

Fig.1. The central optical part of the experiment. obj - microscope oil immersion objective; cov - glass cover (output window); b - focused laser beam; w - side windows; c - sampling volume

The light from emission collection objective, after passing a spectral cut-off glass filter and the diaphragm, is a narrow paraxial beam, allowing to use the multiple internal reflection in the photomultiplier cathode window to increase the efficiency of the detector (photomultiplier FEU-84 with a trialkali cathode). To discriminate against the photomultiplier noise, a beamsplitter with two photomultipliers was used at short channel widths of the correlator ($\theta \leq 25$ μs), while the cross-correlation function between their photocounts were computed. A digital 32-channel clip-correlator was used in our experiments [13].

For excitation a stabilized and attenuated TEM_{∞} beam from an argon ion laser (Spectra-Physics, model 164-05) or a Rhodamine 6G dye laser (Spectra-Physics, model 375) was used. The standard beam focusing technique [14] has been realized with the help of a long-focus lens and an objective 3.7 × 0.11.

For a given sample solution, a sequence of experiments was performed, each doubling the channel width used in the previous experiment. A linear transformation of the data was performed with each following measurement made to fit the correlation curve in common points. Thus the correlation function was obtained over a broad correlation time scale. Although the nonsimultaneous data collection at different time scales is a procedure involving some information loss, caused by imperfection of a single correlator with a limited number of channels, this procedure is essential if one has to resolve the shape and number of all the decay components.

As in other experiments where a monotonic decay function is under study, the information content of a single fluorescence intensity correlation function is quite low. This can cause interpretation difficulties, especially for complicated objects. If the competing models are simple enough to permit estimates of the values of kinetic parameters, it is simple to distinguish between the different processes producing fluorescence intensity fluctuations, because the translational and rotational diffusion and the first-and second-order chemical reaction terms have different shapes and dependences on the sample diameter, on the detection polarization properties and on the concentrations of the reactants.

We have studied the diffusion of the *Escherichia coli* transfer ribonucleic acids labeled with Tetramethylrhodamine isothiocyanate (tRNA-TMR). These solutions do not show a $1/(1 + Bt)$-like time dependences of the correlation function, common for homogeneous dye solutions (see Fig.2). To distinguish between the different possible causes of this phenomenon, we have performed experiments with various concentrations of tRNA-TMR and laser beam radii.

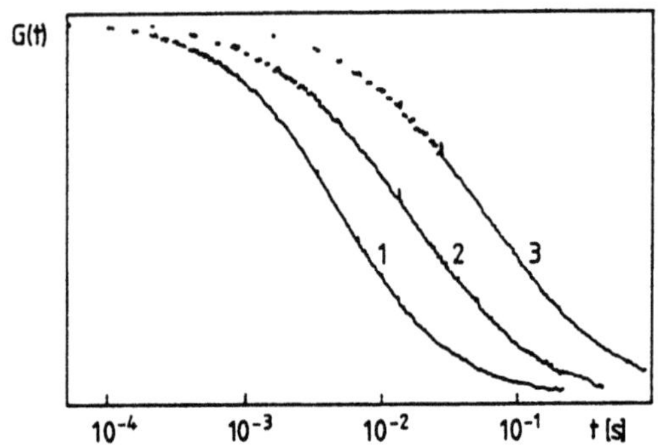

Fig.2. Fluorescence intensity correlation functions, measured at 21°C under 514.5 nm excitation. Curve 1 - for 2×10^{-9} M Rhodamine 6G in water; I_{ex} = = 0.35 mW; mean photon counting rate $\bar{n} = 2.1\times10^{6}$ s^{-1}; the curve fitting gives τ = 4.7 ms, from which the beam radius w = 2.3 µm is calculated. Curve 2 - for 2×10^{-8} M tRNA-TMR in buffer A (see sec. 4); I_{ex} = 0.25 mW; $\bar{n} = 0.6\times10^{6}$ s^{-1}. Curve 3 - the same sample at w = 4.5 µm

No dependence of the correlation function shape on concentration has been noticed. At broader beam the whole correlation curve shifts toward higher correlation times. The conclusion is that the object is of considerable inhomogeneity in diffusion constants, at least during the short time intervals of about 10^{-1} s. The inhomogeneity is larger than expected from the difference of the individual tRNAs and is a subject of further investigation.

Interpretation difficulties have arisen, however, for such modestly complicated objects as the mixtures of some acid dye (Pyronine G, Cresyl Violet or Nile Blue) with DNA or RNA. According to the simple one-step binding model [6], at low dye and high nucleic acid concentrations, two terms in the correlation function are expected. The diffusion term must slow down as the nucleic acid concentration rises, and the reaction term (practically exponential if faster than diffusion) must grow in amplitude and speed up. What is really obtained is the slowing down of the slower part of the correlation function, and growing in amplitude of the faster part of it (see Fig.3). This

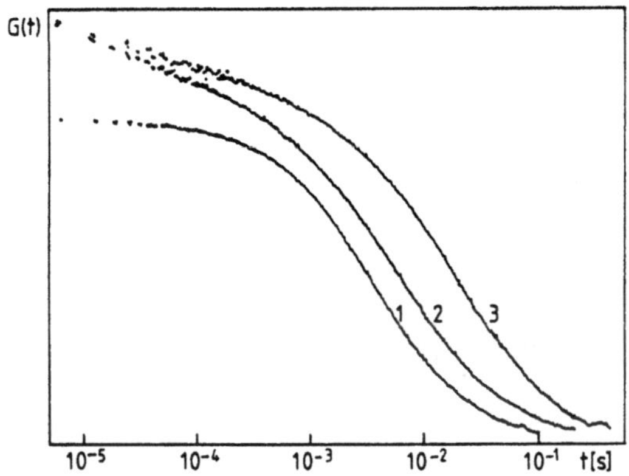

Fig.3. Fluorescence intensity correlation functions, measured at 21°C under 514.5 nm excitation. Curve 1 - for 1.2×10^{-9} M Pyronine G in buffer A; I_{ex} = = 2.0 mW; w = 2.4 μm; $\bar{n} = 1.7\times10^{6}\ s^{-1}$. Curve 2 - for 1.2×10^{-9} M Pyronine G plus 2×10^{-6} M tRNA in buffer A at the same conditions. Curve 3 - the same sample at w = 3.8 μm

allows us to estimate the binding constant [8]. We have never observed a single exponential reaction term in these systems, however. What is obtained is an inhomogeneously broadened band of reaction terms, the slower part of which is cut off by diffusional decay. When the beam radius is enlarged, this band is shifted to larger times. This band is practically structureless in the case of Pyronine G - tRNA mixtures and the fast fluorescence intensity fluctuations corresponding to this band are produced by the first-order (intramolecular) or/and the second-order (intermolecular) chemical reaction.

4. Materials

The dyes Pyronine G and Tetramethylrhodamine isothiocyanate were from Fluka and Sigma, respectively. *Escherichia coli* tRNA was from Olaine (Latvian SSR). For labeling the *E.coli* tRNA was oxidized according to [15], labeled with Tetramethylrhodamine isothiocyanate and purified according to [16].

The sample solutions with tRNA or tRNA-TMR were made in a buffer containing 10 mM tris-HCl, 200 mM KCl, 20 mM MgOAc, 5 mM β-Mercaptoethanol, pH 7.3 (buffer A).

5. Conclusions

The list of advantages of the FC method includes:
- not the time averaged, but dynamic behavior of the sample is studied;
- behavior of specific molecules and their active sites can be studied even in complicated surroundings;
- low quantities of the compounds studied are needed;
- no need to use the macroscopic relaxation techniques (temperature jump, etc.).

As the limitations and possibilities of the FC method are becomming clear and because of its useful principal properties, we believe that the FC spectroscopy will become a standard research technique, used in conjunction with the other methods to study the dynamic properties of microsystems.

References

1. D.Magde, E.L.Elson, W.W.Webb: Phys. Rev. Lett. 29, 705-708 (1972)
2. D.W.Schaefer: Science 180, 1293-1295 (1972)
3. D.Magde, W.W.Webb, E.L.Elson: Biopolymers 17, 361-376 (1978)
4. D.Magde, E.L.Elson, W.W.Webb: Biopolymers 13, 29-61 (1974)
5. M.Ehrenberg, R.Rigler: Chem. Phys. 4, 390-401 (1974)
6. E.L.Elson, D.Magde: Biopolymers 13, 1-27 (1974)
7. R.Rigler: TIBS 2, 252-254 (1977)
8. T.Kändler, P.Kask, P.Piksarv, A.Sirk, E.Lippmaa: ENSV TA Toimetised, Füüs.Matem. (in press, 1982)
9. W.W.Webb: Quart. Rev. Biophys. 9, 49-68 (1976)
10. R.Rigler, P.Grasselli, M.Ehrenberg: Phys. Scr. 19, 486-490 (1979)
11. S.M.Sorscher, J.C.Bartholomew, M.P.Klein: Biochim. Biophys. Acta 610, 28-46 (1980)
12. D.E.Koppel: Phys. Rev. A10, 1938-1945 (1974)
13. A.Sirk, P.Kask, T.Kändler, T.Karu, J.Puskar, E.Lippmaa: ENSV TA Toimetised, Füüs.Matem. 28, 227-232 (1979)
14. D.I.Innes, A.L.Bloom: Spectra-Physics Laser Technical Bulletin, No.5 (1966)
15. P.C.Zamecnick, M.C.Stephenson, J.-F.Scott: Proc. Natl. Acad. Sci. U.S. 46, 811-822 (1960)
16. J.G.J.Bauman, J.Wiegant, P. van Duijn: J. Hifpochem. Cytochem. 29, 227-237 (1981)

Index of Contributors

GPSR Compliance

The European Union's (EU) General Product Safety Regulation (GPSR) is a set of rules that requires consumer products to be safe and our obligations to ensure this.

If you have any concerns about our products, you can contact us on ProductSafety@springernature.com

In case Publisher is established outside the EU, the EU authorized representative is:

Springer Nature Customer Service Center GmbH
Europaplatz 3
69115 Heidelberg, Germany

Batch number: 10620163

Printed by Printforce, the Netherlands